Toyota Celica Owners Workshop Manual

by J H Haynes
Member of the Guild of Motoring Writers

and Alec J Jones BSc(Eng), CEng

Models covered

All Toyota Celica Coupe and Liftback models; ST, XT & GT
1588 cc & 1968 cc

ISBN 1 85010 451 4

© Haynes Publishing Group 1990

All rights reserved. No part of this book may be reproduced or transmitted in any form or by any means, electronic or mechanical, including photocopying, recording or by any information storage or retrieval system, without permission in writing from the copyright holder.

Printed in England *(437-6R6)*

ABCDE
FGHIJ
KLMNO
PQR

Haynes Publishing Group
Sparkford Nr Yeovil
Somerset BA22 7JJ England
0963 40635

Haynes Publications, Inc
861 Lawrence Drive
Newbury Park
California 91320 USA

British Library Cataloguing in Publication Data
Jones, Alec J. (Alec John), *1923–*   Toyota Celica '78 to Jan '82 owners workshop manual.   1. Cars. Maintenance & repair – Amateurs' manuals   I. Title   629.28'722   ISBN 1-85010-451-4

Acknowledgements

Thanks are due to the Toyota Motor Sales Company Limited for the supply of technical information and certain illustrations, to Castrol Limited who supplied lubrication data, and to the Champion Sparking Plug Company who supplied the illustrations showing the various spark plug conditions.

Special thanks are due to all the people at Sparkford who helped in the production of this manual.

About this manual

Its aim

The aim of this manual is to help you get the best value from your vehicle. It can do so in several ways. It can help you decide what work must be done (even should you choose to get it done by a garage), provide information on routine maintenance and servicing, and give a logical course of action and diagnosis when random faults occur. However, it is hoped that you will use the manual by tackling the work yourself. On simpler jobs it may even be quicker than booking the car into a garage and going there twice, to leave and collect it. Perhaps most important, a lot of money can be saved by avoiding the costs a garage must charge to cover its labour and overheads.

The manual has drawings and descriptions to show the function of the various components so that their layout can be understood. Then the tasks are described and photographed in a step-by-step sequence so that even a novice can do the work.

Its arrangement

The manual is divided into fourteen chapters, each covering a logical sub-division of the vehicle. The Chapters are each divided into Sections, numbered with single figures, eg 5; and the Sections into paragraphs (or sub-sections), with decimal numbers following on from the Section they are in, eg 5.1, 5.2, 5.3 etc.

It is freely illustrated, especially in those parts where there is a detailed sequence of operations to be carried out. There are two forms of illustration: figures and photographs. The figures are numbered in sequence with decimal numbers, according to their position in the Chapter – eg Fig. 6.4 is the fourth drawing/illustration in Chapter 6. Photographs carry the same number (either individually or in related groups) as the Section or sub-section to which they relate.

There is an alphabetical index at the back of the manual as well as a contents list at the front. Each Chapter is also preceded by its own individual contents list.

References to the 'left' or 'right' of the vehicle are in the sense of a person in the driver's seat facing forwards.

Unless otherwise stated, nuts and bolts are removed by turning anti-clockwise, and tightened by turning clockwise.

Vehicle manufacturers continually make changes to specifications and recommendations, and these, when notified, are incorporated into our manuals at the earliest opportunity.

Whilst every care is taken to ensure that the information in this manual is correct, no liability can be accepted by the authors or publishers for loss, damage or injury caused by any errors in, or omissions from, the information given.

Introduction to the Toyota Celica

The original Toyota Celica was introduced in July 1971 as a two-door Coupe using a 1588 cc (113 cu in) engine and a four-speed manual gearbox. Development took place until January 1976 when a Liftback version was launched having a 1968 cc (120 cu in) engine.

In January 1978 (earlier in North America) restyled versions of the Coupe and Liftback were introduced using the 1588 cc engine for the Coupe and the 1968 cc engine for the Liftback. Five-speed gearboxes were fitted to all models. North American models are available with a 2189 cc (133.6 cu in) engine, and in 1981 with a 2366 cc (149.4 cu in) engine.

Contents

	Page
Acknowledgements	2
About this manual	2
Introduction to the Toyota Celica	2
Buying spare parts and vehicle identification numbers	5
Tools and working facilities	6
Lubrication chart	8
Routine maintenance	9
General dimensions	12
Jacking and towing	12
Use of English	13
Chapter 1 Engine	14
Chapter 2 Cooling system	67
Chapter 3 Fuel, exhaust and emission control systems	75
Chapter 4 Ignition system	117
Chapter 5 Clutch	131
Chapter 6 Transmission	137
Chapter 7 Propeller shaft	174
Chapter 8 Rear axle	179
Chapter 9 Braking system	186
Chapter 10 Electrical system	200
Chapter 11 Suspension and steering	241
Chapter 12 Bodywork and fittings	261
Chapter 13 Supplement: Revisions and information on later USA models	286
Chapter 14 Supplement: Wiring diagrams for later UK models	306
Safety first!	318
Conversion factors	319
Index	320

Toyota Celica Liftback XT

Toyota Celica Coupé

Buying spare parts and vehicle identification numbers

Buying spare parts

Spare parts are available from many sources, for example Toyota garages, other garages and accessory stores, and motor factors. Our advice regarding spare parts is as follows:

Officially appointed Toyota garages – This is the best source of parts which are peculiar to your car and otherwise not generally available (eg; complete cylinder heads, internal gearbox components, badges, interior trim etc). It is also the only place non-Toyota components may invalidate the warranty. To be sure of obtaining the correct parts it will always be necessary to give the storeman your car's engine and chassis number, and if possible, to take the old part along for positive identification. Remember that many parts are available on a factory exchange scheme – any parts returned should always be clean! It obviously makes good sense to go straight to the specialists on your car for this type of part for they are best equipped to supply you.

Other garages and accessory stores – These are often very good places to buy material and components needed for the maintenance of your car (eg; oil filters, spark plugs, bulbs, fan belts, oils and grease, touch-up paint, filler paste etc). They also sell general accessories, usually have convenient opening hours, often charge lower prices and can usually be found not far from home.

Motor factors – Good factors will stock all the more important components which wear out relatively quickly (eg; clutch components, pistons, valves, exhaust systems, brake cylinders/pipes'hoses'seals/shoes and pads etc). Motor factors will often provide new or reconditioned components on a part exchange basis – this can save a considerable amount of money.

Vehicle identification numbers

Modifications are a continuous and unpublicised process carried out by the vehicle manufacturers so accept the advice of the parts storeman when purchasing a component. Spare parts lists and manuals are compiled upon a numerical basis and individual vehicle numbers are essential to the supply of the correct component.

The engine serial number is stamped on the left-hand side of the cylinder block, close to the engine oil dipstick (photo).

The vehicle serial number is stamped on the engine compartment rear firewall (photo) and on some models this is repeated on top of the instrument panel and on the driver's door post. This identification number is the primary identification number for your Toyota.

On some models, a *Tune-up label* is fixed to the engine rocker cover. On North American models particularly, the information given on this label should be followed in preference to any other information.

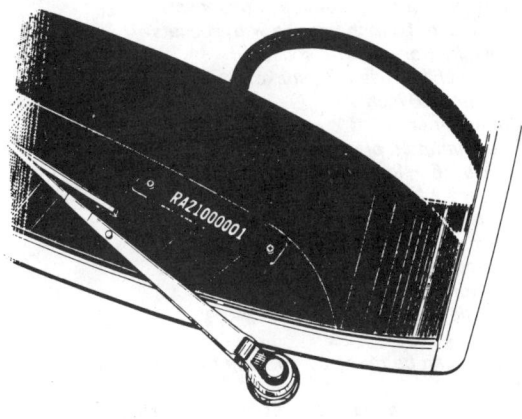

Vehicle number on fascia panel top surface

Engine number

Vehicle identification plate

18R engine tuning label

Tools and working facilities

Introduction

A selection of good tools is a fundamental requirement for anyone contemplating the maintenance and repair of a motor vehicle. For the owner who does not possess any, their purchase will prove a considerable expense, offsetting some of the savings made by doing-it-yourself. However, provided that the tools purchased meet the relevant national safety standards and are of good quality, they will last for many years and prove an extremely worthwhile investment.

To help the average owner to decide which tools are needed to carry out the various tasks detailed in this manual, we have compiled three lists of tools under the following headings: *Maintenance and minor repair*, *Repair and overhaul*, and *Special*. The newcomer to practical mechanics should start off with the *Maintenance and minor repair* tool kit and confine himself to the simpler jobs around the vehicle. Then, as his confidence and experience grows, he can undertake more difficult tasks, buying extra tools as, and when, they are needed. In this way, a *Maintenance and minor repair* tool kit can be built-up into a *Repair and overhaul* tool kit over a considerable period of time without any major cash outlays. The experienced do-it-yourselfer will have a tool kit good enough for most repair and overhaul procedures and will add tools from the *Special* category when he feels the expense is justified by the amount of use to which these tools will be put.

It is obviously not possible to cover the subject of tools fully here. For those who wish to learn more about tools and their use there is a book entitled *How to Choose and Use Car Tools* available from the publishers of this manual.

Maintenance and minor repair tool kit

The tools given in this list should be considered as a minimum requirement if routine maintenance, servicing and minor repair operations are to be undertaken. We recommend the purchase of combination spanners (ring one end, open-ended the other); although more expensive than open-ended ones, they do give the advantages of both types of spanner. All nuts, bolts screws and threads on the Lada are to metric standards.

Combination spanners - 9, 10, 11, 13 & 17 mm
Adjustable spanner - 9 inch
Engine sump/gearbox/rear axle drain plug key (where applicable)
Spark plug spanner (with rubber insert)
Spark plug gap adjustment tool
Set of feeler gauges
Brake adjuster spanner (where applicable)
Brake bleed nipple spanner
Screwdriver - 4 in long x $\frac{1}{4}$ in dia (flat blade)
Screwdriver - 4 in long x $\frac{1}{4}$ in dia (cross blade)
Combination pliers - 6 inch
Hacksaw, junior
Tyre pump
Tyre pressure gauge
Grease gun (where applicable)
Oil can
Fine emery cloth (1 sheet)
Wire brush (small)
Funnel (medium size)

Repair and overhaul tool kit

These tools are virtually essential for anyone undertaking any major repairs to a motor vehicle, and are additional to those given in the *Maintenance and minor repair* list. Included in this list is a comprehensive set of sockets. Although these are expensive they will be found invaluable as they are so versatile - particularly if various drives are included in the set. We recommend the $\frac{1}{2}$ in square-drive type, as this can be used with most proprietary torque wrenches. If you cannot afford a socket set, even bought piecemeal, then inexpensive tubular box spanners are a useful alternative.

The tools in this list will occasionally need to be supplemented by tools from the *Special* list.

Sockets (or box spanners) to cover range in previous list
Reversible ratchet drive (for use with sockets)
Extension piece, 10 inch (for use with sockets)
Universal joint (for use with sockets)
Torque wrench (for use with sockets)
'Mole' wrench - 8 inch
Ball pein hammer
Soft-faced hammer, plastic or rubber
Screwdriver - 6 in long x $\frac{5}{16}$ in dia (flat blade)
Screwdriver - 2 in long x $\frac{5}{16}$ in square (flat blade)
Screwdriver - 1$\frac{1}{2}$ in long x $\frac{1}{4}$ in dia (cross blade)
Screwdriver - 3 in long x $\frac{1}{8}$ in dia (electricians)
Pliers - electricians side cutters
Pliers - needle nosed
Pliers - circlip (internal and external)
Cold chisel - $\frac{1}{2}$ inch
Scriber (this can be made by grinding the end of a broken hacksaw blade)
Scraper (this can be made by flattening and sharpening one end of a piece of copper pipe)
Centre punch
Pin punch
Hacksaw
Valve grinding tool
Steel rule/straight edge
Allen keys
Selection of files
Wire brush (large)
Axle-stands
Jack (strong scissor or hydraulic type)

Special tools

The tools in this list are those which are not used regularly, are expensive to buy, or which need to be used in accordance with their manufacturers' instructions. Unless relatively difficult mechanical jobs are undertaken frequently, it will not be economic to buy many of these tools. Where this is the case, you could consider clubbing together with friends (or a motorists' club) to make a joint purchase, or borrowing the tools against a deposit from a local garage or tool hire specialist.

Tools and working facilities

The following list contains only those tools and instruments freely available to the public, and not those special tools produced by the vehicle manufacturer specifically for its dealer network. You will find occasional references to these manufacturers' special tools in the text of this manual. Generally, an alternative method of doing the job without the vehicle manufacturer's special tool is given. However, sometimes, there is no alternative to using them. Where this is the case and the relevant tool cannot be bought or borrowed you will have to entrust the work to a franchised garage.

Valve spring compressor
Piston ring compressor
Balljoint separator
Universal hub/bearing puller
Impact screwdriver
Micrometer and/or vernier gauge
Dial gauge
Stroboscopic timing light
Dwell angle meter/tachometer
Universal electrical multi-meter
Cylinder compression gauge
Lifting tackle
Trolley jack
Light with extension lead

Buying tools

For practically all tools, a tool factor is the best source since he will have a very comprehensive range compared with the average garage or accessory shop. Having said that, accessory shops often offer excellent quality tools at discount prices, so it pays to shop around.

There are plenty of good tools around at reasonable prices, but always aim to purchase items which meet the relevant national safety standards. If in doubt, ask the proprietor or manager of the shop for advice before making a purchase.

Care and maintenance of tools

Having purchased a reasonable tool kit, it is necessary to keep the tools in a clean serviceable condition. After use, always wipe off any dirt, grease and metal particles using a clean, dry cloth, before putting the tools away. Never leave them lying around after they have been used. A simple tool rack on the garage or workshop wall, for items such as screwdrivers and pliers is a good idea. Store all normal spanners and sockets in a metal box. Any measuring instruments, gauges, meters, etc, must be carefully stored where they cannot be damaged or become rusty.

Take a little care when tools are used. Hammer heads inevitably become marked and screwdrivers lose the keen edge on their blades from time to time. A little timely attention with emery cloth or a file will soon restore items like this to a good serviceable finish.

Working facilities

Not to be forgotten when discussing tools, is the workshop itself. If anything more than routine maintenance is to be carried out, some form of suitable working area becomes essential.

It is appreciated that many an owner mechanic is forced by circumstances to remove an engine or similar item, without the benefit of a garage or workshop. Having done this, any repairs should always be done under the cover of a roof.

Wherever possible, any dismantling should be done on a clean flat workbench or table at a suitable working height.

Any workbench needs a vice: one with a jaw opening of 4 in (100 mm) is suitable for most jobs. As mentioned previously, some clean dry storage space is also required for tools, as well as the lubricants, cleaning fluids, touch-up paints and so on which become necessary.

Another item which may be required, and which has a much more general usage, is an electric drill with a chuck capacity of at least $\frac{5}{16}$ in (8 mm). This, together with a good range of twist drills, is virtually essential for fitting accessories such as wing mirrors and reversing lights.

Last, but not least, always keep a supply of old newspapers and clean, lint-free rags available, and try to keep any working area as clean as possible.

Spanner jaw gap comparison table

Jaw gap (in)	Spanner size
0.250	$\frac{1}{4}$ in AF
0.277	7 mm
0.313	$\frac{5}{16}$ in AF
0.315	8 mm
0.344	$\frac{11}{32}$ in AF; $\frac{1}{8}$ in Whitworth
0.354	9 mm
0.375	$\frac{3}{8}$ in AF
0.394	10 mm
0.433	11 mm
0.438	$\frac{7}{16}$ in AF
0.445	$\frac{3}{16}$ in Whitworth; $\frac{1}{4}$ in BSF
0.472	12 mm
0.500	$\frac{1}{2}$ in AF
0.512	13 mm
0.525	$\frac{1}{4}$ in Whitworth; $\frac{5}{16}$ in BSF
0.551	14 mm
0.563	$\frac{9}{16}$ in AF
0.591	15 mm
0.600	$\frac{5}{16}$ in Whitworth; $\frac{3}{8}$ in BSF
0.625	$\frac{5}{8}$ in AF
0.630	16 mm
0.669	17 mm
0.686	$\frac{11}{16}$ in AF
0.709	18 mm
0.710	$\frac{3}{8}$ in Whitworth; $\frac{7}{16}$ in BSF
0.748	19 mm
0.750	$\frac{3}{4}$ in AF
0.813	$\frac{13}{16}$ in AF
0.820	$\frac{7}{16}$ in Whitworth; $\frac{1}{2}$ in BSF
0.866	22 mm
0.875	$\frac{7}{8}$ in AF
0.920	$\frac{1}{2}$ in Whitworth; $\frac{9}{16}$ in BSF
0.938	$\frac{15}{16}$ in AF
0.945	24 mm
1.000	1 in AF
1.010	$\frac{9}{16}$ in Whitworth; $\frac{5}{8}$ in BSF
1.024	26 mm
1.063	$1\frac{1}{16}$ in AF; 27 mm
1.100	$\frac{5}{8}$ in Whitworth; $\frac{11}{16}$ in BSF
1.125	$1\frac{1}{8}$ in AF
1.181	30 mm
1.200	$\frac{11}{16}$ in Whitworth; $\frac{3}{4}$ in BSF
1.250	$1\frac{1}{4}$ in AF
1.260	32 mm
1.300	$\frac{3}{4}$ in Whitworth; $\frac{7}{8}$ in BSF
1.313	$1\frac{5}{16}$ in AF
1.390	$\frac{13}{16}$ in Whitworth; $\frac{15}{16}$ in BSF
1.417	36 mm
1.438	$1\frac{7}{16}$ in AF
1.480	$\frac{7}{8}$ in Whitworth; 1 in BSF
1.500	$1\frac{1}{2}$ in AF
1.575	40 mm; $\frac{15}{16}$ in Whitworth
1.614	41 mm
1.625	$1\frac{5}{8}$ in AF
1.670	1 in Whitworth; $1\frac{1}{8}$ in BSF
1.688	$1\frac{11}{16}$ in AF
1.811	46 mm
1.813	$1\frac{13}{16}$ in AF
1.860	$1\frac{1}{8}$ in Whitworth; $1\frac{1}{4}$ in BSF
1.875	$1\frac{7}{8}$ in AF
1.969	50 mm
2.000	2 in AF
2.050	$1\frac{1}{4}$ in Whitworth; $1\frac{3}{8}$ in BSF
2.165	55 mm
2.362	60 mm

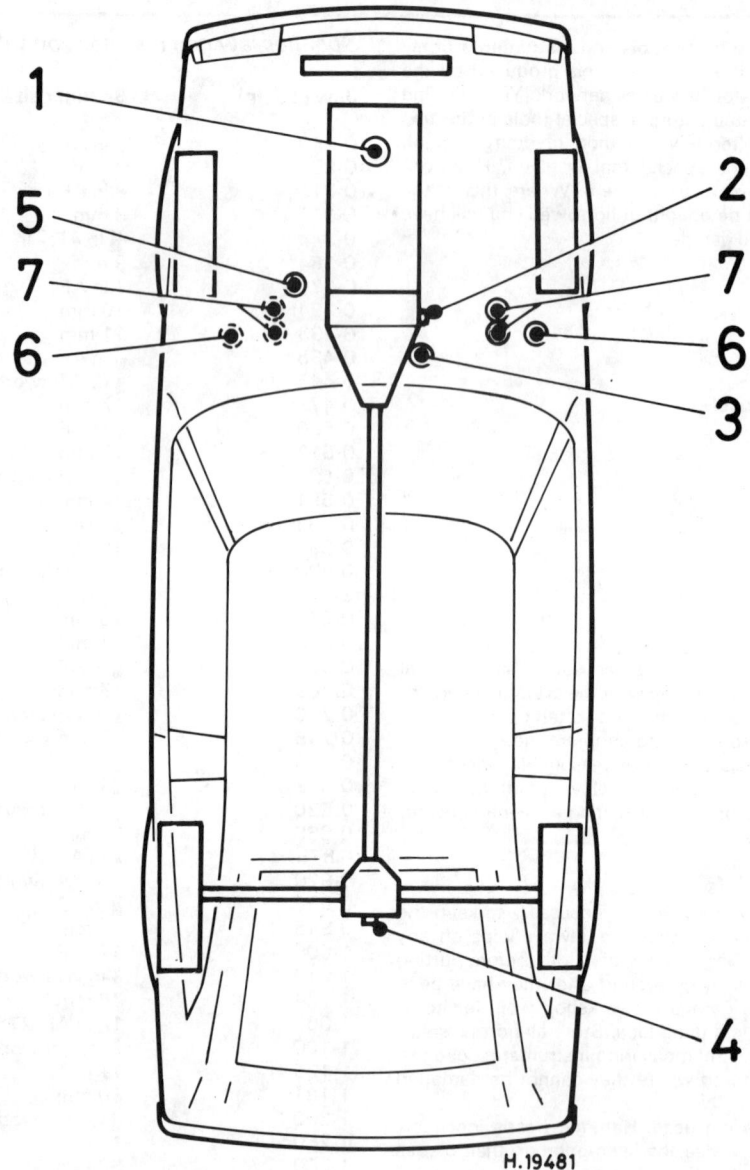

Recommended lubricants and fluids

Component or system	Lubricant type or specification
Engine (1)	SAE 15W/50 multigrade engine oil (API SE)
Manual Gearbox (2)	SAE 90 gear oil (API GL-4)
Automatic transmission (3)	ATF type F
Rear axle (4)	SAE 90 hypoid gear oil (API GL-5)
Manual steering (5)	SAE 90 gear oil (API GL-4)
Power steering	Dexron® ATF
Clutch (6)	DOT 3 or SAE J1703 hydraulic fluid
Braking system (7)	DOT 3 or SAE J1703 hydraulic fluid

Note: The above are general recommendations. Lubrication requirements vary from territory-to-territory and also depend on vehicle usage. Consult the operators handbook supplied with the vehicle.

Routine maintenance

Introduction

Regular maintenance will ensure that your car gives you safety and reliability and at the same time you will get maximum driving enjoyment, maximum fuel economy and long vehicle life. An equally important aspect of regular maintenance is that it will help to prevent breaches of Government regulations relating to vehicle condition and exhaust emission.

The need for periodic lubrication and greasing has now been virtually eliminated, but there are a number of things which still need to be looked at regularly so that breakdowns can be prevented.

Every 250 miles or weekly

Steering
Check tyre pressures
Examine tyres for wear or damage
Check that steering is smooth and accurate

Brakes
Check reservoir fluid level
Check braking efficiency
Try an emergency stop to verify brake adjustment and lining wear

Electrical equipment
Check all bulbs front and rear
Check operation of horns and windscreen wipers
Top-up windscreen washer fluid reservoir

Engine
Check and top-up engine oil
Check and top-up coolant
Check and top-up battery electrolyte level

Every 7500 miles (12 000 km) or 6 months

Engine
Drain engine oil, discard oil filter, fit new filter and fill engine sump with fresh oil
Inspect exhaust system for leaks and damage

Body and transmission
Check the operation of the clutch pedal, brake pedal and handbrake. Adjust their travel if necessary
Check the friction lining thickness of the front brake pads and the condition of the brake discs
Examine the brake pipes and hoses for signs of damage or corrosion
Check the fluid level in the clutch reservoir
Check the level of fluid in the power steering system (if fitted)
Check the oil level in the steering gearbox (manual steering)
Check the condition of the steering balljoints and protective covers
Check the oil level in the differential
Check the oil level in the gearbox, or automatic transmission as appropriate
Check the tightness of the chassis and body nuts and bolts
Clean sill and door drain holes

Checking brake fluid level

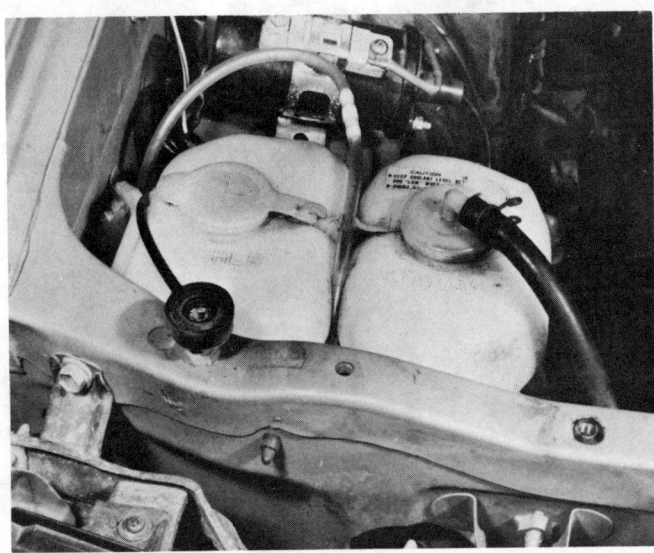

Washer and coolant reservoirs

Topping-up engine oil

Topping-up battery electrolyte

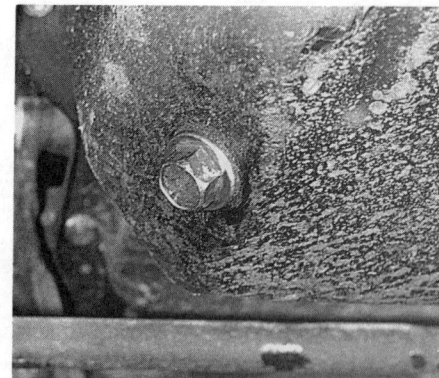
Sump drain plug

Oil filter

Checking clutch fluid level

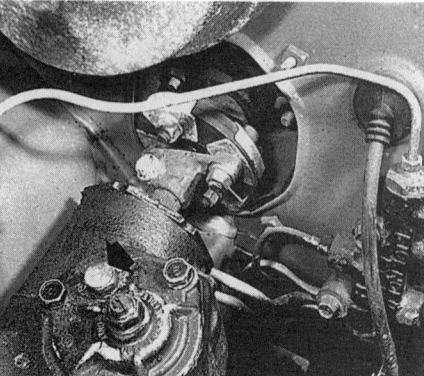
Manual steering box filler plug

Rear axle filler/level plug

Topping-up manual gearbox

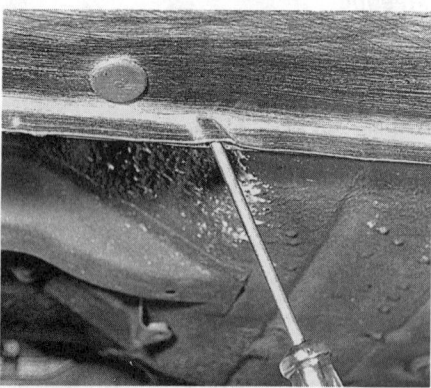
Clearing sill drain hole

Fuel line filter

Gearbox drain plug

Routine maintenance

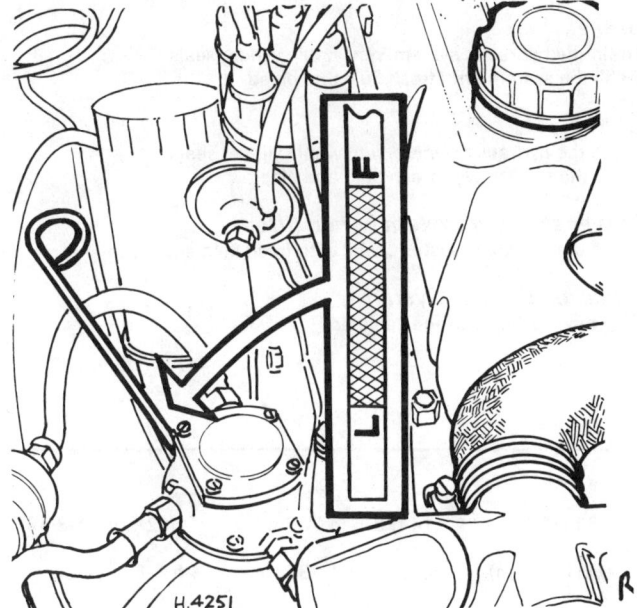

Engine oil dipstick markings

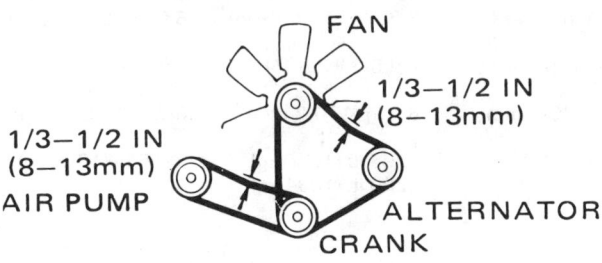

Fan/alternator/air pump drivebelt tensioning diagram

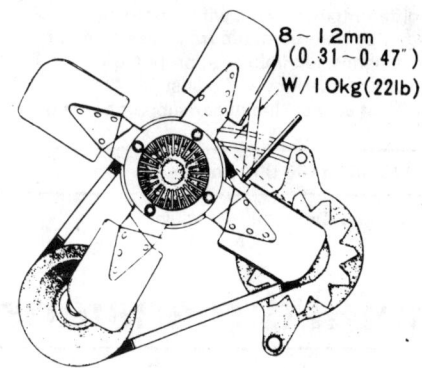

Alternator/fan drivebelt tensioning diagram

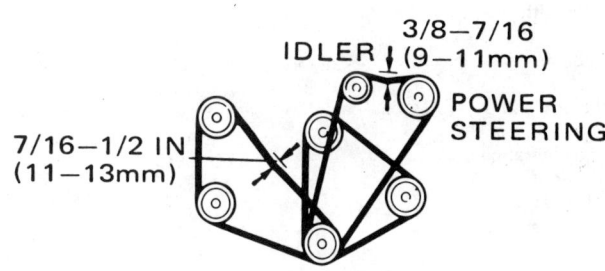

Fan/alternator/air pump/air conditioner/power steering drivebelt tensioning diagram

Every 15 000 miles (24 000 km) or 12 months

In addition to the 6 monthly schedule:

Engine
Adjust the valve clearances
Check the condition and tension of the drivebelts
Check the cooling system hoses for leaks and signs of deterioration
Check the condition of the vacuum hoses and fittings

Fuel system
Check the adjustment of the idle speed and idle mixture
Check for correct operation of the choke system
Inspect the condition of the air filter and clean it
Check the operation of the inlet air temperature control valve (when fitted)
Check the adjustment of the throttle positioner system, (when fitted)
Check the operation of the auxiliary accelerator pump (when fitted)
Check the operation of the deceleration fuel cut system (when fitted)

Ignition system
Discard the spark plugs and fit new ones which have been adjusted to the correct gap
Check the condition of the ignition wiring
Check the ignition timing

Crankcase Ventilation System
Check the conditions of the positive crankcase ventilation system, ensuring that none of the hoses are blocked or damaged

Emission Control System (when fitted)
Inspect and clean the carbon storage canister
Check the condition of the fuel vapour storage system hoses and connections

Electrical system
Examine the wiring harness for signs of damage and check that all connections are sound

Body and transmission
Check the condition of the rear brake drums and friction linings
Check the free-play of the steering wheel
Grease the steering balljoints at the base of the suspension struts
Check the correct operation of seat belt inertia locking mechanism
Check the operation of the seat belt warning system

Every 30 000 miles (48 000 km) or 24 months

In addition to the 12 monthly schedule:

Engine
Fit new drivebelts
Drain the cooling system and re-fill with fresh antifreeze solution of the correct dilution

Fuel system
Fit a new fuel filter element
Fit a new element to the air cleaner
Check the condition of the fuel tank cap, the final lines and all the fuel connections. Fit a new gasket to the fuel tank cap

Ignition system
Check the ignition timing and the operation of the distributor automatic advance

Crankcase Ventilation System
Fit a new PCV valve

Emission Control System (when fitted)
Check the operation of the air injection system

Body and transmission
Drain the differential and refill with fresh oil
Drain the gearbox or automatic transmission and refill with fresh oil, or automatic transmission fluid as appropriate
Grease the wheel bearings and adjust
Grease the front suspension upper support bearing

Every 60 000 miles (96 000 km) or 48 months

In addition to the 24 monthly schedule:

Brakes
Drain the hydraulic system, renew all cylinder seals
Refill the system with fresh fluid and bleed

Clutch
Drain the hydraulic system, renew all cylinder seals
Refill the system with fresh fluid and bleed

Power steering system (when fitted)
Drain the hydraulic system, refill with fresh fluid and bleed

Emission Control System
Fit a new carbon storage canister

General dimensions

	ST	XT	GT
Overall length	14 ft 2·5 in (4·3 m)	14 ft 2·5 in (4·3 m)	14 ft 2·5 in (4·3 m)
Overall height	4 ft 4·0 in (1·32 m)	4 ft 4·0 in (1·32 m)	4 ft 3·8 in (1·32 m)
Overall width	5 ft 4·4 in (1·63 m)	5 ft 4·6 in (1·64 m)	5 ft 4·6 in (1·64 m)
Wheelbase	8 ft 2·4 in (2·499 m)	8 ft 2·4 in (2·499 m)	8 ft 2·4 in (2·499 m)
Front track	4 ft 5·1 in (1·346 m)	4 ft 5·1 in (1·346 m)	4 ft 5·1 in (1·346 m)
Ground clearance	6·7 in (170·2 mm)	6·7 in (170·2 mm)	6·5 in (165·1 mm)
Kerb weight			
Manual	2255 lb (1022 kg)	2295 lb (1040 kg)	2335 lb (1059 kg)
Automatic		2280 lb (1034 kg)	

Jacking and towing

Jacking points
To change a roadwheel, use the body jack fitted under the sill. The jack should be positioned between the notches in the panel seam at the forward, or rear of the panel as appropriate (photo).

For maintenance and repair operations, use a hydraulic screw or trolley jack. To raise the front of the car place the jack under the front crossmember. To raise the rear of the car, place the jack under the centre of the rear axle below the differential housing. Before getting under the car, always supplement the jacks with axle-stands fitted under the bodyframe side members.

Towing
Towing eyes are fitted at both the front and the rear (photos) so that a rope can be attached to tow the vehicle when broken down, or stuck. When a vehicle with automatic transmission is being towed, the transmission must be placed in neutral and the speed limited to 30 mph (48 km/h) and the distance towed to 50 miles (80 km). If the car is to be towed faster, or further than this, the propeller shaft must be disconnected from the differential, to avoid damaging the transmission.

Whenever a vehicle is being towed with its front wheels on the ground, the ignition key must be in the ACC position so that the steering is not locked. The steering lock is not strong enough to hold the front wheels straight while the vehicle is being towed.

Tool kit scissors jack

Front towing eye

Rear towing eye

Use of English

As this book has been written in England, it uses the appropriate English component names, phrases, and spelling. Some of these differ from those used in America. Normally, these cause no difficulty, but to make sure, a glossary is printed below. In ordering spare parts remember the parts list may use some of these words:

English	American	English	American
Accelerator	Gas pedal	Locks	Latches
Aerial	Antenna	Methylated spirit	Denatured alcohol
Anti-roll bar	Stabiliser or sway bar	Motorway	Freeway, turnpike etc
Big-end bearing	Rod bearing	Number plate	License plate
Bonnet (engine cover)	Hood	Paraffin	Kerosene
Boot (luggage compartment)	Trunk	Petrol	Gasoline (gas)
Bulkhead	Firewall	Petrol tank	Gas tank
Bush	Bushing	'Pinking'	'Pinging'
Cam follower or tappet	Valve lifter or tappet	Prise (force apart)	Pry
Carburettor	Carburetor	Propeller shaft	Driveshaft
Catch	Latch	Quarterlight	Quarter window
Choke/venturi	Barrel	Retread	Recap
Circlip	Snap-ring	Reverse	Back-up
Clearance	Lash	Rocker cover	Valve cover
Crownwheel	Ring gear (of differential)	Saloon	Sedan
Damper	Shock absorber, shock	Seized	Frozen
Disc (brake)	Rotor/disk	Sidelight	Parking light
Distance piece	Spacer	Silencer	Muffler
Drop arm	Pitman arm	Sill panel (beneath doors)	Rocker panel
Drop head coupe	Convertible	Small end, little end	Piston pin or wrist pin
Dynamo	Generator (DC)	Spanner	Wrench
Earth (electrical)	Ground	Split cotter (for valve spring cap)	Lock (for valve spring retainer)
Engineer's blue	Prussian blue	Split pin	Cotter pin
Estate car	Station wagon	Steering arm	Spindle arm
Exhaust manifold	Header	Sump	Oil pan
Fault finding/diagnosis	Troubleshooting	Swarf	Metal chips or debris
Float chamber	Float bowl	Tab washer	Tang or lock
Free-play	Lash	Tappet	Valve lifter
Freewheel	Coast	Thrust bearing	Throw-out bearing
Gearbox	Transmission	Top gear	High
Gearchange	Shift	Torch	Flashlight
Grub screw	Setscrew, Allen screw	Trackrod (of steering)	Tie-rod (or connecting rod)
Gudgeon pin	Piston pin or wrist pin	Trailing shoe (of brake)	Secondary shoe
Halfshaft	Axleshaft	Transmission	Whole drive line
Handbrake	Parking brake	Tyre	Tire
Hood	Soft top	Van	Panel wagon/van
Hot spot	Heat riser	Vice	Vise
Indicator	Turn signal	Wheel nut	Lug nut
Interior light	Dome lamp	Windscreen	Windshield
Layshaft (of gearbox)	Countershaft	Wing/mudguard	Fender
Leading shoe (of brake)	Primary shoe		

Chapter 1 Engine

For modifications and information applicable to later USA models, refer to Supplement at end of manual

Contents

Auxiliary driveshaft (18R and 18R-G engines) – servicing	29
Camshaft and camshaft bearings – examination and renovation	26
Connecting rods and bearings – examination and renovation	23
Crankcase ventilation system (2T-B engine)	12
Crankcase ventilation system (18R engine)	17
Crankcase ventilation system (20R engine)	20
Crankshaft and main bearings – examination and renovation	22
Cylinder block – examination and renovation	33
Cylinder bores – examination and renovation	24
Cylinder head and valves – examination, renovation and decarbonising	28
Dismantling the engine – general	8
Driveplate (automatic transmission) – servicing	31
Engine/transmission – refitting	46
Engine/transmission – removal	5
Engine ancillary components – refitting	43
Engine ancillary components – removal	9
Engine dismantling (2T-B engine)	10
Engine dismantling (18R engine)	13
Engine dismantling (18R-G engine)	14
Engine dismantling (20R engine)	18
Engine reassembly (2T-B engine)	35
Engine reassembly (18R engine)	37
Engine reassembly (18R-G engine)	39
Engine reassembly (20R engine)	41
Engine reassembly – general	34
Engine – separation from automatic transmission	7
Engine – separation from manual gearbox	6
Engine to automatic transmission – reconnection	45
Engine to manual transmission – reconnection	44
Examination and renovation	21
Fault diagnosis – engine	48
Flywheel – servicing	30
General description	1
Initial start up after major repair or overhaul	47
Lubrication system, oil pump and filter (2T-B engine)	11
Lubrication system, oil pump and filter (18R engine)	15
Lubrication system, oil pump and filter (18R-G engine)	16
Lubrication system, oil pump and filter (20R engine)	19
Major operations only possible with engine removee	3
Major operations possible with engine in position	2
Method of engine removal	4
Oil seals – renewal	32
Pistons and piston rings – examination and renovation	25
Timing components – examination and renovation	27
Valve clearances (2T-B engine) – adjustment	36
Valve clearances (18R engine) – adjustment	38
Valve clearances (18R-G engine) – adjustment	40
Valve clearances (20R engine) – adjustment	42

Specifications

2T-B engine (used in TA40B model)

General

Type	4 cylinder, in-line OHV
Bore	3·35 in (85·0 mm)
Stroke	2·76 in (70·0 mm)
Capacity	1588 cc (96·9 cu in)
Compression ratio	9·4 : 1
Firing order	1 – 3 – 4 – 2
Ignition timing	12° BTDC at 800 rpm
Valve timing:	
Inlet opens	16° BTDC
Inlet closes	54° ABDC
Exhaust open	58° BBDC
Exhaust closes	12° ATDC

Valve clearances:

	Inlet	Exhaust
Hot	0·008 in (0·20 mm)	0·013 in (0·33 mm)
Cold	0·007 in (0·18 mm)	0·012 in (0·30 mm)

Oil capacity (with filter change)	7·9 Imp pints (4·4 litres)

Cylinder head

Warpage limit	0·002 in (0·05 mm)
Valve seat angle	45°
Valve seat contact width	0·055 in (1·4 mm)

Cylinder block

Warpage limit	0·002 in (0·05 mm)
Bore diameter – standard	3·3465 to 3·3484 in (85·00 to 85·05 mm)
Wear limit	0·008 in (0·2 mm)

Chapter 1 Engine

Pistons
Outside diameter – standard ..... 3·3441 to 3·3461 in (84·94 to 84·99 mm)
Piston oversizes ..... 0·25, 0·5, 0·75, 1·00 mm
Cylinder-to-piston clearance ..... 0·0020 to 0·0028 in (0·05 to 0·07 mm)

Piston ring
Piston ring endgap:
  Compression ring No 1 ..... 0·008 to 0·016 in (0·2 to 0·4 mm)
  Compression ring No 2 ..... 0·004 to 0·012 in (0·1 to 0·3 mm)
  Oil control ring ..... 0·008 to 0·020 in (0·2 to 0·5 mm)
Piston ring to groove clearance:
  Compression ring No 1 ..... 0·0008 to 0·0024 in (0·020 to 0·060 mm)
  Compression ring No 2 ..... 0·0006 to 0·0022 in (0·015 to 0·055 mm)

Connecting rod
Bend limit ..... 0·002 in (0·05 mm)
Sidefloat:
  Standard ..... 0·0063 to 0·0102 in (0·16 to 0·26 mm)
  Limit ..... 0·0118 in (0·3 mm)
Bearing undersizes ..... 0·05, 0·25, 0·50, 0·75 mm

Crankshaft
Bend limit ..... 0·0012 in (0·03 mm)
Endfloat:
  Standard ..... 0·003 to 0·007 in (0·07 to 0·18 mm)
  Limit ..... 0·012 in (0·3 mm)
Journal diameter ..... 2·2825 to 2·2835 in (57·976 to 58·000 mm)
Journal undersizes ..... 0·05, 0·25, 0·50 mm
Crankpin diameter ..... 1·8888 to 1·8898 in (49·976 to 48·000 mm)
Crankpin undersizes ..... 0·25, 0·50, 0·75 mm

Flywheel
Run-out limit ..... 0·004 in (0·1 mm)

Camshaft
Bend limit ..... 0·0012 in (0·03 mm)
Endfloat:
  Standard ..... 0·0028 to 0·0086 in (0·07 to 0·22 mm)
  Limit ..... 0·012 in (0·3 mm)
Bearing clearance:
  Standard ..... 0·0010 to 0·0026 in (0·025 to 0·066 mm)
  Limit ..... 0·004 in (0·1 mm)
Bearing undersizes ..... 0·125, 0·250 mm

Valve lifter (cam follower)
Outside diameter:
  Standard ..... 0·8732 to 0·8739 in (22·179 to 22·199 mm)
  Oversize ..... 0·8752 to 0·8759 in (22·229 to 22·249 mm)

Valves
Head diameter:
  Inlet ..... 1·61 in (41 mm)
  Exhaust ..... 1·42 in (36 mm)
Valve length:
  Standard ..... 4·29 in (109 mm)
  Minimum ..... 4·272 in (108·5 mm)
Stem diameter:
  Inlet ..... 0·3138 to 0·3146 in (7·97 to 7·99 mm)
  Exhaust ..... 0·3138 to 0·3142 in (7·97 to 7·98 mm)
Valve stem to guide clearance:
  Inlet:
    Standard ..... 0·0010 to 0·0024 in (0·025 to 0·060 mm)
    Limit ..... 0·003 in (0·08 mm)
  Exhaust:
    Standard ..... 0·0012 to 0·0026 in (0·030 to 0·065 mm)
    Limit ..... 0·004 in (0·1 mm)
Valve head thickness limit:
  Inlet ..... 0·020 in (0·5 mm)
  Exhaust ..... 0·028 in (0·7 mm)
Valve contact face angle ..... 45°

Valve springs
Free length ..... 1·657 in (42·1 mm)
Installed length ..... 1·484 in (37·7 mm)

Valve rocker shaft
Diameter ..... 0·6287 to 0·6295 in (15·97 to 15·99 mm)

Shaft-to-arm clearance .... 0·0008 to 0·0016 in (0·02 to 0·04 mm)

Manifold
Warpage limit of joint face .... 0·008 in (0·2 mm)

Oil pump
Type .... Trochoid
Rotor tip clearance:
  Standard .... 0·0016 to 0·0063 in (0·04 to 0·16 mm)
  Limit .... 0·010 in (0·25 mm)
Rotor end clearance:
  Standard .... 0·0012 to 0·0035 in (0·03 to 0·09 mm)
  Limit .... 0·006 in (0·15 mm)
Outer rotor-to-body clearance:
  Standard .... 0·0039 to 0·0063 in (0·10 to 0·16 mm)
  Limit .... 0·010 in (0·25 mm)
Relief valve operating pressure .... 50 to 63 lbf in^2 (3·6 to 4·4 kgf/cm^2)

18R engine (used in RA40 models)

General
Type .... 4 cylinder, in-line, single, OHC
Bore .... 3·48 in (88·5 mm)
Stroke .... 3·14 in (80·0 mm)
Capacity .... 1968 cc (120 cu in)
Compression ratio .... 8·5 : 1
Firing order .... 1 – 3 – 4 – 2
Ignition timing .... 7° BTDC at 650 rpm
Valve clearances (hot):
  Inlet .... 0·008 in (0·20 mm)
  Exhaust .... 0·014 in (0·36 mm)
Oil capacity (with filter change) .... 6·8 Imp pints (3·8 litres)

Cylinder head
Warpage limit .... 0·0019 in (0·05 mm)
Valve seat angle .... 45°
Valve seat contact width .... 0·047 to 0·063 in (1·2 to 1·6 mm)

Valves
Overall length (inlet and exhaust) .... 4·437 in (112·7 mm)
Valve stem diameter:
  Inlet .... 0·3138 to 0·3144 in (7·970 to 7·985 mm)
  Exhaust .... 0·3139 to 0·3140 in (7·960 to 7·975 mm)
Valve stem clearance
  Inlet:
    Standard .... 0·0012 to 0·0024 in (0·03 to 0·06 mm)
    Limit .... 0·0032 in (0·08 mm)
  Exhaust:
    Standard .... 0·0016 to 0·0032 in (0·04 to 0·08 mm)
    Limit .... 0·0039 in (0·10 mm)
Valve head thickness limit .... 0·024 in (0·6 mm)

Valve spring
Free length:
  Inner .... 1·736 in (44·1 mm)
  Outer .... 1·830 in (46·5 mm)
Installed length:
  Inner .... 1·476 in (37·5 mm)
  Outer .... 1·634 in (41·5 mm)

Camshaft
Bend limit .... 0·004 in (0·10 mm)
Endfloat:
  Standard .... 0·0056 to 0·0076 in (0·04 to 0·17 mm)
  Limit .... 0·0098 in (0·25 mm)
Bearing clearance:
  Standard .... 0·0012 to 0·0024 in (0·03 to 0·06 mm)
  Limit .... 0·0039 in (0·1 mm)
Journal diameter .... 1·3768 to 1·3780 in (34·97 to 35·00 mm)
Bearing undersizes .... 0·125, 0·25 mm
Valve rocker arm and shaft clearance:
  Standard .... 0·0008 to 0·0020 in (0·02 to 0·05 mm)
  Limit .... 0·0032 in (0·08 mm)

Manifold
Warpage limit of joint face .... 0·016 in (0·4 mm)

Auxiliary driveshaft
Endfloat:
   Standard .... 0·0024 to 0·0051 in (0·06 to 0·13 mm)
   Limit .... 0·012 in (0·3 mm)
Journal diameter:
   Front .... 1·8098 to 1·8106 in (45·96 to 45·98 mm)
   Rear .... 1·6126 to 1·6134 in (40·96 to 40·98 mm)
Running clearance:
   Standard .... 0·0012 to 0·0028 in (0·03 to 0·07 mm)
   Limit .... 0·0032 in (0·08 mm)

Cylinder block
Warpage limit .... 0·002 in (0·05 mm)
Cylinder bore – standard .... 3·4842 to 3·4862 in (88·49 to 88·54 mm)
Wear limit .... 0·008 in (0·2 mm)

Crankshaft
Run-out limit .... 0·0040 in (0·1 mm)
Endfloat:
   Standard .... 0·0008 to 0·0079 in (0·02 to 0·20 mm)
   Limit .... 0·0118 in (0·3 mm)
Crankpin journal clearance:
   Standard .... 0·0008 to 0·0020 in (0·02 to 0·05 mm)
   Limit .... 0·0032 in (0·08 mm)
Bearing undersizes .... 0·05, 0·25, 0·50 mm
Journal diameter – standard .... 2·0857 to 2·0866 in (52·976 to 53·000 mm)
Crank journal clearance:
   Standard .... 0·0008 to 0·0020 in (0·02 to 0·05 mm)
   Limit .... 0·0032 in (0·08 mm)
Bearing undersizes .... 0·05, 0·25, 0·50 mm
Journal diameter – standard .... 2·3613 to 2·3622 in (59·976 to 60·000 mm)

Pistons
Outside diameter – standard .... 3·4819 to 3·4839 in (88·44 to 88·49 mm)
Piston oversizes .... 0·50, 1·00 mm
Cylinder-to-piston clearance .... 0·0020 to 0·0028 in (0·05 to 0·07 mm)

Piston ring
Piston ring endgap:
   Compression ring No 1 .... 0·0039 to 0·0118 in (0·10 to 0·30 mm)
   Compression ring No 2 .... 0·0039 to 0·0118 in (0·10 to 0·30 mm)
   Oil control ring .... 0·008 to 0·020 in (0·2 to 0·5 mm)
Piston ring-to-groove clearance:
   Compression ring No 1 .... 0·0008 to 0·0024 in (0·02 to 0·06 mm)
   Compression ring No 2 .... 0·0008 to 0·0024 in (0·02 to 0·06 mm)

Connecting rod
Big-end side-float:
   Standard .... 0·0063 to 0·0102 in (0·16 to 0·26 mm)
   Limit .... 0·012 in (0·3 mm)
Bearing clearance:
   Standard .... 0·0008 to 0·0012 in (0·02 to 0·05 mm)
   Limit .... 0·0031 in (0·08 mm)
Bearing undersizes .... 0·05, 0·25, 0·50, 0·75, 1·00 mm
Small-end bush clearance:
   Standard .... 0·00020 to 0·00055 in (0·005 to 0·014 mm)
   Limit .... 0·00059 in (0·015 mm)

Flywheel
Run-out limit .... 0·008 in (0·2 mm)

Oil pump
Type .... Trochoid
Rotor tip clearance:
   Standard .... 0·0039 to 0·0059 in (0·10 to 0·15 mm)
   Limit .... 0·008 in (0·2 mm)
Rotor end clearance:
   Standard .... 0·0012 to 0·0028 in (0·03 to 0·07 mm)
   Limit .... 0·0059 in (0·15 mm)
Outer rotor-to-body clearance:
   Standard .... 0·0039 to 0·0063 in (0·10 to 0·16 mm)
   Limit .... 0·008 in (0·2 mm)

18R-G engine (used in RA40NQ models)

General
Type .... 4 cylinder in-line, twin ohc

Bore ... 3·48 in (88·5 mm)
Stroke ... 3·14 in (80 mm)
Capacity ... 1968 cc (120 cu in)
Compression ratio ... 9·7 : 1
Firing order ... 1 – 3 – 4 – 2
Ignition timing:
  Static ... 5° BTDC
  Coolant above 60°C ... 5° BTDC at 1000 rpm
Valve clearances (cold):
  Inlet ... 0·010 to 0·013 in (0·26 to 0·32 mm)
  Exhaust ... 0·012 to 0·015 in (0·31 to 0·36 mm)
Oil capacity (with filter change) ... 8·8 Imp pints (5·0 litres)

Cylinder head
Warpage limit ... 0·0019 in (0·05 mm)
Valve seat angle ... 45°
Valve seat contact width ... 0·047 to 0·063 in (1·2 to 1·6 mm)

Valves
Overall length:
  Inlet ... 4·20 in (106·8 mm)
  Exhaust ... 4·14 in (105·1 mm)
Valve stem diameter:
  Inlet ... 0·3333 to 0·3338 in (8·465 to 8·480 mm)
  Exhaust ... 0·3330 to 0·3337 in (8·460 to 8·475 mm)
Valve stem clearance (standard):
  Inlet ... 0·0008 to 0·0020 in (0·02 to 0·05 mm)
  Exhaust ... 0·0012 to 0·0024 in (0·03 to 0·06 mm)
Valve stem clearance (limit):
  Inlet ... 0·0032 in (0·08 mm)
  Exhaust ... 0·0039 in (0·10 mm)
Valve head thickness limit:
  Inlet ... 0·02 in (0·5 mm)
  Exhaust ... 0·024 in (0·6 mm)

Valve spring
Free length ... 1·795 in (45·6 mm)
Installed length ... 1·535 in (39·0 mm)

Cam follower
Cam follower clearance in cylinder head (COLD):
  Standard ... 0·0008 to 0·0012 in (0·02 to 0·03 mm)
  Limit ... 0·004 in (0·1 mm)

Camshaft
Bend limit ... 0·0012 in (0·03 mm)
Endfloat:
  Standard ... 0·0059 to 0·0138 in (0·15 to 0·35 mm)
  Limit ... 0·0158 in (0·4 mm)
Journal clearance:
  Standard ... 0·0020 to 0·0035 in (0·05 to 0·09 mm)
  Limit ... 0·0059 in (0·15 mm)
Journal diameter – standard ... 1·2572 to 1·258 in (31·934 to 31·950 mm)
Cam height (inlet and exhaust):
  Standard ... 1·786 to 1·790 in (45·37 to 45·47 mm)
  Limit ... 1·77 in (45·0 mm)

Manifold
Warpage limit (inlet and exhaust) ... 0·0039 in (0·1 mm)

Auxiliary driveshaft
Endfloat:
  Standard ... 0·0024 to 0·0051 in (0·06 to 0·13 mm)
  Limit ... 0·012 in (0·3 mm)
Journal diameter:
  Front ... 1·7949 to 1·8012 in (45·59 to 45·75 mm)
  Rear ... 1·5980 to 1·6043 in (40·59 to 40·75 mm)
Running clearance:
  Standard ... 0·0012 to 0·0028 in (0·03 to 0·07 mm)
  Limit ... 0·0032 in (0·08 mm)

Cylinder block
Warpage limit ... 0·0019 in (0·05 mm)
Cylinder bore – standard ... 3·484 to 3·486 in (88·50 to 88·55 mm)
Wear limit ... 0·008 in (0·2 mm)

Chapter 1 Engine

Crankshaft
Run-out limit .... 0·0020 in (0·05 mm)
Endfloat:
    Standard .... 0·0008 to 0·0079 in (0·02 to 0·20 mm)
    Limit .... 0·0118 in (0·3 mm)
Crankpin journal clearance:
    Standard .... 0·0008 to 0·0020 in (0·02 to 0·05 mm)
    Limit .... 0·0032 in (0·08 mm)
Bearing undersizes .... 0·05, 0·25, 0·50 mm
Crank journal clearance:
    Standard .... 0·0008 to 0·0020 in (0·02 to 0·05 mm)
    Limit .... 0·0032 in (0·08 mm)
Bearing undersizes .... 0·05, 0·25, 0·50 mm
Journal diameter – standard .... 2·3613 to 2·3622 in (59·976 to 60·000 mm)

Pistons
Outside diameter – standard .... 3·4819 to 3·4839 in (88·44 to 88·49 mm)
Piston oversizes .... 0·50, 1·00 mm
Cylinder-to-piston clearance .... 0·0020 to 0·0028 in (0·05 to 0·07 mm)

Piston ring
Piston ring endgap:
    Compression ring No 1 .... 0·0039 to 0·0118 in (0·10 to 0·30 mm)
    Compression ring No 2 .... 0·0039 to 0·0118 in (0·10 to 0·30 mm)
    Oil control ring .... 0·008 to 0·020 in (0·2 to 0·5 mm)
Piston ring-to-groove clearance:
    Compression ring No 1 .... 0·0008 to 0·0024 in (0·02 to 0·06 mm)
    Compression ring No 2 .... 0·0008 to 0·0024 in (0·02 to 0·06 mm)

Connecting rod
Big-end side-float:
    Standard .... 0·0063 to 0·010 in (0·16 to 0·26 mm)
    Limit .... 0·012 in (0·3 mm)
Bearing clearance:
    Standard .... 0·0008 to 0·0020 in (0·02 to 0·05 mm)
    Limit .... 0·0032 in (0·08 mm)
Bearing undersizes .... 0·25, 0·50, 0·75, 1·00 mm
Small-end bush clearance:
    Standard .... 0·00020 to 0·00055 in (0·005 to 0·014 mm)
    Limit .... 0·00059 in (0·015 mm)

Flywheel
Run-out limit .... 0·008 in (0·2 mm)

Oil pump
Type .... Trochoid
Rotor tip clearance:
    Standard .... 0·0039 to 0·0059 in (0·10 to 0·15 mm)
    Limit .... 0·008 in (0·2 mm)
Rotor end clearance:
    Standard .... 0·0012 to 0·0028 in (0·03 to 0·07 in)
    Limit .... 0·0059 in (0·15 mm)
Outer rotor-to-body clearance:
    Standard .... 0·0039 to 0·0063 in (0·10 to 0·16 mm)
    Limit .... 0·008 in (0·2 mm)

20R engine (used in RA42 models)

General
Type .... 4 cylinder, in-line, single OHC
Bore .... 3·48 in (88·5 mm)
Stroke .... 3·50 in (89·0 mm)
Capacity .... 2189 cc (133·6 cu in)
Compression ratio .... 8·4 : 1
Firing order .... 1 – 3 – 4 – 2
Ignition timing .... 8° BTDC at 850 rpm
Valve clearances (hot):
    Inlet .... 0·008 in (0·2 mm)
    Exhaust .... 0·012 in (0·3 mm)
Oil capacity (with filter change) .... 8·8 Imp pints (5·0 litres)

Cylinder head
Warpage limit .... 0·0059 in (0·15 mm)
Valve seat angle .... 45°
Valve seat contact width .... 0·047 to 0·063 in (1·2 to 1·6 mm)

Valves
Stem diameter:
- Inlet ... 0·3138 to 0·3144 in (7·97 to 7·99 mm)
- Exhaust ... 0·3136 to 0·3142 in (7·97 to 7·98 mm)

Stem clearance (standard):
- Inlet ... 0·0006 to 0·0024 in (0·02 to 0·06 mm)
- Exhaust ... 0·0012 to 0·0026 in (0·03 to 0·07 mm)

Stem clearance (limit):
- Inlet ... 0·0031 in (0·08 mm)
- Exhaust ... 0·0039 in (0·10 mm)

Valve springs
- Free length ... 1·759 in (45·6 mm)
- Installed length ... 1·594 in (40·5 mm)

Camshaft
Endfloat:
- Standard ... 0·0031 to 0·0071 in (0·08 to 0·18 mm)
- Limit ... 0·0098 in (0·25 mm)

Journal clearance:
- Standard ... 0·0004 to 0·0020 in (0·01 to 0·05 mm)
- Limit ... 0·004 in (0·1 mm)
- Journal diameter ... 1·2984 to 1·2990 in (32·98 to 33·00 mm)

Cam lobe height:
- Inlet ... 1·6783 to 1·6819 in (42·63 to 42·72 mm)
- Exhaust ... 1·6806 to 1·6841 in (42·69 to 42·78 mm)

Rocker arm and shaft
- Rocker shaft diameter ... 0.6287 to 0.6295 in (15.97 to 15.99 mm)
- Shaft-to-arm clearance ... 0·0004 to 0·0020 in (0·01 to 0·05 mm)

Inlet and exhaust manifolds
Warpage limit:
- Inlet ... 0·008 in (0·2 mm)
- Exhaust ... 0·012 in (0·3 mm)

Cylinder block
- Cylinder bore – standard ... 3·4842 to 3·4854 in (88·50 to 88·53 mm)
- Wear limit ... 0·008 in (0·2 mm)

Pistons
- Outside diameter – standard ... 3·4827 to 3·4839 in (88·46 to 88·49 mm)
- Piston oversizes ... 0·50, 1·00 mm
- Cylinder-to-piston clearance ... 0·0012 to 0·0020 in (0·03 to 0·05 mm)

Piston ring endgap:
- Compression rings ... 0·004 to 0·012 in (0·1 to 0·3 mm)
- Piston ring-to-groove clearance – limit ... 0·008 in (0·2 mm)

Connecting rod
Big-end side-float:
- Standard ... 0·0063 to 0·0102 in (0·16 to 0·26 mm)
- Limit ... 0·012 in (0·3 mm)
- Bearing clearance – standard ... 0·0010 to 0·0022 in (0·025 to 0·055 mm)
- Bend limit ... 0·002 in (0·05 mm)

Small-end bush clearance:
- Standard ... 0·0002 to 0·0004 in (0·005 to 0·011 mm)
- Limit ... 0·0006 in (0·015 mm)

Crankshaft
Endfloat:
- Standard ... 0·0008 to 0·0079 in (0·02 to 0·20 mm)
- Limit ... 0·012 in (0·3 mm)

Thrust washer thicknesses:
- Standard ... 0·0787 in (2·0 mm)
- O/S 0·125 ... 0·0811 in (2·06 mm)
- O/S 0·25 ... 0·0839 in (2·13 mm)
- Run-out limit ... 0·004 in (0·1 mm)

Crank journal diameter:
- Standard ... 2·3614 to 2·3622 in (59·98 to 60·00 mm)
- Undersize ... 2·3504 to 2·3508 in (59·70 to 59·71 mm)
- Journal clearance – standard ... 0·0010 to 0·0022 in (0·025 to 0·055 mm)

Flywheel
- Run-out limit ... 0·008 in (0·2 mm)

Oil pump
Type ..... Epicyclic gear
Body clearance:
   Standard ..... 0·0024 to 0·0059 in (0·06 to 0·15 mm)
   Limit ..... 0·008 in (0·2 mm)
Tip clearance – driven gear to crescent:
   Standard ..... 0·0059 to 0·0083 in (0·15 to 0·21 mm)
   Limit ..... 0·012 in (0·3 mm)
Tip clearance – drivegear to crescent:
   Standard ..... 0·0087 to 0·0098 in (0·22 to 0·25 mm)
   Limit ..... 0·012 in (0·3 mm)
Side clearance:
   Standard ..... 0·0012 to 0·0039 in (0·03 to 0·09 mm)
   Limit ..... 0·0059 in (0·15 mm)
Relief valve operating pressure ..... 64 lbf/in² (4·5 kg/cm²)

Torque wrench settings

	lbf ft	Nm

2T-B
	lbf ft	Nm
Crankshaft main bearing cap	60	83
Connecting rod cap	35	48
Camshaft thrust plate-to-cylinder head	10	14
Timing chain cover	10	14
Oil pump-to-cylinder block	15	21
Crankshaft pulley	40	55
Cylinder head	63	87
Camshaft timing sprocket	75	104
Valve rocker support	60	83
Manifold-to-cylinder head	10	14
Cylinder head cover	5·0	7·0
Sump	5·0	7·0
Sump drain plug	25	35
Spark plug	15	21
Flywheel	45	62
Clutch cover	15	21
Clutch housing-to-cylinder block	45	62

18R
	lbf ft	Nm
Cylinder head	87	121
Valve rocker support	15	21
Manifold	35	48
Camshaft bearing cap	15	21
Camshaft timing sprocket	15	21
Auxiliary shaft sprocket	65	90
Crankshaft main bearing cap	75	104
Connecting rod cap	45	62
Sump	5	7·0
Crankshaft pulley	75	104
Flywheel	60	83
Thermo switch	25	35
Clutch bellhousing-to-block	50	69
Driveplate-to-torque converter	20	28
Driveplate-to-crankshaft	45	62
Torque converter housing-to-block	50	69

Otherwise as 2T-B engine

18R-G
	lbf ft	Nm
Cylinder head	63	87
Camshaft bearing cap	15	21
Camshaft timing sprocket	55	76
Auxiliary shaft drive sprocket	45	62
Manifold (Inlet and Exhaust)	10	14
Crankshaft main bearing cap	75	104
Connecting rod cap	48	66
Sump	5	7·0
Crankshaft pulley	72	100
Flywheel	60	83
Thermo vacuum switching valve	25	35

Otherwise as 2T-B engine

20R
	lbf ft	Nm
Cylinder head	63	87
Camshaft bearing cap	15	21
Camshaft sprocket	60	83
Crankshaft main bearing cap	80	111
Connecting rod cap	45	62

Crankshaft pulley	85	118
Flywheel	65	90
Drive plate-to-torque converter	20	28
Driveplate-to-crankshaft	60	83
Torque converter housing-to-engine	50	69

Otherwise as 2T-B engine

1 General description

Variants of three basic engines are used on the range of models covered by this Manual. The smallest engine is of the 2 T-B type which has pushrod operated valves. The middle size of engine is the 18R, which is of single overhead camshaft type. The 18R-C engine used in North American models is basically the same as the 18R of the UK range except for the emission control equipment and the related differences in carburation and ignition equipment. The 18R-G engine is a twin camshaft version of the 18R engine.

Some models in the USA are fitted with the 20R engine which has a single overhead camshaft, and a crossflow cylinder head.

All the engines are of the four cylinder, in-line type with a five bearing crankshaft.

2 Major operations possible with engine in position

1   The following operations can be carried out without removing the engine from the vehicle:

(a) Removal and refitting of the rocker box cover. On vehicles fitted with air conditioning, the clamp which retains the flexible pipe must be removed from the rocker box cover and the pipe pulled away sufficiently to provide clearance for the removal of the cover. On no account disconnect any of the air conditioning system circuit pipes, or unions (if fitted)
(b) Removal and refitting of the valve gear and cylinder head
(c) Removal and refitting of the timing chain and gears
(d) Removal and refitting of the engine front mountings and the transmission rear mounting

2   If a hoist is attached to the engine lifting lugs and the weight of the engine is taken up, the engine mounts can be disconnected and the engine raised a few inches. Disconnect the steering relay rods (Chapter 11) to permit the following operations to be carried out:

(a) Removal and refitting of the sump
(b) Removal and refitting the oil pump
(c) Removal and refitting of the piston/connecting rod assemblies (after removing the sump and cylinder head)

3 Major operations only possible with engine removed

The engine must be removed to carry out the following operations:

(a) Removal of the flywheel, or driveplate (automatic transmission)
(b) Removal and fitting of the crankshaft and main bearings

4 Method of engine removal

1   The engine may be removed with, or without the gearbox/automatic transmission. The procedure given in this Chapter is for the removal of the engine and gearbox, or engine and automatic transmission together. Where the gearbox or automatic transmission is to be left in the vehicle, it must be first disconnected from the engine (see Chapter 6), otherwise the operations are identical.

2   When a vehicle is fitted with automatic transmission and no work is required on the transmission when the engine is to be removed, it is preferable to leave the transmission in place.

3   Make sure that lifting gear of adequate capacity is available before starting to remove an engine. If possible also have a trolley jack available.

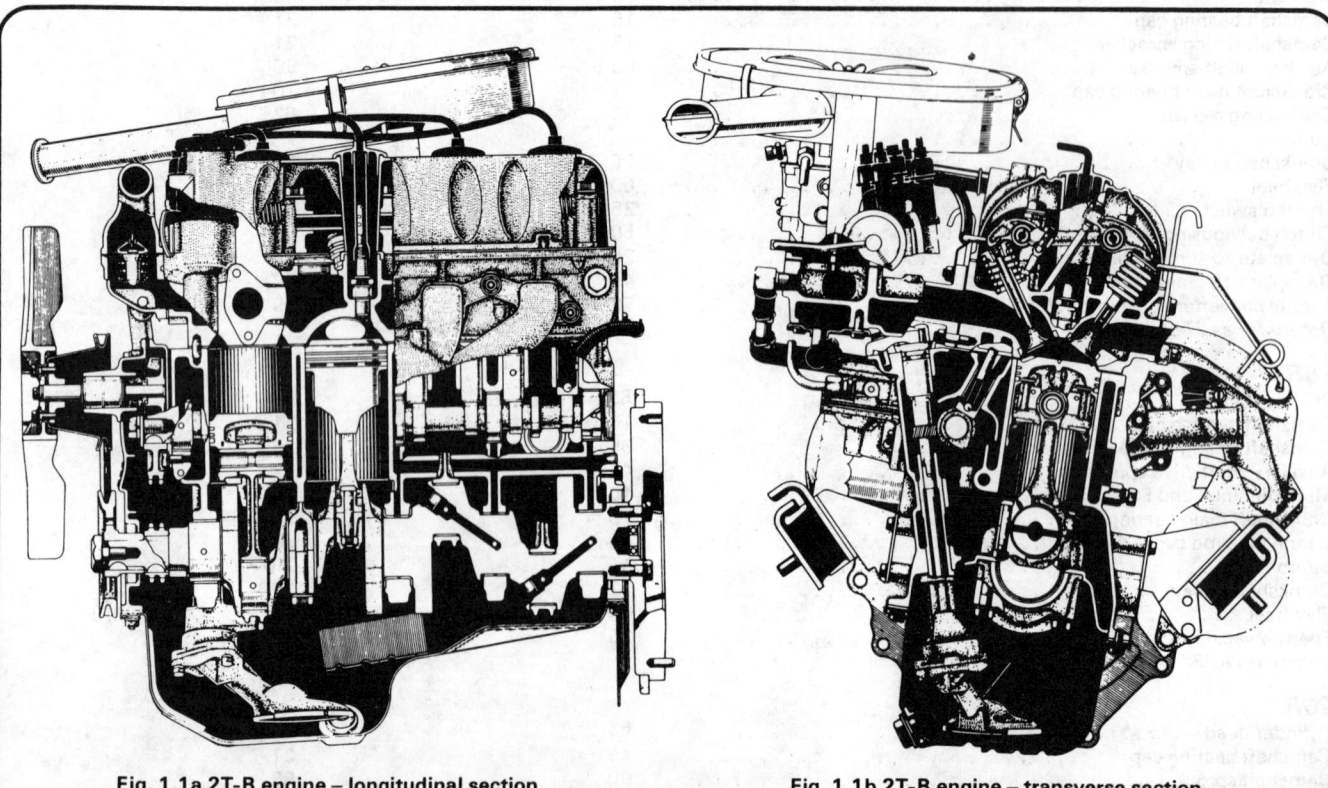

Fig. 1.1a 2T-B engine – longitudinal section

Fig. 1.1b 2T-B engine – transverse section

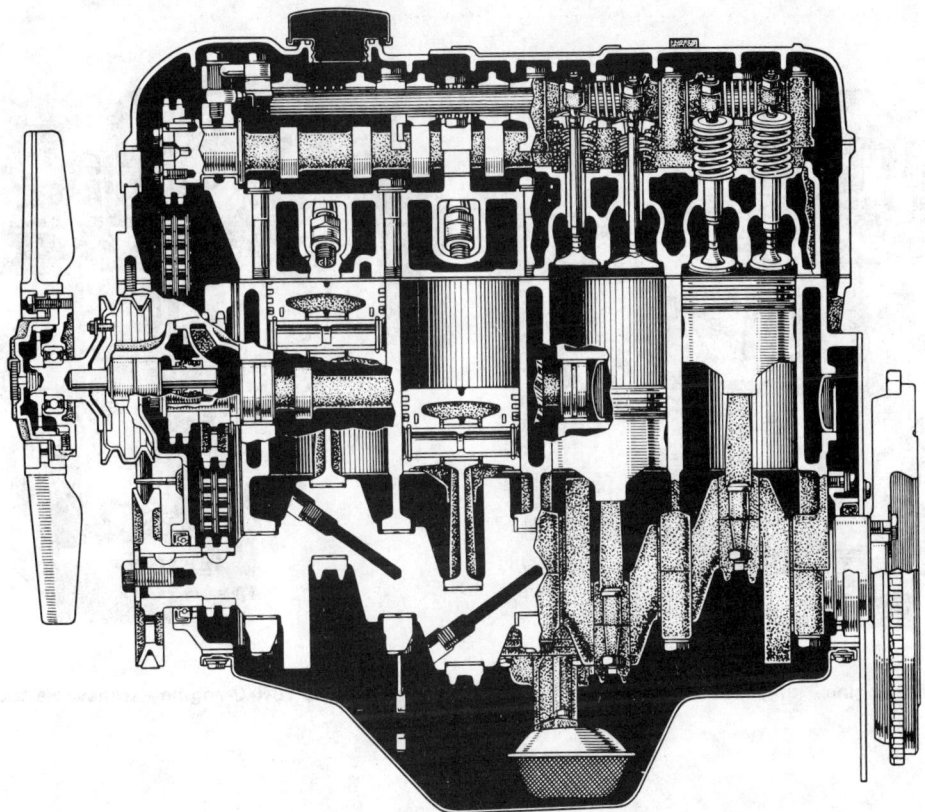

Fig. 1.2a 18R engine – longitudinal section

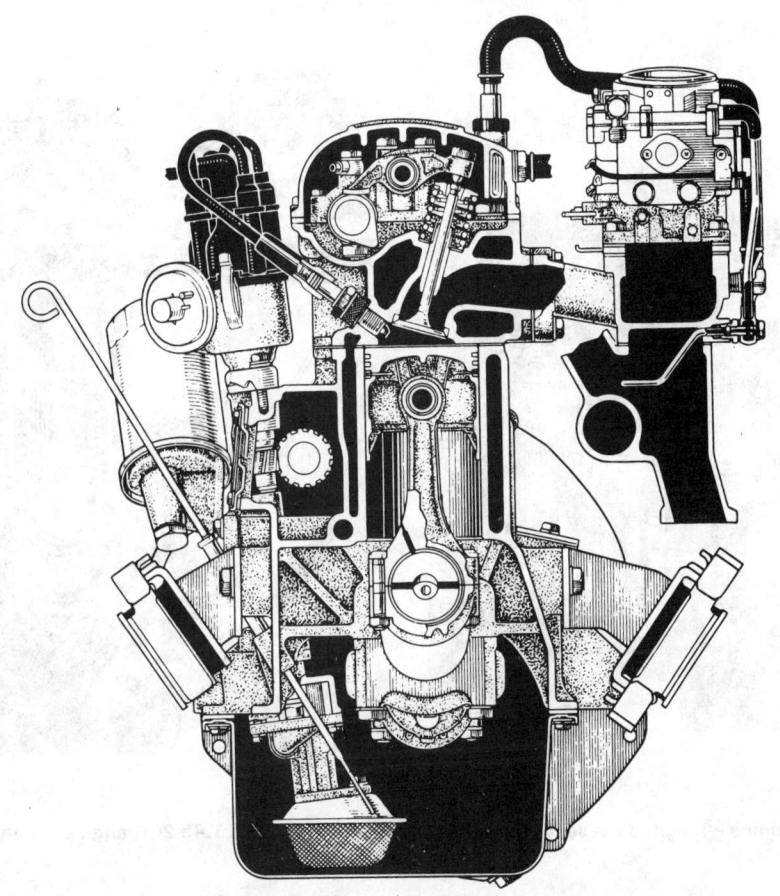

Fig. 1.2b 18R engine – transverse section

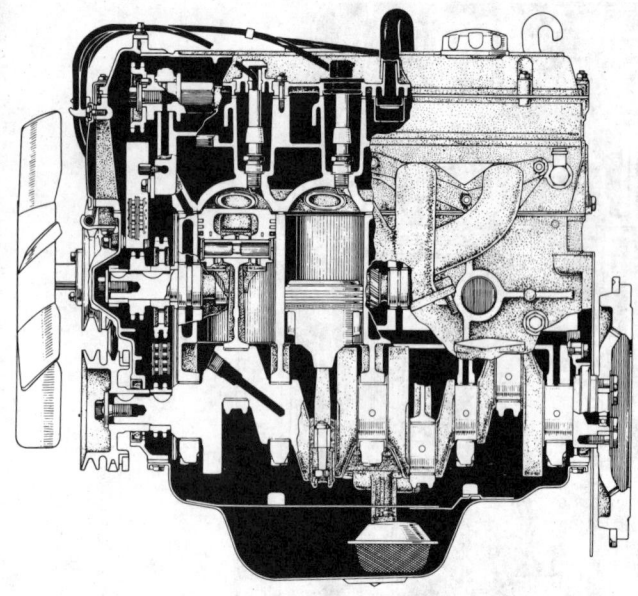

Fig. 1.3a 18R-G engine – longitudinal section

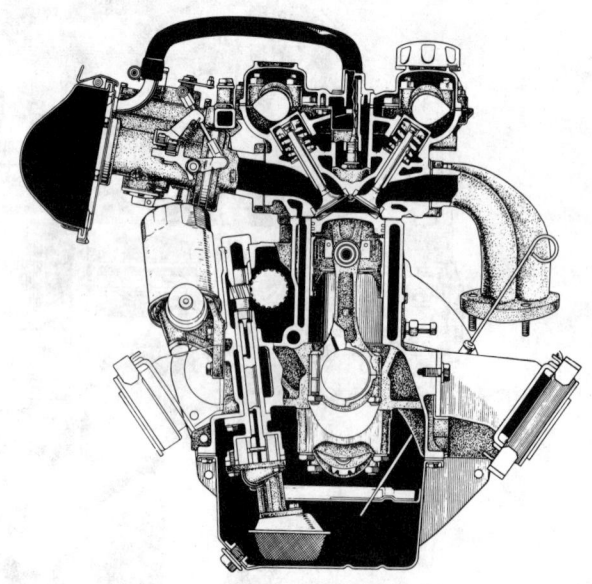

Fig. 1.3b 18R-G engine – transverse section

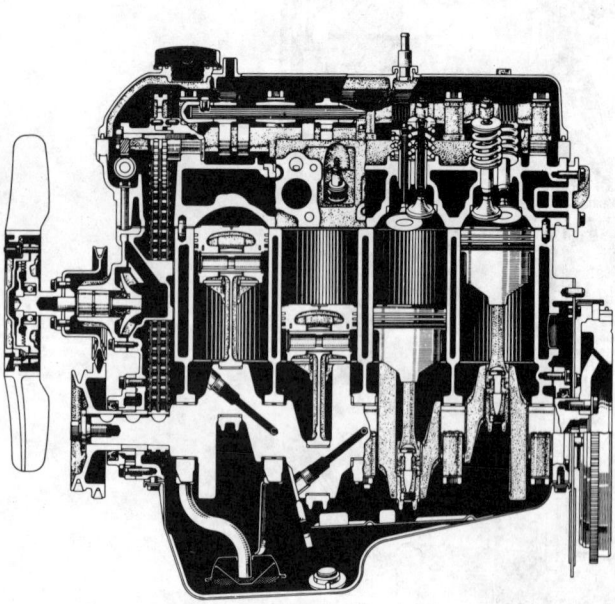

Fig. 1.4a 20R engine – longitudinal section

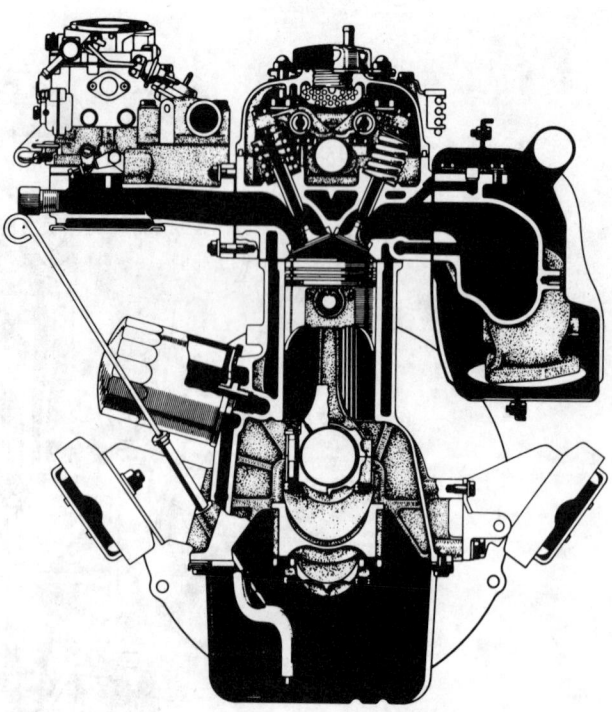

Fig. 1.4b 20R engine – transverse section

5.6a Removing the radiator grille

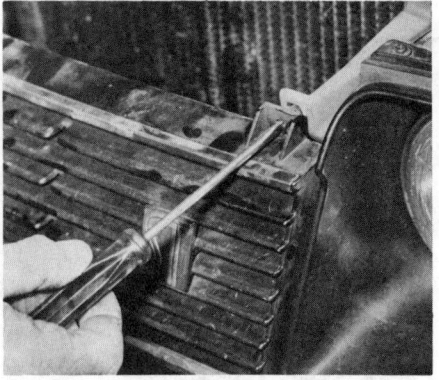
5.6b Removing the radiator grille

5.6c Radiator grille centre support

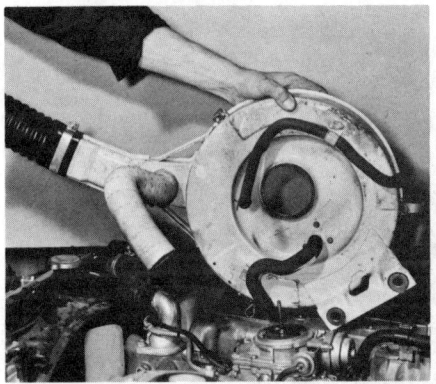
5.8 Removing the air cleaner (18R)

5.9 Fuel hoses removed from fuel pump (18R)

5.14a Alternator connections removed

5.14b Water temperature transmitter

5.14c Oil pressure transmitter

5.14d Typical engine earth connection

5.14e Typical transmission earth connection

5.19 Speedometer cable (note washer)

5.20 Removing the gear lever

5.23 Engine front mounting bolts

5.24a Rear crossmember bolts

5.24b Supporting the engine with blocks

5.25 Lifting out the engine (18R)

5 Engine/transmission – removal

1 Disconnect the lead from the battery negative terminal.
2 Drain the cooling system (see Chapter 3).
3 Drain the engine and transmission lubricant.
4 On cars equipped with air conditioning, the system must be discharged of refrigerant gas so that the pipes which connect to the compressor can be disconnected. This and the later recharging of the system are jobs for the service engineer.
5 Mark the position of the bonnet hinge plates, remove the securing bolts and with the help of an assistant, lift the bonnet from the vehicle.
6 Remove the radiator grille, the upper and lower shields, the fan shrouds, and the grille centre support (photos).
7 Disconnect the upper and lower radiator hoses, unbolt the radiator and remove it from the engine compartment. *On vehicles equipped with air conditioning* the condenser is located in front of the radiator and this too will have to be disconnected (system first having been discharged) and removed. On some models, which are equipped with an oil cooler for the automatic transmission the connecting hoses will have to be removed from the radiator before the latter can be lifted from its location.
8 Remove the air cleaner from the carburettor (photo).
9 Disconnect the fuel hoses from the fuel pump (photo).
10 Disconnect the heater hoses.
11 Disconnect the controls from the carburettor.
12 Disconnect the brake servo unit vacuum hose from the inlet manifold.
13 Unbolt the clutch operating cylinder from the clutch bellhousing, also the hose support bracket, and tie the cylinder up out of the way; there is no need to disconnect the hydraulic line.
14 Disconnect the leads from the following components: starter solenoid, alternator, oil pressure switch, water temperature transmitter, emission control unit and coil. Also disconnect any earth leads (photos).
15 Disconnect the vacuum hoses from the emission control unit. It is important that these are carefully identified to ensure exact refitting in their original positions.
16 Working beneath the car, disconnect the exhaust downpipe from the exhaust manifold.
17 Disconnect the exhaust pipe support bracket from the transmission housing.
18 Unbolt the parking brake cable equalizer and lower the equalizer complete with cables.
19 Disconnect the speedometer drive cable from the transmission housing (photo).
20 *On cars having a manually operated gearbox,* remove the centre console and peel back the flexible rubber boot. Check that the gearshift lever is in neutral and then extract the securing screws from the lever retaining plate and withdraw the lever (photo).
21 *On cars equipped with automatic transmission* disconnect the speed selector rod from the range selector lever on the side of the transmission housing and remove the fluid filler tube and dipstick.
22 Remove the propeller shaft, as described in Chapter 7.
23 Using a suitable hoist and slings positioned securely round the engine, raise the hoist so that it just takes the weight of the engine. Unbolt the engine mountings from the crossmember (photo).
24 Place a jack under the transmission housing and then unbolt and remove the rear mounting support crossmember (photo) and the mounting. Alternatively support the engine by placing a block of wood against the front bulkhead (photo).
25 Remove the jack from below the transmission which will allow it to drop a few inches. Hoist the combined engine/transmission up and out of the engine compartment at a steeply inclined angle (photo).

6 Engine – separation from manual gearbox

1 Remove the starter motor.
2 Unscrew and remove the bolts which connect the clutch bellhousing to the cylinder block.
3 Pull the gearbox from the engine in a straight line supporting the gearbox so that its weight does not hang upon the gearbox input shaft,

even momentarily, whilst it is still engaged in the crankshaft or clutch assembly.

7 Engine – separation from automatic transmission

1   Remove the six torque converter mounting bolts through the access holes in the engine rear plate.
2   Remove the transmission housing attachment bolts and with the transmission supported on a jack to keep it in line with the engine, carefully draw the transmission to the rear until it is clear of the engine. At the same time keep the torque converter pressed to the rear to prevent its disconnection from the oil pump drive slots.

8 Dismantling the engine – general

1   It is best to mount the engine on a dismantling stand but if one is not available, stand the engine on a strong bench at a comfortable working height. Failing this, the engine can be stripped down on the floor.
2   During dismantling, the greatest care should be taken to keep the exposed parts free from dirt. As an aid to achieving this, it is a sound scheme to first clean down the outside of the engine, removing all traces of oil and congealed dirt.
3   Use kerosene or a good grease solvent. The latter will make the job much easier, as, after the solvent has been applied and allowed to stand for a time, a vigorous jet of water will wash off the solvent and all the grease and filth. If the dirt is thick and deeply embedded, work the solvent into it with a brush.
4   Finally wipe down the exterior of the engine with a rag and only then, when it is quite clean, should dismantling begin. As the engine is stripped, clean each part in a bath of kerosene.
5   Never immerse parts with oilways in kerosene, eg; the crankcase, but to clean, wipe down carefully with a solvent dampened rag. Oilways can be cleaned out with wire. If an air line is present all parts can be blown dry and the oilways blown through as an added precaution.
6   Re-use of old engine gaskets is false economy and can give rise to oil and water leaks. To avoid the possibility of trouble after the engine has been reassembled *always* use new gaskets throughout.
7   Do not throw the old gaskets away as it sometimes happens that an immediate replacement cannot be found and the old gasket is useful as a template. Hang up the old gaskets as they are removed on a suitable hook or nail.
8   To strip the engine it is best to work from the top down. The sump provides a firm base on which the engine can be supported in an upright position. When this stage where the sump must be removed is reached, the engine can be turned on its side and all other work carried out with it in this position.
9   Whenever possible, replace nuts, bolts and washers fingertight from wherever they were removed. This helps to avoid later loss and muddle. If they cannot be replaced, lay them out so that it is clear where they came from.

9 Engine ancillary components – removal

1   After the engine has been removed from the car and the transmission has been separated, the ancillary components should be removed before engine dismantling is started.
2   Unbolt the clutch pressure plate assembly (manual transmission) from the flywheel and remove the components of the clutch assembly.
3   When locking tabs are fitted to the flywheel securing bolts, bend them down. Remove the bolts and lift off the flywheel or automatic transmission driveplate from the crankshaft flange.
4   When the car is fitted with air conditioning, remove the air compressor.
5   Remove the dipstick and oil filter assembly, the distributor cap and HT leads and then release the distributor clamp and remove the distributor. Remove the spark plugs.
6   Disconnect the fuel lines from the fuel pump and remove the pump.
7   Remove the alternator and its adjustment strap.
8   Remove the coolant pipes to the automatic choke when fitted, the crankcase ventilation hose and then unbolt and remove the car-

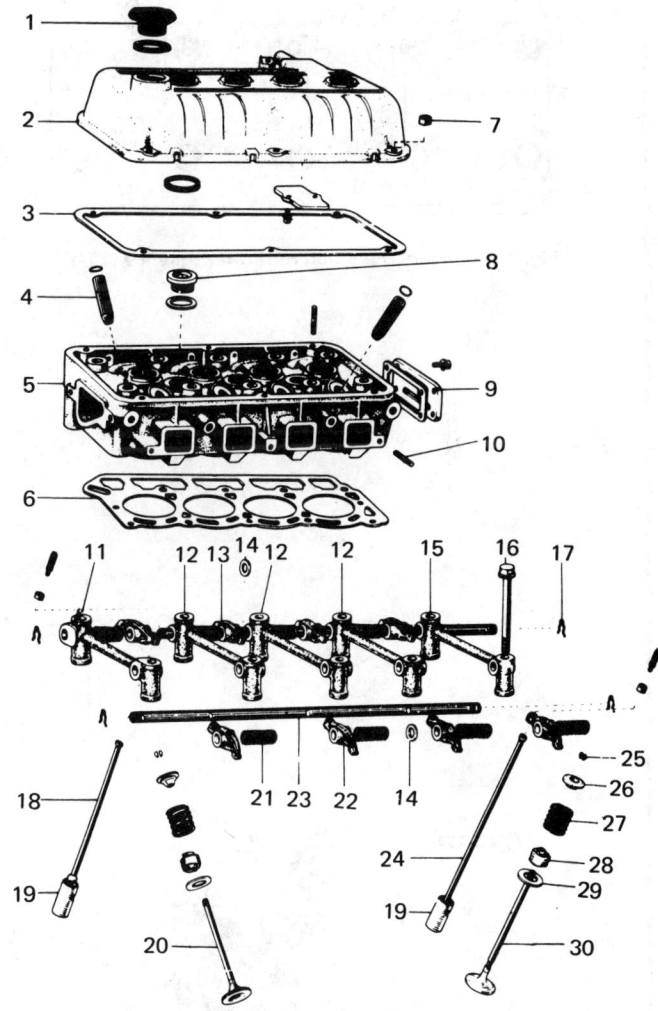

Fig. 1.5 Cylinder head components (2T-B)

1   Oil filler cap
2   Cylinder head cover
3   Cylinder head cover gasket
4   Valve guide (intake)
5   Cylinder head
6   Cylinder head gasket
7   Nut
8   Plug
9   Cylinder head rear plate
10 Stud
11 Valve rocker support
12 Valve rocker support
13 Rocker arm
14 Washer
15 Valve rocker support
16 Bolt
17 Spring clip
18 Valve pushrod
19 Valve lifter (cam follower)
20 Intake valve
21 Compression spring
22 Rocker arm
23 Valve rocker shaft
24 Valve pushrod
25 Valve spring retainer lock
26 Valve spring retainer
27 Valve spring
28 Stem oil seal
29 Plate washer
30 Exhaust valve

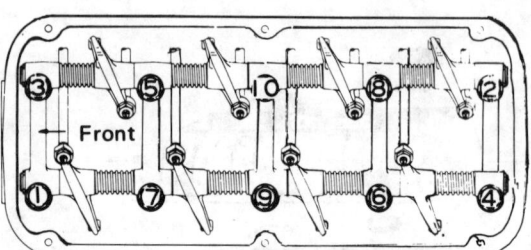

Fig. 1.6 Cylinder head bolt removal sequence (2T-B)

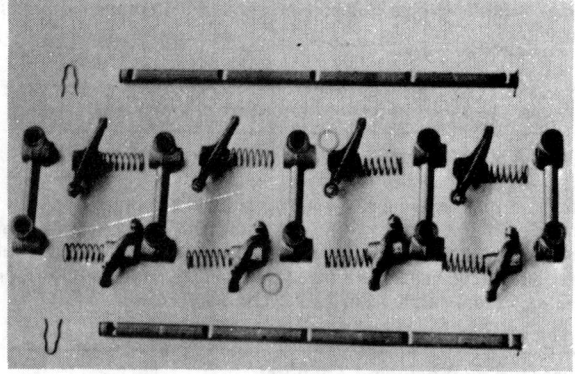

Fig. 1.7 Rocker assembly components (2T-B)

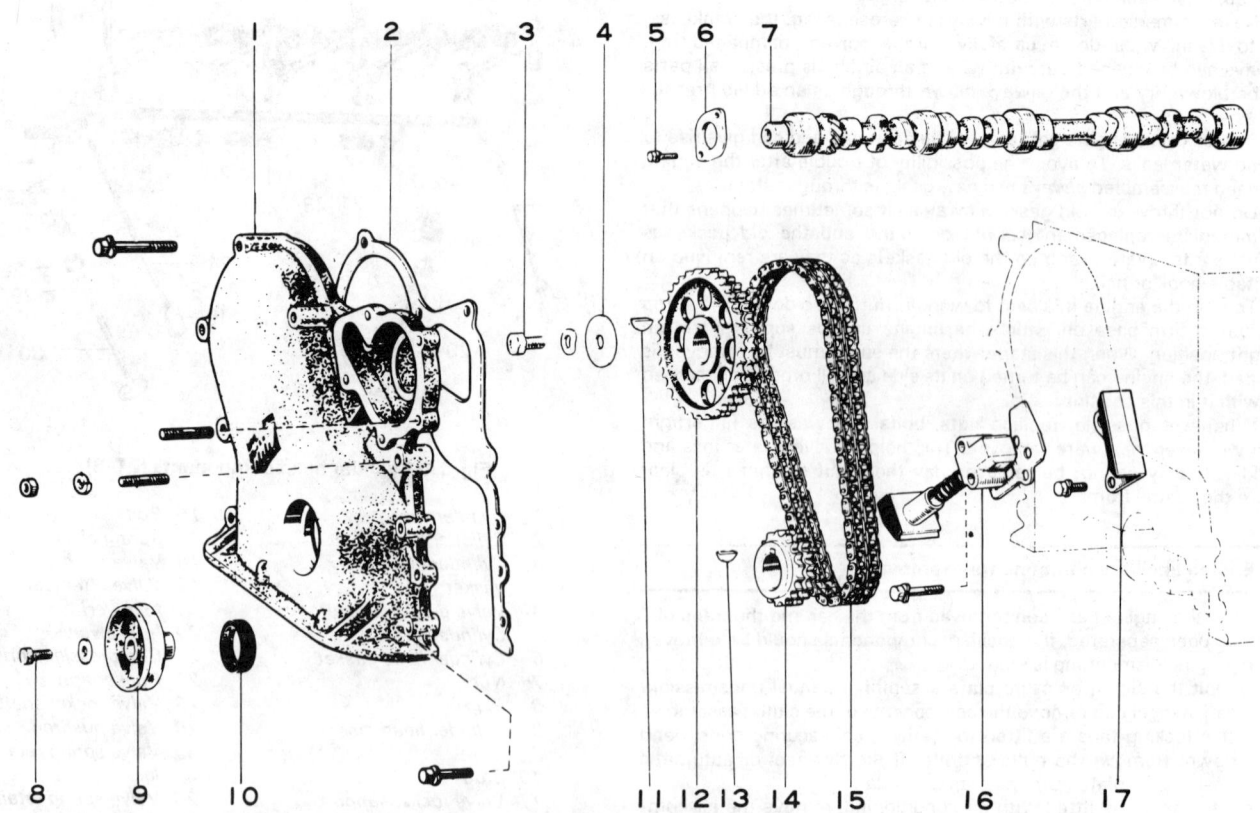

Fig. 1.8 Timing cover and timing gear (2T-B)

1 Timing chain cover	6 Plate	10 Oil seal	14 Crankshaft sprocket
2 Gasket	7 Camshaft	11 Woodruff key	15 Chain
3 Bolt	8 Bolt	12 Camshaft sprocket	16 Chain tensioner assembly
4 Plate washer	9 Crankshaft pulley	13 Woodruff key	17 Chain vibration damper
5 Bolt			

Chapter 1 Engine

burettor(s) from the inlet manifold.
9 Remove the inlet manifold assembly and the exhaust manifold.
10 Remove the water pump.
11 Remove the thermostat housing cover and lift out the thermostat.
12 Remove the rocker box cover.

10 Engine dismantling (2 T-B engine)

Rocker gear

1 Loosen the cylinder head bolts one turn at a time in the sequence shown in Fig. 1.6 until all tension on them has been removed and then remove the bolts completely.
2 Lift the rocker assembly off, noting which is the front of the assembly.
3 Mark or label the four rocker supports with a number, making the front support No. 1.
4 Remove the two spring retainers, take off No 1 support and then the rocker arm and spring of the inlet and exhaust valve. Lay each component out in its exact position as when assembled and make sure that the inlet valve rocker shaft and the exhaust valve rocker shaft are not mixed up.
5 Lift out each of the pushrods in turn and push it through a piece of card with the pushrod number written against it on the card.

Cylinder head

6 Tap the cylinder head with a soft-faced hammer to break the gasket seal and then lift the cylinder head off.
7 Remove the valves by compressing the valve spring with a spring compressor until the collets can be removed. Release the compressor slowly until the spring is no longer under tension and then remove the spring compressor.
8 Remove the valve spring retainer and valve spring, prise off and discard the valve stem oil seal and then lift off the plate washer on which the spring sits. Keep the components of each valve assembly together and label them to make sure that they are put back in their original place.
9 Use a magnet to remove the valve lifters from the cylinder block and place each valve lifter with its associated pushrod or valve assembly.

Camshaft and timing gear

10 Remove the crankshaft pulley bolt after either jamming the flywheel ring gear if the engine is out of the car or the starter removed (engine in car) and separated from the transmission, or by putting the car in gear and giving the spanner on the crankshaft pulley bolt a sharp blow with a heavy hammer.
11 Use a suitable puller to draw the crankshaft pulley off the crankshaft.
12 Lay the engine on its side and remove the sump. If the sump is stuck to the cylinder block after all the bolts have been removed, lever it off carefully with a wide screwdriver.
13 Remove the bolts and nuts securing the timing cover and pull the cover off.
14 Remove the two bolts securing the chain tensioner and lift away the chain tensioner assembly.
15 Remove the bolt from the centre of the camshaft sprocket and ease the crankshaft sprocket, camshaft sprocklet and timing chain forward as an assembly until the crankshaft sprocket is clear of the crankshaft.
16 Remove the two bolts from the camshaft thrust plate and take the thrust plate off, then carefully withdraw the camshaft, keeping it in line with the camshaft bore so that the bearings are not damaged.

Piston and connecting rod assemblies

17 Undo and remove the big-end cap retaining nuts using a socket wrench, and remove the big-end caps one at a time, taking care to keep them in the right order and the right way round. If they are not already marked with the cylinder number, put an appropriate number of centre punch marks onto the bearing cap and also put mating marks on the two parts of each of the big-end bearings, so that the connecting rod and cap will be reassembled correctly.
18 If the big-end caps are difficult to remove, they may be tapped with a soft-faced hammer.
19 Withdraw the pistons and connecting rods upwards out of the

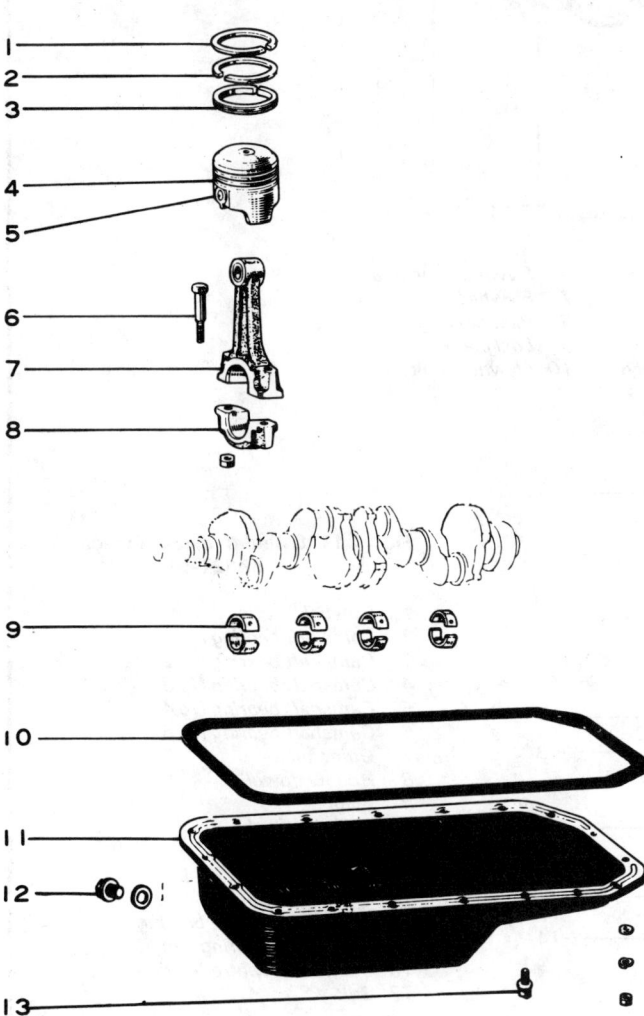

Fig. 1.9 Sump, piston and connecting rod (2T-B)

1 Compression ring (Top)
2 Compression ring (Second)
3 Oil control ring
4 Piston
5 Gudgeon pin
6 Connecting rod bolt
7 Connecting rod
8 Connecting rod big-end cap
9 Connecting rod shell bearing
10 Sump gasket
11 Sump
12 Sump drain plug
13 Bolt and washer

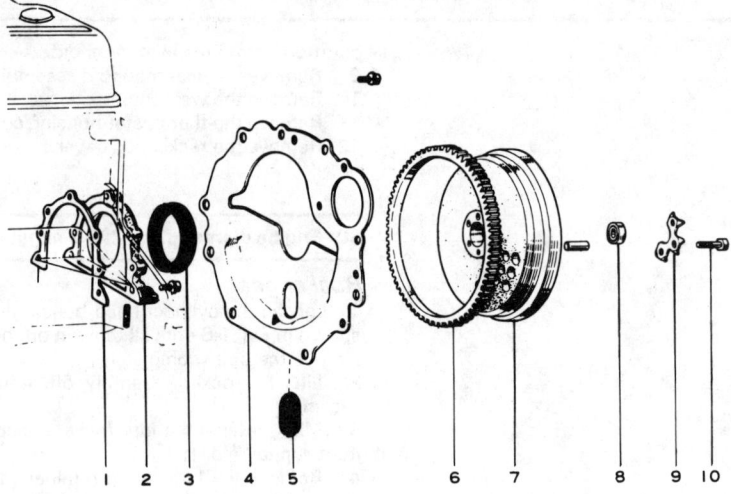

Fig. 1.10 Flywheel and backplate (2T-B)

1 Oil seal retainer gasket
2 Oil seal retainer
3 Oil seal
4 Engine rear plate
5 Plug (automatic transmission only)
6 Flywheel ring gear
7 Flywheel
8 Pilot bearing
9 Lockplate
10 Flywheel bolt

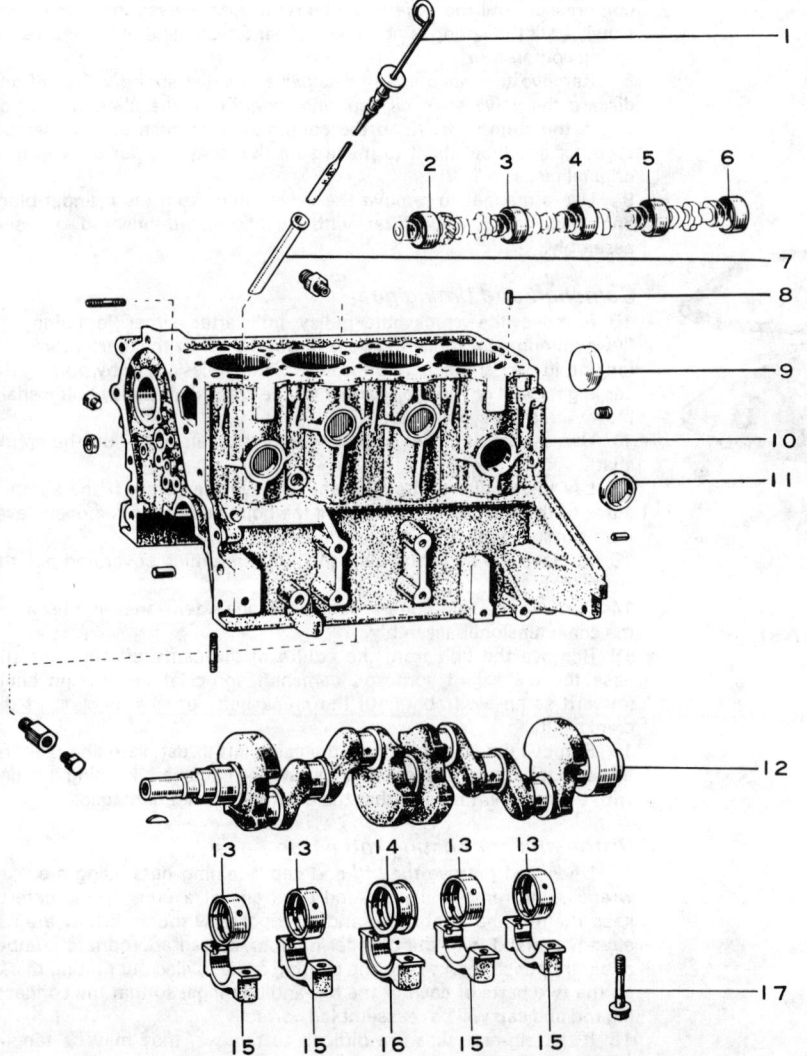

Fig. 1.11 Cylinder block and crankshaft (2T-B)

1 Dipstick
2 Camshaft bearing No 1
3 Camshaft bearing No 2
4 Camshaft bearing No 3
5 Camshaft bearing No 4
6 Camshaft bearing No 5
7 Guide tube
8 Hollow dowel
9 Plug
10 Cylinder block
11 Plug
12 Crankshaft
13 Crankshaft bearing
14 Crankshaft centre bearing
15 Crankshaft bearing cap
16 Crankshaft centre bearing cap
17 Bolt

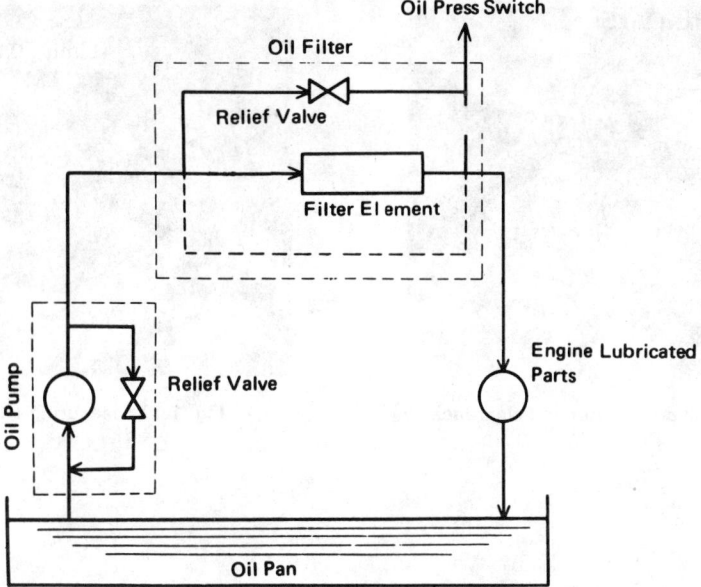

Fig. 1.12 Lubrication system schematic

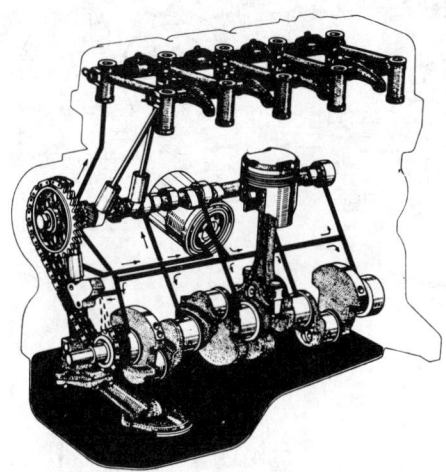

Fig. 1.13 Engine lubrication system

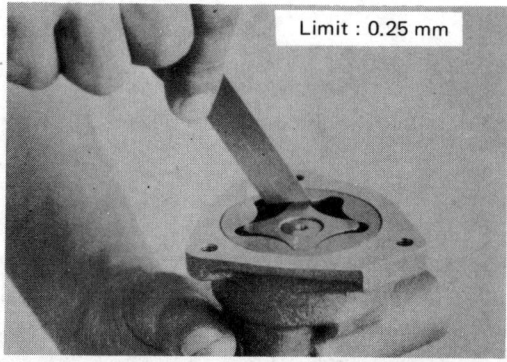

Fig. 1.15 Measuring oil pump rotor tip clearance

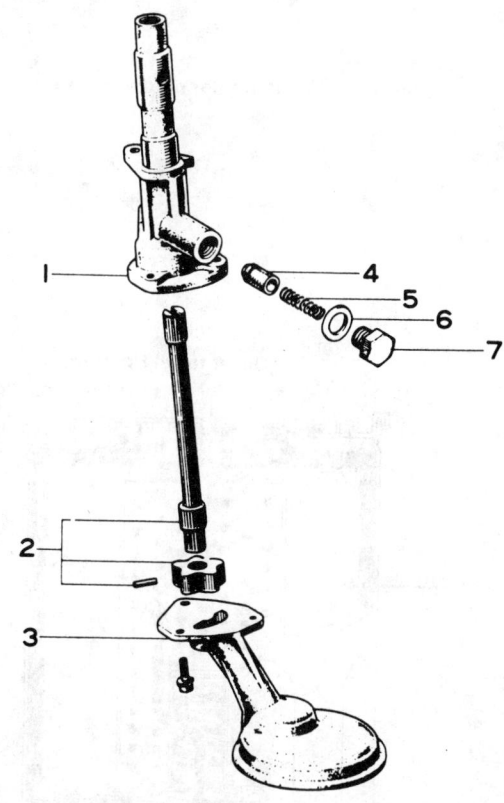

Fig. 1.14 Oil pump components

1 Oil pump body
2 Rotor set
3 Cover assembly
4 Relief valve
5 Spring
6 Gasket
7 Plug

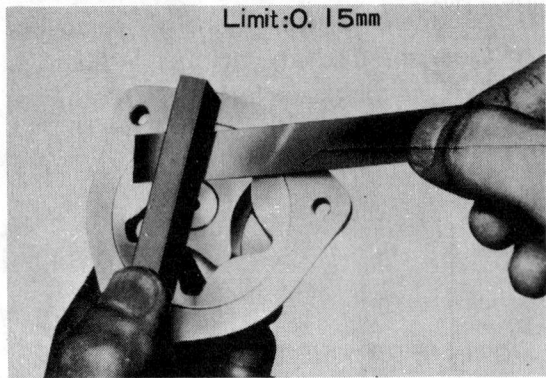

Fig. 1.16 Measuring oil pump rotor end clearance

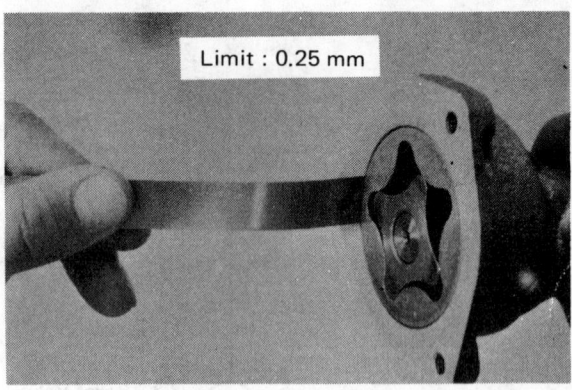

Fig. 1.17 Measuring oil pump rotor clearance

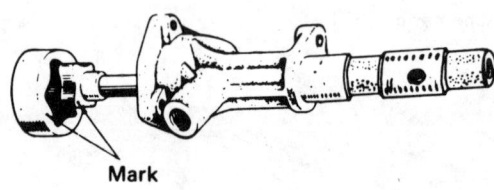

Fig. 1.18 Oil pump rotor mating marks

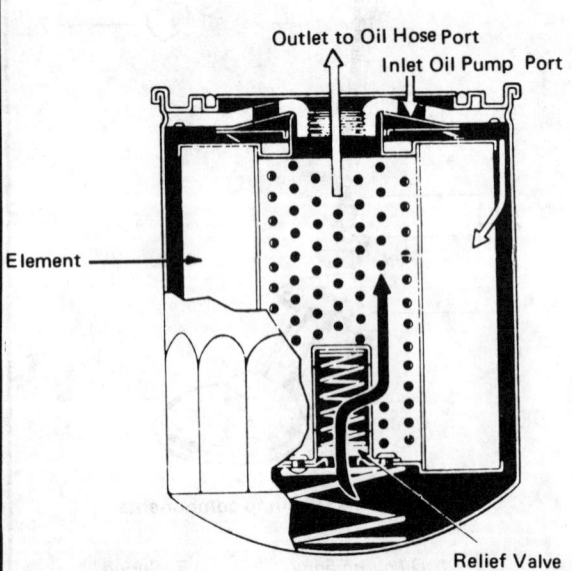

Fig. 1.20 Cartridge oil filter (sectional view)

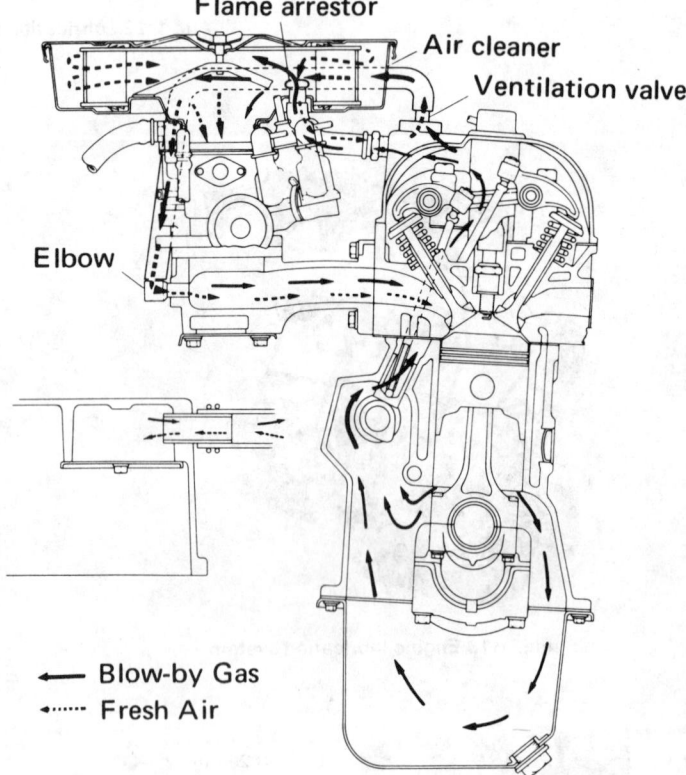

Fig. 1.19 Crankcase ventilation system (2T-B)

cylinder block and re-fit each connecting rod with its cap as soon as it is removed, so that there is no possibility of their being mixed up.
20 On engines which have covered a very high mileage, there may be a pronounced ridge at the top of the cylinder bores because no wear has taken place there. Unless these ridges are scraped away carefully, there is a danger of breaking the piston rings when attempting to push the pistons out of the top of the bores, where the wear step tends to be less pronounced.
21 The gudgeon pins are a press fit in the connecting rod small ends and because of the need for a press and the correct size of mandrel, it is recommended that the removal of the pistons from the connecting rods, if necessary, is left to your Toyota dealer.
22 To remove the piston rings, slide two or three pieces of old feeler gauge between the ring and the piston and space them equidistant around the piston; this will expand the ring just enough for the ring to clear the piston lands. Carefully slide the ring upwards and off the top of the piston. Piston rings are extremely brittle and should not be expanded more than is necessary to just clear the lands. It is equally important that when sliding them up the piston they are always parallel with the top of the piston.
23 If the same rings are being refitted, make sure that the rings are put back in the same groove as the one from which they were removed and that they are the same way up as before.

Crankshaft and main bearings
24 Remove the bolt securing the oil pump to the cylinder block and pull out the oil pump and strainer assembly. If the pump does not pull out, use a soft metal drift and tap it out from the distributor side.
25 Unbolt and remove the crankshaft rear oil seal retainer after first removing the engine rear plate.
26 With the cylinder block resting on the cylinder head jointface, undo and remove the ten bolts securing the main bearing caps to the cylinder block.
27 Make sure that the bearing caps are numbered 1 to 5 on their front faces and also have an arrow pointing towards the front of the engine. If there are no marks, or if they are not distinct, put a *Front* mark and a bearing number on each cap as it is removed.
28 Remove the main bearing caps and their half bearing shells, taking care to keep the bearing shells in the caps.
29 When removing the centre bearing caps, also remove the half thrust washer on each side of the cap and put these thrust washers alongside the bearing caps against the same side of the cap as that from which they were removed.
30 Lift the crankshaft from the crankcase. Remove the bearing shells from the crankcase webs and place each half bearing with its appropriate bearing cap.

11 Lubrication system, oil pump and filter (2 T-B engine)

1 Pressure for the engine lubrication system is generated by a trochoid oil pump which is driven by a gear on the front of the camshaft. Integral with the pump cover is the oil intake pipe and strainer which are submerged in the oil sump.
2 The oil pump normally has a very long life, but in the event of low oil pressure then the oil level is normal and the engine bearings are known to be in good condition, remove the oil pump for inspection.
3 To inspect the pump, remove the sump, remove the pump fixing bolt and washer and pull the pump out of the cylinder block.
4 Unscrew and remove the pressure relief valve.
5 Remove the three bolts securing the pump cover and remove the cover.
6 Remove the two parts of the rotor assembly.
7 Inspect the outside diameter and the top end of the oil pump shaft and if they are worn, or damaged, fit a new shaft.
8 Measure the clearance between the tips of the inner rotor and the outer rotor. If this is greater than 0.010 in (0.25 mm), fit a new rotor set.
9 Measure the clearance between the end of the rotor and the endface of the pump casing. If this exceeds 0.006 in (0.15 mm), fit a new rotor assembly or a pump body if the pump cover shows signs of wear.
10 Measure the clearance between the outer rotor and the pump body and renew either the rotor assembly or the pump body, if the clearance exceeds 0.010 in (10.25 mm).
11 Check that all oil passages are clear and that the oil strainer and components of the relief valve are undamaged.
12 When assembling the pump, note that the two rotors each have a mating mark on their under sides. These marks must be aligned when the rotors are inserted into the pump.
13 After completing the assembly of the pump, immerse the suction end of the pump in a container of engine oil and turn the pump shaft clockwise to check that oil emerges from the pump outlet. Close the outlet hole with the thumb and turn the shaft as before to ensure that a build up of pressure can be felt.
14 When fitting the pump to the engine, use a new gasket.
15 The oil filter is of the full-flow cartridge type and is mounted on the right-hand side of the cylinder block.
16 To replace an oil filter, unscrew the old cartridge, using a strap wrench or grips. In the absence of either of these, punch a long screwdriver through the filter casing and use the screwdriver as a tommy-bar.
17 Clean the filter mating surface on the cylinder block. Smear oil on to the joint ring of a new filter and screw the filter on until the joint ring just touches the cylinder block. Tighten the filter a further one half turn by hand. Do not use any form of wrench or grips to fit a new filter cartridge.

12 Crankcase ventilation system (2 T-B engine)

1 As part of the emission control system (described in Chapter 3), a *Positive Crankcase Ventilation* system is fitted.
2 A ventilation PCV valve in the inlet manifold leads blow-by gas into the manifold, which avoids dirtying the carburettor.
3 At engine idling the amount of blow-by gas is small and manifold vacuum is high. The manifold vacuum tends to close the ventilation valve and reduce the amount of blow-by gas drawn into the inlet manifold.
4 At cruising under light load, the action is similar to that at idling, but as engine load increases, manifold vacuum decreases allowing the valve to open by a greater amount and the increased load also results in an increase in blow-by gas for the inlet manifold to accept.
5 When accelerating under heavy load, there is a further increase in the amount of blow-by gas. The ventilation valve will now tend to be fully open, but if there is more blow-by gas than the valve will pass, the excess will be drawn into the engine through the air cleaner.
6 Because the internal pressure of the crankcase is always lower than the atmospheric pressure, there is no possibility of blow-by gas escaping to the atmosphere.
7 The ventilation valve is fitted to the side of the inlet manifold, beneath the carburettor. Always ensure that the hoses are in good condition and that the hose clips are tight.
8 To test the operation of the valve, start the engine and with it running at idling speed, pinch the hose attached to the valve several times. Each time the hose is pinched, the valve will be heard to close against its seat if the valve is working properly. If this is not so, fit a new valve.
9 If engine performance is poor, especially at low engine speeds, check the operation of the ventilation valve before looking for other faults, or adjusting the engine tuning.

13 Engine dismantling (18R engine)

Rocker gear
1 Unscrew and remove the union bolts and withdraw the oil feed pipe. Also remove the No 2 chain tensioner bolt and spring.
2 Unscrew the rocker shaft pillar bolts in the sequence shown in Fig. 1.23 to ensure that the pressure of the valve springs is relieved gently to prevent distortion of the shaft.
3 Remove the rocker shaft assembly from the cylinder head.
4 The rocker shaft does normally require dismantling unless the heels of the rocker arms are scored or badly worn, or one of the coil springs is broken.
5 If dismantling is essential, first remove the retaining screw which secures the rocker shaft front support pillar to the shaft. The rear pillar is retained to the shaft by means of two securing bolts which engage in cut-outs in the shaft.
6 As each rocker arm, spring and pillar is withdrawn, keep them in strict sequence for refitting.

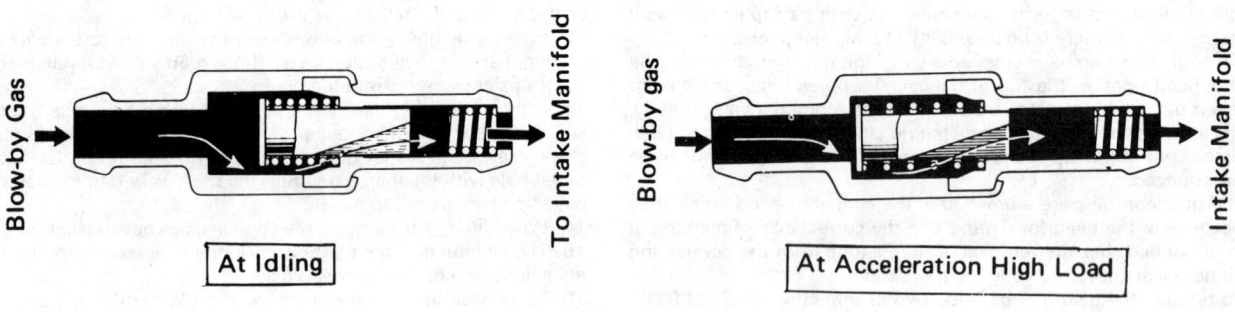

Fig. 1.21 PCV valve operation

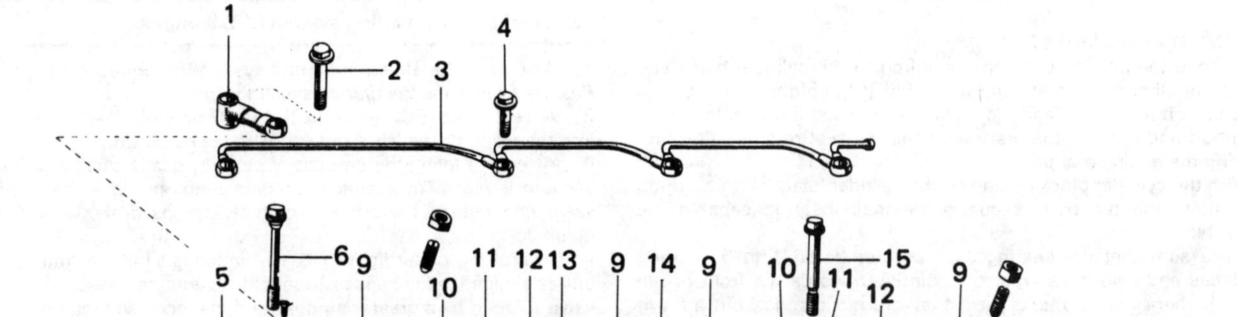

Fig. 1.22 Rocker gear and associated parts (18R)

1 Oil pipe union	8 Bearing	15 Bolt	22 Exhaust valve
2 Union bolt	9 Rocker arm	16 Split collets	23 Inlet valve
3 Oil pipe assembly	10 Rocker arm	17 Valve spring retainer	24 Camshaft sprocket
4 Union bolt	11 Rocker shaft support pillar	18 Valve stem oil seal	25 Dowel pin
5 Rocker shaft	12 Rocker arm	19 Valve (inner) spring	26 Camshaft
6 Bolt	13 Spring	20 Valve (outer) spring	27 Camshaft shell bearings
7 Rocker shaft support pillar	14 Rocker shaft support pillar	21 Plate	

Camshaft

7 Unbolt the sprocket from the end of the camshaft. A semi-circular plastic plug is provided for access to the bolts.
8 Use a hooked piece of wire to support the chain while the gearwheel is removed and to prevent the chain becoming disconnected from the drive sprocket. This is particularly important when the camshaft is being removed with the engine in position in the car and further dismantling is not anticipated.
9 Remove the four camshaft bearing caps together with their respective bearing shells, keeping them in strict sequence for correct refitting (they are usually numbered 1 to 4). Remove the camshaft.

Cylinder head

10 *If the engine is still in the car,* drain the cooling system and disconnect all leads and hoses from the cylinder head; remove the carburettor, and spark plugs, then continue as follows.
11 Unscrew the cylinder head bolts half a turn at a time in the sequence shown in Fig. 1.24.
12 Unscrew and remove the two bolts which secure the top end of the timing chain cover to the cylinder head.
13 Lift the cylinder head straight up to clear the locating dowels. If the cylinder head is stuck, it is permissible to insert a screwdriver at the point shown in Fig. 1.25 to break the seal. Use a piece of wire to support the chain and camshaft sprocket whilst lifting off the head.
14 Each valve should be removed from the cylinder head using the following method:
15 Compress each spring, using a valve spring compressor, until the collets can be removed. Release the compressor slowly, remove it, then remove the retainer, double springs, oil seal and the washer from the valve stem. Finally withdraw the valve from its guide.
16 If, when the valve spring compressor is screwed down, the valve spring retaining cap refuses to free to expose the collets, do not continue to screw down on the compressor as there is a likelihood of bending the valve stem.
17 Gently tap the top of the tool directly over the cap with a light hammer. This will free the cap. To avoid the compressor jumping off the valve spring retaining cap when it is tapped, hold the compressor firmly in position with one hand.
18 It is essential that the valves are kept in their correct sequences unless they are so badly worn that they are to be renewed. If they are going to be kept and used again, place them in a sheet of card having holes numbered 1 to 8 corresponding to the relative positions the valves were in when fitted. Also keep the valve springs, washers etc, in the correct order.

Sump and timing gear

Note: The owner may wish to remove the timing cover with the engine in the car, in which case the cylinder head must first be removed followed by the sump. If the sump is not removed, the sump-to-timing cover gasket will be damaged along the front edge (where the sump-to-timing cover bolts are fitted) when the timing cover is taken off.
However, provided that extreme care is taken, it should be possible to cut the appropriate length from a new gasket to use when reassembling provided that the abutting edges are squarely cut. The use of a silicone/RTV-type gasket sealant should be made at these points to minimize the risk of oil leakage, as well as the use of a normal non-setting sealant on the joint faces.
19 *If the engine is in the car,* drain the engine oil and disconnect the battery negative lead. Remove the cylinder head as described previously in this Section. Remove the radiator.
20 Unbolt and remove the re-inforcement brackets from the rear of the sump.
21 Unscrew and remove the sump securing bolts and detach the sump from the crankcase.
22 Unscrew the crankshaft pulley bolt, with the sump removed, by jamming the crankshaft with a piece of wood to prevent the engine turning as the bolt is loosened. If the pulley is being removed with the sump still fitted, and the car has a *manual gearbox,* engage a gear and apply the handbrake fully to prevent the crankshaft rotating. With automatic transmission, remove the starter and jam the ring gear with a large screwdriver or cold chisel.
23 Remove the crankshaft pulley. It will usually pull straight out but if necessary, remove it by placing two tyre levers behind it or use a puller (there are two holes tapped in the pulley for this purpose).
24 Unbolt and remove the timing cover. Note that the upper bolt is

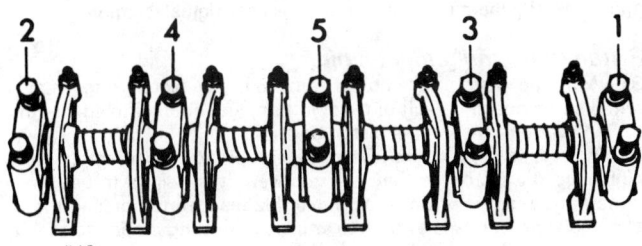

Fig. 1.23 Loosening sequence for rocker assembly (18R)

Fig. 1.24 Loosening sequence for cylinder head bolts (18R)

Fig. 1.25 Prising up the cylinder head (18R)

Chapter 1 Engine

entered from the rear.

25 Remove the camshaft drive chain. The auxiliary driveshaft sprocket and the chain tensioner should now be withdrawn.

26 Remove the crankshaft sprockets, and the second auxiliary driveshaft sprocket, complete with chains as one assembly. Remove the chain damper.

27 Remove the auxiliary driveshaft thrust plate, and withdraw the driveshaft and engine front plate. Unbolt and remove the oil pump assembly from within the crankcase.

Piston/connecting rod assemblies

28 *If the engine is still in the vehicle,* first remove the cylinder head and sump as previously described, then turn the crankshaft so that the pistons are all part way down their bores. Using a bearing scraper, carefully remove as much as possible of the 'wear' ridge at the top of each cylinder bore. This operation is essential to prevent the piston rings breaking as the pistons are extracted through the top of the block.

29 With quick drying paint, mark each piston, connecting rod and big-end bearing cap. Number the components of each assembly 1 to 4 (from the front of the engine) and also the relative positions of the components to each other and to the crankcase, so that if the original assembly is to be fitted, it will be refitted in its exact, previously located position.

30 Unbolt the big-end caps from the connecting rods, then push each piston/connecting rod assembly out through the top of the block. Take great care that the threads of the big-end studs do not score the cylinder bores during this operation. If the bearing shells are to be used again, identify them in respect of their exact original location.

Piston rings and gudgeon pins

31 With the piston assemblies removed, the piston rings may be removed by opening each of them in turn, just enough to enable them to ride over the lands of the piston body.

32 In order to prevent the lower rings dropping into an empty groove higher up the piston as they are removed, it is helpful to use two or three narrow strips of tin or old feeler blades inserted behind the ring at equidistant points, and then to employ a twisting motion to slide the ring from the piston.

33 To remove a gudgeon pin, first extract the circlips (one at each

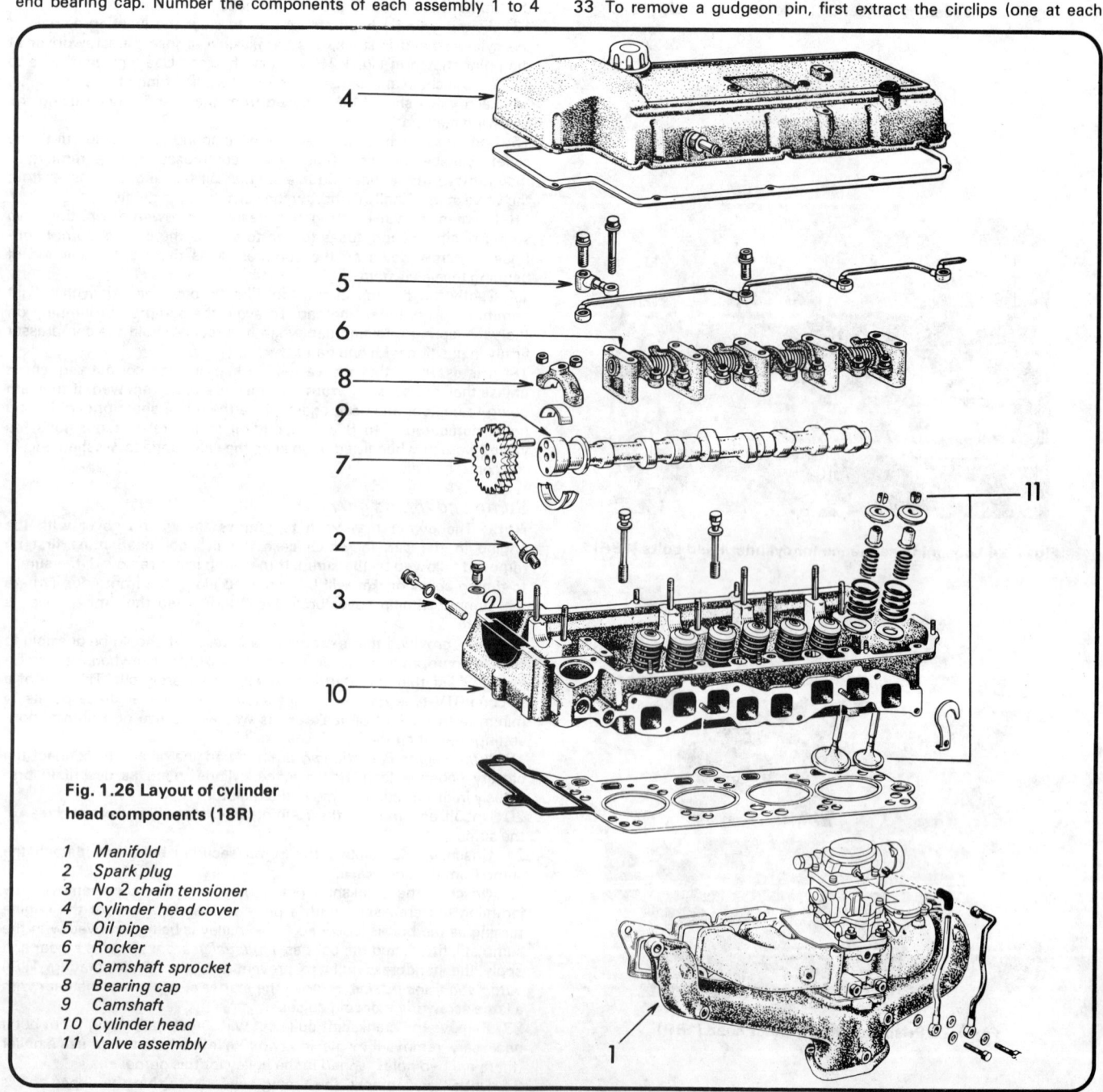

Fig. 1.26 Layout of cylinder head components (18R)

1 Manifold
2 Spark plug
3 No 2 chain tensioner
4 Cylinder head cover
5 Oil pipe
6 Rocker
7 Camshaft sprocket
8 Bearing cap
9 Camshaft
10 Cylinder head
11 Valve assembly

Chapter 1 Engine

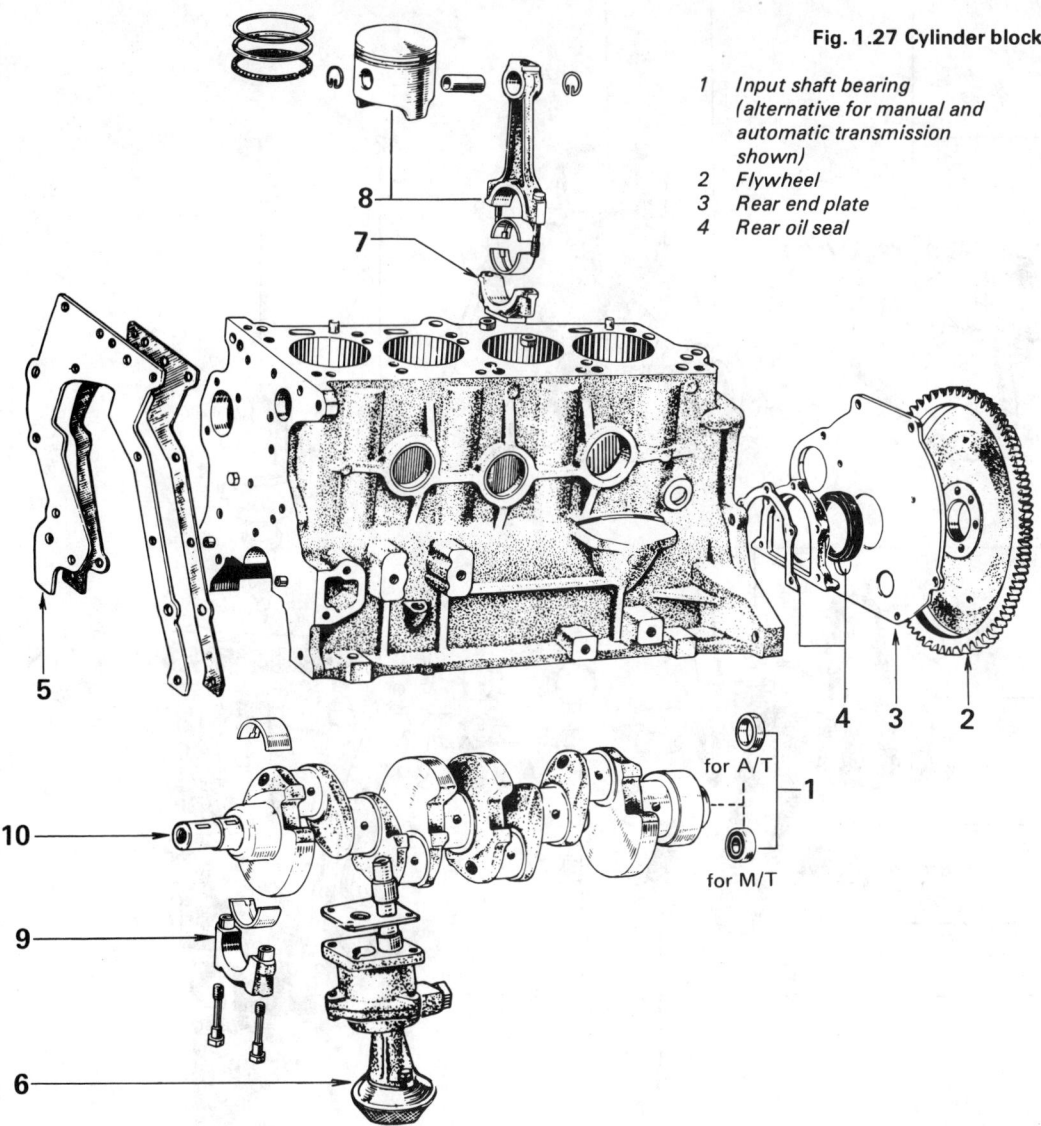

Fig. 1.27 Cylinder block components (18R)

1 Input shaft bearing (alternative for manual and automatic transmission shown)
2 Flywheel
3 Rear end plate
4 Rear oil seal
5 Front end plate
6 Oil pump
7 Connecting rod cap
8 Piston and connecting rod
9 Main bearing cap
10 Crankshaft

end) and then immerse the piston in hot water at a temperature of approximately 140°F (600°C). After a few minutes, the gudgeon pin will be able to be pushed out of the piston and connecting rod with finger pressure only.
34 Mark each gudgeon pin as it is removed with the piston sequence number (use masking tape) so that it can be refitted in its original location.

Crankshaft and main bearings
35 Unbolt and remove the crankshaft rear oil seal retainer.
36 Mark each of the main bearing caps with quick-drying paint (numbered 1 to 5 from the front of the engine), making sure that the caps are also marked as to which way round they are to be refitted. Some caps are marked with a triangle, the apex of which points to the front of the engine, and they are already numbered, but check before removing them.
37 Unscrew the main bearing cap bolts and remove the caps complete with shell bearings. The centre main bearing shells incorporate thrust flanges.
38 Lift the crankshaft from the crankcase. If the bearing shells are to be used again, identify them in respect of exact original location.
39 If not already removed, the input shaft bearing (clutch pilot bearing) can be extracted from the rear end of the crankshaft.

14 Engine dismantling (18R-G engine)

1 With the exception of the cylinder head and timing gear the dismantling of the 18R-G engine is the same as for the 18R.

Camshafts
2 Remove the vibration damper and the tensioner from the timing chain.
3 Set No 1 cylinder to TDC on the compression stroke and the cam lobes for No 1 cylinder will then both be pointing outwards (Fig. 1.29).
4 Put mating marks on each of the gears to indicate the position of the pins in the camshaft sprocket flange (Fig. 1.30).
5 Unscrew the nuts on the end of the camshaft. Remove the nut, washer and camshaft sprocket. If the engine is not to be dismantled further, support the timing chain to prevent it from becoming disengaged from the lower drive sprocket.
6 Remove the cap of No 1 camshaft bearing and then gradually loosen the other cap nuts in two or three stages in the sequence shown in Fig. 1.31.
7 Before lifting off the bearing caps, mark each one with its number and also indicate which side of it faces the front.
8 As an added precaution lay them out in order after removing them, making sure that the caps for each camshaft are kept separately.

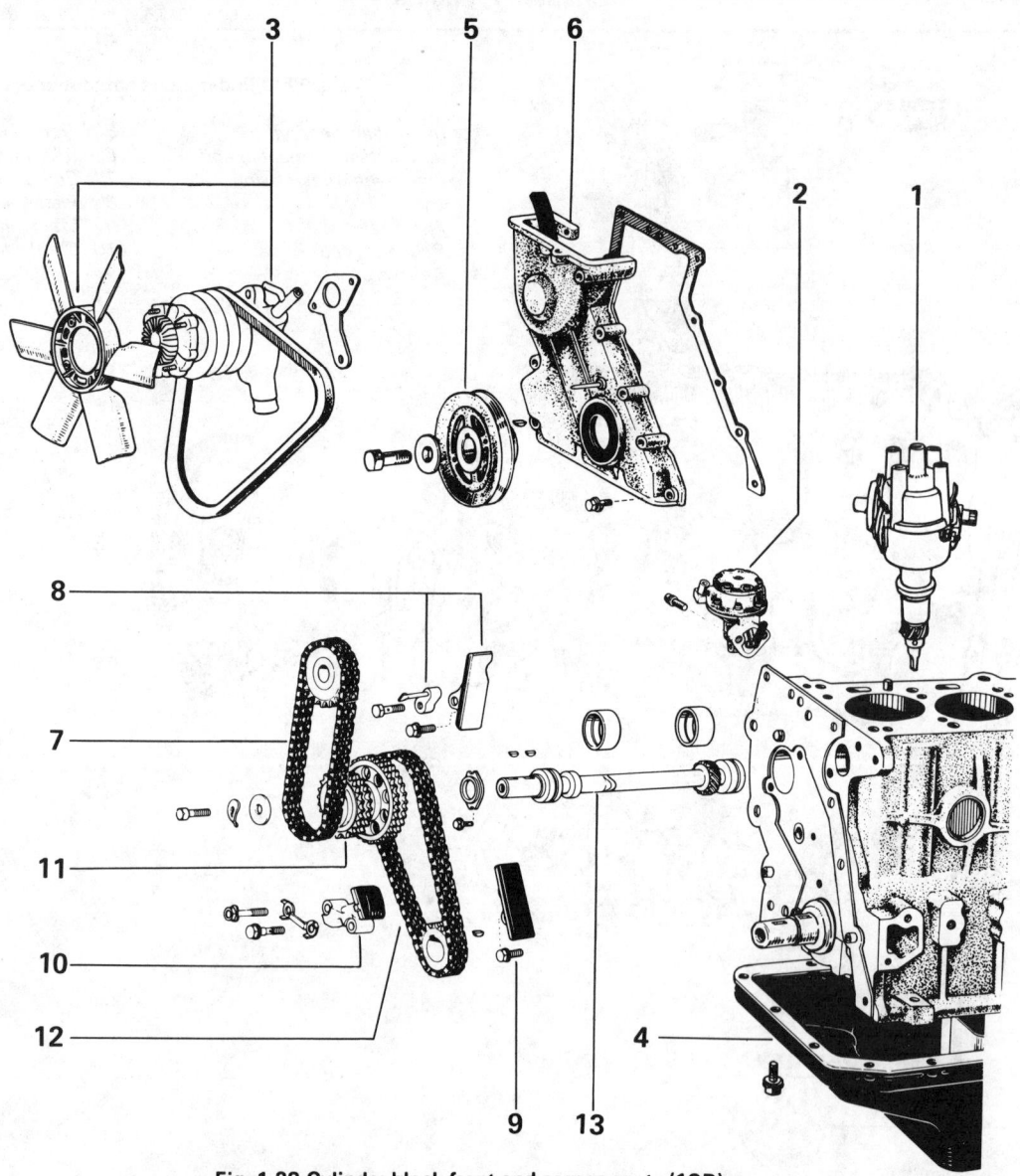

Fig. 1.28 Cylinder block front end components (18R)

1. Distributor
2. Fuel pump
3. Fan and water pump
4. Sump
5. Crankshaft pulley
6. Timing cover
7. No 2 timing chain
8. No 2 chain damper and oil jet
9. No 1 chain damper
10. No 1 chain tensioner
11. Camshaft sprocket
12. No 1 timing chain and sprocket
13. Auxiliary driveshaft

Fig. 1.29 Positions of camshaft lobes and crankshaft pulley notch (No 1 piston at TDC) (18R-G)

Fig. 1.30 Mating marks on camshaft drive pins and sprockets (18R-G)

Fig. 1.31 Camshaft bearing cap loosening sequence (18R-G)

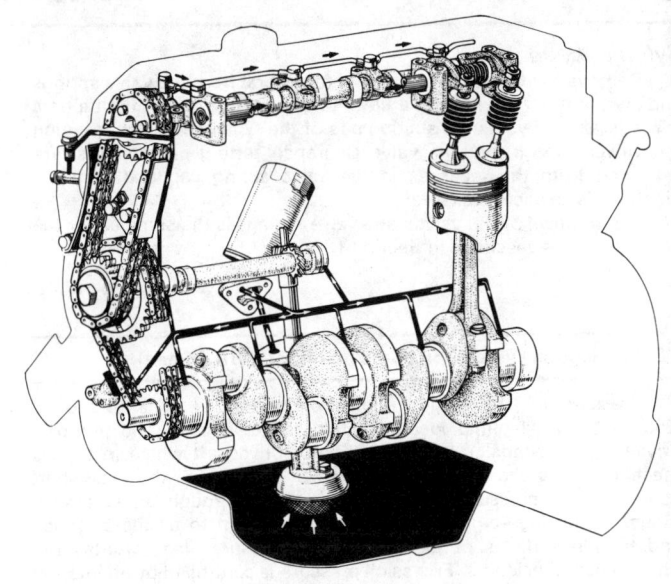

Fig. 1.32 Engine lubrication system (18R)

Fig. 1.33 Oil pump (18R) – exploded view

1 Gasket
2 Oil strainer
3 Relief valve
4 Pump cover
5 Rotor assembly

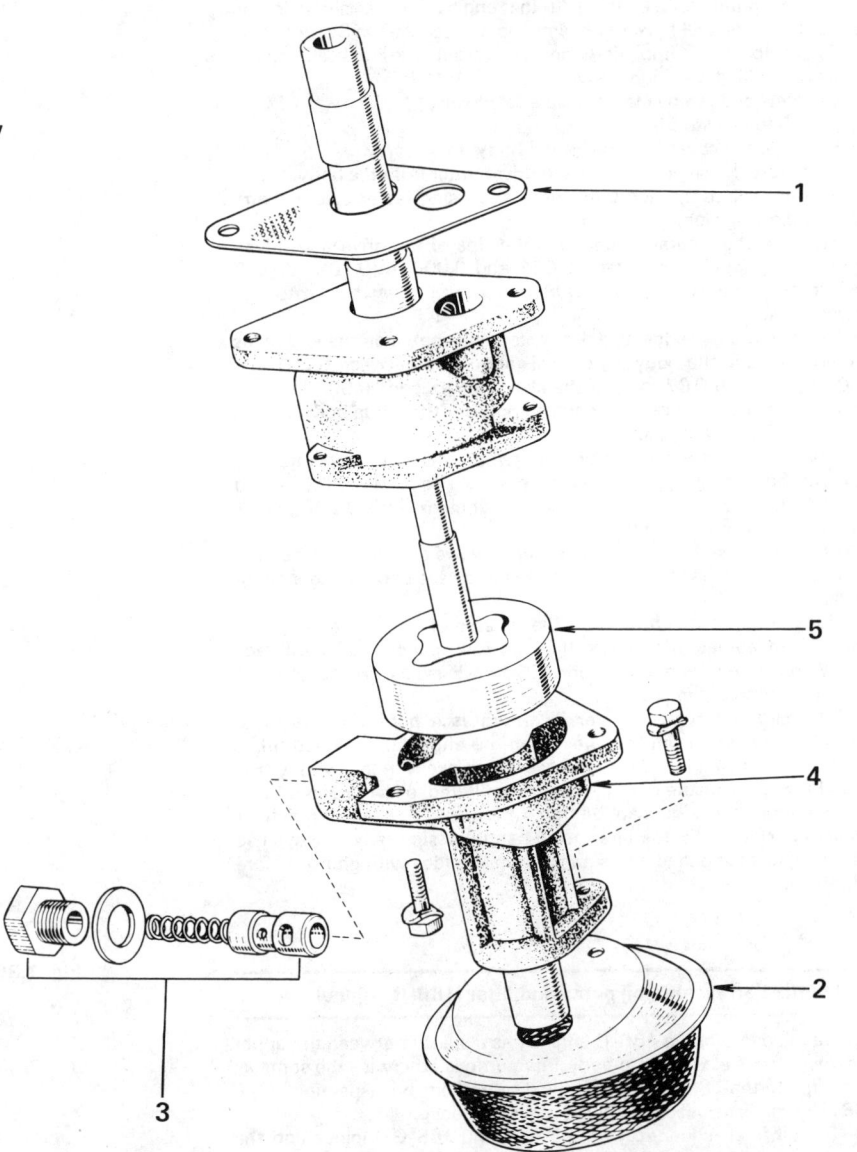

Cylinder head

9 Remove the bucket tappets (cam followers) from the valve springs and lay them out in order on a piece of paper. Removal of the tappets is made easier by using a suction pad of the type used when grinding the valves. Check that the valve clearance setting pads (shims) are extracted from the recesses in the valve spring cap and carefully identified with each bucket tappet.

10 The removal of the valves and valve springs is the same as for the 18R engine, as described in Section 13.

15 Lubrication system, oil pump and filter (18R engine)

1 Pressure for the engine lubrication system is generated by a trochoid type oil pump located within the crankcase. The pump is driven by an extension of the distributor driveshaft which in turn is meshed with a short driveshaft driven by chain from the crankshaft sprocket. The pressurised oil is first passed through an externally mounted cartridge-type disposable oil filter, then to all the bearings and friction surfaces of the engine. Oil pressure also actuates the timing chain tensioners. Excess oil pressure is controlled by an internal relief valve within the oil pump.

2 The oil pump normally has a very long life but in the event of low oil pressure (not due to worn bearings or lack of oil) being observed, remove the pump for servicing. If the engine is in position in the vehicle, the sump will have to be removed as described in Section 3.

3 To service the pump, first remove it from the crankcase (three screws) by pulling it straight down.

4 Unscrew and remove the pressure relief valve.

5 Unbolt the oil strainer.

6 Separate the cover from the pump body (three screws).

7 Withdraw the oil pump shaft and driven rotor from the body.

8 Examine all components for wear and using feeler blades, carry out the following clearance tests.

9 Measure the clearance between the tips of the drive and driven rotors. This should be between 0.004 and 0.006 in (0.10 and 0.15 mm). If the clearance exceeds 0.008 in (0.2 mm) renew both rotors as a matched set.

10 Using a straight-edge, check the clearance between the end faces of the rotors and the body flange. This should be between 0.001 and 0.003 in (0.03 to 0.07 mm). If the clearance exceeds 0.006 in (0.15 mm) the rotors should be renewed and possibly the pump body as well to achieve the correct tolerance.

11 Finally, measure the clearance between the outer rotor and the inside of the pump body. The clearance should be between 0.004 and 0.006 in (0.10 and 0.16 mm). If the clearance exceeds 0.008 in (0.2 mm), renew the pump body.

12 Check the condition of the relief spring valve and the valve head. It is a good policy to renew the spring if the engine is undergoing a major overhaul.

13 Reassembly and refitting are the reverse of the dismantling and removal procedures, but ensure that the rotor punch marks will face downwards when the pump is refitted. Always use a new gasket when fitting the pump to the crankcase.

14 The cartridge type oil filter incorporates a non-return valve to prevent oil draining from the filter when the engine is switched off. A bypass valve is built into the filter base which opens in the event of the filter clogging to ensure normal (though unfiltered) oil circulation.

15 The filter cartridge can be removed using a chain wrench or special filter strap. Fit the filter using hand pressure only. Always use the new gasket supplied and smear its sealing face with grease before tightening.

16 Lubrication system, oil pump and filter (18R-G engine)

1 The oil pump of the 18R-G engine has a spacer between the upper and lower parts of the pump body. The pump is otherwise the same as that fitted to the 18R engine and the dismantling and inspection of the 18R-G pump is as described in the previous Section.

2 The oil filter is identical on the 18R and 18R-G engines and the only difference in the lubrication system is the oil supply to the second shaft.

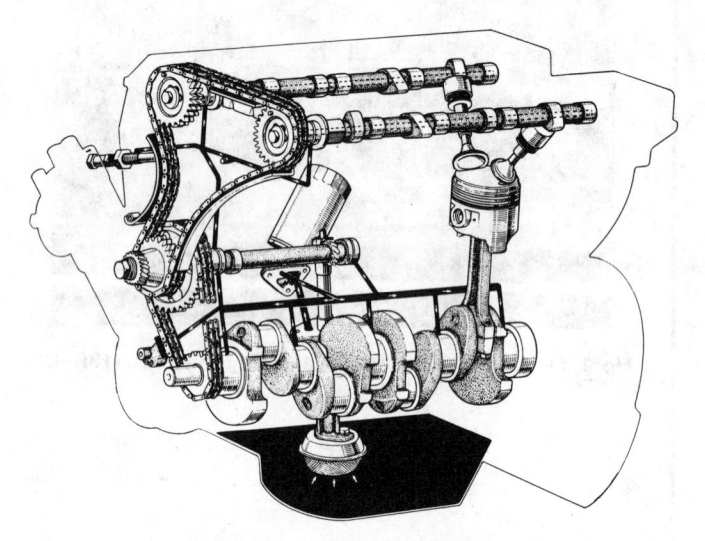

Fig. 1.34 Engine lubrication system (18R-G)

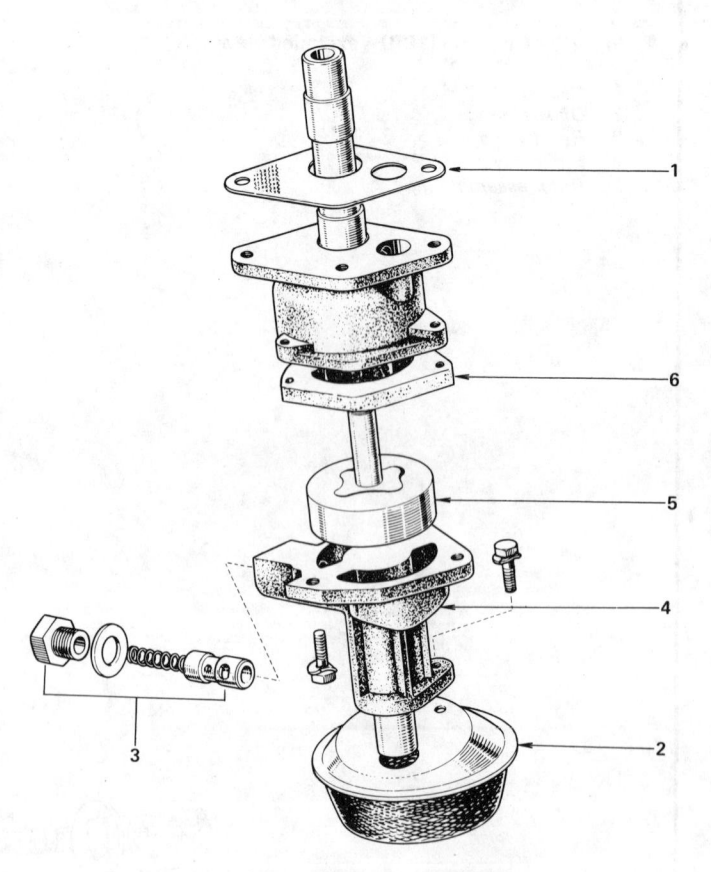

Fig. 1.35 Oil pump (18R-G) – exploded view

1 Gasket
2 Oil strainer
3 Relief valve
4 Pump cover
5 Rotor assembly
6 Spacer

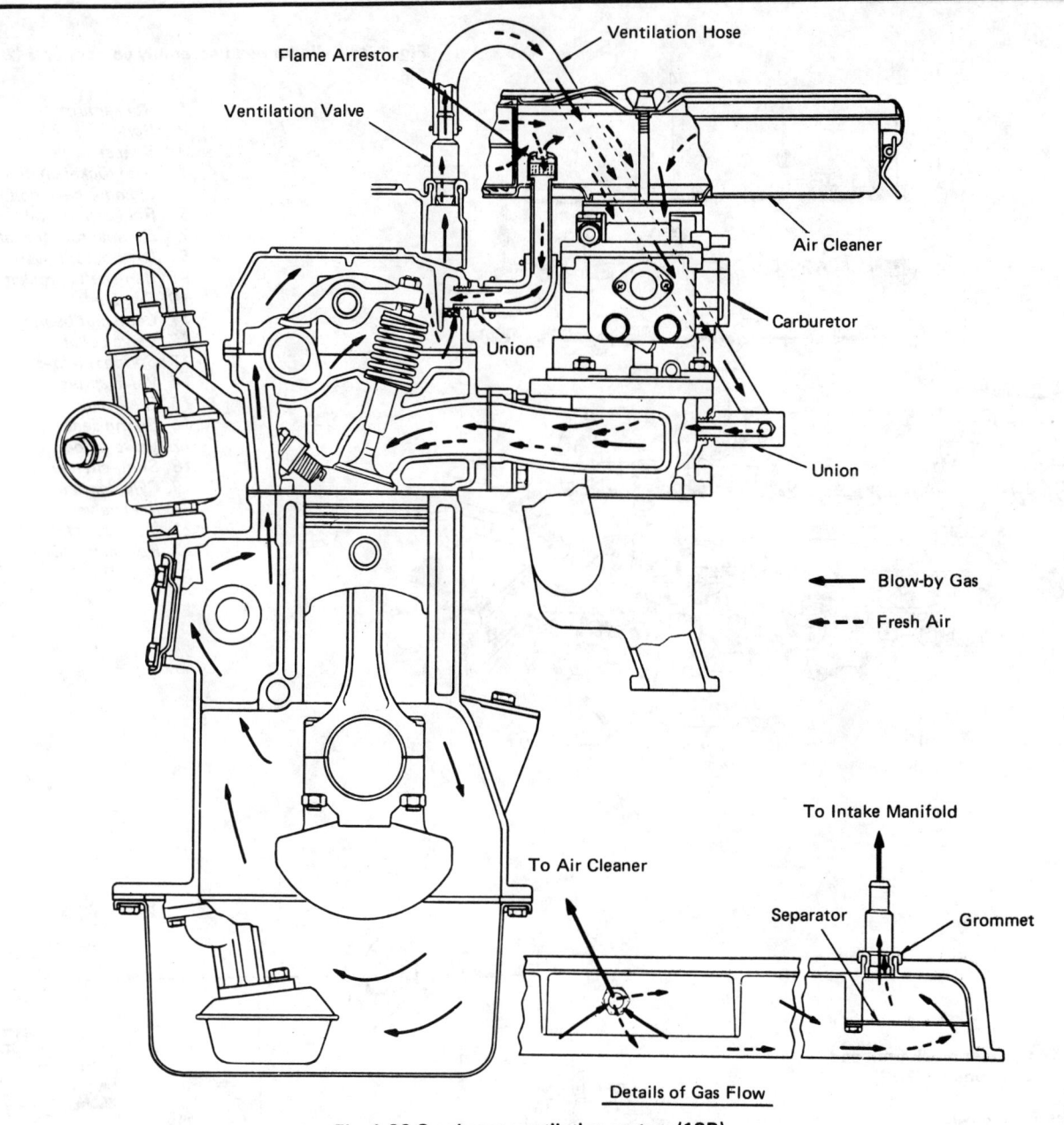

Fig. 1.36 Crankcase ventilation system (18R)

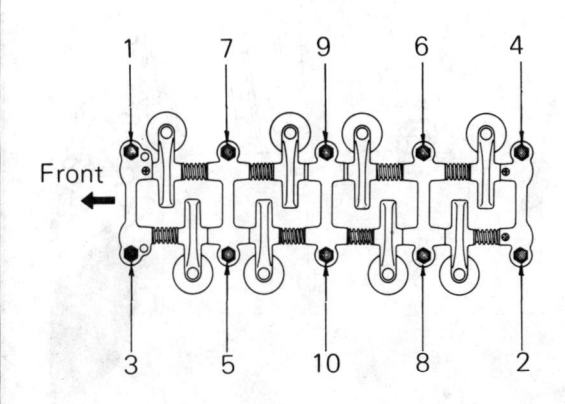

Fig. 1.37 Cylinder head bolt loosening sequence (20R)

Fig. 1.38 Cylinder head assembly components (20R)

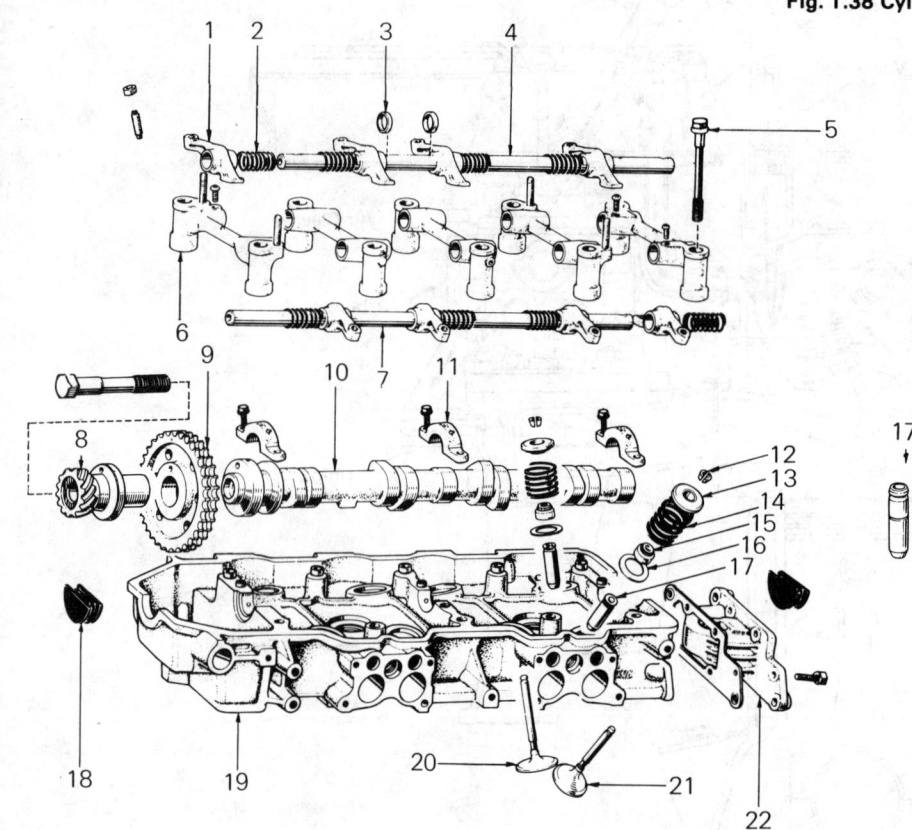

1. Rocker arm
2. Spring
3. Spacer
4. Inlet rocker shaft
5. Cylinder head bolt
6. Rocker pedestal
7. Exhaust rocker shaft
8. Distributor drivegear
9. Camshaft sprocket
10. Camshaft
11. Camshaft bearing cap
12. Valve collets
13. Spring retainer
14. Valve spring
15. Stem seal
16. Spring seat
17. Valve guide
18. Semi-circular seal
19. Cylinder head
20. Inlet valve
21. Exhaust valve
22. Rear cover (EGR cooler)

Fig. 1.39 Cylinder block front end components (20R)

1. Distributor drivegear
2. Cam sprocket
3. Chain cover assembly
4. Damper No 2
5. Damper No 1
6. Crankshaft pulley
7. Pump drive spline
8. Crankshaft sprocket
9. Chain tensioner
10. Chain

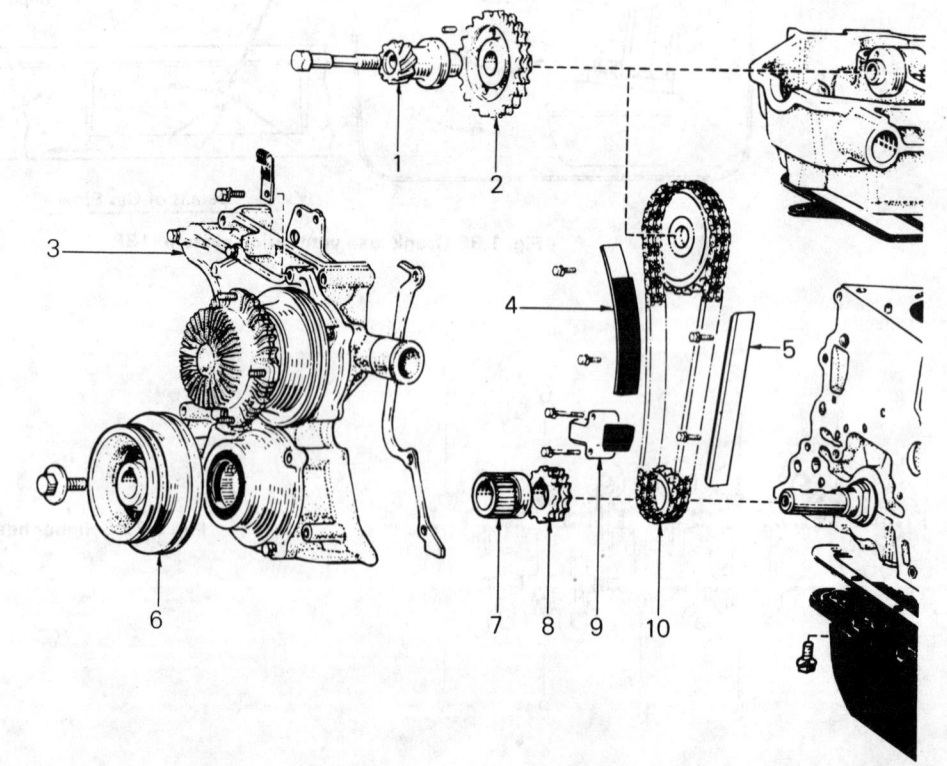

17 Crankcase ventilation system (18R engine)

1 As part of the emission control system (described in Chapter 3), a positive crankcase ventilation system is fitted.
2 Every 12 000 miles (20 000 km) check the operation of the PCV valve. To do this, let the engine run at idling speed and first pinch and then release the hose just above the valve, at the same time, listening for the sound of the valve seating. If it does not close, or is sluggish in operation, remove it and wash it thoroughly in fuel. Other indications of a faulty PCV valve are evidence of oil in the air cleaner and rough idling.
3 Every 24 000 miles (38 000 km) renew the PCV valve and at all times make sure that the connecting hoses and clips are secure and in good condition, also the O-ring seal of the oil filler cap (photo).

18 Engine dismantling (20R engine)

Cylinder head, rocker shafts and camshaft

1 *If the engine is in the vehicle,* disconnect the battery negative lead, all hoses, controls and leads from the rocker cover, cylinder head and carburettor. Disconnect the exhaust downpipe, drain the cooling system and remove the air cleaner and the distributor, then continue as follows.
2 Unbolt and remove the rocker cover. Remove the rubber semi-circular plug from the front edge of the cylinder head and remove the bolt which is exposed.
3 Using a quick drying paint, put alignment marks on the timing chain and camshaft sprocket.
4 Withdraw the distributor drivegear but leave the cam sprocket and chain undisturbed.
5 Remove the chain cover bolt from directly in front of the cam sprocket. *This must be done before attempting to remove the cylinder head bolts.*
6 Unscrew the cylinder head bolts in the sequence shown in Fig. 1.37 to prevent warpage of the cylinder head.
7 Withdraw the rocker assembly which is secured by the cylinder head bolts (now removed). The rocker pillars are located on dowels and both ends of the assembly must be levered upwards simultaneously to prevent distortion.
8 Push the cam sprocket forward off its mounting flange and allow it to rest on the ends of the chain dampers. Remove the cylinder head by lifting both ends simultaneously as this too is located on dowels.
9 *If this work is being carried out while the engine is still in the vehicle,* once the cylinder head is removed, it is recommended that the engine oil is drained and discarded as it will have become contaminated with coolant. Failure to observe this recommendation may result in corroded crankcase components.
10 Remove the EGR valve from the cylinder head.
11 Unbolt and remove the inlet manifold complete with carburettor.
12 Remove the thermostatic valve.
13 Unbolt and remove the exhaust manifold heat insulator.
14 With the heat insulator removed, unbolt and withdraw the exhaust manifold.
15 Unscrew and remove the spark plugs.
16 At this stage, measure the camshaft endfloat, using a feeler blade. If it exceeds 0.0098 in (0.25 mm), the complete cylinder head assembly will have to be renewed.
17 Remove the camshaft bearing caps and lift out the camshaft.
18 Remove the valve from the cylinder head using a valve spring

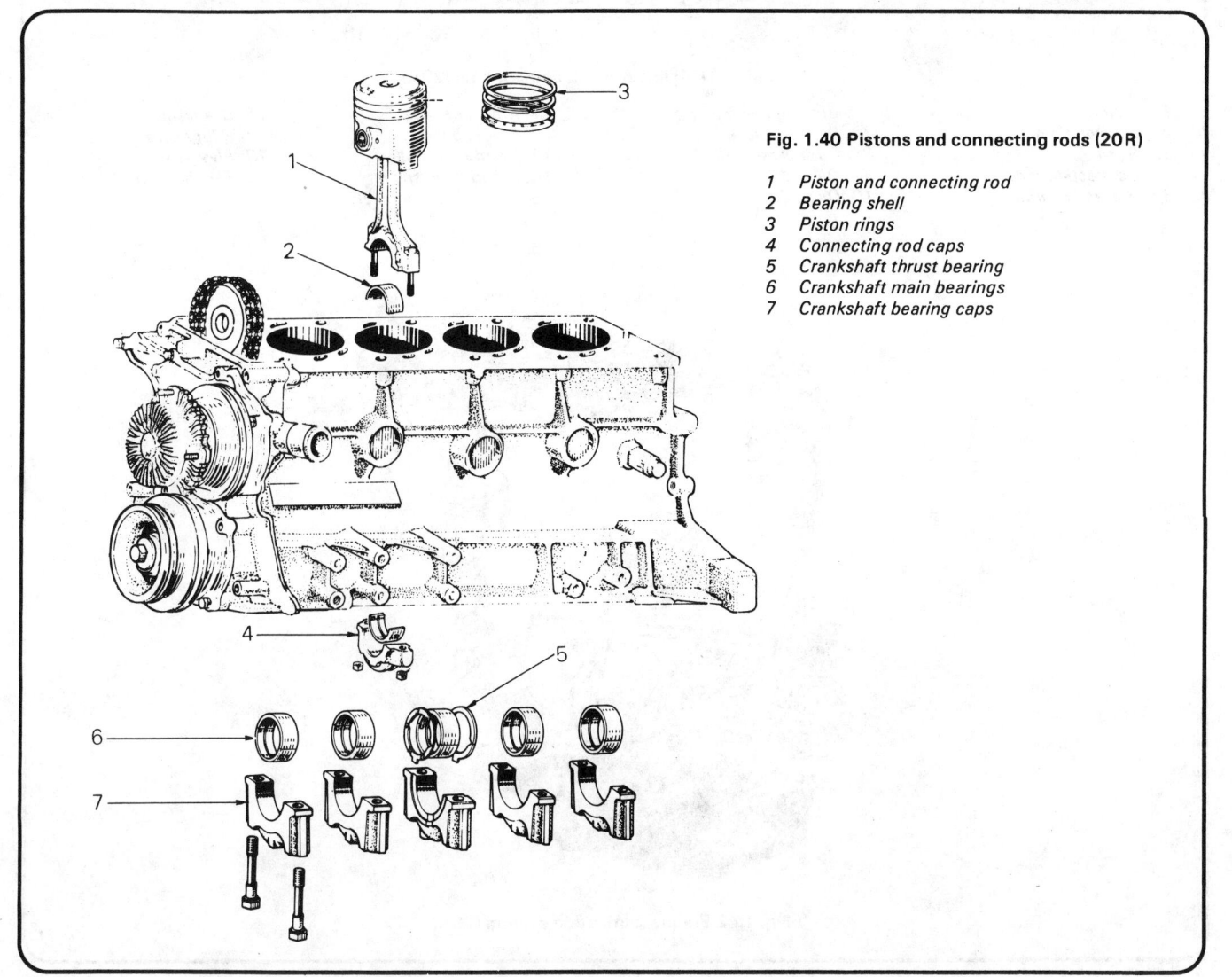

Fig. 1.40 Pistons and connecting rods (20R)

1 Piston and connecting rod
2 Bearing shell
3 Piston rings
4 Connecting rod caps
5 Crankshaft thrust bearing
6 Crankshaft main bearings
7 Crankshaft bearing caps

44

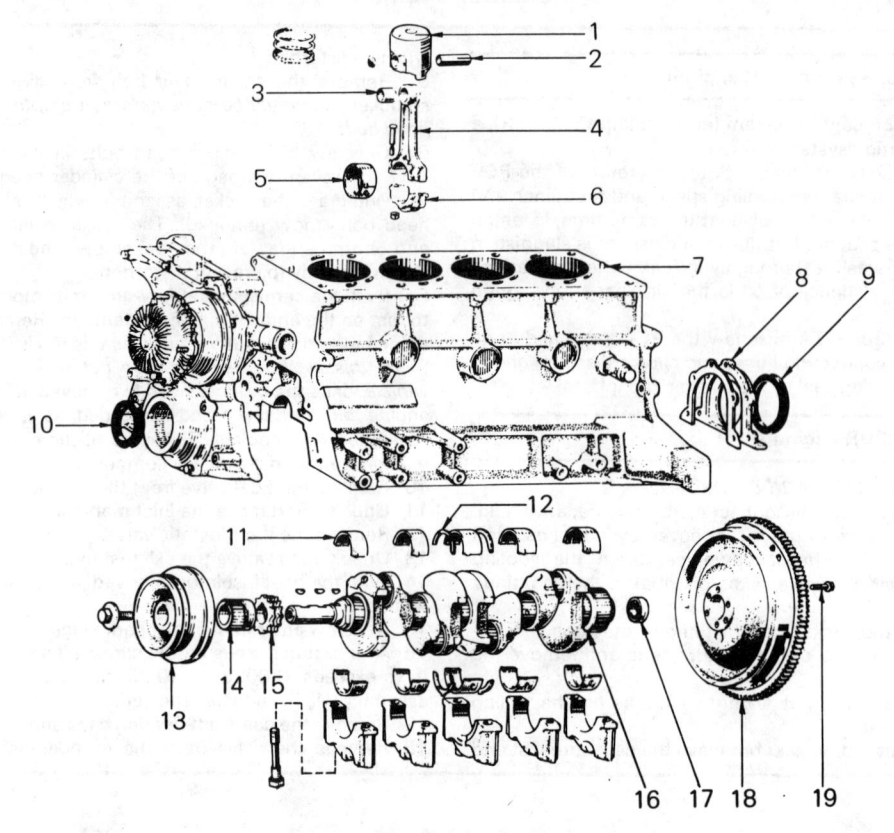

Fig. 1.41 Cylinder block components (20R)

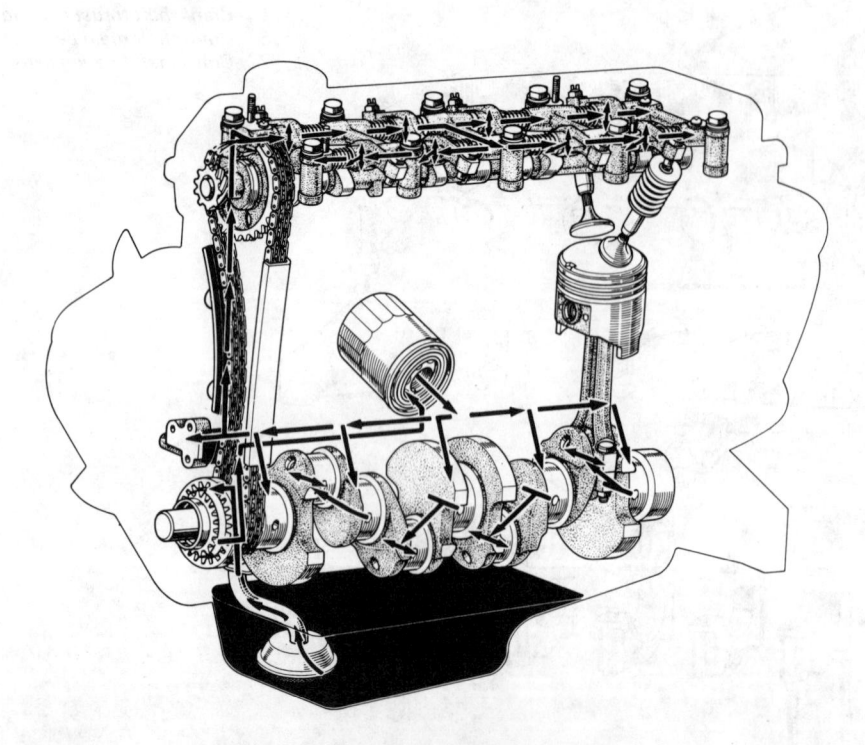

1 Piston	6 Big-end bearing cap	11 Main bearing	16 Crankshaft
2 Gudgeon pin	7 Cylinder block	12 Thrust bearing	17 Pilot bearing
3 Bush	8 Oil seal retainer	13 Crankshaft pulley	18 Flywheel
4 Connecting rod	9 Oil seal	14 Pump drive spline	19 Flywheel bolt
5 Big-end bearing	10 Oil seal	15 Crankshaft sprocket	

Fig. 1.42 Engine lubrication system (20R)

Fig. 1.43 Oil pump (20R) – exploded view

1 Spring
2 Relief valve
3 Pump
4 Inner gear
5 Outer gear
6 O-ring
7 Crankshaft drivegear

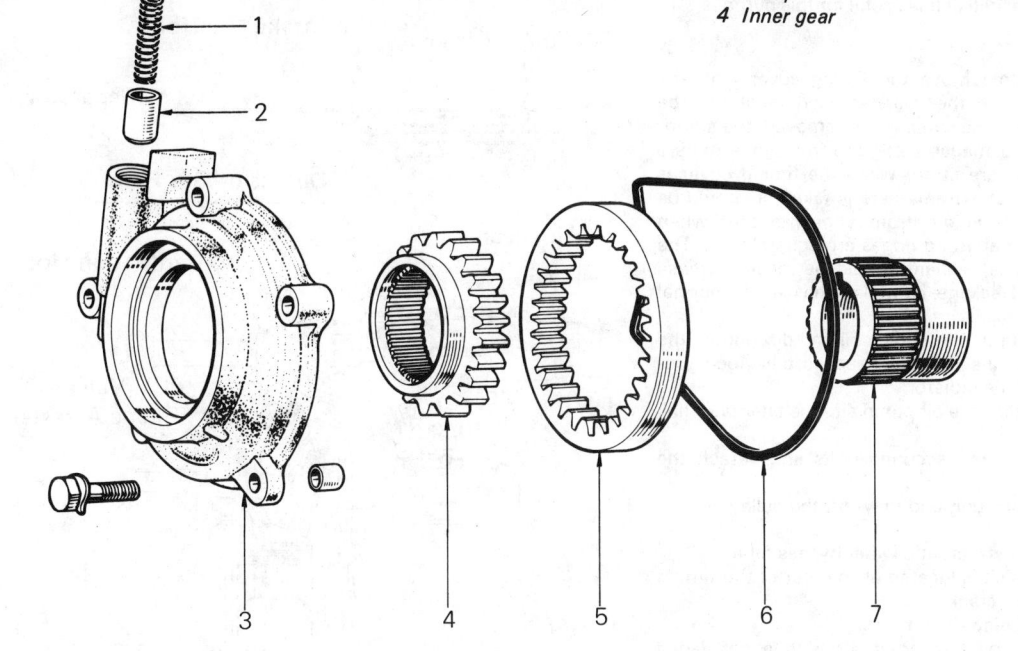

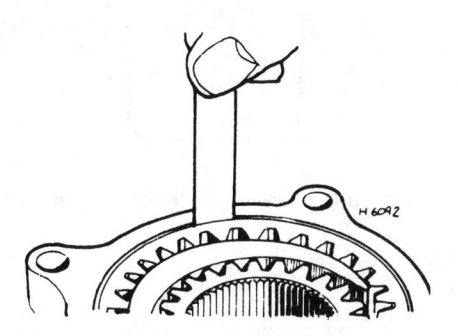

Fig. 1.44 Checking oil pump outer gear clearance (20R)

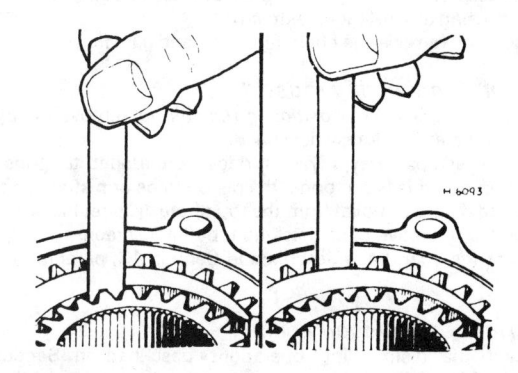

Fig. 1.45 Checking oil pump tip to crescent clearances (20R)

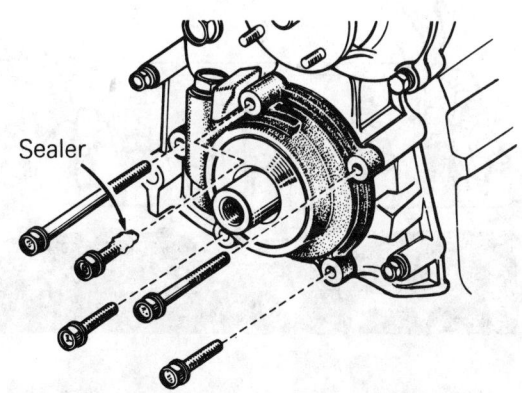

Sealer

Fig. 1.46 Oil pump securing bolts (20R)

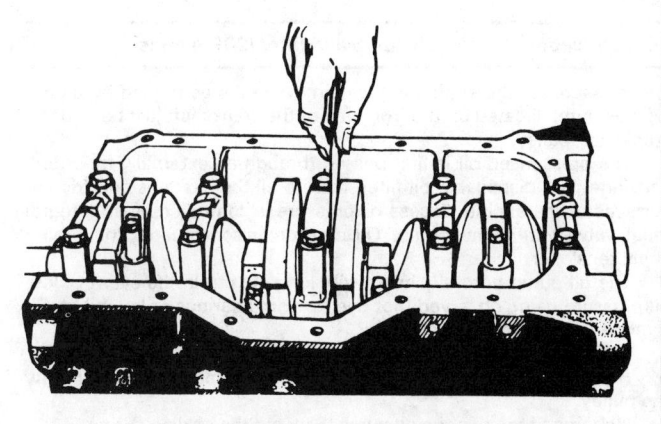

Fig. 1.47 Measuring crankshaft endfloat

compressor. Extract the split collets and withdraw the valve spring retainers, the springs, valves, seals and seats.
19 Keep the valves in their original fitted sequence. A piece of card with holes numbered 1 to 8 punched in it is useful for this purpose.

Sump and timing gear
Note: The owner may wish to remove the timing cover with the engine in the car, in which case the cylinder head must first be removed followed by the sump. If the sump is not removed, the sump-to-timing cover gasket will be damaged along the front edge (where the sump-to-timing cover bolts are fitted) when the timing cover is taken off. However, provided that extreme care is taken, it should be possible to cut the appropriate length from a new gasket when reassembling, provided that the abutting edges are squarely cut. The use of a silicone/RTV-type gasket sealant should be made at these points to minimise the risk of oil leakage, as well as the use of a normal non-setting sealant on the joint faces.
20 *If the engine is in the car,* drain the engine oil and disconnect the battery negative lead. Remove the cylinder head as described previously in this Section. Remove the radiator.
21 Remove the drivebelts. Detach the air pump and the alternator link from the timing cover.
22 Unscrew and remove the sump securing bolts and detach the sump from the crankcase.
23 Remove the crankshaft pulley bolt and draw off the pulley using a suitable extractor.
24 Remove the two bolts which secure the water bypass tube.
25 Remove the single bolt which is located at the rear of the timing cover on the left-hand side of the crankcase.
26 Unscrew and remove the timing cover bolts.
27 Remove the timing cover by tapping it off with a soft-faced hammer.
28 Remove the timing chain together with the camshaft sprocket.
29 If necessary, the crankshaft sprocket and oil pump drive can be withdrawn using a two-legged extractor.
30 Unbolt and remove the chain tensioner and guides.

Pistons and connecting rods
31 *If the engine is still in position in the vehicle,* remove the cylinder head and sump as previously described.
32 Carefully scrape away any wear ridge from around the tops of the cylinder bores. If this is not done, the piston rings or piston itself could be damaged during removal from the top of the cylinder block.
33 Unbolt and remove the oil pick-up tube/filter screen.
34 Repeat the operations described in Section 13, paragraphs 28 to 30.

Piston rings and gudgeon pins
35 Repeat the dismantling operations described in Section 13, paragraphs 31 to 34 inclusive.

Crankshaft and main bearings
36 Repeat the dismantling operations described in Section 13 paragraphs 35 to 39.

19 Lubrication system, oil pump and filter (20R engine)

1   Pressure for the engine lubrication system is generated by a gear type oil pump located on the front end of the crankshaft just behind the crankshaft pulley.
2   The pressurised oil is first passed through an externally mounted, cartridge type disposable oil filter, then to all the bearings and friction surfaces of the engine. Excess oil pressure is controlled by an integral relief valve within the pump. Oil pressure also actuates the timing chain tensioner.
3   The oil pump normally has a very long life but in the event of low oil pressure being observed (not due to worn bearings or low oil level), remove the oil pump for servicing.
4   Remove the sump and oil pick-up tube/strainer.
5   Remove the drivebelts from the crankshaft pulley and withdraw the pulley.
6   Unbolt and remove the oil pump. Pick out the sealing O-ring.
7   Withdraw the pump drivegear from the front end of the crankshaft.
8   Unscrew and remove the relief valve components and then withdraw the inner and outer gears from the pump body.

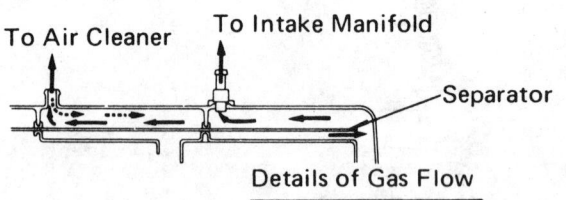

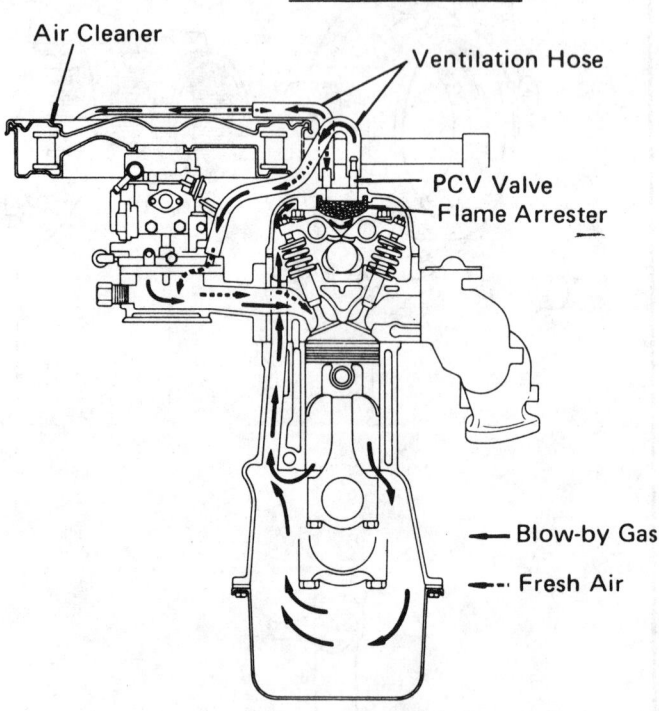

Fig. 1.48 Crankcase ventilation system (20R)

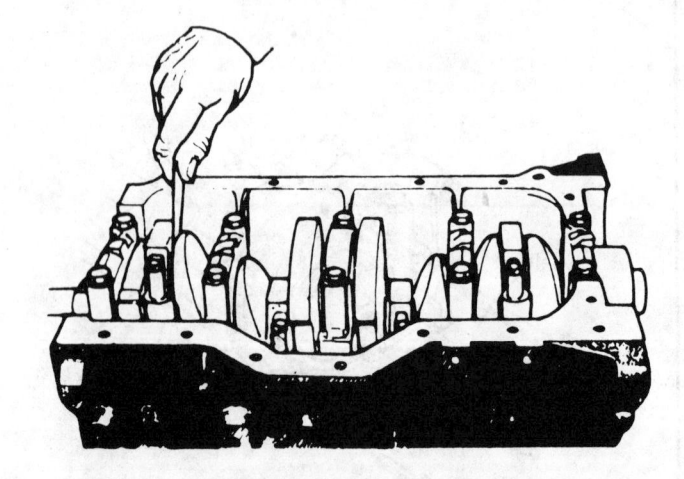

Fig. 1.49 Measuring big-end bearing sidefloat

Chapter 1 Engine

9 Clean and inspect all components for damage.
10 Using a feeler blade, check the clearance between the outer gear and the pump body. This should not exceed 0.008 in (0.2 mm) otherwise renew one or both components.
11 Now measure the tip to crescent clearance between the gears which should not exceed 0.012 in (0.3 mm), when all gears are in position within the pump body.
12 Using a straight-edge and a feeler blade, measure the gear endfloat within the pump body. This should not exceed 0.0059 in (0.15 mm).
13 Renew the oil seal if there have been signs of oil seepage from this part of the engine. Drive out the defective seal with a piece of tubing and refit the new one in the same way.
14 Reassembly is a reversal of dismantling but make sure that the punch marks are visible when the gears are refitted.
15 When refitting the oil pump, renew the O-ring and apply jointing compound to the threads of the uppermost securing bolt.

20 Crankcase ventilation system (20R engine)

1 Refer to Section 17 but note the difference in layout of the system on this engine (see Fig. 1.48).

21 Examination and renovation – general

With the engine stripped down and all parts thoroughly cleaned, it is now time to examine everything for wear. The following items should be checked and where necessary renewed or renovated as described in the following Sections. The information applies to all engine types except where specifically annotated.

22 Crankshaft and main bearings – examination and renovation

1 Examine the crankpin and journal surfaces for signs of scoring or scratches. Check the ovality of the crankpins at several different positions using a micrometer. If more than 0.001 in (0.0254 mm) out of round, the crankshaft will have to be reground. Check the journals in the same manner.
2 If it is necessary to regrind the crankshaft and to fit new bearings, your Toyota dealer will decide how much to grind off and he will supply new oversize shell bearings to suit. Details of regrinding tolerances and bearings are given in Specifications.
3 If the crankshaft is in good condition and requires no attention, it is always worthwhile renewing the bearing shells at the same time of a major overhaul. Renew them with ones of the same size as the originals.
4 It is possible (although difficult, and not really recommended), to renew the main bearing shells while the engine is still in the car. To do this, remove the sump and then detach one of the main bearing caps.
5 Renew the bearing shell in the cap.
6 Insert a flat-headed screw in the crankshaft journal oil hole and carefully turn the crankshaft. The head of the screw will push the second bearing shell from its seat. Install the new shell in a similar way, having first oiled it liberally.
7 The crankshaft endfloat should be checked now by temporarily installing all the main bearing caps and shells and tightening to the specified torque.
8 Push and pull the crankshaft in both longitudinal directions and measure the total endfloat with feeler blades or a dial gauge. The endfloat should be as stated in Specifications according to engine type.
9 Where the endfloat is incorrect, renew the centre main bearing shells which incorporate the thrust washers (18R). On the 2T and 20R engines, semi-circular thrust washers are used (photo).
10 When carrying out the operations just described, make sure that the arrows on the main bearing caps point towards the front of the engine and that the caps and shells are located in their original sequence.
11 The clutch input shaft spigot bearing is located in the centre of the rear mounting flange on the crankshaft. Renew if worn, greasing its reverse side before fitting.

23 Connecting rods and bearings – examination and renovation

1 Big-end bearing failure is indicated by a knocking from within the

17.3 Oil filler cap O-ring seal (18R)

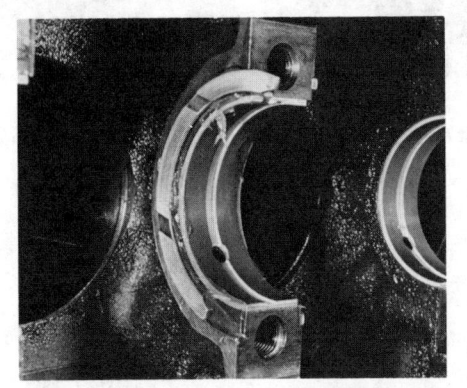

22.9a Crankshaft thrust washer (2T-B)

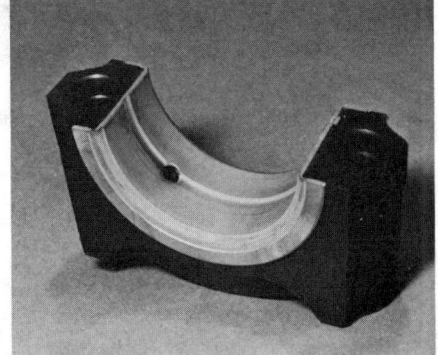

22.9b Bearing incorporating thrust washer (18R)

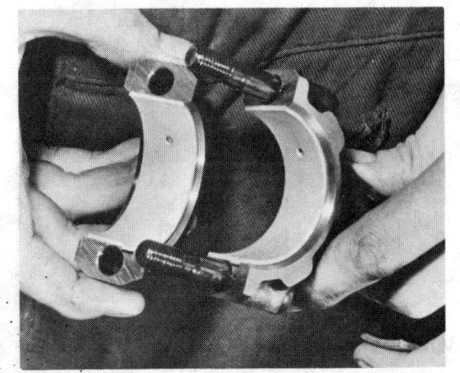

23.2a Big-end bearing shell and oil hole (2T-B)

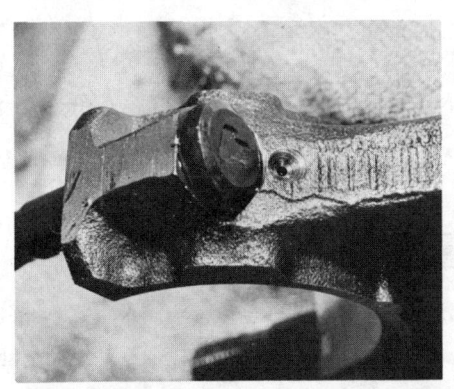

23.2b Connecting rod oil jet hole (2T-B) arrowed

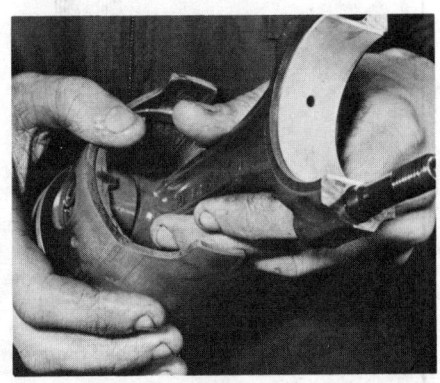

23.2c Big-end bearing shell and oil hole (18R)

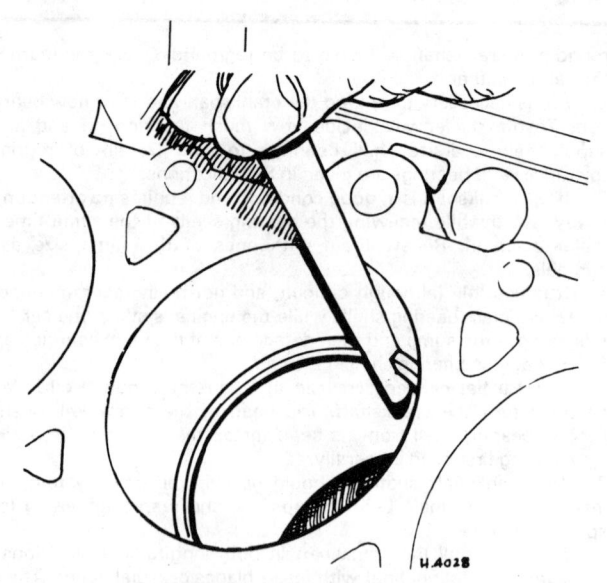

Fig. 1.50 Measuring piston ring end gap

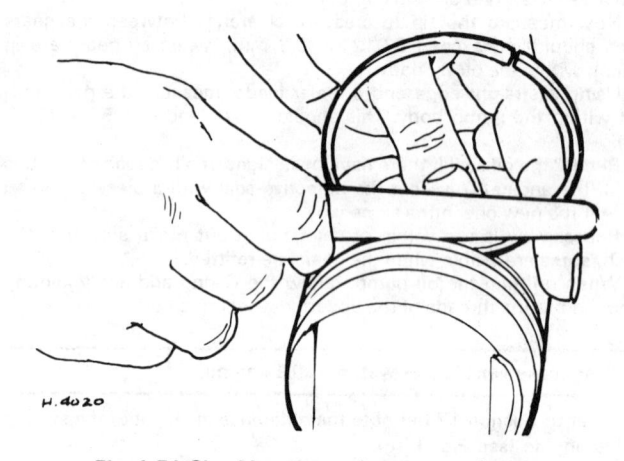

Fig. 1.51 Checking piston ring groove clearance

Fig. 1.52 Measuring camshaft thrust clearance (2T-B)

Fig. 1.53 Measuring camshaft thrust clearance (18R and 20R)

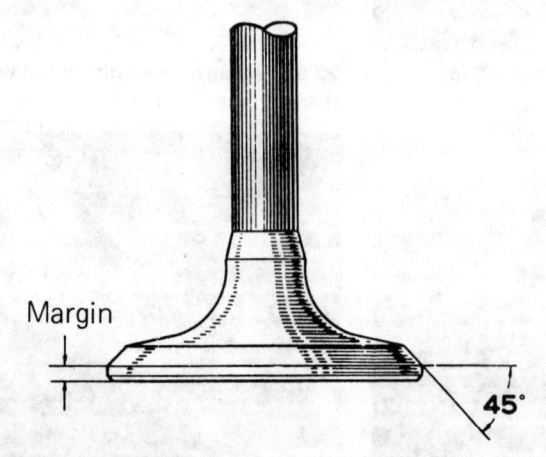

Fig. 1.54 Valve head margin measurement

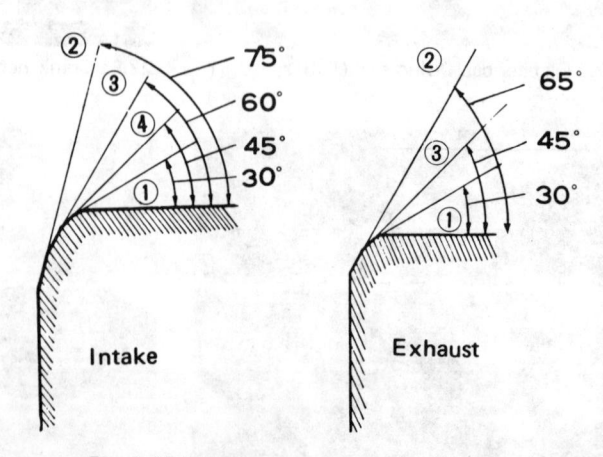

Fig. 1.55 Valve seat cutting angles (2T-B)

crankcase and a slight drop in oil pressure.
2   Examine the big-end bearing surfaces for pitting and scoring. Renew the shells in accordance with the sizes specified in Specifications. Where the crankshaft has been reground, the correct oversize big-end shell bearings will be supplied by the engineer (photos).
3   Fit each connecting rod to its respective crankpin and, using a feeler blade, check the side-float. If this exceeds the tolerance in Specifications, the connecting rod will have to be renewed.
4   Check each small-end bush for wear or scoring. Each gudgeon pin should be a push fit in its bush using thumb pressure only. If the bush is worn it will have to be pressed out and a new one fitted, ensuring that the oil holes of the bush and the connecting rod coincide. As the bush will have to be reamed after fitting, this is probably a job best left to your Toyota dealer.

24 Cylinder bores – examination and renovation

1   The cylinder bores must be examined for taper, ovality, scoring and scratches. Start by carefully examining the top of the cylinder bores. If they are at all worn, a very slight ridge will be found on the thrust side. This marks the top of the piston ring travel. The owner will have a good indication of the bore wear prior to dismantling the engine, or removing the cylinder head. Excessive oil consumption accompanied by blue smoke from the exhaust is a sure sign of worn cylinder bores and piston rings.
2   Measure the bore diameter just under the ridge with a micrometer and compare it with the diameter at the bottom of the bore, which is not subject to wear. If the difference between the two measurements is more than 0.008 in (0.2 mm) then it will be necessary to fit special pistons and rings or to have the cylinder rebored and fit oversize pistons. If no micrometer is available remove the rings from a piston and place the piston in each bore in turn about halfway down the bore. If a 0.0012 in (0.03 mm) feeler gauge, slid between the piston and cylinder wall, requires less than a pull of between 2.2 and 5.5 lbs (1.0 and 2.5 kg) to withdraw it, using a spring balance, then remedial action must be taken. Oversize pistons are available as listed in Specifications.
3   These are accurately machined to just below the indicated measurements so as to provide correct running clearances, in bores taken out to the exact oversize dimensions.
4   If the bores are slightly worn, but not so badly worn as to justify reboring them, then special oil control rings and pistons can be fitted which will restore compression and stop the engine burning oil. Several different types are available and the manufacturer's instructions concerning their fitting must be followed closely.
5   If new pistons or rings are being fitted and the bores have not been reground, it is essential to slightly roughen the hard glaze on the sides of the bores with fine glass paper so the new piston rings will have a chance to bed in properly.
6   If the cylinder bores have been bored out beyond the limit so that the maximum oversize pistons available cannot be fitted, then sleeves can be supplied which after fitting and boring will accept standard sized pistons. This again is a job for your Toyota dealer or motor engineering works.

25 Pistons and piston rings – examination and renovation

1   If the original pistons are to be refitted, carefully remove the piston rings as described in Section 13, paragraphs 31 onwards.
2   Clean the grooves and rings free from carbon, taking care not to scratch the aluminium surfaces of the pistons.
3   If new rings are to be fitted, then order the top compression ring to be stepped to prevent it impinging on the 'wear ring' which will almost certainly have been formed at the top of the cylinder bore.
4   Before fitting the rings to the pistons, push each ring in turn down to the bottom of its respective cylinder bore (use an inverted piston to do this so that the ring is kept square in its bore) and then measure the piston ring end gap. This should be as shown in Specifications. If the gap is incorrect, carefully grind the ends of the ring.
5   Now test each ring in its groove in the piston for side clearance using a feeler blade. If the clearance exceeds that specified, renew the piston as it will be the groove that is worn.
6   Where necessary a piston ring which is slightly tight in its groove may be rubbed down holding it perfectly squarely on an oilstone or a sheet of fine emery cloth laid on a piece of plate glass. Excessive tightness can only be rectified by having the grooves machined out.
7   The gudgeon pin should be a push fit into the piston when heated in water to a temperature of 140°F (60°C). If it appears slack, then both the piston and gudgeon pin should be renewed.

26 Camshaft and camshaft bearings – examination and renovation

2 T-B and 18R engines

1   Check the camshaft journals for scoring or grooves and then measure each journal at several different points to detect any taper or out of round. If the difference between the measurements exceeds 0.0004 in (0.01 mm) the camshaft must be reground and oversize shell bearings fitted. This is a job for your Toyota dealer.
2   With the camshaft on the cylinder head complete with shell bearings and caps and the cap bolts tightened as specified, check the camshaft endfloat. This should be as shown in Specifications, otherwise renew the bearing shell which incorporates the thrust flanges.
3   Firstly examine the camshaft lobes for scoring or wear. Using a micrometer, check the overall lengths of the inlet and exhaust valve cam lobes and compare them with those specified. If they are worn, renew the camshaft complete.

18R-G and 20R engines

4   Check the camshaft for wear, as described in the preceding paragraphs but where bearing running clearance or camshaft endfloat is found to be excessive, then the cylinder head will have to be renewed as the camshaft runs directly in bearings and caps which are in-line machined, no detachable bearing shells are fitted.

27 Timing components – examination and renovation

1   Examine all the sprocket teeth for wear or 'hooked' appearance and renew if necessary.
2   Wash the timing chains thoroughly in paraffin and examine for wear or stretch. If the chain is supported at both ends so that the rollers are vertical then a worn chain will take on a deeply bowed appearance while an unworn one will dip slightly at its centre point.
3   Check the chain tensioners and guides for wear, and renew the slippers if they are cut or grooved.

28 Cylinder head and valves – examination, renovation and decarbonising

1   Examine the heads of the valves for pitting and burning, especially the heads of the exhaust valves. The valve seatings should be examined at the same time. If the pitting on the valve and seat is very slight the marks can be removed by grinding the seats and valves together with coarse and then fine, valve grinding paste.
2   Where bad pitting has occurred to the valve seats, it will be necessary to recut them and fit new valves. In practice, it is very seldom that the seats are so badly worn that they require renewal. Normally it is the valve that is too badly worn to use again, and the owner can easily purchase a new set of valves and match them to the seats by grinding. Where the seat has to be recut, a 45° cutter should be used, and only the minimum amount of metal removed to provide a satisfactory finish. Ensure that the valve head margin (see Fig. 1.54) is not less than 0.024 in (0.6 mm), then apply a little engineers blue to the valve seating surfaces and check the contact area. The seat contact should be in the middle of the valve face with a width of 0.047 to 0.063 in (1.2 to 1.6 mm). If the seating is too high, a 15° or 30° and 45° cutter should be used to correct it; if the seating is too low a 75° or 60° and 45° cutter should be used to correct it (see Figs. 1.55, 1.56 and 1.57).
3   Valve grinding where the seats do not have to be recut, is carried out as follows: Smear a trace of coarse carborundum paste on the seat face and apply a suction grinder tool to the valve head. With a semi-rotary motion, grind the valve head to its seat, lifting the valve occasionally to redistribute the grinding paste. When a dull matt, even surface finish is produced on both the valve seat and the valve, wipe off the paste and repeat the process with fine carborundum paste, lifting and turning the valve to distribute the paste as before. A light

Chapter 1 Engine

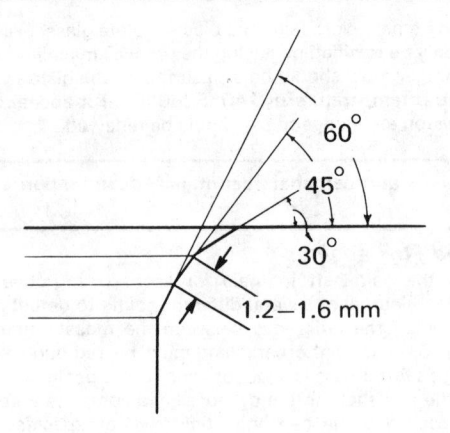

Fig. 1.56 Valve seat cutting angles (18R)

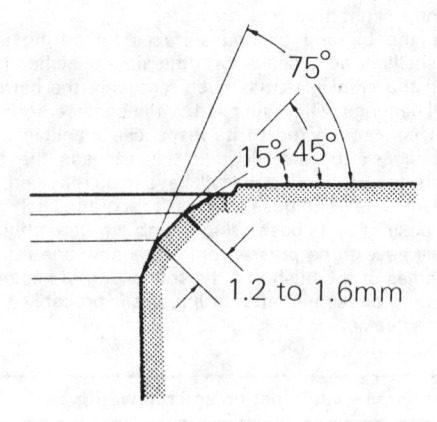

Fig. 1.57 Valve seat cutting angles (20R)

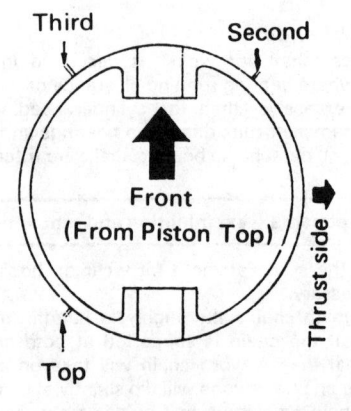

Fig. 1.58 Ring end gap staggering (2T engine)

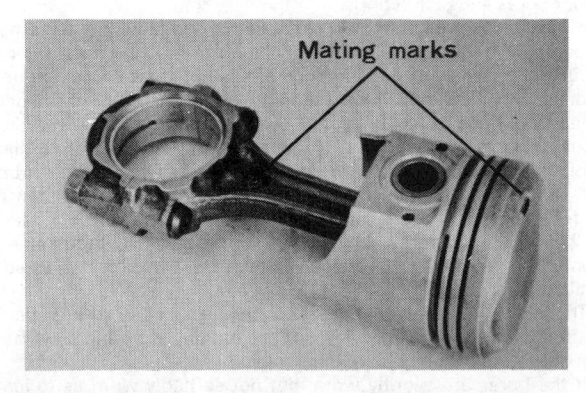

Fig. 1.59 Piston and connecting rod FRONT marks (2T-B)

spring placed under the valve head will greatly ease this operation. When a smooth unbroken ring of light grey matt finish is produced, on both valve and valve seat faces, the grinding operation is complete.

4 Scrape away all carbon from the valve head and the valve stem. Carefully clean away every trace of grinding compound, taking care to leave none in the ports or in the valve guides. Clean the valves and valve seats with a paraffin soaked rag then with a dry rag, and finally, if an air line is available, blow the valves, valve guides and valve ports clean.

5 Wear in the valve guides can best be checked by inserting a new valve and testing for rocking movement in all directions. The clearance between the guide and valve stem must not exceed that shown in Specifications.

6 To renew a valve guide, first check whether the existing one has a snap-ring retaining it. Where this is so, use a hammer and a brass drift to snap the valve guide off, then drive out the remaining part.

7 Where there is no snap-ring, drive out the valve guide with a suitable drift.

8 All new valve guides have snap-rings, and the guide is simply driven in until the snap-ring locates in the groove. The guide must now be reamed to obtain the specified clearance.

9 The valve springs should be compared with their specified free lengths. Renew the springs as a set if they differ from their specified new length or have been in operation more than 24 000 miles (38 000 km). Always renew the valve stem oil seals.

10 With the cylinder head removed, use a blunt scraper to remove all trace of carbon and deposits from the combustion spaces and ports. Scrape the cylinder head free from scale or old pieces of gasket or jointing compound. Clean the cylinder head by washing it in paraffin and take particular care to pull a piece of rag through the ports and cylinder head bolt holes. Any grit remaining in these recesses may well drop onto the gasket or cylinder block mating surface as the cylinder head is lowered in position and could lead to a gasket leak after reassembly is complete.

11 With the cylinder head clean, test for distortion if a history of coolant leakage has been apparent. Carry out this test using a straight edge and feeler gauges or a piece of plate glass. If the surface shows any warping in excess of 0.002 in (0.05 mm), then the cylinder head will have to be resurfaced which is a job for a specialist engineering company.

12 Clean the piston and top of the cylinder bores. If the pistons are still in the block, then it is essential that great care is taken to ensure that no carbon gets into the cylinder bores, as this could scratch the cylinder walls or cause damage to the piston and rings. To ensure this does not happen, first turn the crankshaft so that two of the pistons are at the top of their bores. Stuff a rag into the other two bores or seal them off with paper and masking tape. The waterways should also be covered with small pieces of masking tape to prevent particles of carbon entering the cooling system and damaging the water pump.

13 Press a little grease into the gap between the cylinder walls and the two pistons which are to be worked on. With a blunt scraper carefully scrape away the carbon from the piston crown, taking great care not to scratch the aluminium. Also scrape away the carbon from the surrounding lip of the cylinder wall. When all carbon has been removed scrape away the grease which will now be contaminated with carbon particles, taking care not to press any into the bores. To assist prevention of carbon build-up the piston crown can be polished with a metal polish. Remove the rags or masking tape from the other two cylinders and turn the crankshaft so that the two pistons which were at the bottom are now at the top. Place a rag or masking tape in the cylinders which have been decarbonised and proceed as just described.

Chapter 1 Engine

29 Auxiliary driveshaft (18R and 18R-G engines) – servicing

1 The auxiliary driveshaft bearings should be inspected for scoring or scratches.
2 The correct running clearances between the shaft and bearing is between 0.0010 and 0.0026 in (0.025 and 0.066 mm). Where the clearance exceeds 0.003 in (0.08 mm) the bearings must be renewed.
3 To do this, remove the plug at the back of the rear shaft bearing and using a suitably stepped mandrel, drive out the old and insert the new bearings.
4 Check the endfloat of the driveshaft; this must not exceed 0.012 in (0.03 mm). If it does, renew the thrust plate to provide the standard endfloat of between 0.002 and 0.005 in (0.06 and 0.13 mm).

30 Flywheel – servicing

1 Examine the clutch driven plate contact surface of the flywheel for scoring or grooves. If they are deep or tiny cracks are visible, it is recommended that the flywheel is renewed.
2 Check the starter ring gear for cracks or chipped teeth.
3 If the ring gear is damaged on a 20R engine, the flywheel must be renewed.
4 If the ring gear is damaged on an 18R engine, either obtain a replacement flywheel complete with ring gear, or proceed as follows:
5 Either split the ring with a cold chisel after making a cut with a hacksaw blade between two teeth, or use a soft-headed hammer (not steel) to knock the ring off, striking it evenly and alternately at equally spaced points. Take great care not to damage the flywheel during this process.
6 Heat the new ring in either an electric oven to about 200°C (392°F).
7 Hold the ring at this temperature for five minutes and then quickly fit it to the flywheel so the chamfered portion of the teeth faces the gearbox side of the flywheel.
8 The ring should be tapped gently down onto its register and left to cool naturally when the connection of the metal on cooling will ensure that it is a secure and permanent fit. Great care must be taken not to overheat the ring (indicated by the ring turning light metallic blue) as if this happens the temper of the ring will be lost.

31 Driveplate (automatic transmission) – servicing

1 Examine the starter ring gear for worn or broken teeth; where these are evident, renew the driveplate complete.
2 Check the torque converter securing bolt holes for elongation and if apparent, renew the driveplate.

32 Oil seals – renewal

1 During a major overhaul, always discard the old oil seals and fit new ones during reassembly.
2 Renew the timing cover and crankshaft rear oil seals on 18R engines. Renew the oil pump oil seal and crankshaft rear oil seal on 20R engines.

33 Cylinder block – examination and renovation

1 Examine the crankcase and cylinder block for cracks, especially around bolt holes and between the cylinders.
2 Probe waterways and oil galleries to ensure that they are not blocked.
3 Check the security and condition of the core plugs. To renew a core plug, first drill a hole in its centre and lever it out. If it is particularly stubborn, tap a thread in the hole and screw in a bolt, using a piece of tubing and a large washer to act as a point of leverage and extract the plug as the bolt is tightened.
4 Where the cooling mixture has frozen due to the use of a weak anti-freeze mixture, it is quite likely that one or more of the core plugs will have been partially dislodged from their seats by the expansion of the ice. In such an event, drive the plug fully home or better still, renew it. The engine side cover can be removed to gain access to the threaded type core plug located behind the oil pump driveshaft on 18R engines.

34 Engine reassembly – general

1 To ensure maximum life with minimum trouble from a rebuilt engine, not only must everything be correctly assembled but everything must be spotlessly clean, all the oilways must be clear, locking washers and spring washers must always be fitted where indicated and all bearing and other working surfaces must be thoroughly lubricated during assembly.
2 Before assembly begins renew any bolts or studs, the threads of which are in any way damaged and whenever possible use new spring washers.
3 Apart from your normal tools, a supply of clean rags, an oil can filled with engine oil, a new supply of assorted spring washers, a set of new gaskets and a torque spanner, should be collected together.

35 Engine reassembly (2T-B engine)

Crankshaft and main bearings

1 Ensure that the crankcase is clean and that all oilways are clear. A thin drill is useful for cleaning oilways, but it is preferable to use compressed air if it is available. Treat the crankshaft in the same way and then inject some engine oil into the crankshaft oilways.
2 Wipe the crankcase bearing surfaces clean and then fit the bearing shells to them (photo). The centre bearing has a thrust washer fitted to each side of it and these washers should be smeared in grease to keep them in place. Take care that the tab on each of the bearings is fully recessed into the groove of its seating.
3 If new bearings are being fitted they will have been coated with protective grease and this should be removed before fitting them.
4 Wipe the crankshaft bearing caps and fit the bearings to them in a similar manner to the crankcase bearings. Fit the thrust washers to each side of the centre bearing cap (photo), keeping the washers in place with grease. The grooves in the thrust washers should face outwards.
5 Apply oil liberally to the crankshaft journals and to the upper and lower bearing shells. Lower the crankshaft into position, taking care not to displace the thrust washers and making sure that the crankshaft is the right way round.
6 Fit the main bearing caps, making sure that they are in their proper positions and are the same way round as when they were removed.
7 Fit the bearing cap bolts and tighten the cap bolts one at a time to the specified torque setting.
8 After tightening each bearing cap, check that the crankshaft can be rotated by hand. If it is excessively stiff, check that the bearing shells have been fitted properly and grit is not trapped between the shell and cap.
9 Check the endfloat of the crankshaft by levering the crankshaft against one of the thrust washers and then check that there is a clearance of 0.003 to 0.007 in (0.07 to 0.18 mm) between the other thrust washer and the crankshaft journal shoulder. If the endfloat is incorrect, different thickness thrust washers must be fitted to bring it within the specified tolerance.

Pistons and connecting rods

10 If the piston rings have been removed, fit the rings to the pistons. Note that one side of the ring has a code letter and number stamped on it and place this side uppermost when fitting the ring. After fitting the rings, oil them and the grooves liberally and space the ring gaps as shown in Fig. 1.58.
11 Check that the mark on the piston crown to denote its front is on the same side of the connecting rod as the cast mark on the connecting rod (Fig. 1.59). Fit a piston ring compressor to the piston and insert each piston into its correct bore with the mark on the piston crown facing the front of the engine (photo). When the ring compressor is in contact with the cylinder block, tap the piston crown with the wooden handle to push the piston into the cylinder bore fully.
12 Fit the bearing to the connecting rod big-end, rotate the crankshaft so that the crankpin is at its lowest point and engage the connecting rod big-end with the crankpin.
13 Fit a bearing shell to the connecting rod cap, lubricate the bearing and crankpin and fit the cap (photo).

35.2 Fitting a crankshaft upper bearing (2T-B)

35.4 Centre crankshaft bearing cap and thrust washers (2T-B)

35.11 Refitting a piston (2T-B)

35.13 Fitting a connecting rod cap (2T-B)

35.17 Fitting the crankshaft rear oil seal (2T-B)

35.21 Fitting the camshaft thrust plate (2T-B)

35.25 Timing chain and sprockets correctly fitted (2T-B)

35.27 Timing chain tensioner fitted (2T-B)

35.28 Timing chain damper fitted (2T-B)

35.30 Refitting the front cover (2T-B)

35.32 Refitting the oil pump (2T-B)

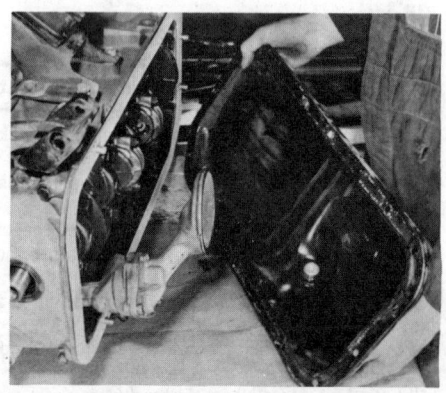

35.34 Refitting the sump (2T-B)

35.35 Stabiliser bracket refitted (2T-B)

35.36 Engine backplate fitted (2T-B)

35.37 Flywheel bolts and locking tabs (2T-B)

35.38 Inserting a valve into its guide (2T-B)

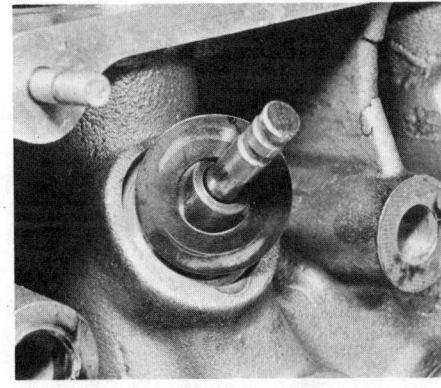

35.39a Plate washer fitted to valve guide

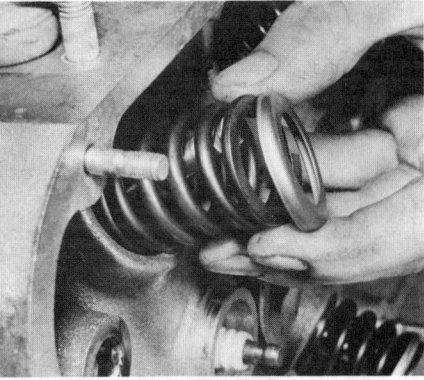

35.39b Fitting a valve spring and shield

35.40 Fitting the split collets

35.41a Fitting a valve lifter

35.41b Fitting a cylinder head gasket

35.41c Lowering the cylinder head

35.42a Fitting the pushrods

35.42b Fitting the rocker assembly

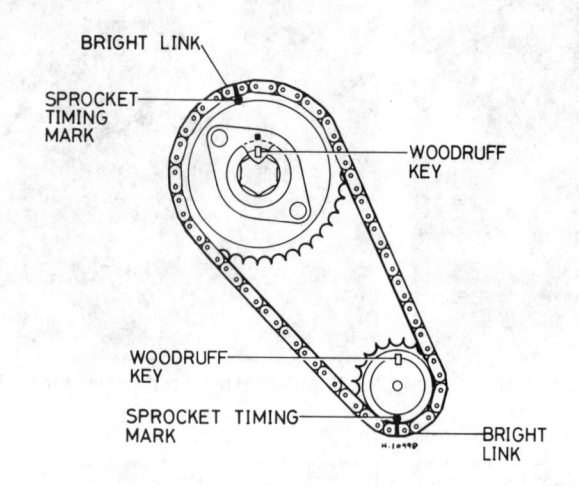

Fig. 1.60 Timing marks (2T-B)

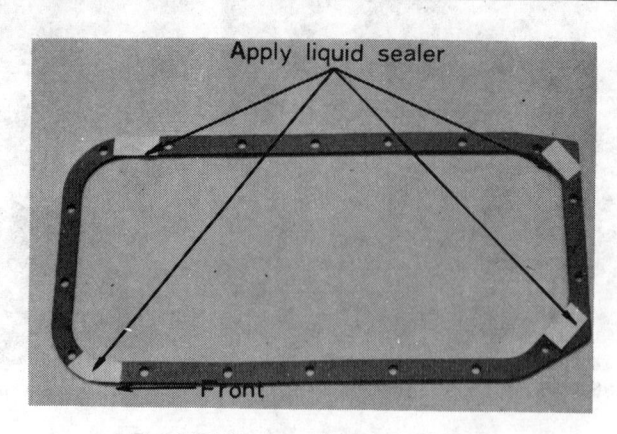

Fig. 1.61 Sump sealing points (2T-B)

Fig. 1.62 Cylinder head bolt tightening sequence (2T-B)

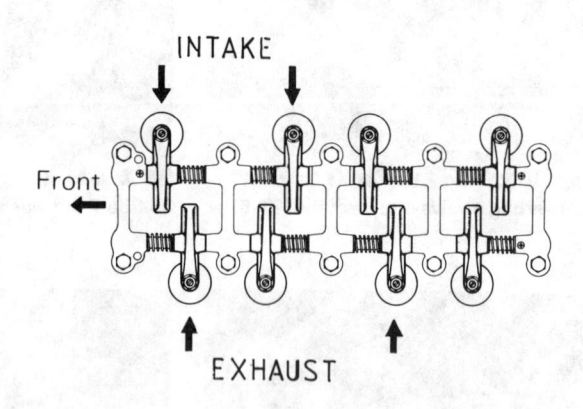

Fig. 1.63 Adjusting valve clearances (2T-B) first stage

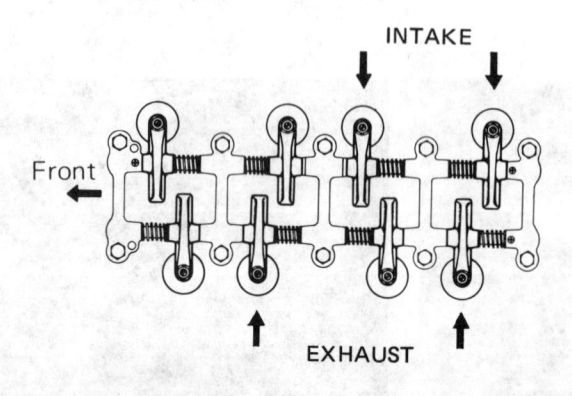

Fig. 1.64 Adjusting valve clearances (2T-B) second stage

Fig. 1.65 Main bearing cap tightening sequence (18R)

14 Check that the correct bearing cap has been fitted and that it is the correct way round, then fit the cap retaining nuts and tighten them to the specified torque.
15 After fitting each piston and tightening the cap bolts, rotate the crankshaft to ensure that the connecting rod bearing is not excessively tight.

Crankshaft rear oil seal

16 Fit a new oil seal to the seal housing, fitting it so that the lip of the seal is towards the front of the engine.
17 Fit a new gasket over the dowels of the cylinder block, smear oil over the seal lip and fit the seal assembly on to the end of the crankshaft, taking care not to damage the seal lip during fitting (photo).
18 Fit the retaining bolts to the seal housing and tighten them successively and in diagonal sequence to the specified torque.

Camshaft and timing gear

19 Wipe the camshaft journals with a clean rag and then lubricate them with engine oil.
20 Carefully insert the camshaft, keeping it central with the bore in the cylinder block, so as not to damage the bearings.
21 Fit the camshaft thrust plate and secure it with its two bolts and spring washers (photo).
22 Fit the Woodruff key to the crankshaft nose and then slide the crankshaft sprocket onto the crankshaft, making sure that the timing mark on the sprocket is facing outwards.
23 Line up the keyways of the crankshaft and camshaft as shown in Fig. 1.60.
24 Fit the timing chain to the camshaft sprocket with one of the bright links of the chain opposite the timing mark on the sprocket.
25 While holding the camshaft sprocket with the chain hanging from it, fit the chain to the crankshaft sprocket so that the other bright link of the chain is opposite the timing mark on the crankshaft sprocket (photo).
26 It should now be possible to fit the camshaft sprocket to the camshaft. Fit the bolt and washer to the end of the camshaft and tighten the bolt.
27 Fill the timing chain tensioner with oil and refit the tensioner. Secure it in position with two bolts and spring washers (photo).
28 Fit the timing chain damper, fit the two bolts and spring washers and tighten the bolts (photo).
29 Fit a new oil seal to the timing cover and smear the lip of the seal with oil.
30 Ensure that the faces of the timing cover and its mating surface are clean. Fit a new gasket, fit the timing cover and insert the bolts (photo).
31 Tighten the timing cover bolts sequentially in diagonal pairs. Fit the crankshaft pulley, taking care not to damage the timing cover oil seal. Insert the pulley fixing bolt and tighten it to the specified torque.

Oil pump and sump

32 Using a new gasket, fit the oil pump assembly (photo) into its hole in the crankcase and secure it in position with a bolt and spring washer.
33 Trim off any of the timing cover or crankshaft oil seal gasket which is proud of the sump joint face and apply jointing compound to the points indicated in Fig. 1.61.
34 Fit a new sump gasket, fit the sump (photo), and tighten the securing bolts and nuts to the specified torque.
35 Fit the stabiliser brackets to the sides of the crankcase and secure them with two bolts and washers (photo).

Flywheel and backplate

36 Fit the engine backplate over the dowels of the cylinder block (photo).
37 Fit the flywheel (or driveplate if the vehicle has automatic transmission). Fit the securing bolts and tab washers. Tighten the bolts to the specified torque and secure them in position by bending up the tab washers (photo).

Cylinder head and valves

38 Oil the valve guides and insert each valve into the guide from which it was removed (photo).
39 Over each valve stem fit the plate washer (photo), valve stem, oil seal, valve spring, shield (photo), and valve retainer in that order.

40 Compress the valve spring, insert the split collets (photo) and release the spring compressor. Give the top of the valve stem a sharp tap with a soft-faced hammer to ensure that the collets have seated properly.
41 Fit the valve lifters to the cylinder block (photo), fit a new cylinder head gasket (photo) and carefully lower the cylinder head onto the block (photo), fitting it over the dowels.
42 Insert the pushrods (photo), lower the rocker assembly onto the cylinder head (photo) and engage each pushrod with its rocker arm. Note that later cylinder heads are fitted with a guide pin to locate a valve rocker support; when using an old rocker assembly with a new cylinder head the dowel should be removed. All other component combinations are fully compatible. Insert the cylinder head bolts and tighten them in several stages in the order shown in Fig. 1.62.

36 Valve clearances (2T-B engine) – adjustment

1 With the rocker box cover removed, turn the crankshaft pulley until the piston of No 1 cylinder is at TDC on its compression stroke. This will be when the notch on the crankshaft pulley is opposite the TDC mark on the timing scale and the inlet valve of No 1 cylinder has closed.
2 Check the clearance between the pad of the rocker arm and the stem of the valve, using a feeler gauge. If the clearance is not as given in Specifications slacken the locknut and turn the adjuster screw until the correct clearance is obtained. While holding the screw to prevent it from turning, tighten the locknut and then re-check the clearance (photos). Without turning the crankshaft check the clearance of the following valves:

Inlet side	1 and 2
Exhaust side	1 and 3

3 Turn the crankshaft one complete revolution and check the clearances of the remaining valves

Inlet side	3 and 4
Exhaust side	2 and 4

The valves are numbered from the front of the engine, inlet valves on the left when viewed from the crankshaft pulley end.

37 Engine reassembly (18R engine)

Crankshaft and main bearings

1 Locate the main bearing shells in their crankcase recesses and lubricate them with engine oil.
2 Carefully lower the crankshaft into position.
3 Fit the main bearing caps complete with shell bearings, noting that the caps (previously numbered 1 to 5) should have their arrows pointing towards the front of the engine (photo).
4 Note that the centre bearing incorporates the thrust washers (photo).
5 Tighten the main bearing cap bolts to the specified torque in two stages in the order shown in Fig. 1.65.
6 Bolt on the crankshaft rear oil seal retainer complete with new seal and gasket (photo). Tighten the securing bolts to the specified torque. Check that the crankshaft turns freely.

Pistons, rings and connecting rods

7 Assemble the piston to the connecting rod so that the marks on the connecting rod and the piston crown are in alignment. These marks face the front of the engine when fitted.
8 Connect the two components by pushing in the gudgeon pin by thumb pressure only (immerse the piston in hot water if necessary). Fit new circlips, one at each end of the gudgeon pin.
9 Fit the rings to the pistons, using the same method as for removal. It is vital that the rings are fitted in the correct order with their tapers running the correct way. This will be achieved if the ring markings face upwards. Stagger the piston ring gaps as indicated in Fig. 1.66.
10 Lubricate the piston rings and the piston bore liberally, fit a piston ring compressor to the piston and place the assembly into a cylinder bore with the mark on the piston crown towards the front of the engine (photo). When the ring compressor meets the block surface, tap the piston/connecting rod assembly into the cylinder bore using the handle

36.2a Measuring valve clearance

36.2b Adjusting valve clearance

37.3 Main bearing cap front mark (18R)

37.4 Centre crankshaft bearing shell (18R)

37.6a Rear oil seal retainer (18R)

37.6b Crankshaft rear oil seal retainer gasket (18R)

37.10 Fitting a piston (18R)

37.11 Installing a big-end cap (18R)

37.14 Fitting the oil pump (18R)

37.15 Fitting the front plate and gasket (18R)

37.17 Timing chain and lower tensioner (18R)

37.27 Fitting the engine rear plate (18R)

Chapter 1 Engine   57

37.28 Tightening the flywheel bolts (18R)

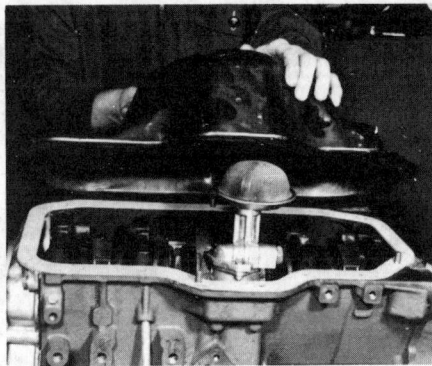

37.31 Fitting the sump (18R)

37.33 Inner and outer valve springs (18R)

37.34 Completed valve assembly (18R)

37.39 Cylinder head gasket in position (18R)

37.45 Camshaft bearing cap in position (18R)

37.51 Fitting the rocker shaft assembly (18R)

38.3 Adjusting valve clearances (18R)

38.4 Fitting the rocker cover (18R)

of a hammer.
11 Rotate the crankshaft so that the crankpin is at the lowest point, engage the connecting rod big-end with the crankpin and then fit the big-end bearing cap so that the marks made prior to dismantling on the rod and cap are adjacent and on the same side (photo).
12 Tighten the big-end cap nuts to the specified torque.
13 Repeat the procedure with the other three piston/connecting rod assemblies, then check that the crankshaft turns smoothly.

Oil pump and timing gear
14 Fit the oil pump to the crankcase using a new joint gasket (photo).
15 Fit the front plate complete with new gasket, and secure it with the single bolt which is adjacent to the oil pump driveshaft hole (photo).
16 Insert the auxiliary driveshaft and the thrust plate, tightening to the specified torque.

17 Fit the lower chain tensioner, so that the projection on the slipper is visible when fitted. Also fit the lower chain damper (photo).
18 Rotate the crankshaft and auxiliary driveshaft so that their keyways are both pointing towards the top (cylinder head) face of the block (Fig. 1.67). Check that the keys are in position.
19 Locate the crankshaft and auxiliary driveshaft sprockets inside the run of the chain so that the timing marks are aligned as shown in Fig. 1.68, then carefully fit the assembly. Don't use too much force tapping on the auxiliary driveshaft sprocket, or the rear plug in the cylinder block may be forced out.
20 Fit the timing cover gasket, sticking it in position with jointing compound.
21 Fit the upper chain damper and oil jet bolt ensuring it is correctly located, and the slipper.
22 Fit the camshaft drive sprocket on the auxiliary driveshaft and tighten the retaining bolt to the specified torque. Use a wooden bar to

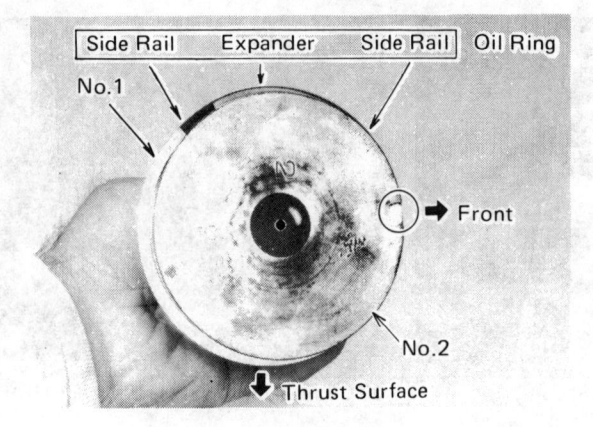

Fig. 1.66 Piston ring end gap staggering (18R engine)

Fig. 1.67 Setting the keyways of crankshaft and auxiliary driveshaft

Fig. 1.68 Sprockets and timing chain alignment (18R engine)

Fig. 1.69 Fitting No 1 chain and camshaft sprocket (18R)

Fig. 1.70 Alignment of No 2 chain and camshaft sprocket (18R)

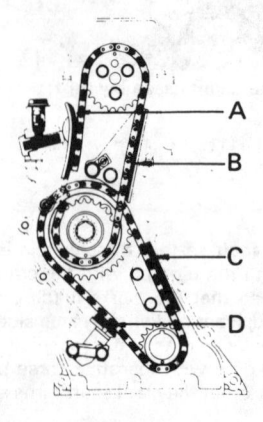

Fig. 1.71 Timing chains, tensioners and dampers (18R)

A Upper (No 2) tensioner slipper   C Lower (No 1) damper
B Upper (No 2) damper   D Lower (No 1) tensioner

Chapter 1 Engine

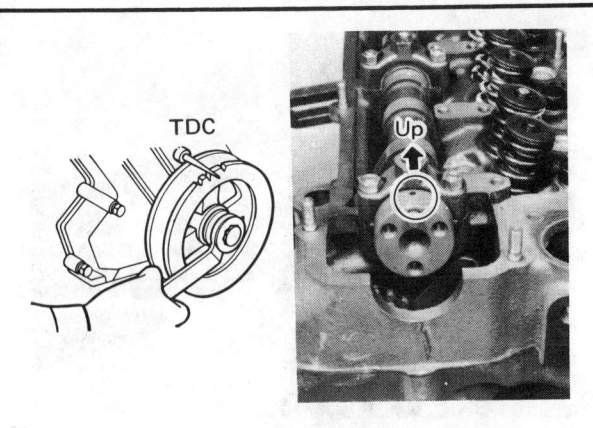

Fig. 1.72 Position of crankshaft and camshaft for valve timing setting (18R)

Fig. 1.73 Position of timing chain and sprocket for valve timing setting (18R)

Fig. 1.74 Tightening order of camshaft bearing bolts (18R)

wedge the crankshaft against the crankcase to prevent it from rotating.
23 Position the camshaft drive chain on the auxiliary driveshaft sprocket so that the two bright links are as shown in Fig. 1.70, then fit the sprocket to the shaft. If the engine is in the car, or the correct way up, use a hooked piece of wire to keep the chain engaged with the sprocket teeth pending fitting the timing cover and cylinder head.
24 Fit the timing cover, using gasket cement on its mating surfaces and on the right-hand upper securing bolt. Tighten the timing cover bolts to the specified torque.
25 Fit the crankshaft pulley (a piece of tubing may be used to drive it into position). Tighten the pulley securing bolt, with its washer, to the specified torque.
26 Fit a new input shaft bearing at the rear end of the crankshaft.

Flywheel (or driveplate – automatic transmission)
27 Bolt on the engine rear plate (photo).
28 Refitting either the flywheel or driveplate is the reverse of the removal procedure, but tighten the securing bolts to the specified torque, and bend up the tabs on the locking plates (photo).

Sump
29 Ensure that the mating faces of the sump and crankcase are quite clean and free from old pieces of gasket.
30 Smear the crankcase flange with jointing compound and stick a new gasket into position.
31 Smear the sump flange with jointing compound and bolt into position. Do not overtighten the sump securing bolts (photo).

Cylinder head and valves
32 Install the first valve into its respective guide, having first lubricated its stem with engine oil.
33 Fit the valve stem washer, oil seal, double valve springs and cap. Note that the closer coils of the spring are fitted nearer the cylinder head (photo).
34 Fit the valve spring retainer, and then compress the valve springs with a compressor and fit the split collets. Gently release the compressor and check that the collets are correctly seated (photo).
35 Repeat the operations on the remaining valves making sure that each valve is returned to its original guide or, if a new valve has been fitted, into the seat into which it has been ground.
36 When all the valves have been reassembled into the cylinder head, tap the end of each valve stem using a block of hardwood and a hammer in order to settle the valve components.
37 Check that the surfaces of the cylinder head and block are scrupulously clean.
38 Smear the top of the block with a thin film of non-setting gasket cement, making sure that none runs down into the oil or water passages or the bolt holes.
39 Lay a new gasket carefully into position on the block (photo).
40 Smear the face of the cylinder head with a film of non-setting gasket cement and then lower the head straight down onto the block so that the positioning dowels engage first time. Do not slide the head about to position it as this will damage the gasket.
41 Make sure that the threads of the cylinder head bolts are clean and screw them in finger tight.
42 Tighten the bolts progressively ($\frac{1}{2}$ turn at a time) and in the reverse sequence to that indicated for dismantling to the specified torque. The timing chain should have been pulled through the aperture in the cylinder head with the hooked wire.

Camshaft and rocker gear
43 Fit the camshaft lower bearing shells into their recesses on the top of the cylinder head.
44 Lubricate the bearings with engine oil and lower the camshaft into position.
45 Fit the bearing caps, complete with shell bearings. The caps should be numbered 1 to 4 (counting from the front of the engine) and the flat portion of their upper bosses must face the front (photo).
46 Tighten the camshaft bearing cap bolts to the specified torque.
47 Rotate the crankshaft by means of the pulley bolt until No 1 piston is at TDC. This is indicated by the timing mark on the pulley and timing chain cover (refer to Chapter 3 if necessary).
48 Turn the camshaft so that the dowel pin and punch marks are uppermost. Pull the timing chain upwards with the hooked piece of wire previously used to retain it on its drive sprocket.

Fig. 1.75 Checking valves 1, 2, 3 and 5 (18R)

Fig. 1.76 Checking valves 4, 6, 7 and 8 (18R)

Fig. 1.77 Cylinder head sealing points (18R–G)

Fig. 1.78 Position of timing chain after fitting cylinder head (18R–G)

Fig. 1.79 Cylinder head bolt tightening sequence (18R–G)

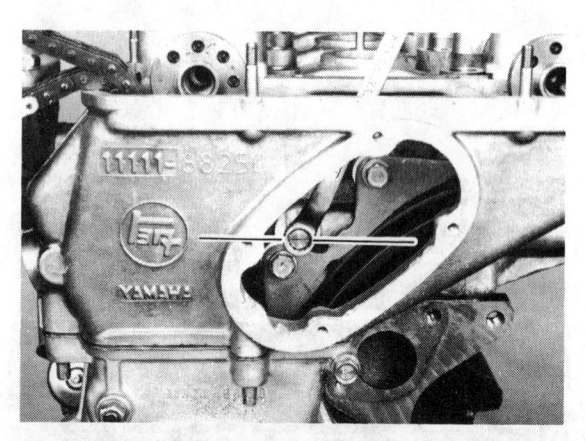

Fig. 1.80 Correct position of slot on oil nozzle (18R–G)

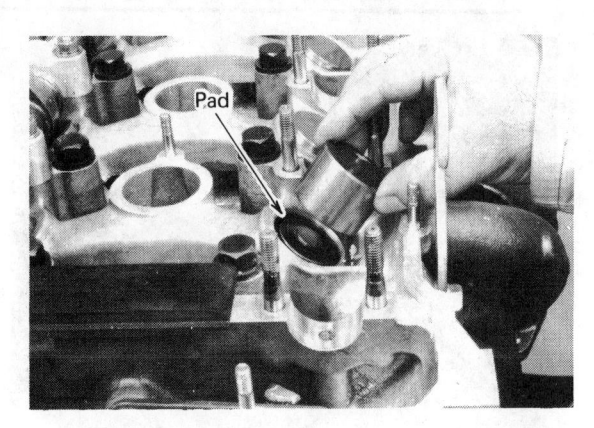

Fig. 1.81 Fitting the valve lifters and pads

Fig. 1.82 Camshafts with slots (arrowed) uppermost (18R-G)

Fig. 1.83 Camshaft bearing cap arrows to front of engine (18R-G)

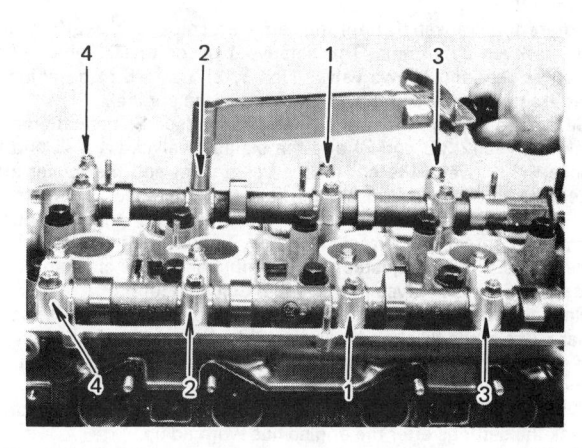

Fig. 1.84 Camshaft bearing cap tightening sequence (18R-G)

Fig. 1.85 Position of timing chain and sprockets for valve timing (18R-G)

Fig. 1.86 Camshaft sprocket dowel pin bolt and lockplate (18R-G)

Chapter 1 Engine

49 Engage the camshaft sprocket within the upper loop of the chain so that the timing marks are aligned as shown in Fig. 1.73, then fit the sprocket ensuring that the dowel engages in the sprocket hole.
50 Tighten the camshaft sprocket bolts to the specified torque.
51 Fit the rocker shaft assembly but make sure that each of the adjusting screws is backed off or damage may occur. If the rocker gear has been dismantled, refit the components in their original order and make sure that the rocker shaft support pillars have their front facing markings correctly set (photo).
52 Tighten the pillar bolts progressively and evenly to the specified torque in the order shown.
53 Ensure that the oilways in the oil feed assembly are unobstructed, then locate the assembly correctly and tighten the bolts to the specified torque.

38 Valve clearances (18R engine) – adjustment

Note: Valve clearances should be adjusted when the engine is hot. Where the adjustment is made with a cold engine (eg, during rebuild), it must be checked again, after running up to normal operating temperature.

1 If the engine is in the car, it will be necessary to first remove the rocker cover. Take careful note of where any hoses and electrical connections are fitted. For information on removal of the air cleaner, refer to Chapter 3.
2 Rotate the crankshaft until the camshaft sprocket dowel hole and timing mark are uppermost. This sets No 1 piston at TDC on its compression stroke, and allows valves Nos 1, 2, 3 and 5 to be adjusted (valves are numbered 1 to 8 from the front of the engine).
3 Select the appropriate size of feeler gauge (see Specifications) for the inlet valves (2,3,6 and 7) and the exhaust valves (1, 4, 5 and 8), and insert it between the end of the valve stem and the rocker arm. The feeler should be a firm sliding fit if the clearance is correct. To adjust, loosen the nut on the adjusting screw, then tighten or loosen the screw to obtain the correct clearance. Tighten the nut and recheck the clearance. Having adjusted the first four valves, the crankshaft can be rotated 360° and valves 4, 6, 7 and 8 adjusted (photo).
4 Refit the rocker cover using a new gasket, don't forget the semi-circular plugs at the front and rear of the block, if removed (photo).
5 If the engine is in the car, refit the ancillaries which have been removed.
6 Where adjustment has been made with a cold engine, don't forget to check the settings after the engine has warmed up.

39 Engine reassembly (18R-G engine)

Crankshaft and main bearings
1 Carry out the operations described in Section 37, paragraphs 1 to 6.

Pistons, rings and connecting rods
2 Carry out the operations described in Section 37, paragraphs 7 to 13.

Oil pump and timing gear
3 Carry out the operations described in Section 37, paragraphs 14 to 26.

Flywheel (or driveplate – automatic transmission)
4 Carry out the operations described in Section 37, paragraphs 27 and 28.

Sump
5 Carry out the operations described in Section 37, paragraphs 29 to 31.

Cylinder head and valves
6 Lubricate the valve guides and the lips of the valve stem seals.
7 Fit the valve spring seats and then press the valve stem seals on to the valve guides by hand.
8 Insert the valves, taking care not to damage the lips of the valve stem seals. Fit the valve springs and the spring retainers.
9 Compress each valve spring in turn, using a spring compressor and

Fig. 1.87 Adjusting the chain tensioner (18R–G)

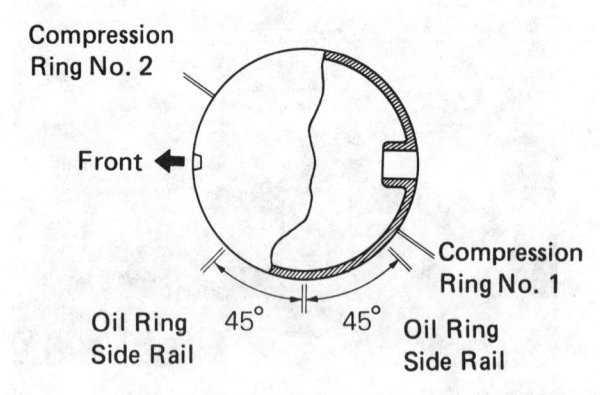

Fig. 1.87A Piston ring end gap setting diagram (20R)

Fig. 1.88 Correct position of bright link on crankshaft sprocket (20R)

Chapter 1 Engine

fit the split collets. After fitting the collets and removing the spring compressor tap the top of the valve stem lightly with a soft-faced hammer to ensure that the collets have seated properly.

10 Apply jointing compound to the three points indicated in Fig. 1.77 and fit the cylinder head. Make sure that the timing chain is secured so that it cannot come off the bottom sprocket (Fig. 1.78).

11 Insert the cylinder head bolts and tighten them in stages, progressively in the order shown in Fig. 1.79.

12 Fit the chain vibration damper and then the oil nozzle. The slot in the end of the oil nozzle must be positioned horizontally (Fig. 1.80). Fit the cylinder head front cover and gasket.

13 Fit the valve lifters (cam followers) and adjusting pads.

14 Fit the camshafts, taking care that the left and right-hand shafts are returned to their original positions. The camshafts should be positioned with the slot at their front end uppermost.

15 Taking care to keep the timing chain engaged with the bottom sprocket, turn the crankshaft pulley 90° anti-clockwise, to lower the pistons and prevent the possibility of interference between the valves and the piston crowns.

16 Fit camshaft bearing caps Nos 2 to 5, taking care to ensure that every cap is returned to its original position and that the arrow on the cap points to the front of the engine.

17 Tighten bearing caps Nos 2 to 5 in stages to the specified torque (Fig. 1.84).

18 Fit the caps to No 1 bearing of each camshaft and tighten the cap nuts.

19 Check that the valve lifters of No 2 and No 4 exhaust valves protrude above the cylinder head by the same amount (about 2 mm).

20 Use feeler gauges to measure the gap between the surface of the valve lifter and the lowest point of the cam for each of the exhaust valves and note the value. Use a pair of grips to turn the camshaft and position the cams.

21 Check that the valve lifters of Nos 3 and No 4 inlet valves protrude above the cylinder head by the same amount. Measure and record the valve clearances for the inlet valves.

22 If the clearance of any of the valves is not within the specification, adjust the clearance as described in the following Section, then turn the camshafts to their slot uppermost positions and return the crankshaft to TDC for No 1 cylinder.

23 Taking care to keep the timing chain engaged with the bottom sprocket, engage the camshaft sprockets in the chain so that the timing mark on each sprocket is opposite the plated link of the chain (Fig. 1.85).

24 Fit the bolts and washers to the sprockets and screw the bolts a few threads into the camshafts.

25 Fit the pin into the hole in the camshaft sprocket which was marked on dismantling and engage the pin into the end of the camshaft. If necessary the camshaft may be turned a fraction of a degree to engage the pin, but the amount turned must be less than three quarters of a degree. Hold the pin in place by pushing the washer forward and then screw the bolt in finger tight.

26 Turn the crankshaft a small amount clockwise to take up the slack in the chains and pins and then tighten the camshaft sprocket bolts to the specified torque.

27 Fit the vibration damper to the top of the cylinder head.

28 Screw in the chain tensioner until it is firm and then back it off 0.020 to 0.040 in (0.5 to 1.0 mm).

40 Valve clearances (18R-G engine) – adjustment

1 The valve clearances must be measured when the engine is cold. Remove the rocker box cover and turn the crankshaft pulley until the piston of No 1 cylinder is at TDC on its compression stroke. This will be when the notch of the crankshaft pulley is aligned with the TDC mark on the timing scale and the inlet valve of No 1 cylinder has closed.

2 Check that the slots in the camshaft flanges behind No 1 camshaft bearing are uppermost.

3 With the camshafts in this position, measure the clearance between the cam and the valve lifter on the following valves only and record the clearances.

Inlet side	*1 and 2*
Exhaust side	*1 and 3*

4 Turn the crankshaft pulley one complete revolution in the normal direction of rotation (clockwise) so that No 4 cylinder is at TDC on its compression stroke.

5 With the camshafts in this position, measure the clearances of the following valves.

Inlet side	*3 and 4*
Exhaust side	*2 and 4*

6 If any of the clearances are outside the limits given in Specifications, proceed as follows.

7 Turn the crankshaft pulley until No 1 cylinder is at TDC on its compression stroke.

8 Put mating marks on each camshaft sprocket to indicate the hole to which the pin is fitted. Also mark the link of the timing chain which is opposite the timing mark on each of the sprockets.

9 Remove the timing chain tensioner. Remove the cylinder head front cover and remove the timing chain damper.

10 Remove the bolt and washer from the camshaft sprocket and disconnect the sprocket from the camshaft. Take care that the timing chain does not drop down and become disengaged from the crankshaft sprocket. Either secure the chain out of the way, or wire it to the camshaft sprocket.

11 Remove the cap from No 1 camshaft bearing.

12 Gradually loosen the remaining camshaft bearings in stages in the order shown in Fig. 1.31, then remove the nuts, the bearing caps and the camshaft. Lay the bearing caps out in their correct order and if both camshafts are removed label them and their bearing caps to indicate the side of the engine to which they are fitted.

13 Use a suction tool or magnet to remove the valve lifters of the valves which require adjustment; label the lifters to indicate their position on the engine. Remove the pads (shims) from the top of the valve stems and place them with their valve lifters.

14 Calculate the thickness of pad required by the following method. Using a micrometer, measure the thickness of the pad removed from the lifter. Work out how much the new pad (shim) must be thicker or thinner to bring the valve clearance within the specified tolerance. For example if the valve clearance is *too small* by 0.002 in (0.005 mm) then the new pad will have to be *thinner* by this amount.

15 Pads are available in 41 sizes by 0.002 in (0.05 mm) increments from 0.039 in (1.0 mm) to 0.118 in (3.0 mm). Choose a pad which is as near as possible to the calculated value. Fit the pad and refit the valve lifter. Recheck the valve clearance.

16 When all the pads requiring alteration have been changed, fit the camshaft into its journals on the cylinder head and complete the assembly of the camshafts and timing gear as described in paragraphs 14 and then 16 to 28 of Section 39.

41 Engine reassembly (20R engine)

Crankshaft and main bearings

1 Carry out the operations described in Section 37, paragraphs 1 to 6.

Pistons, rings, connecting rods, sump and flywheel

2 Carry out the operations described in Section 37, paragraphs 7 to 13.

3 Fit the oil pick-up tube and screen, and the sump, using a new gasket and applying jointing compound as shown for the 2T engine (Fig. 1.61).

4 Fit the flywheel (or driveplate – automatic transmission) and tighten the bolts to the specified torque.

Timing gear, chain and cover

5 Fit the chain guides and chain tensioner.

6 Rotate the crankshaft until the Woodruff key at its front end is pointing upwards (ie, towards the cylinder head).

7 Engage the timing chain so that its single bright link is opposite the mark on the crankshaft sprocket (Fig. 1.88).

8 Engage the camshaft sprocket within the loop of the timing chain so that the mark on the sprocket is between the two bright links on the chain (Fig. 1.89).

9 Fit the oil pump drivegear onto the front end of the crankshaft.

10 Fit the timing cover gasket, and then gently turn the camshaft sprocket in an anti-clockwise direction to remove any slack from the chain.

11 Fit the timing cover complete with oil pump onto the locating dowels; refit the bolts and tighten to the specified torque.

Fig. 1.89 Correct position of bright links on camshaft sprocket (20R)

Fig. 1.90 Sealing compound application points (20R)

Fig. 1.91 Camshaft position for setting valve timing (20R)

Fig. 1.92 Chain cover bolt (20R)

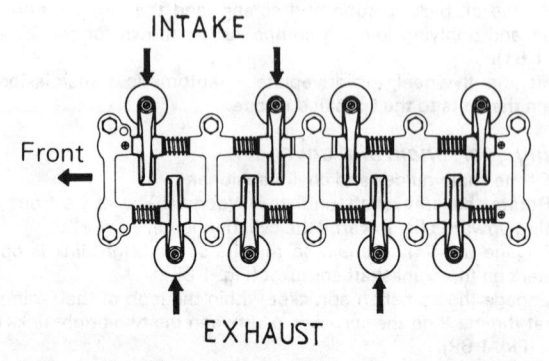

Fig. 1.93 First four valves to be adjusted (20R)

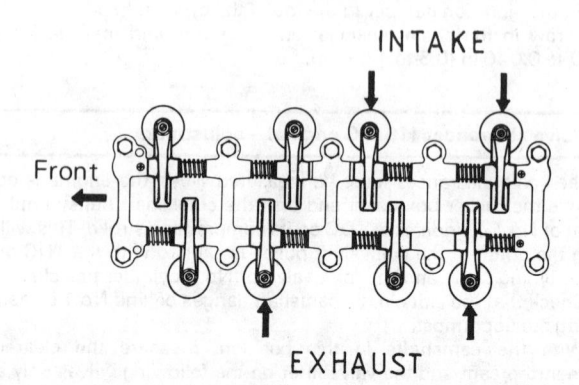

Fig. 1.94 Second four valves to be adjusted (20R)

12 Fit the crankshaft pulley. When tightening the retaining bolt do not turn the crankshaft, but either jam one of the crankshaft webs with a block of wood or refit the flywheel and jam the starter ring gear.

Cylinder head

13 Refer to Section 37 paragraphs 32 to 36, but note that single valve springs are used on this engine.
14 Fit the camshaft and the bearing caps, in their correct sequence with arrows pointing to the front of the engine. Make sure that all the bearing surfaces are oiled before reassembly. Tighten the cap bolts to the specified torque wrench settings.
15 Fit the thermostatic valve, the inlet manifold, the EGR valve, the exhaust manifold, and the heat insulator to the cylinder head.
16 Make sure that the surfaces of the cylinder head and block are quite clean; smear jointing compound at the locations indicated in Fig. 1.90, then fit a new cylinder head gasket.
17 Fit the cylinder head onto its locating dowels.
18 Without moving the head, rotate the camshaft so that the dowel on its flange and the timing marks are at the top. This is the No 1 TDC position.
19 Apply tension upwards to retain the timing chain and sprocket in engagement, then turn the crankshaft by means of its pulley retaining bolt until the hole in the camshaft sprocket is in alignment with the flange dowel on the camshaft.
20 If the rocker gear has been dismantled, reassemble it, making sure that the arrows on the shaft support pillars face towards the front of the engine when fitted.
21 Fit the rocker assembly, and insert the combined cylinder head/rocker pillar bolts, tightening them to the specified torque in the reverse sequence to that shown in Fig. 1.37.
22 Screw in the chain cover bolt (Fig. 1.92).
23 Fit the camshaft sprocket/chain assembly to the dowel on the camshaft flange, and then push on the distributor drivegear and tighten the securing bolt to the specified torque.

42 Valve clearances (20R engine) – adjustment

The procedure for valve clearance adjustment on the 20R engine is similar to that given for the 18R engine in Section 38. Fig. 1.93 shows the first four valves to be adjusted; Fig. 1.94 shows the second four valves to be adjusted. Note that no separate figures are given for a cold engine so the hot adjustment figures must be used.

43 Engine ancillary components – refitting

1 The components which have to be refitted will vary according to the model of engine and the market for which the vehicle was intended and will be the reversal of the operations in Section 9.
2 Do not attempt to hurry the fitting of these items and take care to ensure that nothing is forgotten, or fitted incorrectly. Pay particular attention to the details, such as fixing clips and the routing of pipes and cables.

44 Engine to manual gearbox – reconnection

This is the reverse of the separation procedure – see Section 6.

45 Engine to automatic transmission – reconnection

1 This is the reverse of the separation procedure given in Section 7, but observe the following points:
2 Check that the two projections on the torque converter key with the slots in the transmission fluid pump impeller, particularly if the torque converter has been partially withdrawn during dismantling operations.
3 Make sure that the driveplate is bolted to the torque converter with the marks made prior to dismantling in alignment.
4 Tighten all bolts to the specified torque, and check the security of the transmission drain plug.

46 Engine/transmission – refitting

1 Locate slings or chains round the engine and support the weight of the combined unit on suitable lifting tackle. Where a fixed hoist is being used, raise the power unit and roll the car under it.
2 Lower the unit into the engine compartment ensuring that nothing fouls during the operations.
3 With the front engine mountings roughly aligned, jack-up the transmission so that the rear crossmember and mounting can be installed. Remove the jack.
4 With the hoist still supporting the weight of the engine, the engine/transmission can be moved fractionally so that the front mountings can be aligned and bolted up.
5 Refit the propeller shaft making sure that the rear driving flanges have their marks (made before dismantling) in alignment.
6 Reconnect the gearchange or selector mechanism according to transmission type.
7 Reconnect the speedometer cable to the transmission housing.
8 Reconnect the reverse light switch leads.
9 Reconnect the exhaust downpipe to the manifold and secure the support bracket to the transmission housing.
10 Reconnect the vacuum hoses to the emission control unit.
11 Reconnect the emission control electrical leads.
12 Reconnect the choke and throttle controls.
13 Reconnect all electrical leads.
14 Bolt the clutch slave cylinder to the bellhousing (manual gearbox).
15 Reconnect the brake servo pipe to the inlet manifold.
16 Reconnect the heater hoses.
17 Reconnect the fuel inlet pipe to the fuel pump.
18 Refit the air cleaner.
19 Refit the radiator and connect the top and bottom hoses.
20 On vehicles with air conditioning, install the condenser in front of the radiator and reconnect the pipes and hoses both to the condenser and the compressor pump.
21 Check that any other item of equipment on your particular car has been or can now be fitted.
22 Refit the radiator grille, upper and lower shields and the fan shrouds.
23 Refit the bonnet.
24 Refill the cooling system.
25 Refill the engine with oil.
26 Refill the gearbox or automatic transmission unit.
27 Connect the lead to the battery negative terminal.

47 Initial start-up after major repair or overhaul

1 Start the engine and check for oil or water leaks. None should be apparent if new gaskets have been used throughout and the specified torque wrench settings adhered to.
2 Where an air conditioning system is installed, have the system professionally recharged with refrigerant gas.
3 Run the vehicle until normal operating temperature is reached and check the following:

(a) *Carburettor and emission control settings (Chapter 3).*
(b) *Ignition timing (Chapter 4).*
(c) *Check the torque of the cylinder head bolts one at a time (unscrew each bolt a quarter-turn and retighten to specified figure and in correct sequence). Check them again after 500 miles (800 km).*
(d) *Check the valve clearances as described earlier for 2T-B, 18R and 20R engines but this time with the engine HOT, (refer to Specifications for the limits).*
(e) *Recheck all oil levels and top-up the engine oil to make up for the amount absorbed by the new filter element.*

48 Fault diagnosis – engine

Symptom	Cause
Engine will not turn over when starter switch is operated	Flat battery Bad battery connections Bad connections at solenoid switch and/or starter motor Starter motor jammed Defective solenoid Starter motor defective
Engine turns over normally but fails to start	No spark at plugs No fuel reaching engine Too much fuel reaching the engine (flooding)
Engine starts but runs unevenly and misfires	Ignition and/or fuel system faults Incorrect valve clearances Burnt out valves Worn out piston rings
Lack of power	Ignition and/or fuel system faults Incorrect valve clearances Burnt out valve Worn out piston rings
Excessive oil consumption	Oil leaks from crankshaft rear oil seal, timing cover gasket and oil seal rocker cover gasket, sump gasket, sump plug washer Worn piston rings or cylinder bores resulting in oil being burnt by engine Worn valve guides and/or defective valve stem seals
Excessive mechanical noise from engine	Wrong valve to rocker clearances Worn crankshaft bearings Worn cylinders (piston slap) Slack or worn timing chain and sprockets
Poor idling	Leak in inlet manifold gasket Perforated or leaking PCV connecting pipe Perforated or leaking brake servo pipe

Note: *When investigating starting and uneven running faults, do not be tempted into snap diagnosis. Start from the beginning of the check procedure and follow it through. It will take less time in the long run. Poor performance from an engine in terms of power and economy is not normally diagnosed quickly. In any event, the ignition and fuel systems must be checked first before assuming any further investigation needs to be made.*

In addition to the foregoing, reference should also be made to the fault finding chart for emission control equipment which is to be found at the end of Chapter 3. Such a fault can have an immediate effect upon engine performance.

Chapter 2 Cooling system

Contents

Antifreeze and corrosion inhibiting mixtures	5
Cooling system – draining	2
Cooling system – filling	4
Cooling system – flushing	3
Drivebelts – tensioning and renewal	10
Fault diagnosis – cooling system	11
General description	1
Radiator – removal, inspection and refitting	6
Thermostat – removal, testing and refitting	7
Water pump and fluid coupling – overhaul	9
Water pump – removal and refitting	8

Specifications

System type .... Pump assisted, thermo-syphon pressurised with thermostatic control and radiator with fan cooling

Coolant capacity
- 2TB .... 6.8 Imp quarts, 8.1 US quarts, 7.8 litres
- 18R .... 7.5 Imp quarts, 8.5 US quarts, 8 litres
- 20R .... 7.5 Imp quarts, 8.5 US quarts, 8 litres

Radiator pressure cap setting .... 12.8 lbf/in^2 (0.9 kgf/cm^2)

Thermostat
Type .... Wax pellet

	Starts to open	Fully open
2T and 18R temperate climate	177/182°F (80.5/83.5°C)	203°F (95°C)
2T and 18R cold climate	188/193°F (86.5/89.5°C)	212°F (100°C)
20R	179.6°F (82°C)	203°F (95°C)
Valve travel	0.31 in (8.0 mm)	

1 General description

The cooling system consists of a fan cooled radiator, a centrifugal pump, thermostat and interconnecting hoses. The car heater is also connected to the water circuit of the cooling system. The cooling system is pressurised by means of a spring-loaded cap on the radiator and this, combined with the use of an ethylene glycol and water mixture in the system reduced the tendency of the coolant to boil.

The principle of operation of the system is that coolant from the bottom of the radiator, where the temperature is lowest, is passed through the bottom hose to the water pump which circulates it round the cylinder block and combustion chambers where it absorbs heat produced by the burning of the fuel in the engine. When the engine has reached its correct operating temperature, a thermostatic valve opens and allows the heated coolant to pass to the top of the radiator. In passing through the radiator from the top to the bottom, the coolant is cooled by the combined effect of air passing through the radiator as a result of the forward motion of the car and air being drawn through the radiator by the engine driven fan mounted behind it.

There are variations in detail and in component layout on the different engines used. Some models have a fluid coupling incorporated in the fan hub which limits fan speed at high engine speeds with a consequent reduction in fan noise and power loss by the fan. A variant of this is a temperature sensitive cooling fan which incorporates a bi-metallic strip (photo). This responds to engine compartment temperature and controls a valve within the fluid coupling to vary the amount of coupling slip and hence the fan speed to maintain optimum cooling efficiency.

2 Cooling system – draining

1 Stand the car on level ground and if the coolant is not to be re-used, it is convenient to position the car so that the coolant can run to a drain gulley.
2 If the engine is cold, the radiator filler cap can be turned anti-clockwise and removed without precautions. If the engine is hot, the cooling system will be under pressure, so the filler cap should be

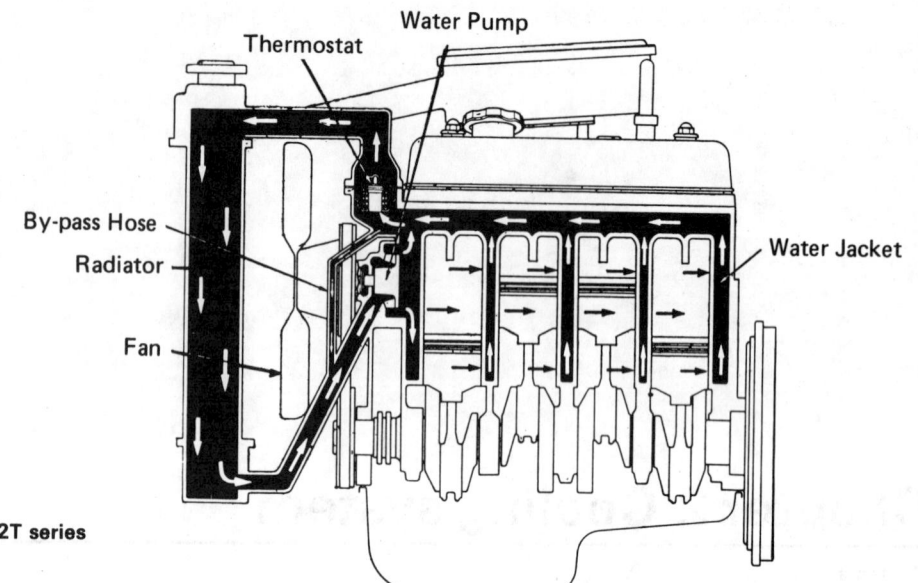

68

Fig. 2.1 Cooling system schematic – 2T series engine

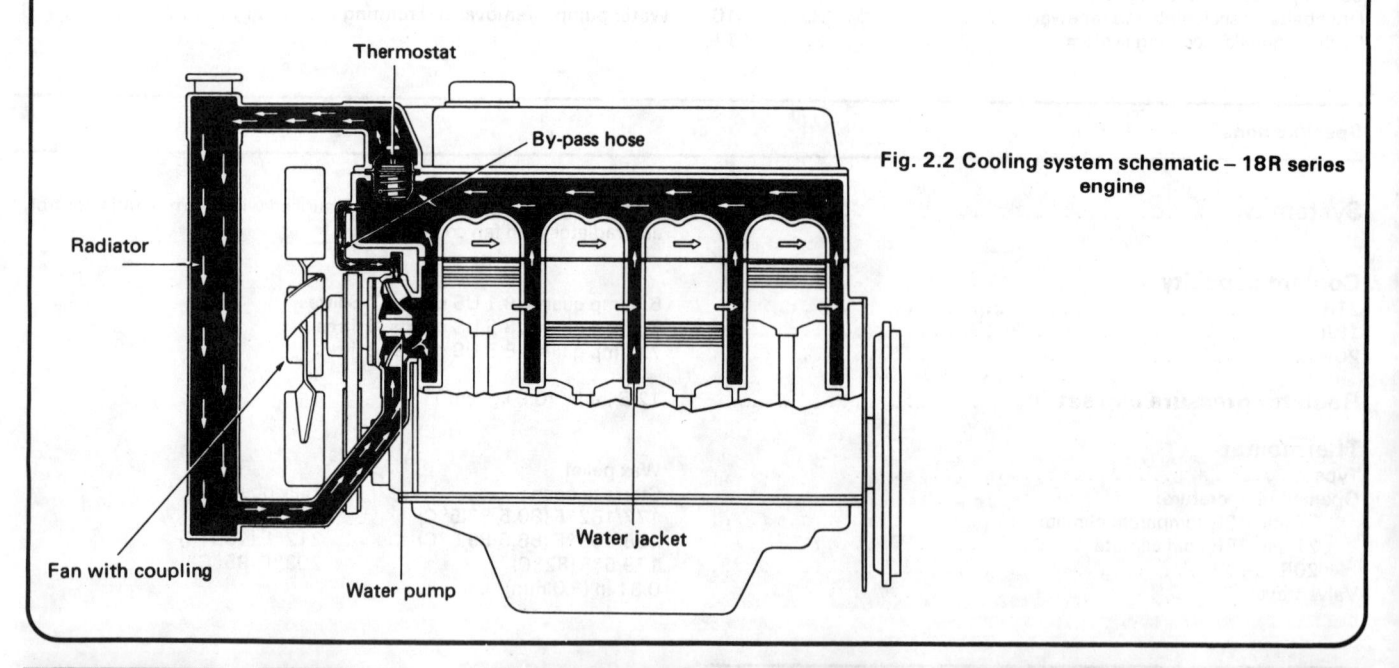

(Labels: Thermostat, Water Pump, By-pass Hose, Radiator, Fan, Water Jacket)

Fig. 2.2 Cooling system schematic – 18R series engine

(Labels: Thermostat, By-pass hose, Radiator, Fan with coupling, Water pump, Water jacket)

1.3 Bi-metallic strip of temperature sensitive fan

2.3 Radiator drain tap

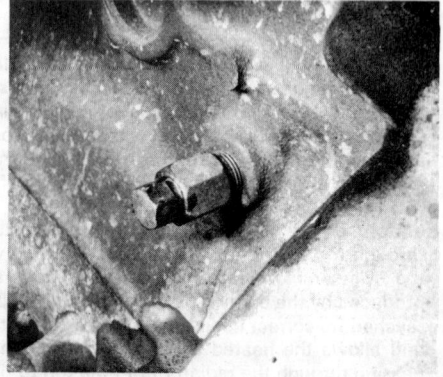

2.4 Cylinder block drain plug

Chapter 2 Cooling system

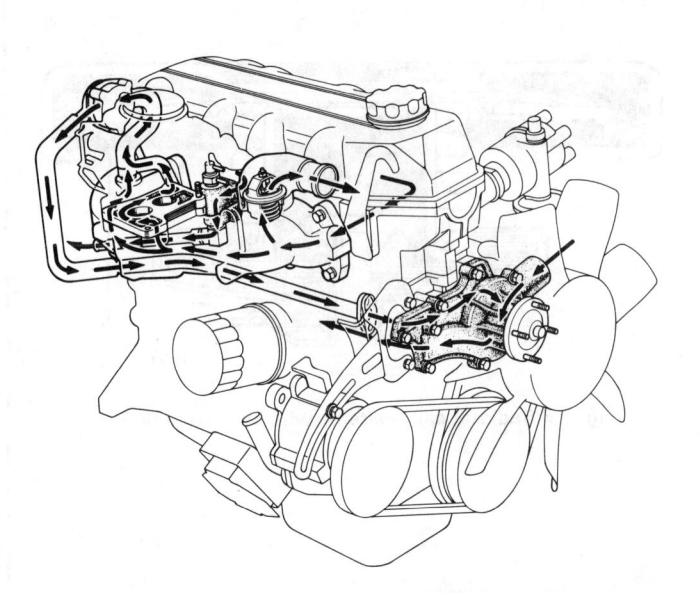

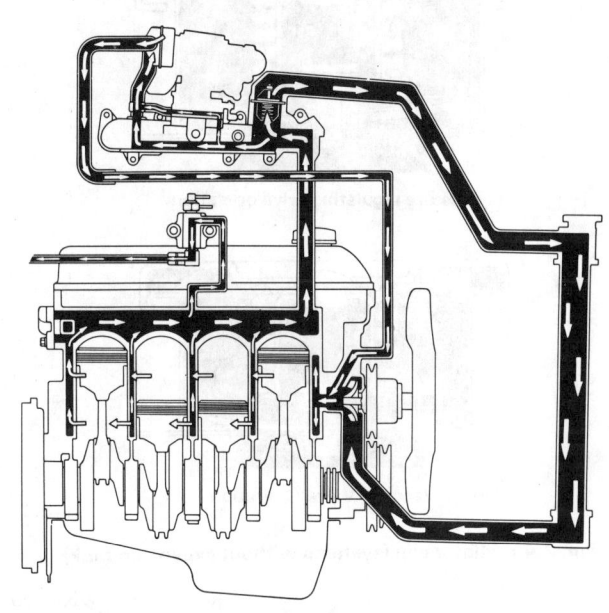

Fig. 2.3 Cooling system schematic – 20R series engine

turned very slowly so that the pressure in the system is released gradually. The cap should be covered with a cloth to protect your hand from escaping steam. If the engine is very hot, the drop in pressure caused by removing the radiator cap may cause the coolant to boil and it is not advisable to drain the cooling system unless the engine has cooled to well below its normal operating temperature.
3 Place the heater control to the *Hot* position.
4 If the coolant is to be re-used, place a clean two gallon container underneath the radiator drain tap and open the tap (photo).
5 When the radiator has drained, move the container to the left-hand side of the engine and unscrew the cylinder block drain plug (photo). Replace and tighten the plug when the cylinder block has drained.

3 Cooling system – flushing

1 After a very high mileage, the radiator and waterways in the engine may become partially or even totally blocked by scale and sediment and the efficiency of the cooling system will be reduced. When this condition is reached and the engine has a tendency to overheat, or when the coolant is a dark, rusty colour, the system should be flushed.
2 Set the heater control to the *Hot* position, open the radiator drain tap and remove the cylinder block drain plug.
3 Remove the radiator filler cap and place a hosepipe in the filler neck. Allow water to flow through the system until the liquid flowing from both the drain holes is clean. If the engine is hot when it is drained, do not immediately flush it with cold water. Either use hot water initially, or allow the engine to cool.
4 If the radiator is heavily sedimented it is better to remove the radiator and turn it upside down. Place the hose in the neck for the bottom hose and allow the water to flow out of the filler neck.
5 In extreme cases, the use of a proprietary descaling fluid may be necessary. In this event be very careful to follow the manufacturers instructions.

4 Cooling system – filling

1 Set the heater control to *Hot*.
2 Ensure that the radiator drain tap is closed and that the cylinder block drain plug is screwed in and tightened.
3 Check that all hoses are in place and that their clamps are tightened firmly.
5 Run the engine at idling speed and as the level of coolant falls, top

it up until no further coolant can be added.
6 Stop the engine. Fit the radiator cap and pour coolant into the coolant reservoir until the reservoir is half full. Put the cap on the reservoir.
7 Run the engine until it reaches normal operating temperature and check the cooling system for leaks.

5 Antifreeze and corrosion inhibiting mixtures

1 In addition to giving protection from frost, the addition of antifreeze raises the temperature at which the coolant boils. Antifreeze also contains corrosion inhibitors which prolong the life of the cooling system.
2 The use of an alcohol based antifreeze is not recommended, but if it is used, the system should be drained at the end of the winter and refilled using clean water plus corrosion inhibitor. It is preferable to use an ethylene glycol mixture and this may be used all the year round. Long life brands of ethylene glycol are available and their manufacturers advise that these may be left in the cooling system for two and sometimes three years.
3 Before adding antifreeze to the system, check all hose connections and make sure that the hoses are in good condition because antifreeze will leak from places which are otherwise watertight.
4 The quality of antifreeze which should be used for various levels of protection is given in the table, expressed as a percentage of the total coolant volume.

Antifreeze volume	Protection to
25%	–26°C (–15°F)
30%	–33°C (–28°F)
35%	–39°C (–38°F)

5 When topping-up a cooling system containing antifreeze, do not use water only because this will dilute the antifreeze and lower the degree of protection. Always use an antifreeze and water mixture in the same proportions as the coolant already in the system.

6 Radiator – removal, inspection and refitting

1 Drain the coolant from the system, collecting it for re-use if it is still in good condition.
2 Remove the lower shield, to give access to the bottom hose clip (photo) and remove the radiator shroud (photo).

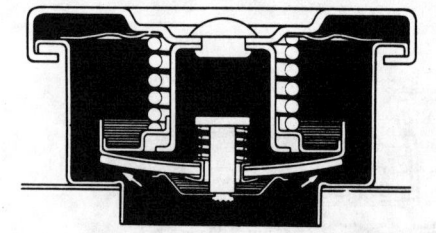

Pressure regulating valve operation

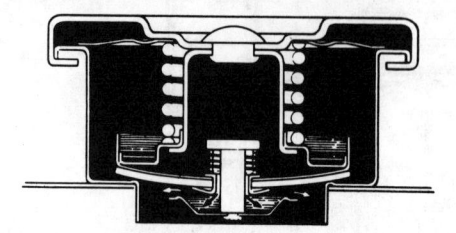

Vacuum Valve Operation

Fig. 2.4 Radiator cap (systems without expansion tank)

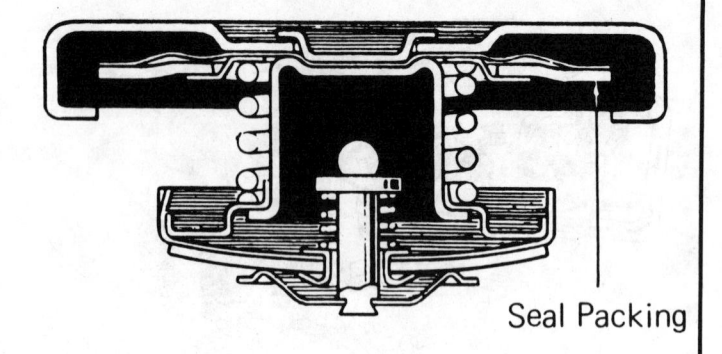

Seal Packing

Fig. 2.5 Radiator cap (systems with expansion tank)

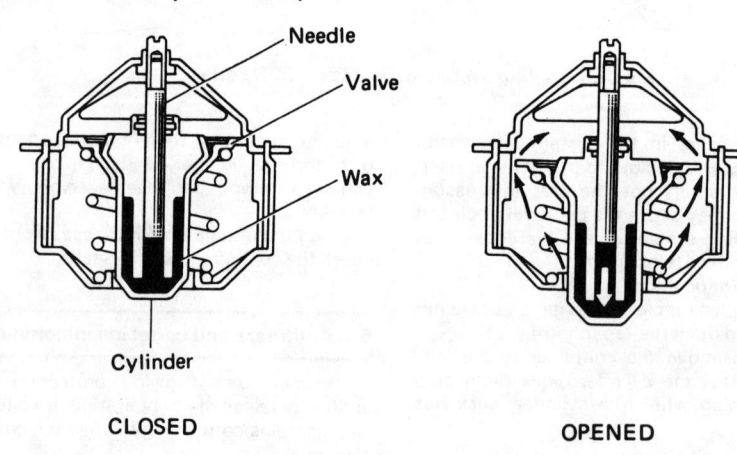

Needle
Valve
Wax
Cylinder

CLOSED   OPENED

Fig. 2.6 Thermostat operation

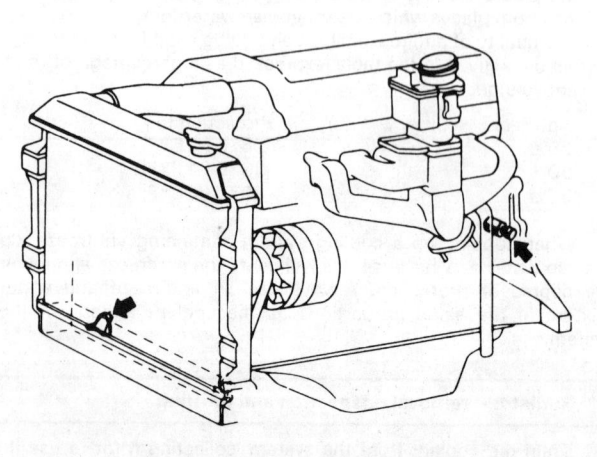

Fig. 2.7 Radiator and cylinder block draining points

Fig. 2.8 Testing the thermostat

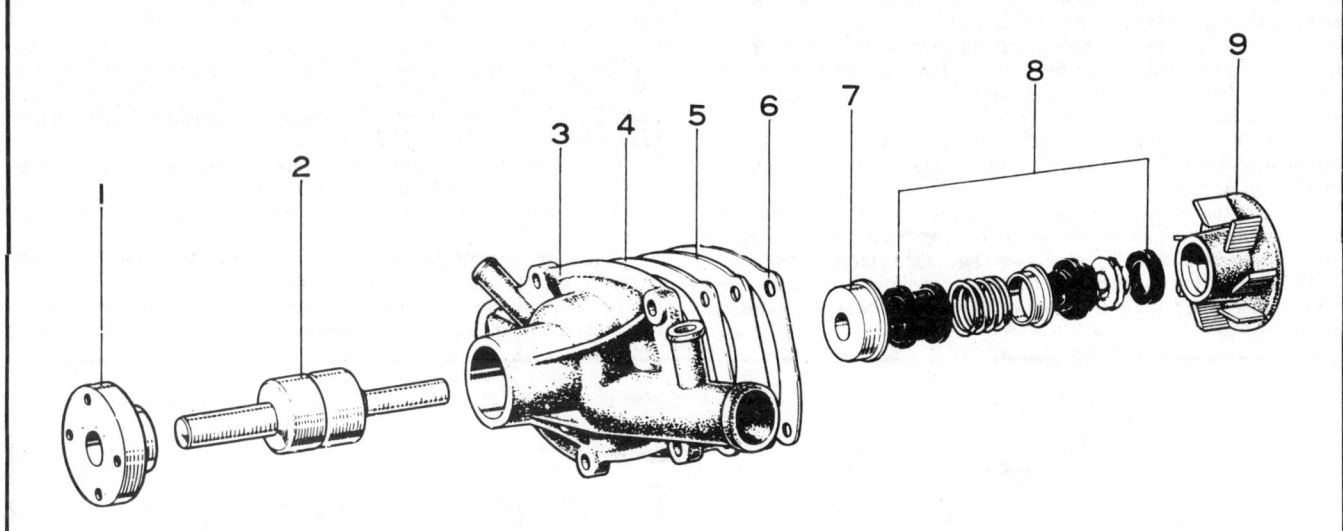

Fig. 2.9 Water pump (2T series) – exploded view

1 Pulley mounting flange
2 Bearing/spindle assembly
3 Body
4 Gasket
5 Cover plate
6 Gasket
7 Seal cover
8 Seal assembly
9 Impeller

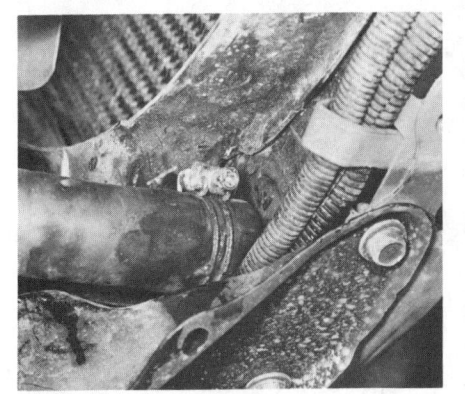

6.2a Radiator bottom hose clip

6.2b Removing the radiator shroud

6.5 Radiator fixing bolt (arrowed)

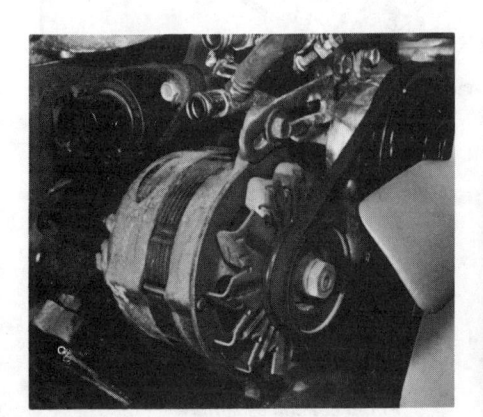

8.3 Bypass hose (18R series)

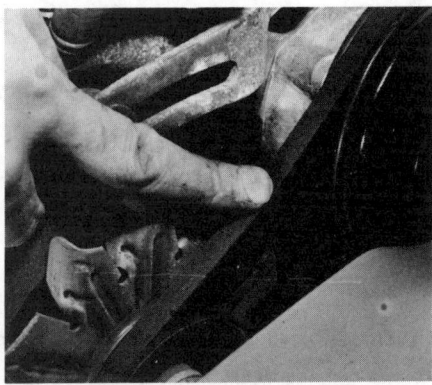

10.2 Checking drivebelt tension

10.3 Alternator link bolt

Chapter 2 Cooling system

3 On cars fitted with automatic transmission, disconnect and plug the oil cooler pipes.
4 On cars equipped with an air conditioning system the condenser is mounted in front of the radiator. It must not be disconnected, or damaged during the removal of the radiator.
5 Having completed the necessary preliminary work, disconnect the top and bottom hoses from the radiator. Remove the four radiator fixing bolts (photo) and lift the radiator out, being careful not to damage the fan.
6 With the radiator removed from the car, brush off any accumulated debris from the fins and reverse flush the radiator as described in Section 3.
7 Examine the radiator for signs of leaks, which can usually be seen as areas which show a coloured stain from the dye in the antifreeze. If the radiator is defective, its repair is best left to a radiator specialist.
8 Before refitting the radiator, check the condition of the top and bottom hoses and fit new ones if the existing ones are hard, or cracked.

7 Thermostat – removal, testing and refitting

1 Drain enough coolant from the radiator to drop the level below the thermostat housing joint face. If the header tank of the radiator is empty and the tops of the cooling tubes are exposed when viewed through the filler neck, it is an indication that enough has been drained.
2 The thermostat is beneath the housing to which the inner end of the radiator top hose is connected and the housing is secured by two bolts. To gain access to the bolts on the 20R engine it is first necessary to remove the air cleaner (refer to Chapter 3 if necessary).
3 Disconnect the top hose from the thermostat housing cover, remove the two fixing bolts and tap the housing cover with a soft-faced hammer to break the joint. Lift off the cover and lift the thermostat out.
4 Clean the thermostat seating and remove all traces of the old gasket from both joint faces of the housing.
5 A faulty thermostat can cause overheating, or slow engine warm up and an ineffective car heater, depending on whether the thermostat

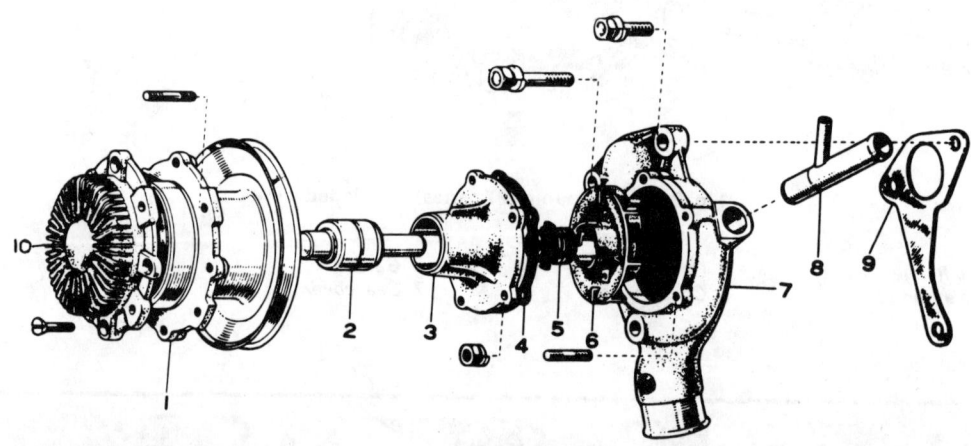

Fig. 2.10 Water pump (18R series) – exploded view (without temperature sensitive cooling fan)

1 Fan fluid coupling
2 Shaft/bearing assembly
3 Cover
4 Cover gasket
5 Seal assembly
6 Impeller
7 Body
8 Union
9 Gasket
10 Fluid coupling case cover

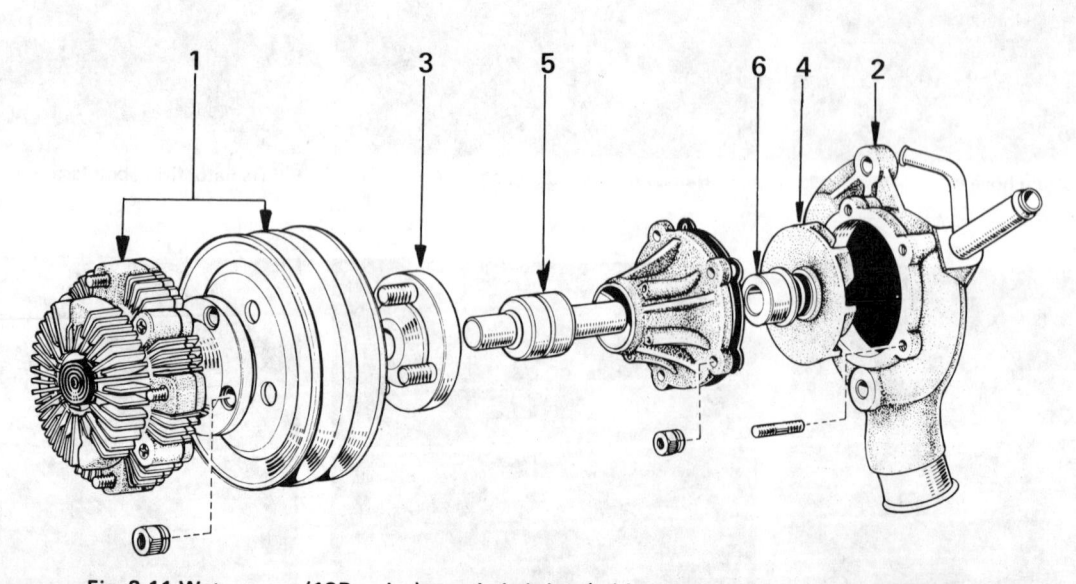

Fig. 2.11 Water pump (18R series) – exploded view (with temperature sensitive cooling fan)

1 Tempered coupling and pulley
2 Pump body
3 Pump flange
4 Pump rotor
5 Pump bearing assembly
6 Pump seal assembly

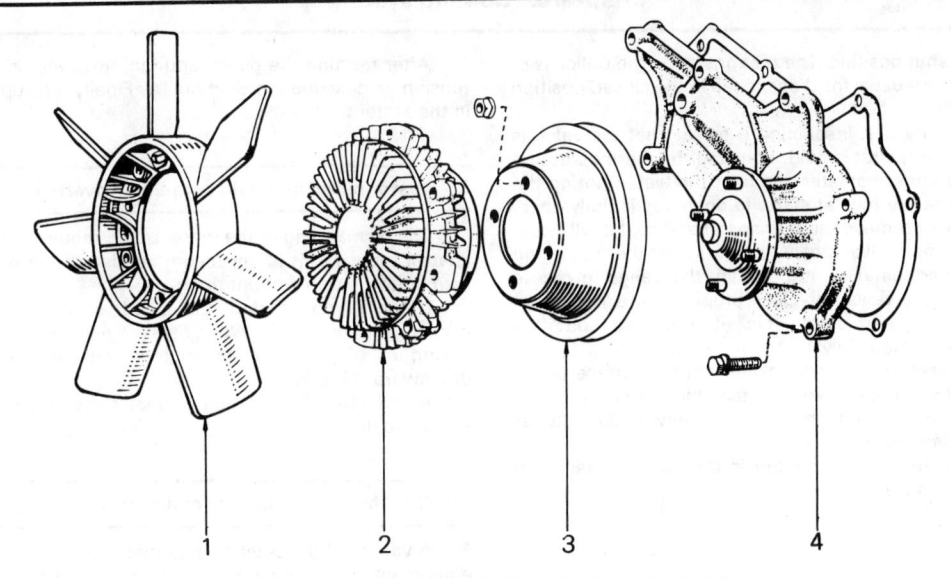

Fig. 2.12 Water pump (20R series) – exploded view

1 Fan   2 Fluid coupling   3 Pulley   4 Water pump

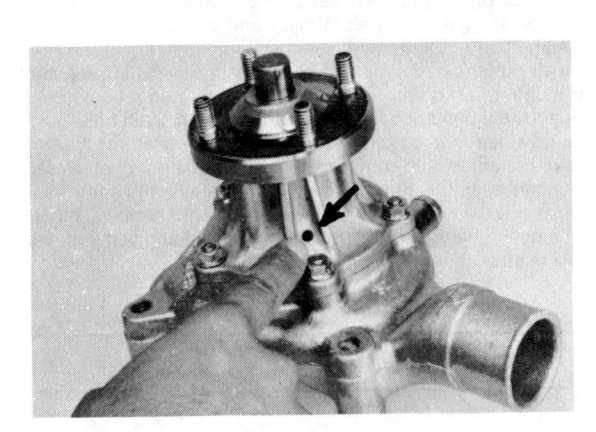

Fig. 2.13 Drain hole (18R series) arrowed

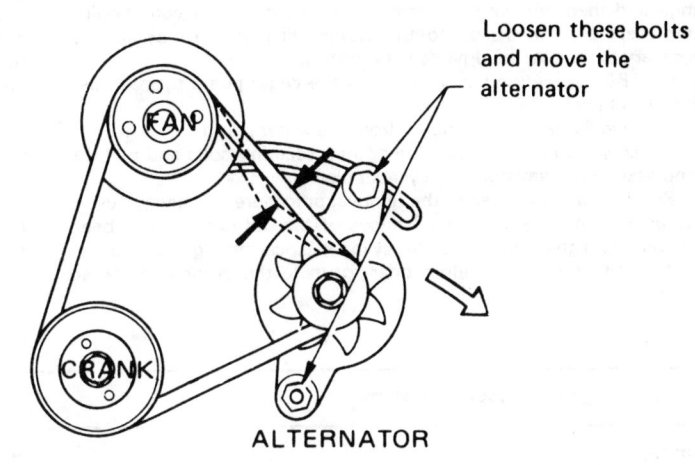

Fig. 2.14 Drivebelt tension adjustment – alternator fitted

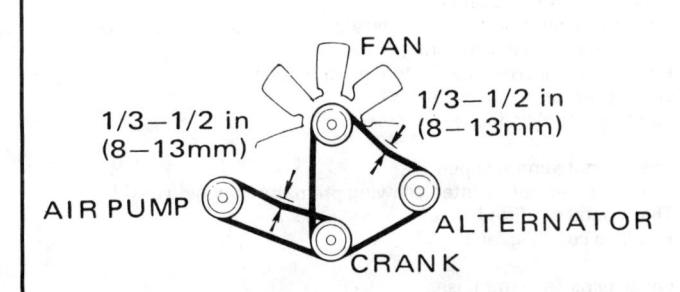

Loosen these bolts and move the alternator

Fig. 2.15 Drivebelt tension – air pump and alternator fitted

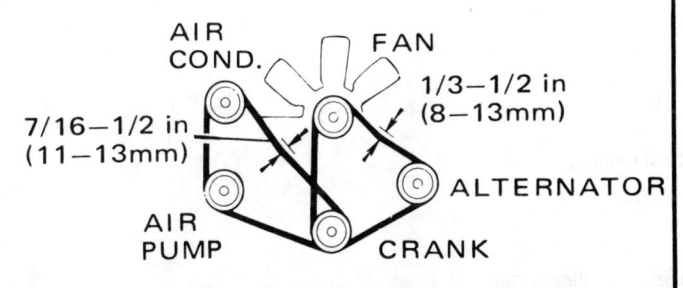

Fig. 2.16 Drivebelt tension – air pump, air conditioner and alternator fitted

fails in the open or the shut position. The thermostat on the Celica is of the wax type and it is more usual for these to fail in the closed position and cause overheating.

6 Test the thermostat by first inspecting it to see that the valve is fully closed when the thermostat is cold. Immerse the thermostat in a pan of cold water with a thermometer and heat the water, noting the temperature at which the thermostat starts to open and is dully open. Allow the water to cool, or remove the thermostat and cool it with cold water to check that the valve closes properly. If the operating temperature of the thermostat is not within the range given in Specifications, or the valve does not close properly, fit a new thermostat. Never re-fit a faulty thermostat – it is better to leave it out if no replacement is available immediately.

7 Refitting is the reverse of removal. Ensure that the surface of the thermostat marked TOP is uppermost and that the periphery of the thermostat flange fits snugly into the housing. Fit a new gasket and re-fit and tighten the housing cover.

8 Re-fit the top hose, top-up the coolant in the system, then start the engine and check for leaks.

8 Water pump – removal and refitting

1 Drain the cooling system and remove the radiator.
2 Slacken the alternator mountings and push the alternator towards the engine, so that the drivebelt can be eased over the alternator pulley flange and then removed. On models fitted with an air conditioning compressor or an emission control system air pump, it may also be necessary to remove the drivebelts from these.
3 On 18R engine models, disconnect the water pump bypass hose (photo) and the heater hose.
4 Remove the fan and fan pulley from the water pump.
5 Unscrew and remove the water pump securing bolts and pull the pump assembly away from the cylinder block.
6 Refitting is the reverse of the removal procedure, but always use a new gasket and ensure that all traces of the old gasket have been removed from the pump and the cylinder block mating faces. Insert one bolt into the pump before offering up to the cylinder block, to provide a location for the pump.

7 After refitting the pulley and fan, re-fit the drivebelt and adjust its tension as described in Section 10. Finally, top-up the level of coolant in the system.

9 Water pump and fluid coupling – overhaul

1 The dismantling of the water pump requires the use of a press and special tools and it is not recommended that any attempt is made to dismantle a pump without them.
2 If the pump bearing has failed, or the seal is leaking, fit an exchange impeller unit. When fitting a new impeller unit into the pump casing of the 18R series engines, ensure that the drain hole is downwards (Fig. 2.13).
3 A defective fluid coupling should be replaced complete by an exchange unit.

10 Drivebelts – tensioning and renewal

1 A variety of drivebelt arrangements are used, dependent upon the equipment fitted and the variants are shown in Fig 2.14, 2.15 and 2.16.
2 Drivebelt tension should be checked by pressing down (photo) at the points shown and a check made to see that the deflection is within limits.
3 To adjust the alternator/fan belt tension, loosen the alternator mounting and link bolts (photo) and lever the alternator away from the engine until the required tension is obtained. Tighten the bolts and re-check the tension. It is preferable to use a piece of wood as a lever, but if a metal bar or long screwdriver is used, take care that if the lever does not press against the alumimiun and castings of the alternator and damage them
4 The air pump/air conditioning compressor drivebelt is adjusted in a similar manner, but do not lever against the air pump body. Press the lever against the head of one of the pump body through bolts.
5 If a drivebelt is to be removed completely, loosen the facing bolts and move the adjustable component to give maximum belt slack. Ease the drivebelt over the rim of one of the pulleys and then lift it off. Refitting a drivebelt is done in the reverse order to removal, but always tension a new drivebelt to the minimum dimension given and re-check the tension after about 500 miles (800 km).

11 Fault diagnosis – cooling system

Symptom	Reason/s
Overheating	Insufficient water in cooling system
	Drivebelt slipping (accompanied by a shrieking noise on rapid engine acceleration)
	Radiator core blocked or radiator grille restricted
	Kinked hose, impeding flow
	Thermostat not opening properly
	Ignition advance and retard incorrectly set (accompanied by loss of power, and perhaps, misfiring)
	Carburettor incorrectly adjusted (mixture too weak)
	Oil level in sump too low
	Fan fluid coupling faulty
Cool running	Thermostat jammed open
	Incorrect thermostat fitted allowing premature opening of valve
	Thermostat not fitted
	Fan fluid coupling faulty
Loss of cooling water	Loose clips on water hoses
	Hoses perished and leaking
	Radiator core leaking
	Thermostat gasket leaking
	Radiator pressure cap spring worn or seal ineffective
	Blown cylinder head gasket
	Cylinder wall or head cracked
	Incorrect pressure cap

Chapter 3 Fuel, exhaust and emission control systems

For modifications and information applicable to later USA models, refer to Supplement at end of manual

Contents

Accelerator linkage	36	Carburettor (18R-G) – general description	19
Air cleaner – element renewal	2	Carburettor (18R-G) – removal and refitting	20
Air injection (AI) system	30	Catalytic converter (CCO) system	31
Automatic hot air intake (HAI) system	26	Choke breaker (CB) system	33
Carburettor (2TB) – adjustments after installation	13	Deceleration fuel cut system	35
Carburettor (2TB) – adjustments during assembly	12	Emission control – general	25
Carburettor (2TB) – dismantling, overhaul and reassembly	11	Exhaust gas recirculation (EGR) system	29
Carburettor (2TB) – general description	9	Exhaust system – removal and refitting	24
Carburettor (2TB) – removal and refitting	10	Fault diagnosis – emission control system	38
Carburettor (18R and 20R) – adjustments after installation	18	Fault diagnosis – fuel system	37
Carburettor (18R and 20R) – adjustments during assembly	17	Fuel contents gauge and sender unit	8
Carburettor (18R and 20R) – dismantling, overhaul and reassembly	16	Fuel evaporative emission control (EVAP) system	34
Carburettor (18R and 20R) – general description	14	Fuel filter – renewal	3
Carburettor (18R and 20R) – removal and refitting	15	Fuel pump – general description	4
Carburettor (18R-G) – adjustments after installation	22	Fuel pump – removal, overhaul and refitting	6
Carburettor (18R-G) – dismantling, overhaul and reassembly	12	Fuel pump – testing	5
		Fuel tank – removal and refitting	7
		General description	1
		High altitude compensation (HAC) system	32
		Inlet and exhaust manifolds	23
		Spark control (SC) system	28
		Throttle positioner (TP) system	27

Specifications

Fuel tank capacity
13.4 Imp gallons, 16.1 US gallons, 61 litres

Fuel octane rating (RON)
18R engines .... 90 octane
2T-B, 18R-G engines .... 98 octane
20R .... 91 octane – unleaded

Fuel pump
Type:
  2T-B, 18R, 18R-G .... Mechanical
  20R .... Electric submersible

Mechanical pump output:
  Capacity .... 900 cc/min (54.9 cu in/min) at camshaft speed of 3000 rpm
  Delivery pressure .... 2.8 to 4.3 psi (0.2 to 0.3 kg/cm^2)
Electrical pump output:
  Capacity .... over 1.2 litres/min
  Delivery pressure .... 2.1 to 4.3 psi (0.15 to 0.3 kg/cm^2)

Carburettor (2T-B)
Primary main jet diameter .... 0.0315 in (0.80 mm)
Secondary main jet diameter .... 0.0520 in (1.32 mm)
Primary slow jet diameter .... 0.0185 in (0.47 mm)
Secondary slow jet diameter .... 0.0217 in (0.55 mm)
Power jet diameter .... 0.0185 in (0.47 mm)

Pump jet diameter ... 0.0197 in (0.50 mm)
Accelerator pump stroke ... 0.12 in (3 mm)
Float level from upper surface – needle
valve closed ... 0.148 in (3.5 mm)
Float lip-to-needle clearance – float lowered ... 0.047 in (1.2 mm)
Primary throttle valve angle:
  Closed ... 7°
  Fully open ... 90°

Carburettor (18R)
Float level from upper surface – needle valve closed ... 0.39 to 0.43 in (10.0 to 11.0 mm)
Float lip-to-needle clearance – float lowered ... 0.039 to 0.047 in (1.0 to 1.2 mm)
Kick-up – secondary throttle valve-to-body clearance ... 0.004 to 0.012 in (0.1 to 0.3 mm)
Secondary throttle touch angle ... 57 to 61°
Fast idle (first throttle valve to body clearance):
  Automatic choke ... 0.032 in (0.81 mm)
  Manual choke ... 0.039 in (1.01 mm)
Unloader angle (from bore) ... 50°
Accelerator pump stroke ... 0.16 in (4.0 mm)
Choke valve fully closed temperature ... Below 77°F (25°C)
Choke breaker:
  Automatic choke ... 19°
  Manual choke ... 16°

Carburettor (18R-G)
Float level from upper surface – needle valve closed ... 0.62 to 0.71 in (16 to 18 mm)
Float level adjustment per turn of screw ... 0.07 in (1.8 mm)
Accelerator pump – discharge time ... 0.8 to 1.1 secs

Carburettor (20R)
Carburettor part number ... 21100-38010   21100-38030
21100-38020   21100-38060
21100-38040

Main jet diameter:
  Primary ... 0.0476 in (1.21 mm)   0.0469 in (1.19 mm)
  Secondary ... 0.0697 in (1.77 mm)
Primary slow jet diameter ... 0.0201 in (0.51 mm)
Power jet diameter ... 0.020 in (0.5 mm)
Float raised position ... 0.20 in (5 mm)
Float lowered position ... 0.04 in (1 mm)
Primary throttle valve full open angle ... 90°
Secondary throttle valve full open angle ... 90°
Kick-up ... 0.008 in (0.2 mm)
Fast idle ... 0.047 in (1.2 mm)
Unloader ... 50°
Choke opener ... 50°
Choke breaker ... 40°
Throttle positioner:
  Manual transmission ... 0.024 in (0.6 mm)
  Automatic transmission ... 0.020 in (0.5 mm)
Accelerator pump stroke ... 0.177 in (4.5 mm)

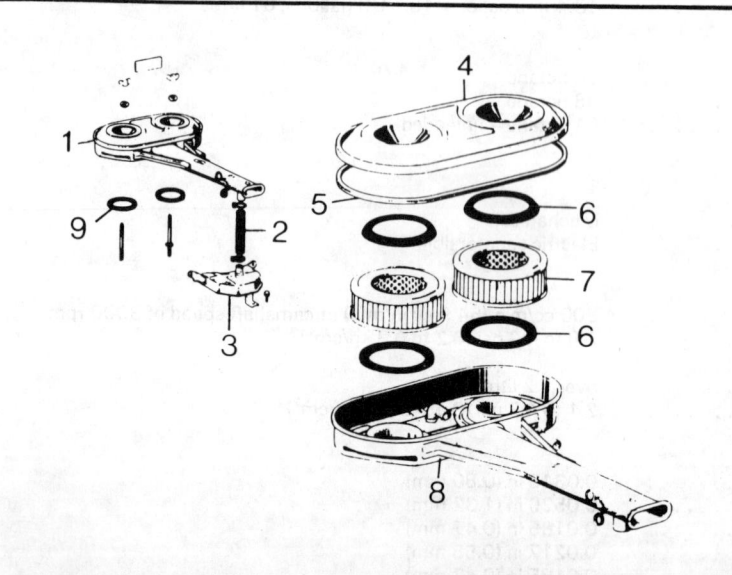

Fig. 3.1 Twin carburettor air cleaner

1 Air cleaner assembly
2 Hose
3 Air cleaner inlet
4 Air cleaner cover
5 Air cleaner case gasket
6 Filter element gasket
7 Air cleaner filter element
8 Air cleaner case
9 Air cleaner to carburettor gasket

Chapter 3 Fuel, exhaust and emission control systems

1 General description

All models have a rear mounted fuel tank and on the 20R engined models the tank contains a submersible, electrically driven fuel pump. All other engine variants covered by this manual have a mechanically driven fuel pump.

With the exception of the 18R-G, the carburettors are of the twin-choke, down-draught type. The 18R-G has twin side-draught carburettors. A single carburettor is fitted to the 18R and 20R and there are twin carburettors on the 2T-B. Because of the differences in carburettor type and the need to fit emission control equipment for some territories, there are differences in the type of air cleaner, but all are of the renewable paper element type.

The type of cold start service which is fitted may be a manual choke or a heat sensing automatic type, depending upon which engine is fitted to the vehicle. The type of choke used is described with the carburettor appropriate to each engine.

2 Air cleaner – element renewal

1 The basic procedure for fitting a new element is the same for models of air cleaner.
2 Remove the central wingnut – in the case of twin carburettor models there is a wingnut in the centre of each element – and push down the clips which hold the rim of the cover in place.
3 Lift the cover off and remove the dirty element carefully so that no dirt falls into the carburettor air intake.
4 If the element is not very dirty it may be blown clean by directing a jet of compressed air from the inside of the element, but a heavily contaminated element should be discarded.
5 Fit the new, or cleaned element, making sure that the sealing rings are in good condition and are seated properly.
6 Fit the air cleaner cover, lining up the arrow on the lid with the arrow on the intake spout and making sure that the cover gasket is in good condition and is seated properly.
7 Screw on the centre wing nut and snap the lip clips into place.
8 If the air cleaner has a lever indicating *Summer* and *Winter* positions, check to see that the lever is in the position which is appropriate to the season (photo).
9 Never run the engine without an element in the air cleaner, because if the engine backfires and there is no air cleaner element the engine intake may ignite.
10 If in doubt about whether a filter element is serviceable, fit a new one. A dirty element causes excessive fuel consumption and poor engine performance.

3 Fuel filter – removal

1 All models have a disposable fuel filter fitted in the engine compartment, the type of filter varying with the type of engine which is fitted.
2 On engines with a mechanical fuel pump, the filter is fitted in the petrol feed line to the fuel pump. On the 20R engine, which has an electric pump in the fuel tank, the petrol line filter is in the feed to the carburettor and supplements the strainer which forms part of the petrol pump assembly.
3 Every 24 000 miles fit a new filter by placing a container under the filter to collect any spilled fuel, then removing the fuel pipes from the filter. When the filter has a screwed connection, fit a spanner to the hexagon on each part of the union in order to unscrew it.
4 If the filter case is cracked, or if the filter is suspected of being excessively dirty, fit a new filter without waiting for the appropriate maintenance interval.

4 Fuel pump – general description

Mechanical type

1 There are two different mechanical pumps used on the range of models covered by this manual, one having a fuel return connection (photo) in addition to the fuel suction and fuel output pipes, but their principle of operation is the same.

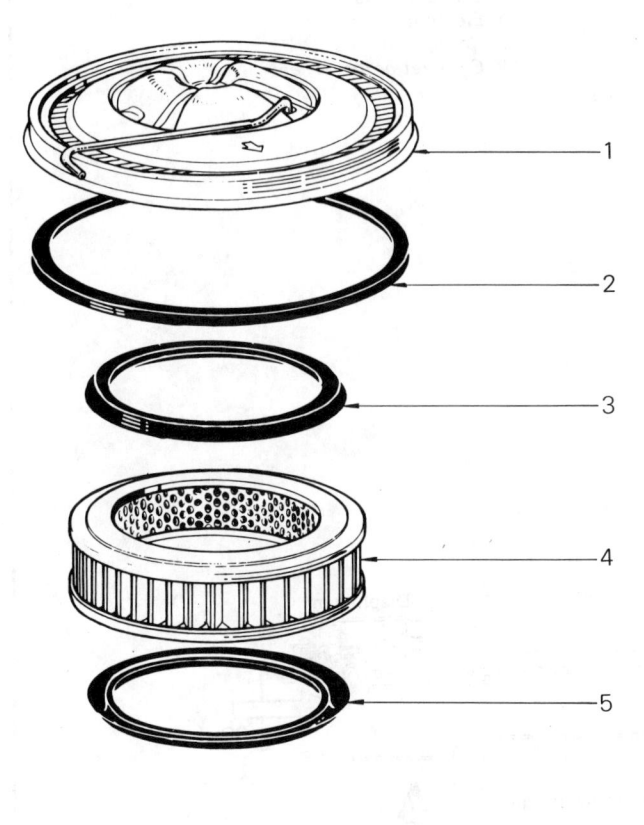

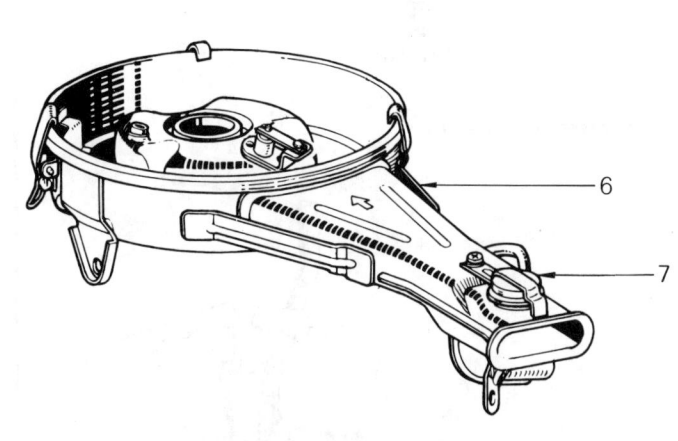

Fig. 3.2 Thermostatic air cleaner

1 Cover
2 Seal
3 Seal
4 Element
5 Seal
6 Case
7 Diaphragm assembly

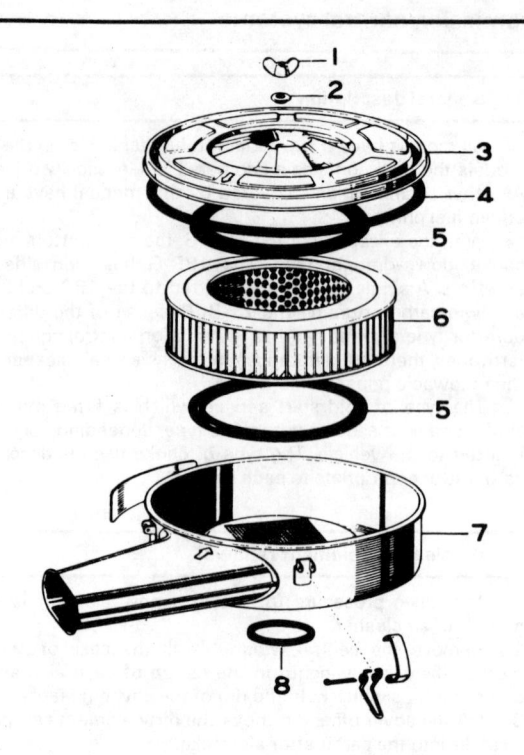

Fig. 3.3 Single carburettor air cleaner

1 Wing nut
2 Seal
3 Cover
4 Cover seal
5 Element seals
6 Element
7 Casing
8 Carburettor throat seal

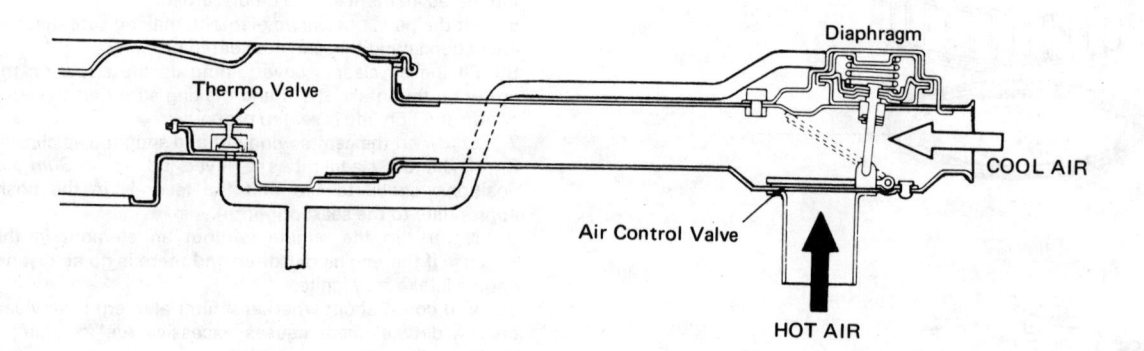

Fig. 3.4 Thermostatic air cleaner (sectional view)

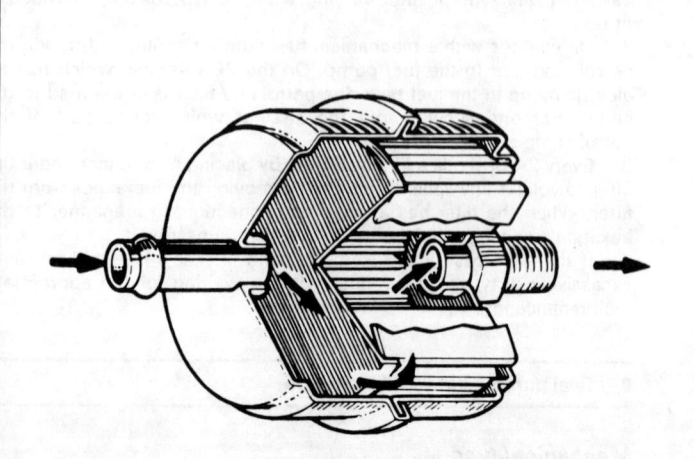

Fig. 3.5 Typical in-line fuel filter (sectional view)

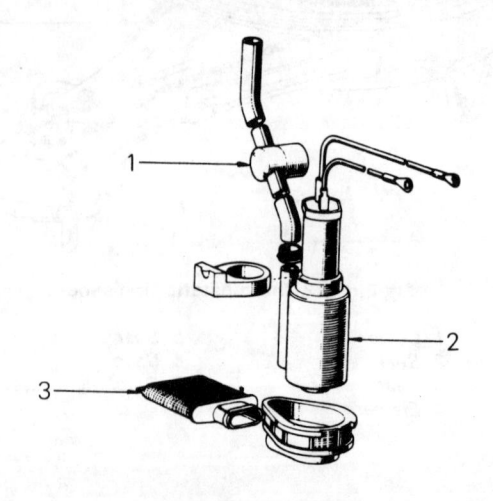

Fig. 3.6 Electric fuel pump

1 Relief valve
2 Pump
3 Filter

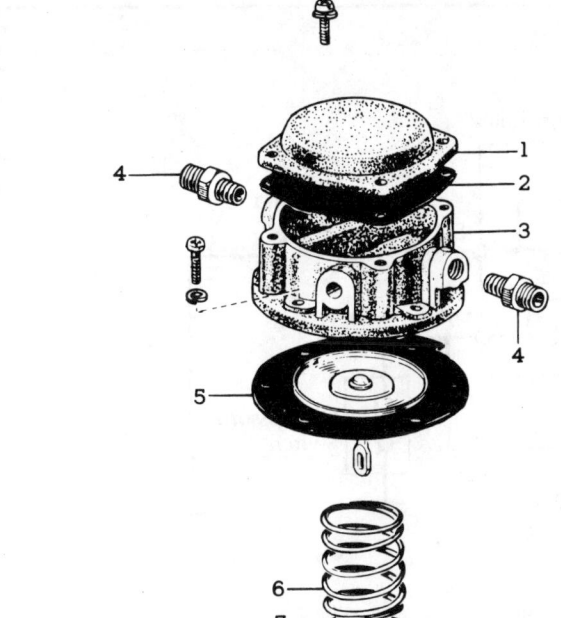

Fig. 3.7 Mechanical fuel pump (exploded view)

1 Cover
2 Gasket
3 Upper body
4 Union
5 Diaphragm/rod assembly
6 Spring
7 Oil seal retainer
8 Oil seal
9 Lower body
10 Rocker arm link
11 Spring
12 Rocker arm
13 Pivot pin
14 Gaskets
15 Insulator

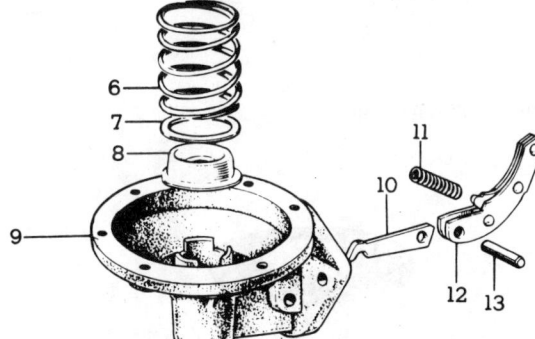

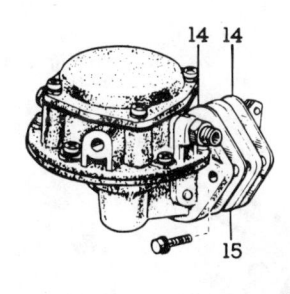

2.3 Air cleaner lid raised

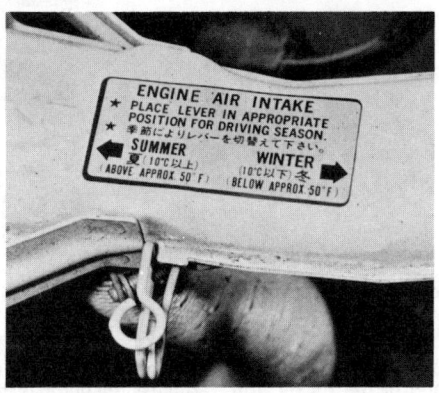

2.8 Air intake selector lever

4.1 Fuel pump with fuel return connection

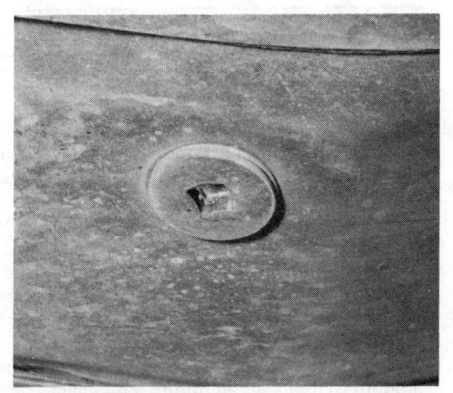

7.1 Fuel tank drain plug

7.6 Fuel tank support strap hinge

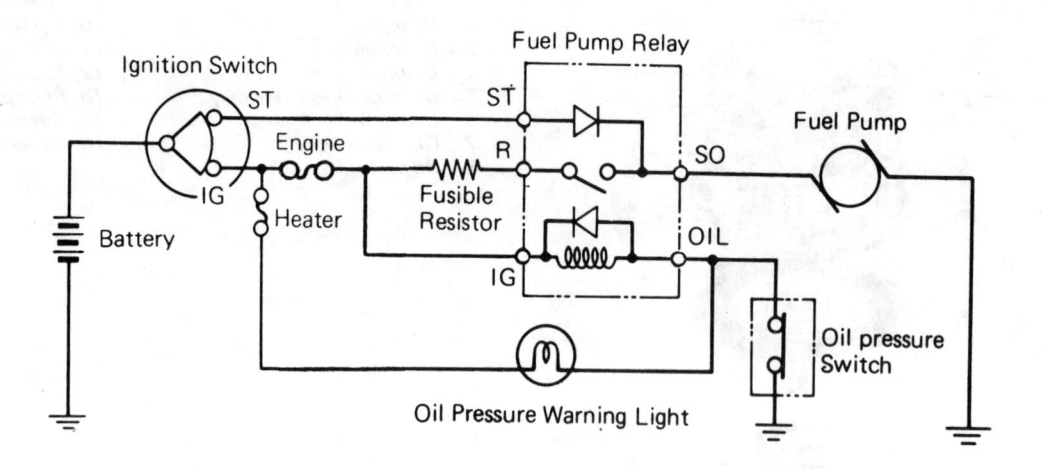

Fig. 3.8 Electric fuel pump circuit

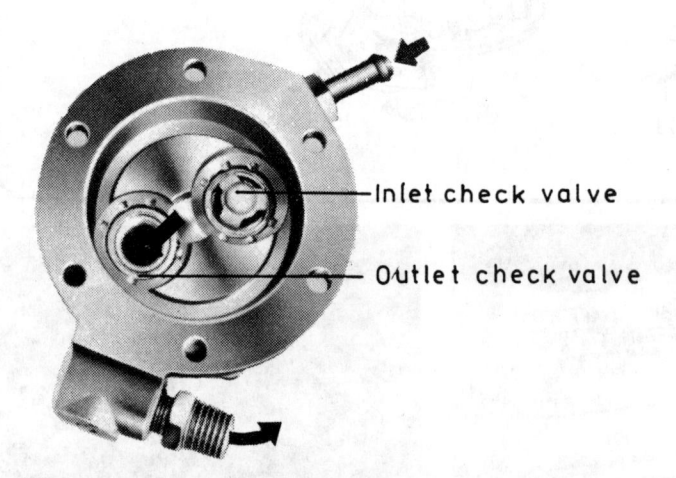

Fig. 3.9 Pump valve operation

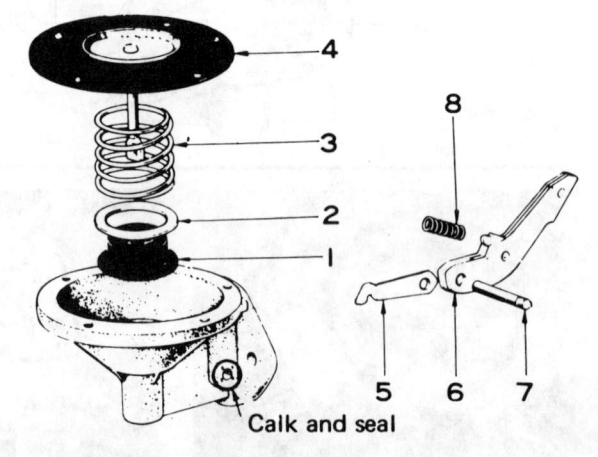

Fig. 3.10 Pump assembly sequence

2 The pump has a spring loaded diaphragm, which is connected to one end of a pivotted rocker arm. The other end of the rocker arm bears against an eccentric on the camshaft so that rotation of the camshaft causes the diaphragm to move.
3 When the movement of the cam pushes the rocker arm towards the pump, the diaphragm is pulled downwards against its return spring, creating suction in the pump fuel chamber which causes the inlet valve to open, admitting fuel to the chamber.
4 When the cam moves away from the rocker arm, the diaphragm return spring is able to expand and push the diaphragm upwards. The inlet valve of the pump closes and the outlet valve opens allowing the force of the spring against the diaphragm to eject the fuel through the pump outlet connection.
5 If the pump does not have a fuel return connection and the carburettor float chamber becomes full, the float chamber needle valve closes and the pump is unable to eject any more fuel. This results in the diaphragm being pressed down against its spring until the carburettor needle valve opens to admit more fuel. If the pump has a fuel return connection the pump output is maintained and excess fuel is returned to the fuel tank.

Electric pump

6 The electric pump is a motor driven unit, mounted in the fuel tank which operates continuously if the engine is turning and engine oil pressure is present.

7 The pump is of the submersible type and must never be operated unless the pump is immersed in petrol.
8 The pump is protected by a fusible resistor in its electrical circuit and must not be connected up without it.

5 Fuel pump – testing

Mechanical pump

1 If it is suspected that the pump is faulty either because the engine will not run, or if the fuel level visible through the carburettor sight glass is low, first check that there is adequate fuel in the tank and that the fuel filter is not blocked.
2 Release the hose clamp on the fuel inlet pipe to the carburettor and place the pipe end in a rigid container.
3 Disconnect the ignition coil, so that the engine will not fire and operate the starter switch to crank the engine for several revolutions. Check to see that there is a strong spurt of fuel into the container at regular intervals.
4 To check that the pump is within specification, measure the amount of fuel ejected during a measured time with the engine turning at normal cranking speed i.e. with the battery in good condition and fully charged. The correct rate of discharge and discharge pressure are given in Specifications.

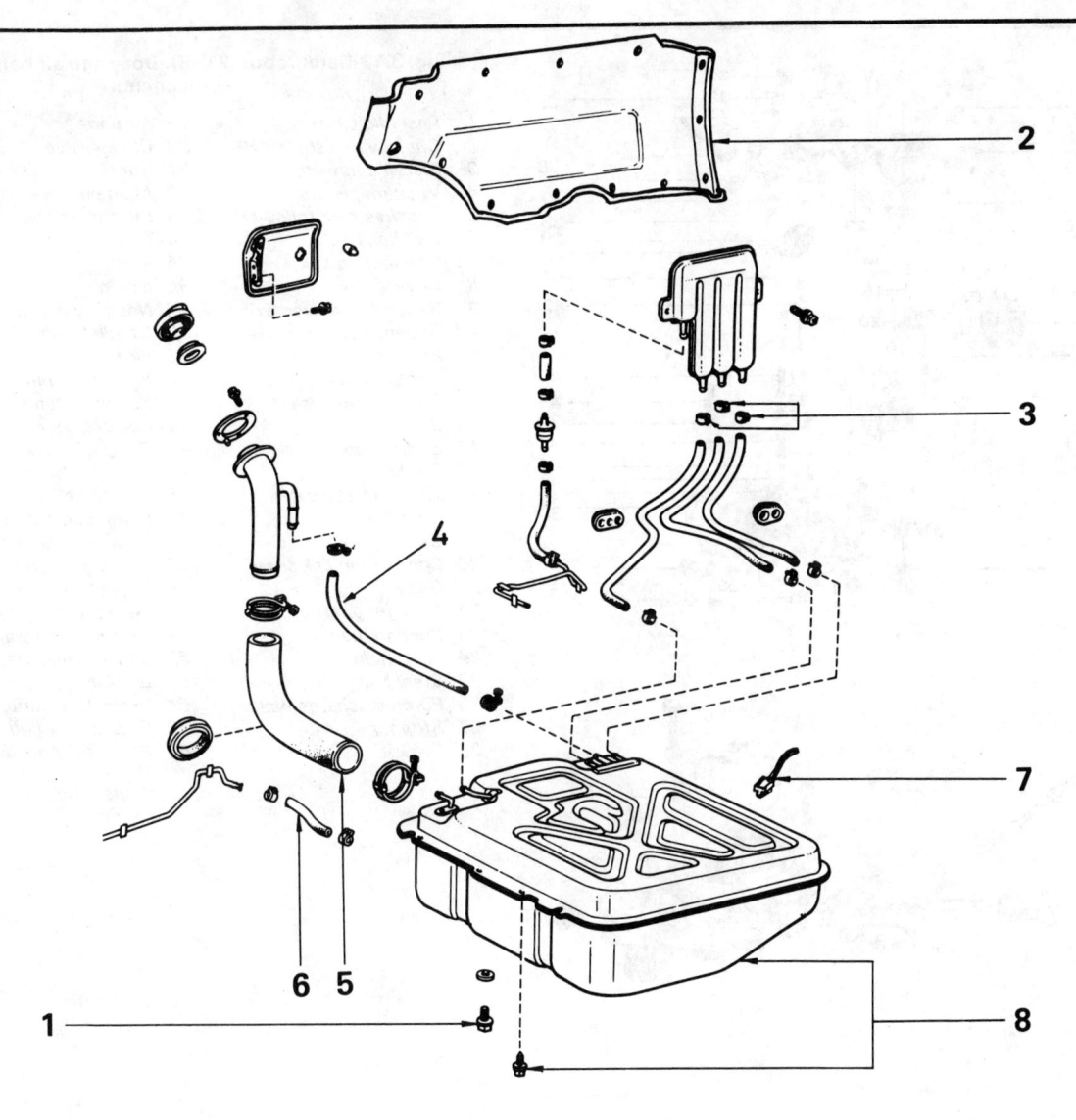

Fig. 3.11 Fuel tank assembly (exploded view)

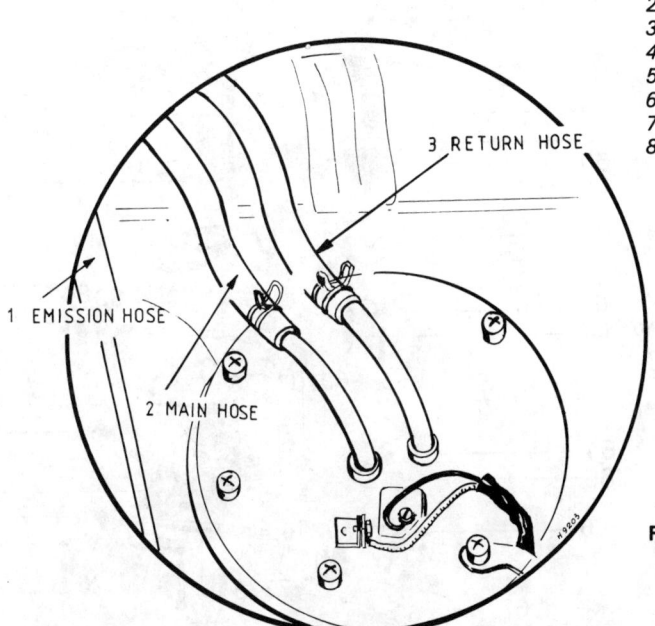

1 Drain plug
2 Luggage compartment trim
3 Fuel evaporator hose connections
4 Fuel tank breather hose
5 Fuel filler hose
6 Fuel feed hose
7 Fuel gauge sender connection
8 Fuel tank and fixing screws

Fig. 3.12 Electric fuel pump connections

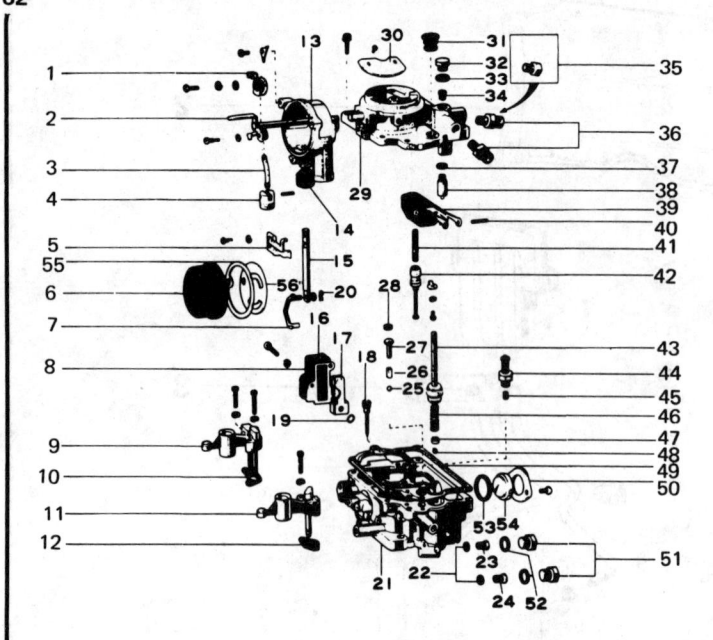

82

Fig. 3.13 Carburettor (2T–B). Body and air horn components

1	Fast idle cam	29	Air horn
2	Choke valve shaft	30	Choke valve
3	Piston connector	31	Boot
4	Vacuum piston	32	Main passage plug
5	Fast idle cam follower	33	Inlet strainer gasket
6	Coil housing	34	Strainer
7	Connecting link	35	Plug
8	Cover	36	Union
9	Second small venturi	37	Needle valve seat gasket
10	Second venturi gasket	38	Needle valve
11	First small venturi	39	Float
12	Venturi gasket	40	Float lever pin
13	Thermostat case	41	Power piston spring
14	Boot	42	Power piston
15	Sliding rod	43	Pump plunger
16	Gasket	44	Power valve sub-assembly
17	Thermostatic valve	45	Power jet
18	Slow jet	46	Pump damping spring
19	O-ring	47	Retainer, check ball
20	Connecting link snap	48	Ball, steel
21	Body	49	Gasket, air horn
22	Main jet gasket	50	Clamp, level gauge
23	Second main jet	51	Plug main passage*
24	First main jet	52	Gasket, main passage
25	Steel ball	53	Gasket
26	Pump discharge weight	54	Glass, level gauge
27	Stopper	55	Gasket, coil housing
28	Gasket	56	Plate, coil housing

*Some plugs also retain a spring

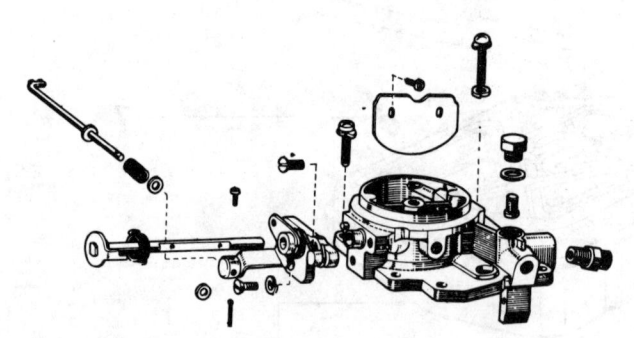

Fig. 3.14 Carburettor (2T–B). Flange assembly components

57	Cap, diaphragm housing	74	Shaft, first throttle
58	Spring, diaphragm	75	Valve, first throttle
59	Gasket, diaphragm	76	Shim, throttle shaft
60	Diaphragm	77	Ring, retainer
61	Housing diaphragm	78	Flange
62	Support, back spring	79	Spring
63	Spring, back	80	Screw, idle adjusting
64	Lever, second throttle	81	Screw, throttle adjusting
65	Collar	82	Shaft, second throttle
66	Lever, first throttle	83	Spring, diaphragm relief
67	Arm, first throttle shaft	84	Lever, diaphragm
68	Bolt	85	Gasket, body flange
69	Spring	86	Link, pump connecting
70	Lever, fast idle adjusting	87	Spring, pump
71	Collar	88	Screw, pump arm set
72	Spring	89	Lever, pump
73	Valve, second throttle		

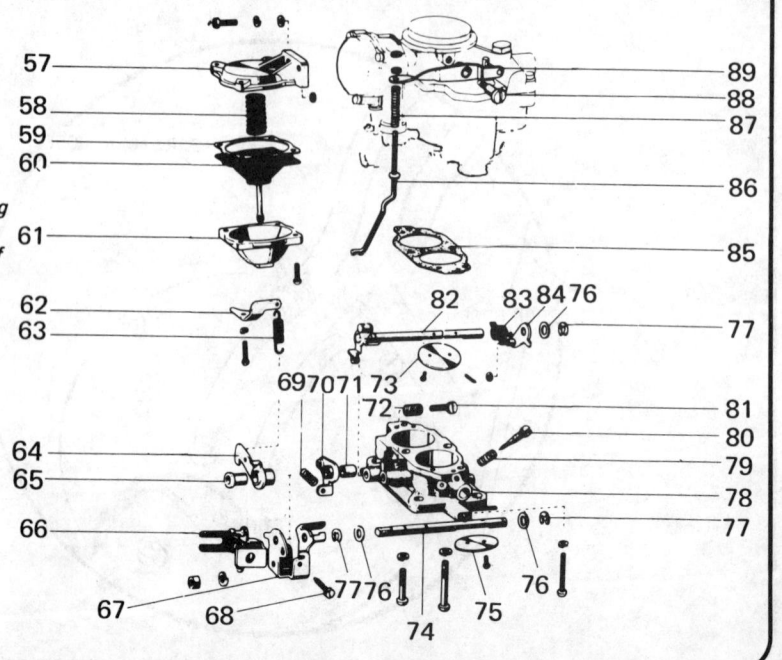

Chapter 3 Fuel, exhaust and emission control systems

Electrical pump

5 The electrical circuit of the fuel pump is designed to ensure that the pump does not work unless the engine is producing oil pressure and a faulty oil pressure switch will prevent the pump operating.
6 If there is no evidence of the oil pressure switch being faulty, test the operation of the pump by disconnecting the lead to the oil pressure switch and then turn the ignition switch to the start position, when the pump should operate. If the pump does not operate smoothly and quietly, fit a new pump. If the pump does not operate at all, the pump relay, resistor, or pump may be faulty and each of these should be checked.
7 To check the discharge capacity of the pump, disconnect the ignition coil so that the engine will not fire. Release the hose clamp on the fuel inlet pipe to the carburettor and place the pipe end in a rigid container.
8 Operate the pump as described previously and measure the volume of fuel discharged over a timed interval. Compare this with the rate of flow given in Specifications.

6 Fuel pump – removal, overhaul and refitting

Mechanical pump

1 Disconnect the inlet and outlet hoses from the fuel pump and either fit caps to the hose ends or plug them to prevent loss of fuel and the ingress of dirt.
2 Undo and remove the nuts and spring washers, (or bolts) securing the fuel pump to the engine and lift the pump away. Remove the gasket and clean the pump joint face of the cylinder block.
3 Before starting to dismantle the pump, clean its outside with solvent and lint-free cloth and wipe the pump dry. Put mating marks on the pump body and pump cover so that they will be reassembled correctly. Also put mating marks on the upper and lower pump body.
4 Undo and remove the screws securing the pump cover and remove the cover and gasket.
5 Undo and remove the screws securing the pump upper body to the lower body and separate the two parts. If the pump diaphragm sticks to either of its mating surfaces, free it with a sharp knife, taking care not to cut the diaphragm.
6 Use a pin punch to tap out the rocker arm pin, lift out the rocker arm and recover the anti-rattle spring.
7 Depress the centre of the diaphragm and then unhook and remove the rocker arm link.
8 Carefully lift the diaphragm and its spring out of the lower body.
9 Do not remove the valves unless they are known to be defective. The valves are peened in and if it is necessary to change a valve, remove the peening and lift the valves out after making a note of which way they are fitted.
10 To remove the oil seal from the lower body, make a note of which way up it is fitted and then prise out the seal and its retainer with a screwdriver. Do not disturb the oil seal unless a new one is available for refitting.
11 Inspect the pump body for signs of cracks and check that the threads in the tapped holes are in good condition. Check the rocker arm, link and pin for wear and obtain new parts if necessary. Examine the diaphragm and check that it is not cracked, or split. If the diaphragm shows any signs of deterioration, discard it and obtain a new one.
12 If new valves are being fitted, carefully clean the valve recesses, fit the valves, making sure that they are placed the correct way up to perform their inlet, or outlet function (Fig 3.9) and then peen the edges of the recess with a small, sharp, punch to keep the valves in place.
13 Fit the shaft oil seal and its retainer.
14 Fit the diaphragm spring and then insert the diaphragm pull rod through the spring and the oil seal.
15 Insert the rocker arm link and hook its end into the eye of the diaphragm pull rod, then refit the rocker arm into the lower body and insert the anti-rattle spring.
16 Line up the hole in the rocker arm with the hole in the pump body and then insert the rocker arm pin. After inserting the pin, peen the body at four places around the pin, using a centre punch and then apply some proprietary locking compound to each end of the pin.
17 Fit the pump cover to the upper body, using a new gasket and taking care to ensure that the mating marks made before dismantling are lined up. This will make certain that the inlet and outlet chamber separating walls of the pump and cover are aligned.
18 After fitting the pump cover and before fitting the upper body to the lower body, place the mouth over the suction port and check that the valve opens when air is blown in and closes when air is sucked out. Repeat the process for the discharge port to see if the valve works the other way, opening when air is sucked out.
19 After checking the correct operation of the valves, line up the mating marks of the upper and lower bodies, align the holes in the upper and lower bodies and in the diaphragm and insert and tighten the retaining screws.
20 On completion of assembly, move the rocker arm by hand to check that it operates properly and re-fit the pump to the engine, using a new gasket.

Electrical pump

21 Remove the lining from the floor of the luggage compartment. Remove the four screws from the fuel pump cover and lift the cover off to gain access to the pump.
22 Disconnect the battery leads and then separate the fuel pump leads from the wiring harness.
23 Label the three fuel hoses and the pipes to which they are connected and remove the hoses, being careful not to spill any fuel which drains out of them.
24 Remove the six pump fixing screws and lift the pump and filter assembly out of the tank.
25 When refitting the pump, use a new gasket and ensure that the hoses are fitted to the correct pipe and that the hose clips are tight.
26 Re-connect the pump leads and fit the cable clips.
27 The electric fuel pump is a sealed unit and cannot be dismantled for overhaul.

7 Fuel tank – removal and refitting

1 Make sure that the vehicle is in a well ventilated area free from sparks and sources of ignition, then disconnect the battery leads. Remove the tank drain plug (photo) and drain the fuel into a container suitable for the storage of petrol. Because of the danger of accumulations of petrol vapour, draining the tank should not be done over an inspection pit or near a drain gulley.
2 From inside the luggage compartment remove the inside trim from the fuel tank side quarter and disconnect the hoses from the fuel evaporative separator, after making a note of the connection from which each hose is removed.
3 Disconnect the fuel tank breather hose from the filler pipe.
4 From beneath the vehicle disconnect the fuel tank inlet hose from the side of the tank and the fuel main hose.
5 Disconnect the fuel gauge sender unit.
6 Remove the bolts from the front end of the two fuel tank support straps (photo) and while supporting the tank, swing the two straps downwards.
7 Remove the fuel tank from beneath the vehicle.
8 The repair of fuel tanks is a specialist job and should not be attempted by anyone not having facilities to steam clean the tank and make it safe from any risk of explosion.

8 Fuel contents gauge and sender unit

1 The testing of these units is described in Chapter 10, together with the remainder of the vehicle's instrumentation.

9 Carburettor (2T-B) – general description

1 Twin carburettors are used and they are of the dual barrel downdraught type.
2 Different types of cold start device may be fitted to engines intended for different markets and there may also be variations resulting from the exhaust emission requirements of different territories.

10 Carburettor (2T-B) – removal and refitting

1 Remove the air cleaner as described previously.

Chapter 3 Fuel, exhaust and emission control systems

2 Disconnect the throttle linkage from the carburettor.
3 Disconnect the manual choke cable, or the leads to the electrically heater automatic choke.
4 Disconnect the carburettor fuel pipes and vacuum connections.
5 Unscrew and remove the carburettor mounting nuts and washers and lift the assembly from the inlet manifold.
6 Remove the gasket and clean the mating flanges.
7 If only one carburettor is being removed, the connecting links between the two throttles and chokes must be disconnected.
8 Refitting is the reversal of removal, but always use a new flange gasket and ensure that both mating faces are clean and completely free of any pieces of the old gasket.

11 Carburettor (2T-B) – dismantling, overhaul and reassembly

1 The carburettor should not be dismantled unnecessarily. If a carburettor has had prolonged use and is likely to be badly worn, it is better to purchase a new, or exchange unit which has been tested and calibrated, rather than try and obtain a lot of small replacement parts.
2 Before starting to dismantle a carburettor it is vital to have a clean work place, screwdrivers and spanners which are the correct fit on the parts to be removed and several clean containers in which to put the parts of each sub-assembly.
3 Clean the outside of the carburettor before starting to dismantle it.

Air horn parts
4 Unscrew and remove the accelerator pump arm pivot bolt, then detach the arm and connecting link.
5 Remove the fast idle connector and throttle positioner if fitted.
6 Extract the air horn securing screws and remove the air horn and gasket.
7 If the air horn is to be dismantled, push out the float pivot pin, remove the float and the components of the fuel inlet needle valve.
8 Remove the power piston, spring and pump plunger.
9 Remove the plug, gasket and fuel inlet strainer.

Automatic choke
10 If an automatic choke is fitted, remove the housing, gasket and housing plate.
11 Do not dismantle the assembly further unless it is essential. If it is necessary, file off the peened part of the choke valve set screws, remove the screws and draw out the choke shaft.

Body
12 Turn the carburettor body upside down and remove the small circular sealing ring, the pump discharge weight and steel ball.
13 Remove the return spring and diaphragm from the diaphragm lever.
14 Remove the four recessed set screws and separate the flange from the carburettor body.

15 To dismantle the body, extract the primary venturi screws and venturi, then the secondary venturi.
16 Remove the thermostatic valve.
17 Use a pair of tweezers to extract the check ball retainer from the bottom of the pump bore. Turn the body upside down and let the steel ball drop out.
18 Remove the plugs, jets, fuel level sight glass and other components if necessary.
19 Remove the diaphragm from the carburettor body. To remove the diaphragm, spring and gasket, remove the four screws securing the cover, making sure to mark the relative positions of the diaphragm, cover and diaphragm housing.

Flange parts
20 Remove the idle adjusting screw.
21 Do not disturb the throttles unless necessary. To remove them, file off the peened part of the valve set screws, remove them and discard them.
22 Mark the throttle plates so that they can be re-fitted the same way up and the same way round, then lift them out and withdraw the throttle rods.
23 Check all parts for damage and wear. Blow out the jets and air passages to clear them, but do not attempt to use wire to clean jets and orifices, because this can distort them and have a bad effect on fuel consumption.
24 Reassemble by reversing the dismantling operations. Wash all parts in clean petrol before reassembling. Use new gaskets and as each unit is reassembled, check that any sliding or rotating parts move smoothly. If screws have been peened and filed to remove the peening, they should be discarded and new screws fitted. The new screws should be peened after tightening.

12 Carburettor (2T-B) – adjustments during assembly

Float level
1 Adjust the float level by bending the float lips.
2 To adjust the top level, hold the air horn upside down and allow the float to hang under its own weight. Check the gap between the tip of the float and the air horn, which should be 0.138 in (3.5 mm). Adjust the gap by bending float lip A (Fig 3.17A).
3 Lift the float and check the gap between the needle valve pin and the float lip (Fig 3.17B). The correct gap is 0.047 in (1.2 mm) and can be adjusted by bending float lips B.
4 When both the top and bottom level positions have been adjusted properly, the float will maintain the fuel at the level marked on the float chamber glass when the engine is running.

Throttle valve opening
5 Open each of the throttle valves fully and check that they are

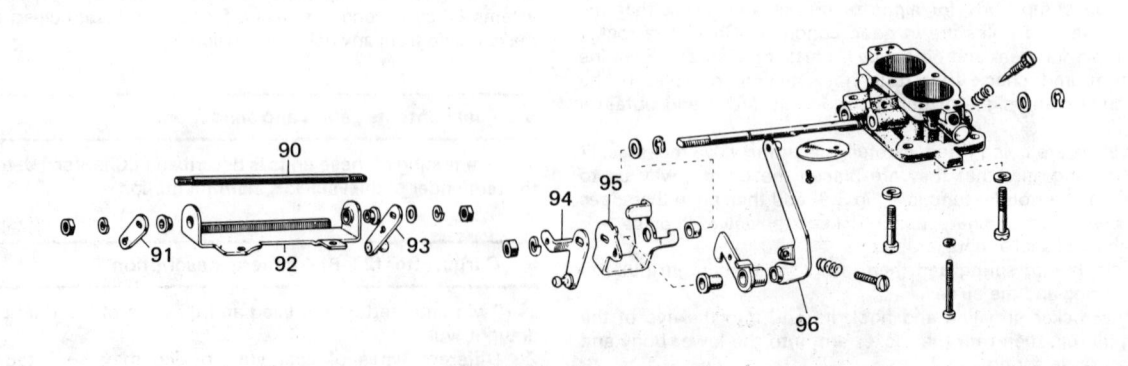

Fig. 3.15 Carburettor (2T–B). Choke and throttle levers

90 Choke connector shaft
91 Lever
92 Bracket
93 Choke lever
94 First throttle lever
95 First throttle arm
96 First adjusting lever

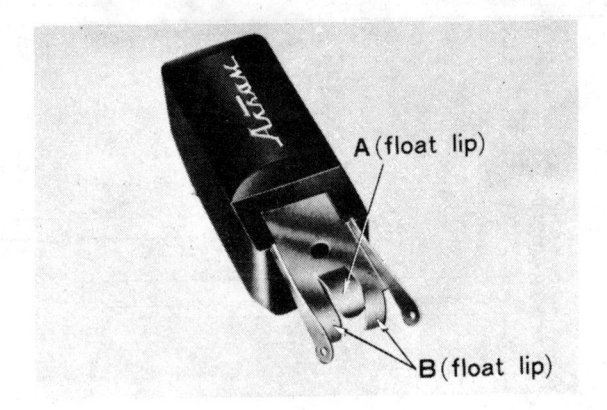

Fig. 3.16 Float lips

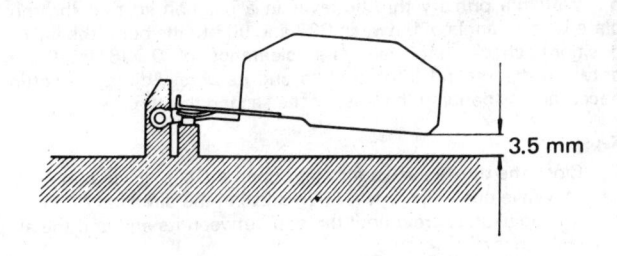

Fig. 3.17A Float adjustment (needle valve closed)

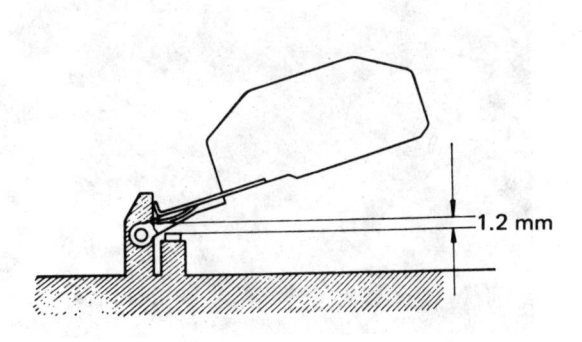

Fig. 3.17B Float adjustment (nib-to-needle clearance)

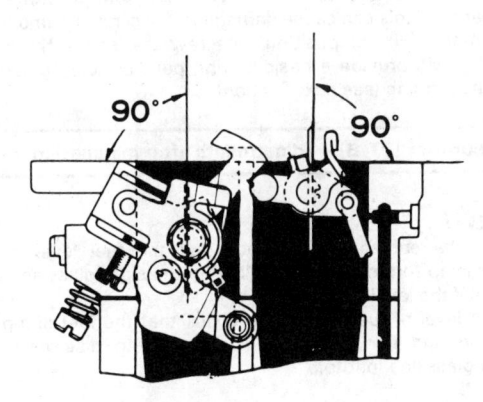

Fig. 3.18 Throttle valve opening adjustment

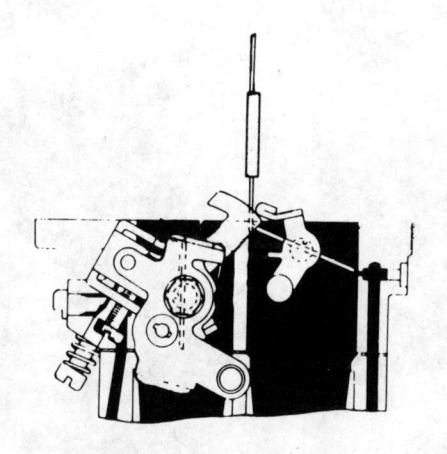

Fig. 3.19 Kick-up adjustment

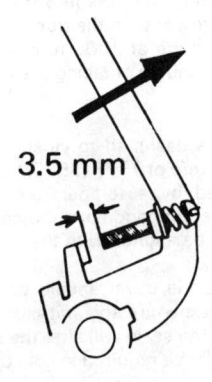

Fig. 3.20 Fast idle adjustment

Chapter 3 Fuel, exhaust and emission control systems

perpendicular to the face of the flange. Adjust the setting, if necessary, by bending the throttle levers.

Kick-up
6 With the primary throttle lever in a position so that the throttle plate is at an angle of between 62° and 90° to the bore (the fully open position), check that there is a clearance of 0.008 in (0.2 mm) between the second throttle plate and its bore. Adjust the setting, if necessary, by bending the lever of the second throttle.

Fast idle
7 Close the throttle valve fully and open the choke valve to its fullest extent. While pushing the lever (Fig 3.20) in the direction of the arrow, turn the adjusting screw until the gap between its end and the stop is 0.15 in (3.5 mm).

Accelerator pump
8 Check that the accelerator pump stroke (Fig 3.21) is 0.12 in (3 mm).
9 Adjust if necessary by bending the offset part A, of the connecting link.

Idle adjusting screw
10 Screw in the idle adjusting screw fully, but do not screw it in tightly because this can cause damage to the screw tip and its seating.
11 From the fully in position, unscrew the screw three complete turns. This will provide a basic setting pending idle adjustment with the engine running (see next Section).

13 Carburettor (2T-B) – adjustments after installation

Float level
1 With the engine idling, check that the fuel level in the float chamber is up to the level of the line on the sight glass and adjust the float level if the level of fuel is too high.
2 If the level of fuel is too low, check that the fuel pump output is satisfactory and if so, adjust the float setting to raise the fuel level to the sight glass line (photo).

Idle speed
3 Run the engine until the coolant temperature in 75° to 80°C (167° to 176°F).
4 Connect a tachometer and a vacuum gauge unless the car is already fitted with them.
5 Ensure that the air cleaner is fitted and that the PCV system hose has been installed. Then turn the throttle adjusting screw and the idle mixture adjusting screw to obtain maximum vacuum at the specified idling speed. On later carburettors, the mixture screw is sealed. Break off the plastic seal only if essential (photo).

CO adjustment
6 In territories with legislation on CO concentration use a CO meter to check the exhaust emission.
7 Ensure that the sampling system is free from leaks, because this will give a reading which is lower than the true concentration of CO.
8 Run the engine for a minute at 2000 rpm before taking a sample and measure within 1 to 3 minutes of racing the engine.

Automatic choke
9 The automatic choke is designed to close the choke valve plate fully at an ambient temperature of 77°F (25°C).
10 The valve is controlled by a temperature sensitive bi-metallic spring and for normal operating conditions the centre line on the thermostat coil housing should be aligned with the centre of the graduated scale on the body (Fig 3.23).
11 In extremes of climate the cover should be turned clockwise to produce a weaker cold start mixture and anit-clockwise to give a richer mixture. One graduation of the scale will alter the setting by 9°F (5°C). A richer mixture will normally be required in cold climates.

14 Carburettor (18R and 20R) – general description

The carburettors used on the two engines are very similar, both

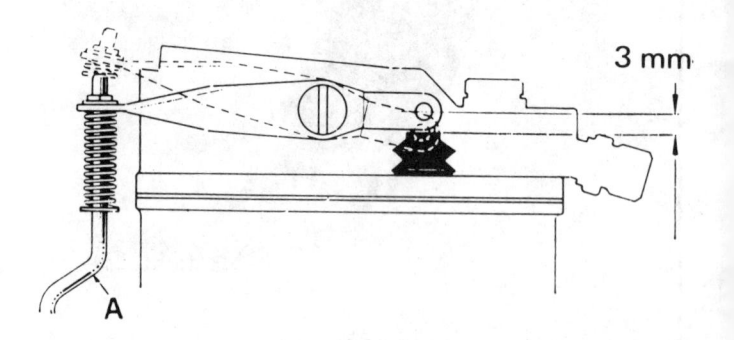

Fig. 3.21 Accelerator pump adjustment

A Connecting link

Fig. 3.22 Idle adjusting screws (arrowed)

A Throttle speed
B Mixture

Fig. 3.23 Automatic choke adjustment

Zero setting arrowed

Chapter 3 Fuel, exhaust and emission control systems

being of the dual barrel downdraught type. Both types use an automatic choke mechanism; the 18R has an electrically actuated bi-metallic spring which is controlled by carburettor ambient temperature, whereas the 20R engine has the choke controlled by the coolant temperature of the engine.

15 Carburettor (18R and 20R) – removal and refitting

1 Remove the air cleaner assembly.
2 Disconnect the fuel pipes from the carburettor float chamber.
3 Disconnect the throttle linkage from the carburettor.
4 On the 20R engine partially drain the cooling system (refer to Chapter 2 if necessary) and detach the coolant hoses from the choke.
5 Remove the electrical connection from the anti-run-on valve.
6 Mark the vacuum connections to the carburettor to identify the pipes to which they are connected and then disconnect them.
7 Remove the flange mounting nuts and lift the carburettor off.
8 Refitting is the reverse of removal, but make sure that the mating flanges are clean; and use a new gasket. After refitting, check the idle speed and mixture, as well as the idle CO if appropriate.

16 Carburettor (18R and 20R) – dismantling, overhaul and reassembly

1 The carburettor should not be dismantled unnecessarily, If a carburettor has had prolonged use and is likely to be badly worn, it is better to purchase a new, or exchange unit which has been tested and calibrated, rather than try and obtain a lot of small replacement parts.
2 Before starting to dismantle a carburettor, it is vital to have a clean work place, screwdrivers and spanners which are the correct fit on the parts to be removed and several clean containers in which to put the parts of each sub-assembly.
3 Clean the outside of the carburettor before starting to dismantle it.

Air horn parts
4 Remove the screw or clip from the pump pivot arm and disconnect the connecting link from the throttle shaft lever.
5 Remove the two connecting links (20R) and loosen the air horn screws in stages in diagonal order.
6 Remove the choke opener (20R), the air horn screws and lift away the air horn and gasket.
7 If the air horn is to be dismantled, remove the float pivot pin and float; remove the needle valve, spring, plunger and seat.
8 Pull out the pump plunger. Remove the power piston retainer, the power piston and spring.

Automatic choke (18R)
9 Remove the choke housing, the choke lever, the thermostat case and its gasket.
10 Unhook the choke breaker rod and remove the choke breaker assembly.
11 Do not dismantle the assembly further unless it is essential. If it is necessary, file off the peened part of the choke valve set screws, remove the screws and draw out the choke shaft.

Automatic choke (20R)
12 Unscrew and remove the three housing set screws, remove the water and coil housing, the housing plate and its gasket.
13 Remove the shaft screw and the three body screws and take off the choke lever and housing body.
14 Remove the choke breaker assembly with the relief lever and link.
15 Do not dismantle the assembly further unless it is essential. If it is necessary, file off the peened part of the choke valve set screws, remove the screws and draw out the choke shaft.

Body (18R)
16 Remove the two air bleed jets and the slow jet.
17 Remove the main air bleed and then the pump discharge weight and outlet valve.
18 Remove the auxiliary accelerator pump (AAP) outlet valve and the power valve.
19 Remove the two main jets and the two small venturis.
20 Remove the three screws from the AAP and take off the cover,

13.2 Carburettor fuel level sight glass

13.5 Mixture screw, sealed on later type carburettor

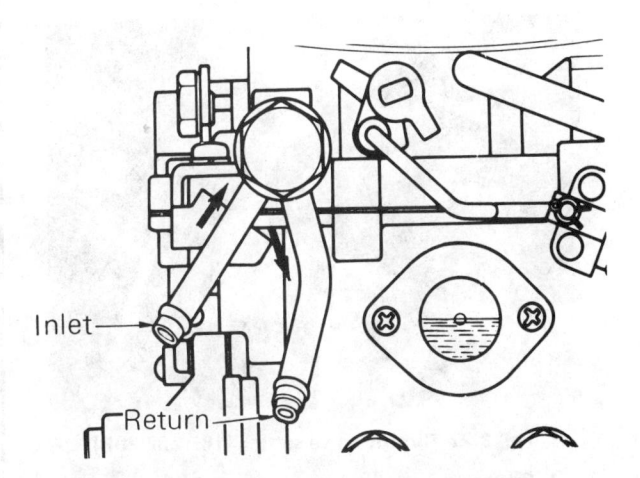

Fig. 3.24 Fuel inlet and return pipes of carburettor

Fig. 3.25 Air horn components (18R and 20R)

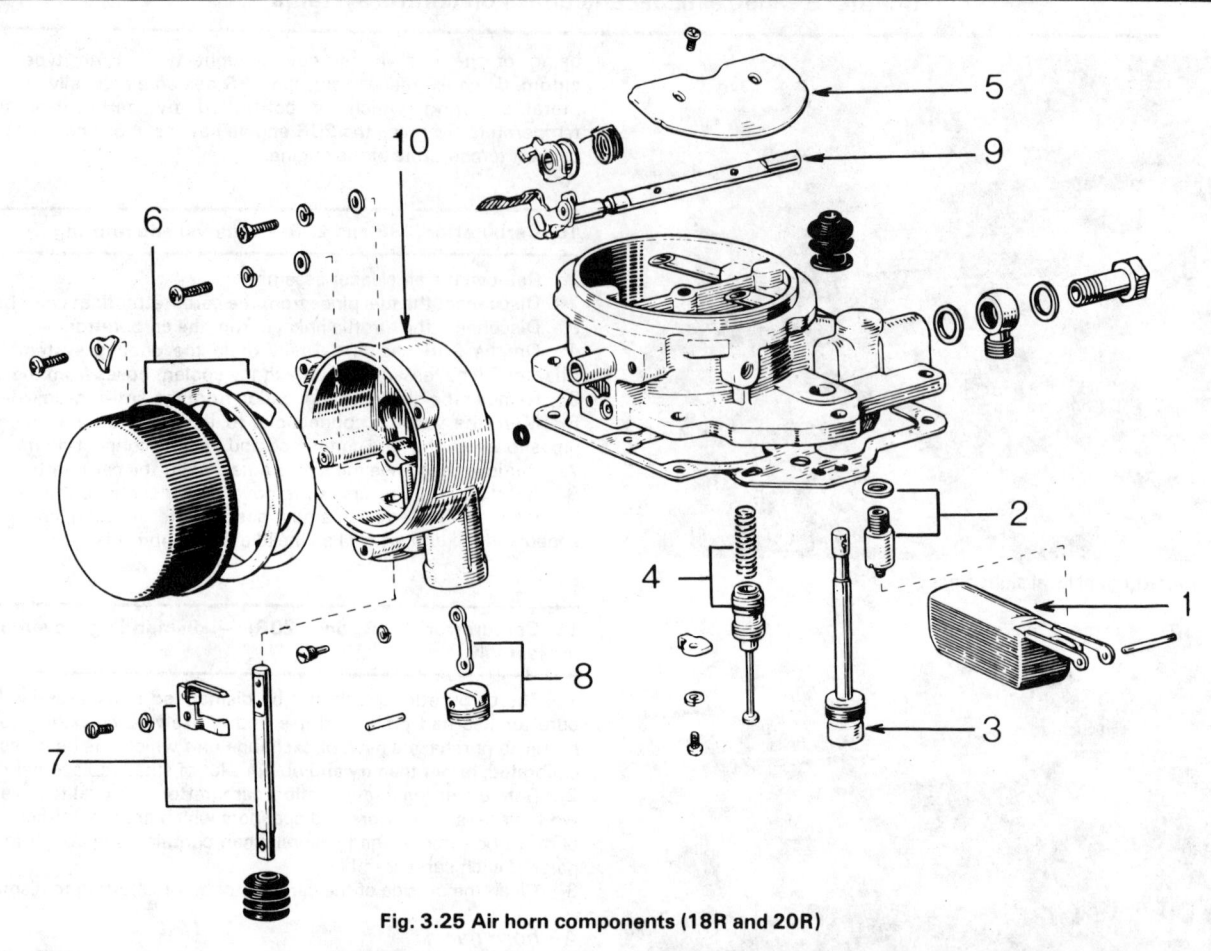

1 Float
2 Needle valve assembly
3 Pump plunger
4 Power piston and spring
5 Choke valve
6 Choke coil housing
7 Sliding rod and fast idle cam follower
8 Vacuum piston and connector
9 Choke spindle
10 Thermostat case

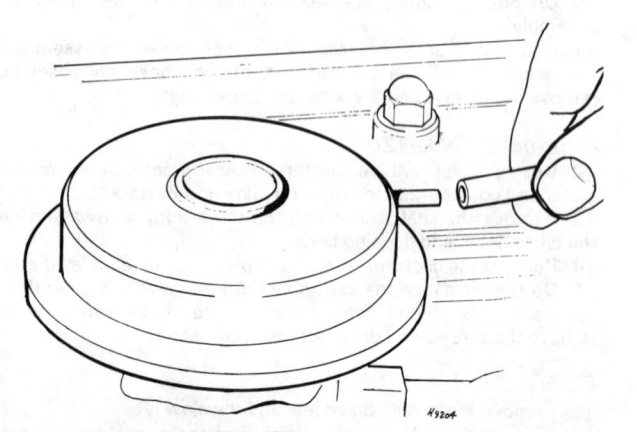

Fig. 3.26 Disconnecting the carburettor from the EGR valve

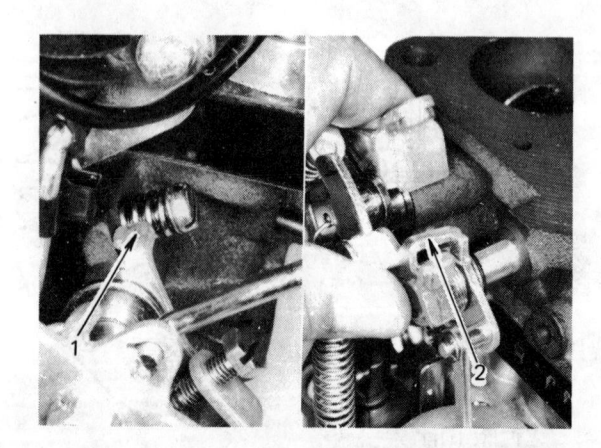

Fig. 3.27 Throttle valve setting (18R and 20R)

1 Primary stop
2 Secondary stop

Fig. 3.28 Kick up adjustment (18R and 20R)

1 Secondary throttle lever

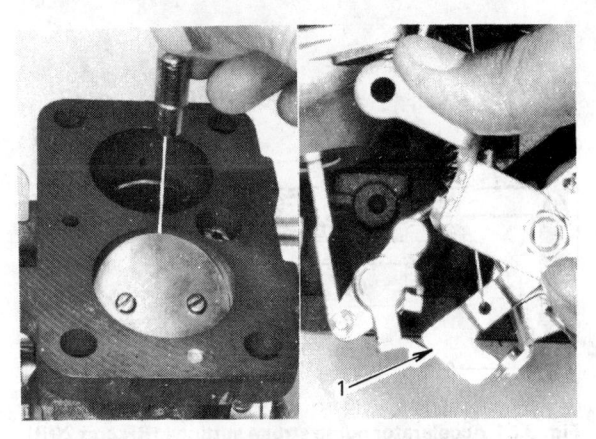

Fig. 3.29 Fast idle setting (18R and 20R)

1 Fast idle screw

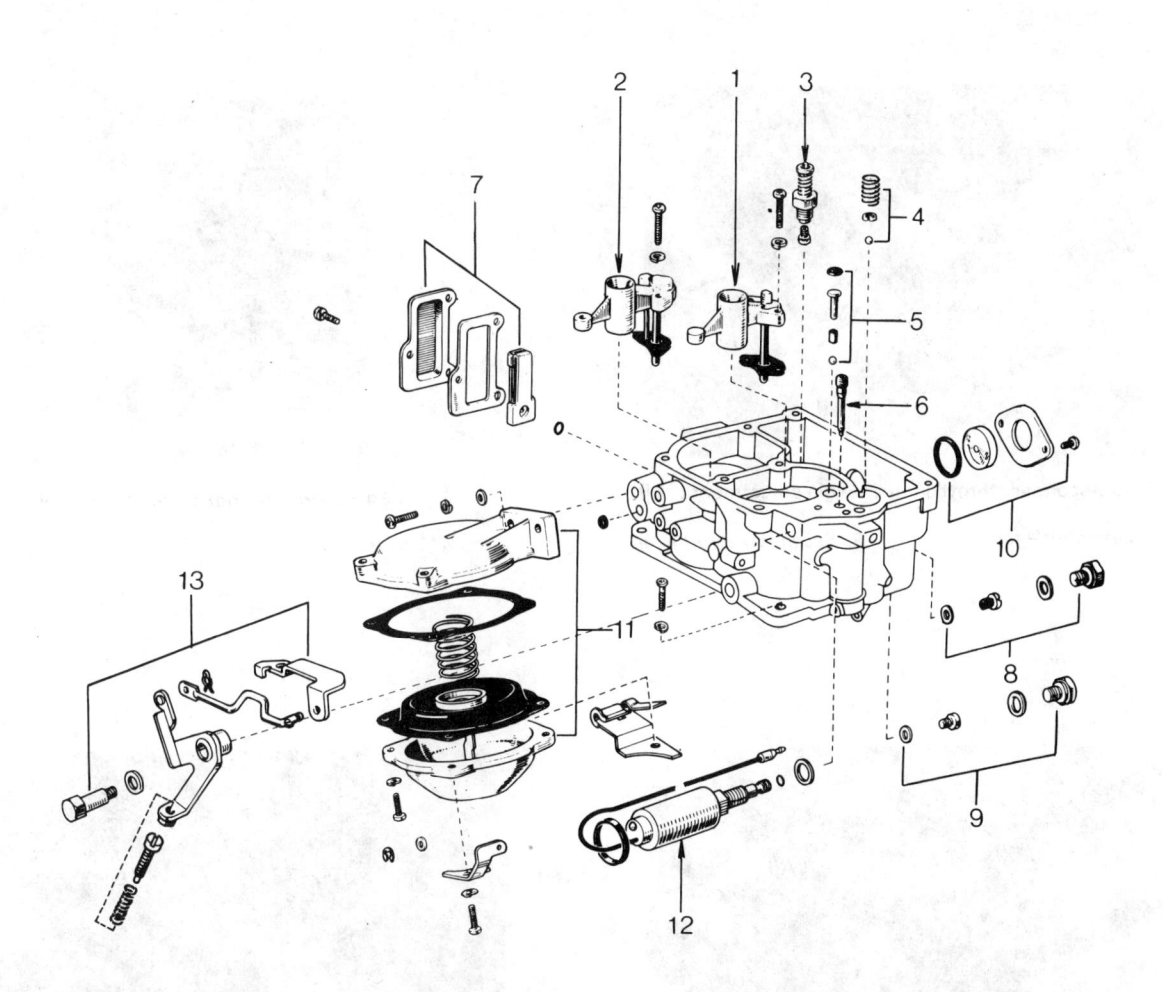

Fig. 3.30 Body components (18R and 20R)

1 Primary small venturi
2 Secondary small venturi
3 Power valve
4 Pump damping spring and small steel ball
5 Pump discharge weight and large steel ball
6 Slow jet
7 Thermostatic valve
8 Primary main jet*
9 Secondary main jet*
10 Level glass
11 Diaphragm assembly
12 Anti-run-on solenoid valve
13 Throttle positioner lever

* Some plugs also retain a spring

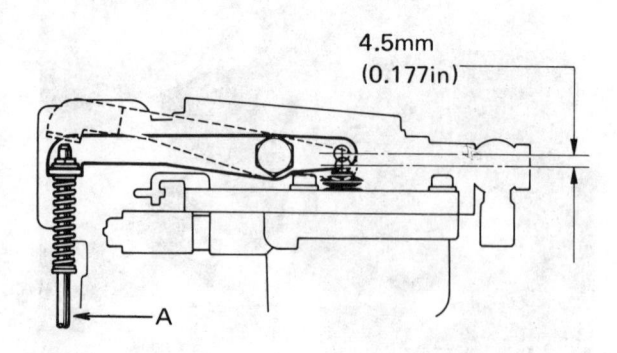

Fig. 3.31 Accelerator pump stroke setting (18R and 20R)

A Adjustment point

Fig. 3.32 Idle mixture adjusting screw (18R and 20R)

Fig. 3.33 Secondary throttle touch angle adjustment (18R)

A Primary throttle tab   B Secondary throttle tab

Fig. 3.34 Automatic choke adjustment (18R)

A Throttle lever

Fig. 3.35 Automatic choke unloader (18R)

A Adjustment tab

Fig. 3.36 Automatic choke breather (18R)

A Adjustment tab

Chapter 3 Fuel, exhaust and emission control systems

gasket and diaphragm.
21 Unscrew and remove the solenoid valve and take out its plunger.
22 Disconnect and remove the second throttle valve diaphragm.
23 Remove the fast idle cam.

Body (20R)
24 Remove the two screws from each of the two small venturis and lift the venturis out.
25 Remove the pump jet, O-ring, spring and ball. Use a pair of tweezers to remove the retainer and the ball.
26 Remove the slow jet and power valve.
27 Remove the two plugs giving access to the main jets and screw out the main jets.
28 Remove the cover of the thermostatic valve, take out the thermostatic valve and O-ring, but do not attempt to dismantle the thermostatic valve.
29 Remove the sight glass cover and O-ring seal.
30 Remove the throttle positioner securing screws and remove the throttle positioner and link.
31 Remove the AAP inlet plug and ball. Remove the outlet plug, spring and ball.
32 Remove the three screws from the AAP cover, take the cover off and remove the gasket, spring and diaphragm.
33 Remove the spring between the diaphragm assembly and the throttle lever. Remove the two screws securing the diaphragm assembly, remove the assembly and recover the O-ring from the joint face.
34 Remove the solenoid valve and the fast idle cam.
35 Remove the three screws securing the carburettor body and insulator to the flange and separate the parts.

Flange parts
36 Remove the idle mixture adjusting screw, the throttle lever, spring and collars. Remove the throttle positioner lever.
37 Do not disturb the throttles unless necessary. To remove them, file off the peened part of the valve set screws, remove them and discard them.
38 Mark the throttle plates so that they will be re-fitted the same way up and the same way round, then lift them out and withdraw the throttle rods.
39 Check all parts for damage and wear. Blow out the jets and air passages to clear them, but do not attempt to use wire to clean jets and orifices, because this can distort them and have a bad effect on fuel consumption.
40 Wash all parts with clean petrol before reassembling. Use new gaskets and as each unit is reassembled, check that any sliding or rotating parts move smoothly. If screws have been peened and filed to remove the peening, they should be discarded and new screws fitted. The new screws should be peened after tightening. Reassemble by reversing the dismantling operations.

17 Carburettor (18R and 20R) – adjustments during assembly

Throttle valve opening
1 Open each of the throttle valves fully and check that they are perpendicular to the face of the flange. Adjust the setting, if necessary by bending the throttle levers.

Kick-up
2 With the primary throttle lever fully open and the kick arm about to open the secondary throttle, check the clearance between the second throttle plate and its bore.
3 The correct clearance is:

18R	0.004 to 0.012 in (0.1 to 0.3 mm)
20R	0.008 in (0.2 mm)

Adjust the clearance, if necessary, by bending the secondary throttle lever.

Fast idle
4 With the choke valve fully closed, check the clearance between the bore and the primary throttle valve.
5 Adjust to the specified clearance by turning the fast idle adjusting screw:

18R (manual choke)	0.039 in (1.01 mm)
20R	0.047 in (1.2 mm)

Accelerator pump
6 Check the stroke of the accelerator pump and if necessary adjust it by bending the offset part A of the connecting link (Figs 3.31).
7 The correct accelerator pump stroke is:

18R	0.16 in (4.0 mm)
20R	0.177 in (4.5 mm)

8 After making any adjustment, make a check to see that the linkage operates smoothly and freely.

Idle adjusting screw (basic setting after overhaul)
9 Screw in the idle adjusting screw fully, but do not screw it in tightly, because this can cause damage to the screw tip and its seating.
10 From the fully in position, unscrew the screw by the following amount:

18R	3 turns from fully closed position
20R	1¾ turns from fully closed position

Second throttle touch angle (18R)
11 Open the first throttle until its throttle valve lever is about to start opening the second throttle, then measure the angle of the first throttle, which should be between 57° and 61°. The simplest way of checking is to cut a piece of cardboard with an angle of 61° and trim it to fit as necessary.
12 After getting the piece of cardboard trimmed to fit, measure its angle and then if the throttle requires adjustment, bend the tab of the primary throttle lever.

Fast idle adjustment – automatic choke (18R)
13 Loosen the three clamping screws of the automatic choke housing and turn the coil housing so that the choke valve is fully closed.
14 Open the first throttle valve slightly, to check that the hooked part of the throttle lever engages the fast idle cam, then measure the clearance between the first throttle and its bore, which should be 0.032 in (0.81 mm).
15 Adjust the clearance if necessary by turning the fast idle screw.
16 After completing the check, turn the choke housing back to its original position and tighten the clamp screws.

Unloader – automatic choke (18R)
17 Close the choke valve fully as described in paragraph 12.
18 Open the first throttle valve fully and check the angle of opening of the choke valve, which should be 50°.
19 Adjust the angle of opening by bending the tab A in Fig 3.35.
20 Reset the choke valve housing as in paragraph 15.

Choke breaker – automatic choke (18R)
21 Close the choke fully as described in paragraph 12.
22 Connect a piece of rubber tubing to the vacuum diaphragm and suck the end of the hose to move the diaphragm to its fullest extent.
23 When the diaphragm is at the limit of its travel, measure the angle of opening of the choke valve.
24 If the angle of opening differs from the specified value of 19°, level the choke tab A in Fig 3.36 to adjust the angle.
25 After completing the check, turn the choke housing back to its original position and tighten the clamp screws.

Manual choke (18R)
26 Close the choke fully by turning the choke lever.
27 Connect a piece of rubber tube to the diaphragm and suck as in the case of the automatic choke, again measuring the maximum angle of opening of the choke.
28 For the manual choke the maximum angle of opening is 16° and is adjusted as in paragraph 24.

Float level (18R)
29 With the air horn pasket removed, hold the air horn upside down and let the float hang down.
30 Measure the clearance between the tip of the float and the surface of the air horn (Fig 3.37), which should be between 0.39 and 0.43 in

Fig. 3.37 Float (valve closed) measurement (18R)

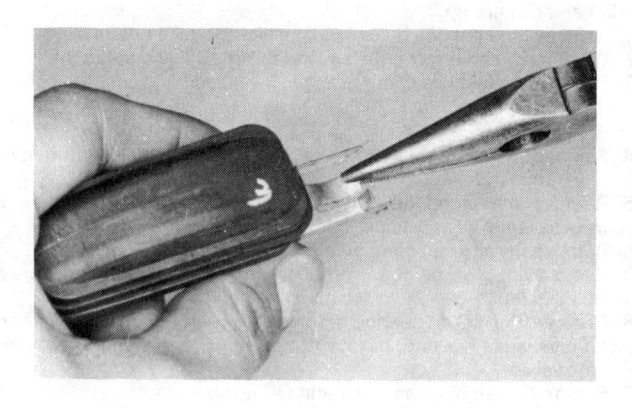

Fig. 3.38 Bending the float nib to adjust (18R)

Fig. 3.39 Adjusting the float nib-to-needle clearance (18R)

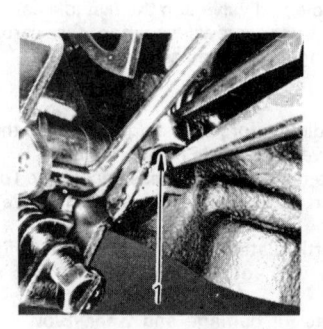

Fig. 3.40 Choke unloader (20R)

1 Fast idle lever

Fig. 3.41 Choke opener (20R)

1 Choke opener rod
2 Choke opener link

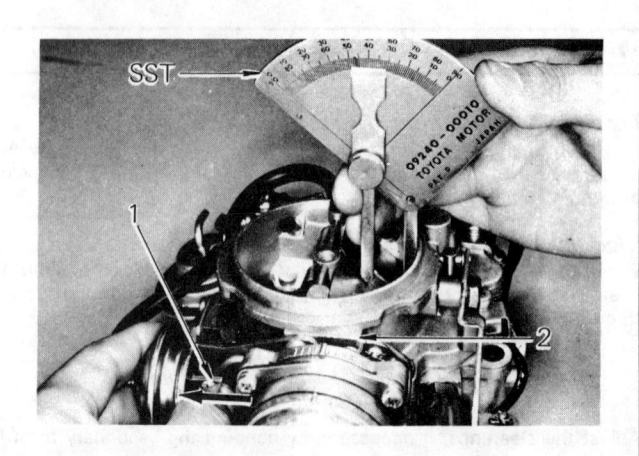

Fig. 3.42 Choke breaker adjustment (20R)

1 Choke breaker rod   2 Relief lever

Fig. 3.43 Throttle positioner (20R)

1 Adjusting screw   2 Throttle lever tab

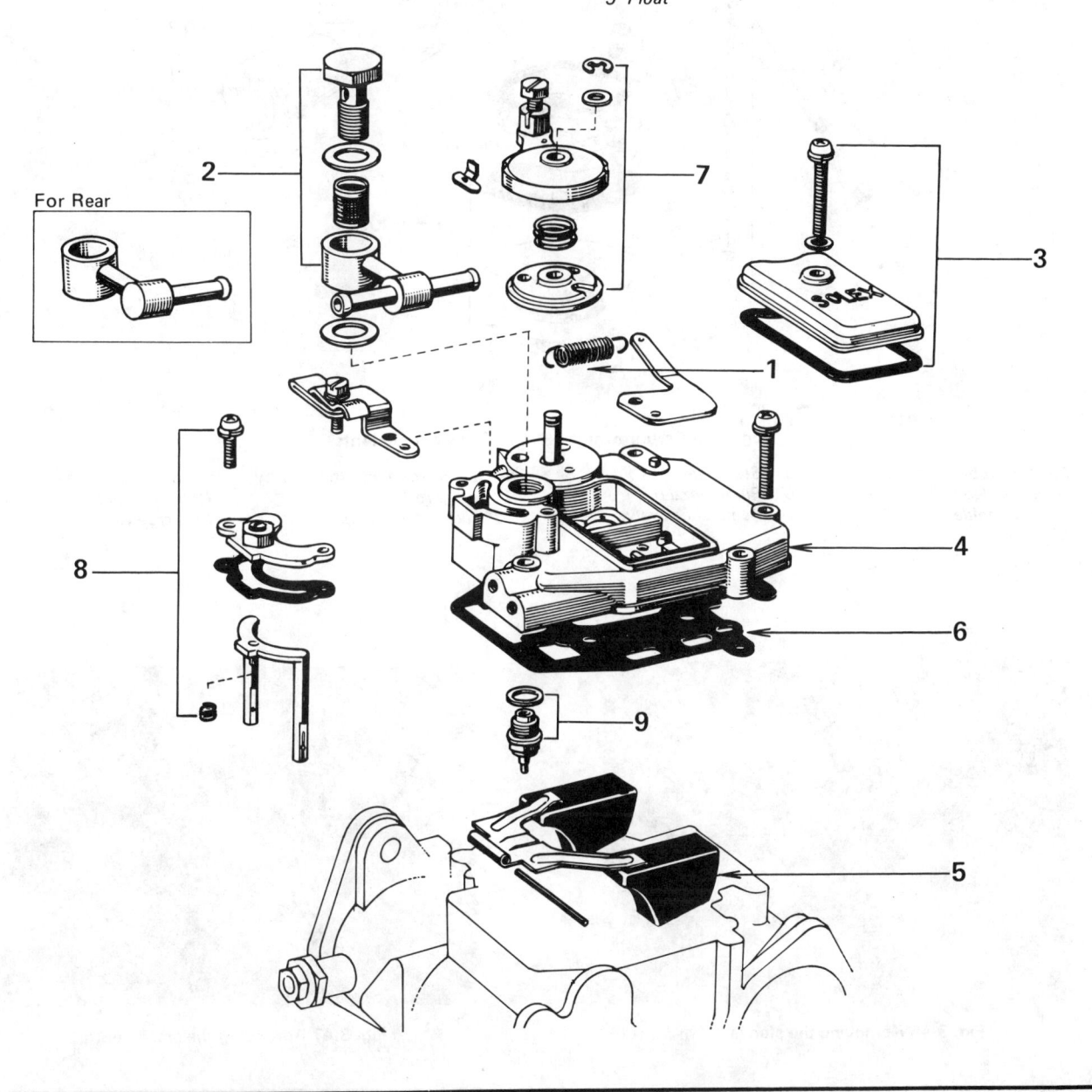

Fig. 3.44 Carburettor (18R–G) – Bowl cover assembly

1 Choke return spring
2 Fuel union
3 Jet chamber cover
4 Bowl cover
5 Float
6 Gasket
7 Choke plate
8 Float level adjuster
9 Needle valve assembly

For Rear

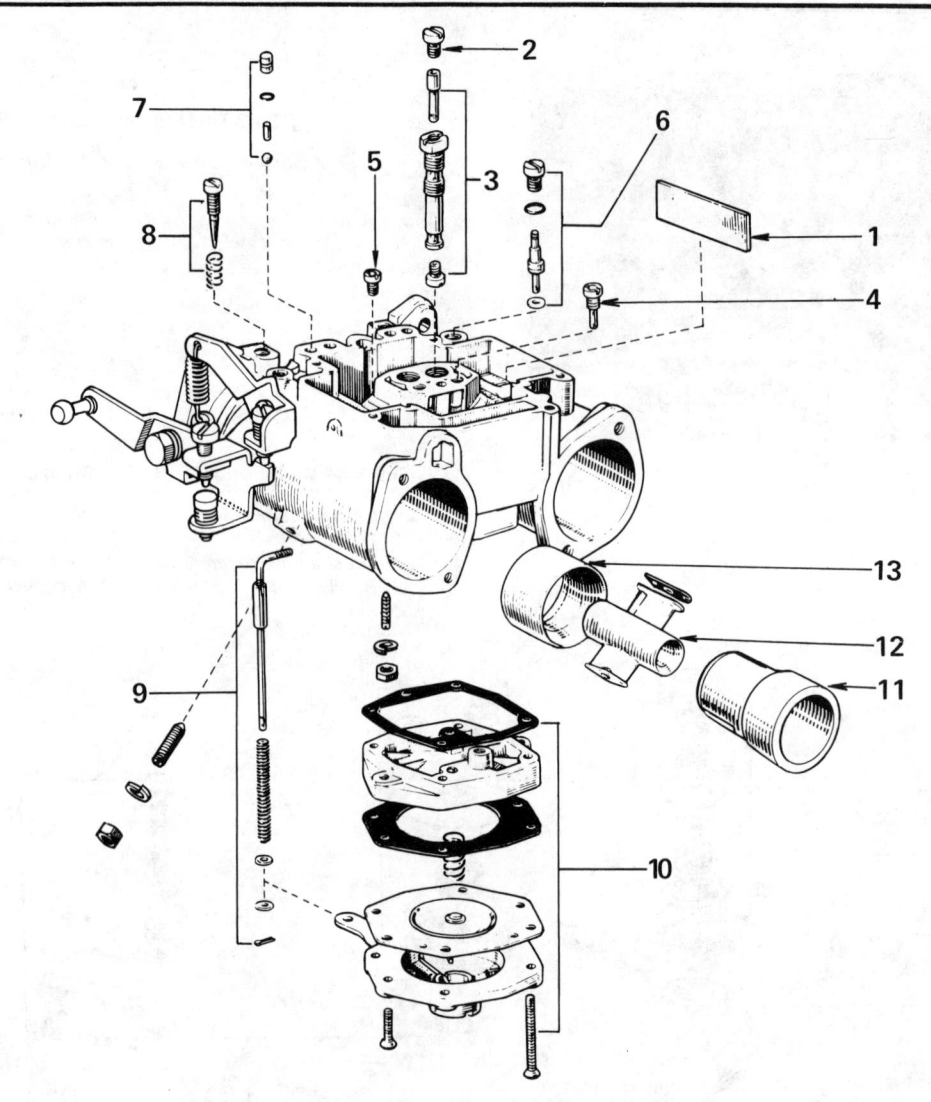

Fig. 3.45 Carburettor (18R–G) – Body components

1 Float chamber plate	5 Starter jet	8 Idle mixture adjusting screw	11 Sleeve
2 Main air bleed jet	6 Pump nozzle	9 Pump rod	12 Small venturi
3 Main jet holder	7 Pump check valve	10 Accelerator pump	13 Large venturi
4 Slow jet			

Fig. 3.46 Removing the sleeve

Fig. 3.47 Removing the small venturi

Chapter 3 Fuel, exhaust and emission control systems

(10.0 to 11.0 mm).
31 Adjust the clearance if necessary, by bending the centre tab of the float (Fig 3.38).
32 Lift the float and use a gauge to check the clearance between the needle valve plunger and the centre float tap. If the measurement is outside the range 0.039 to 0.047 in (1.0 to 1.2 mm), bend the tabs of the outer float lips (Fig 3.39).

Unloader (20R)
33 With the primary throttle valve fully open, check the angle of the choke valve, which should be inclined at 50° to the bore.
34 Adjust the angle if necessary by bending the fast idle lever (Fig 3.40).

Choke opener (20R)
35 Push in the choke opener rod as far as possible and check the angle of the choke valve (Fig 3.41).
36 If the angle is not 55°, adjust it to this angle by bending the choke opener link.

Choke breaker (20R)
37 Push in the choke breaker rod fully (Fig 3.42), to open the choke valve and measure the choke valve angle.
38 If the angle is not 40°, adjust it to this angle by bending the relief lever.

Throttle positioner (20R)
39 Screw in the throttle positioner adjustment screw until it just makes contact with the throttle lever tab.
40 Turn the screw until the clearance between the primary throttle valve and bore is:

Manual transmission models 0.024 in (0.6 mm)
Automatic transmission models 0.020 in (0.5 mm)

Float level (20R)
41 With the air horn gasket removed, hold the air horn upside down and let the float hang down.
42 Measure the clearance between the top of the float and the surface of the air horn, which should be 0.020 in (5 mm).
43 To alter the clearance, bend the centre tab of the float.
44 Lift the float and use a gauge to check that the clearance between the needle valve plunger and the centre float tab is 0.04 in (1 mm).
45 Adjust the clearance by bending the tabs on the float's two outer lips.

18 Carburettors (18R and 20R) – adjustment after installation

1 Having installed the carburettor on models which have a return fuel hose from the carburettor, check that the inlet and return fuel pipes have been connected correctly.
2 The adjustments which should be made are the same as those detailed in Section 13.

19 Carburettor (18R-G) – general description

Twin carburettors are used and they are of the dual barrel sidedraught Solex type. The cold start device is a manually operated choke plate mounted on top of the float chamber lid assembly.

20 Carburettor (18R-G) – removal and refitting

1 Remove the air cleaner, as described previously.
2 Disconnect the throttle linkage from the carburettor.
3 Disconnect the choke cable from the choke lever.
4 Disconnect the carburettor fuel pipe and vacuum connection.
5 Unscrew and remove the carburettor mounting nuts and washers and lift the assembly away from the inlet manifold.
6 Remove the gasket and clean the mating flanges.
7 If only one carburettor is being removed, the connecting links between the two throttles and chokes must first be disconnected.
8 Refitting is the reversal of removal, but always use a new flange gasket and ensure that both mating faces are clean and completely free of any pieces of old gasket.

21 Carburettor (18R-G) – dismantling, overhaul and reassembly

1 The carburettor should not be dismantled unnecessarily. If a carburettor has had prolonged use and is likely to be badly worn, it is better to purchase a new, or exchange unit which has been tested and calibrated, rather than try and obtain a lot of replacement parts.
2 Before starting to dismantle a carburettor it is vital to have a clean work place, screwdrivers and spanners which are the correct fit on the parts to be removed and several clean containers in which to put the parts of each sub-assembly.
3 Clean the outside of the carburettor before starting to dismantle it.

Bowl cover
4 Unhook and remove the choke return spring.
5 Remove the banjo bolt from the fuel union and remove the union and the washer from each side of it. Note that the union on the front carburettor has two fuel pipe connections and that of the rear carburettor only one.
6 Remove the screw and washer from the centre of the jet chamber cover and lift off the cover and its gasket.
7 Remove the screws and washers securing the float bowl cover and take off the cover and gasket.

Fig. 3.48 Removing the large venturi

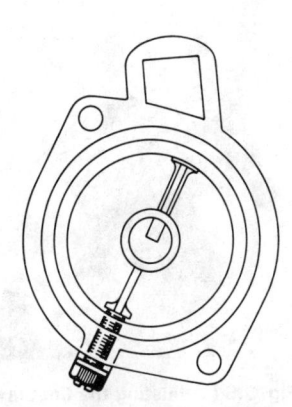

Fig. 3.49 Correct position of small venturi

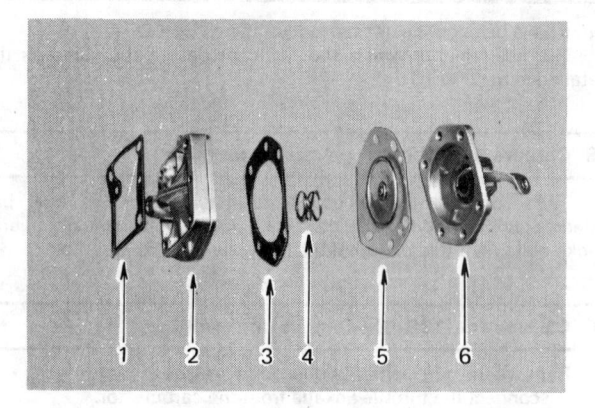

Fig. 3.50 Order of assembly of accelerator pump parts

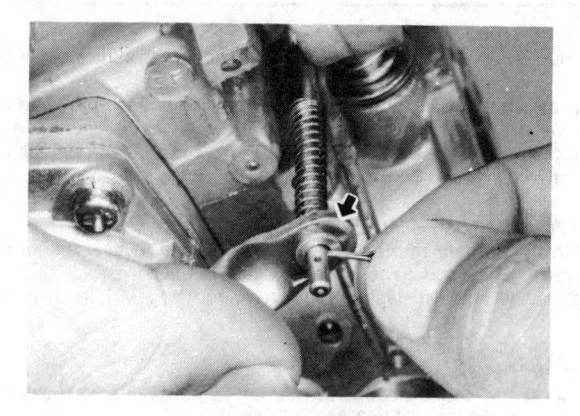

Fig. 3.51 Inserting the accelerator pump split-pin

Fig. 3.52 Idle mixture adjusting screws

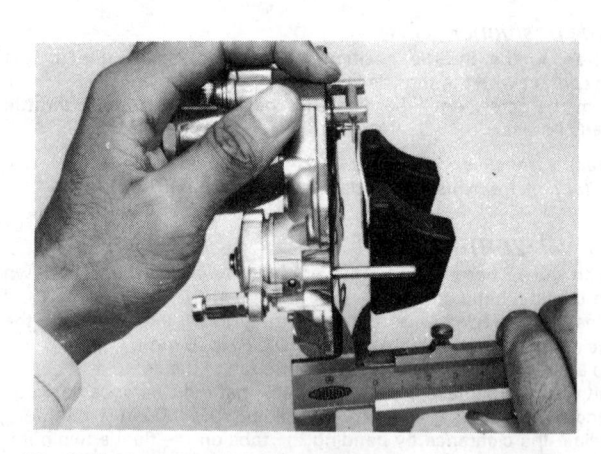

Fig. 3.53 Measuring the float needle valve closed position

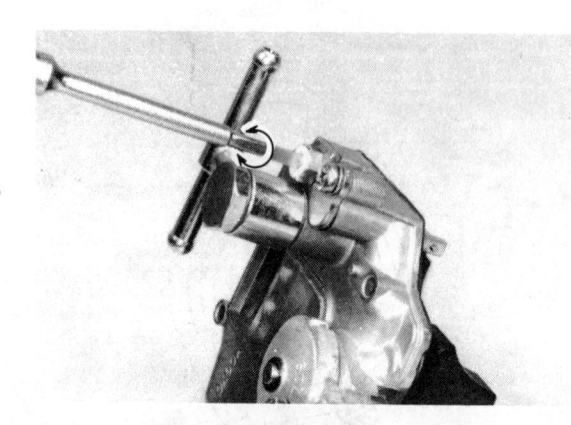

Fig. 3.54 Adjusting the float level

Fig. 3.55 Removing a jet holder assembly

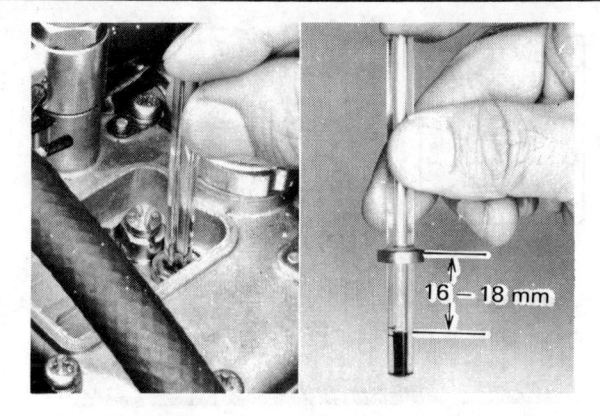

Chapter 3 Fuel, exhaust and emission control systems

Fig. 3.56 Measuring the float chamber fuel level

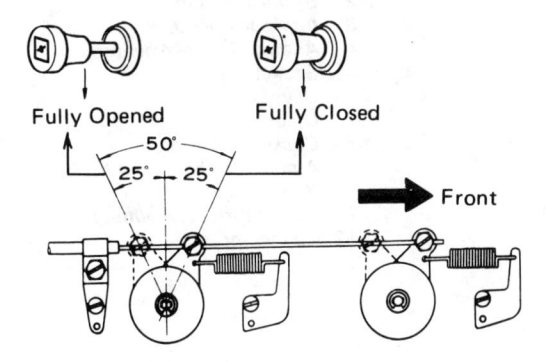

Fig. 3.57 Choke cable adjustment

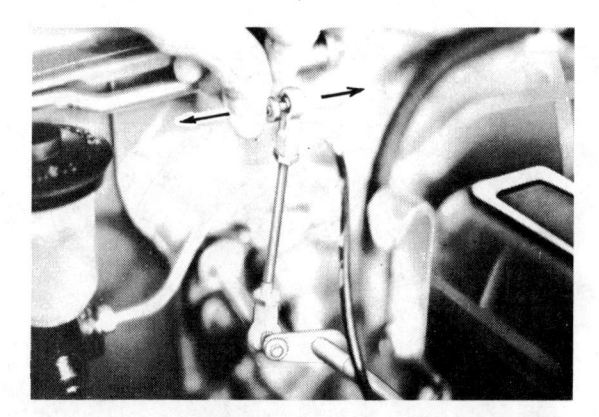

Fig. 3.58 Disconnecting the throttle rod

8 Pull out the float hinge pin and remove the float, then remove the float chamber cover gasket.
9 Remove the C-clip and washer from the cover of the choke assembly and remove the cover, spring and choke plate.
10 Remove the two screws from the cover plate on which the float adjusting screw is mounted. Do not disturb the centre screw and locknut. Lift off the cover plate and gasket then remove the float adjusting stay. Take care not to lose the very small spring which locates on the peg of the float adjusting stay.
11 Unscrew and remove the needle valve assembly and gasket.

Body

12 Withdraw the float chamber from its slot.
13 Unscrew and remove the main air bleed jet and jet holder.
14 Unscrew and remove the slow jet and starter jet.
15 Unscrew and remove the pump check valve, gasket and seating.
16 Unscrew and remove the idle mixture adjusting screw and its spring.
17 Withdraw the split-pin from the pump end of the accelerator pump rod. Disconnect the rod from the pump lever and recover the two washers and the spring.
18 Unscrew and remove the screws from the accelerator pump cover. Remove the cover, diaphragm, spring, gasket base plate and base plate gasket.
19 Remove the sleeve, small venturi and large venturi in that order (Fig 3.46, 3.47 and 3.48).
20 Wash all the parts in clean petrol. Blow out all the jets and air passages to clear them, but do not attempt to use wire to clean jets and orifices, because this can distort them and have a bad effect on fuel consumption.
21 Check the bowl cover for cracks or damaged threads and see that the starter pipe is clear and undamaged.
22 Check that the petrol filter on the banjo bolt is not clogged, damaged or corroded.
23 Examine the choke plate for damage and excessive wear of its sliding surfaces.
24 Examine the needle valve and ensure that there is no ridge worn on its taper. Check that the valve seating is undamaged.
25 Check the float for damage, the float lever hinge pin and mating surfaces for wear.
26 Check the body for damage, wear and adhering carbon.
27 Check the throttle shafts and bearings for wear, ensuring that the throttles close fully.
28 Ensure that all jets are clear and that their threads, screwdriver slots and seatings are undamaged.
29 Examine the idle mixture adjusting screw and ensure that its taper is smooth and its tip undamaged.
30 Check that the accelerator pump body is in good condition, the pump nozzle is clear and undamaged and that the diaphragm is not perished, or damaged.
31 Reassembly is a reverse of dismantling, using new gaskets and noting the following points.
32 When refitting the small venturi (Fig 3.49), use the longest of the three screws securing the venturis and sleeve.
33 Assemble the accelerator pump components in the order shown in Fig 3.50 and fit the split-pin into the third hole from the tip of the rod (Fig 3.51).
34 Screw each idle mixture screw (Fig. 3.52) in fully, but not tightly and then unscrew it 1½ turns as a basic setting.
35 With the bowl cover held vertically and the float hanging down (Fig. 3.53), measure the distance from the tip of the float to the face of the bowl cover. This should be 0.63 in (16 mm) and can be adjusted by unlocking the locknut and turning the adjusting screw (Fig. 3.54). After making the adjustment, keep the screw in position with a screwdriver while re-tightening the locknut.

22 Carburettor (18R-G) – adjustments after installation

Float level

1 If it is suspect that the level of petrol in the float chamber is incorrect, check the level in the following way.
2 Start the engine and with it running at about 1000 rpm take out one of the main jet holders as an assembly (Fig. 3.55), after removing the jet chamber cover.

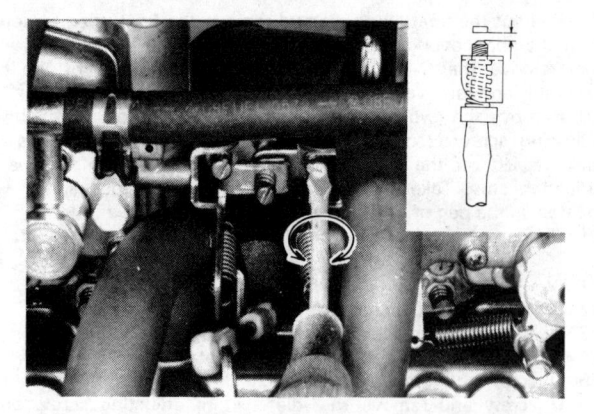

Fig. 3.59 Rear idle speed adjusting screw clear of throttle lever

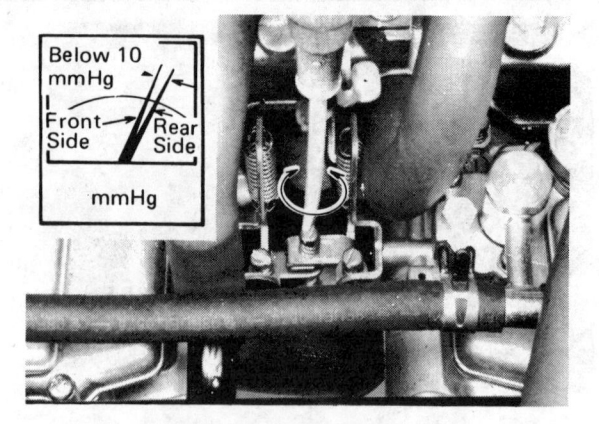

Fig. 3.60 Adjusting the synchronising screw

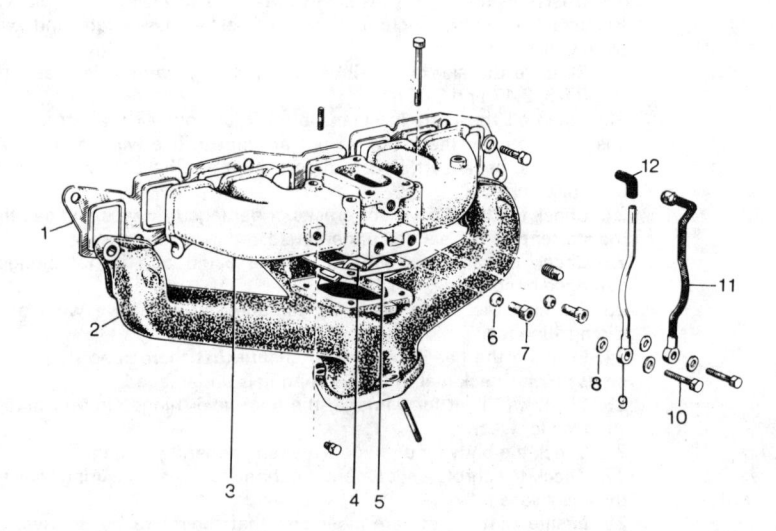

Fig. 3.61 Manifolds (18R/18R-C)

1 Gasket
2 Exhaust manifold
3 Intake manifold
4 Automatic choke heater pipe
5 Gasket
6 Olive
7 Union
8 Gasket
9 Automatic choke inlet heater pipe
10 Bolt
11 Automatic choke outlet heater pipe
12 Connector

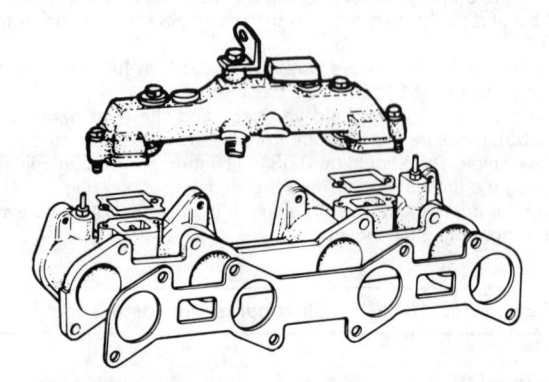

Fig. 3.62 Inlet manifold and balance pipe (18R-G)

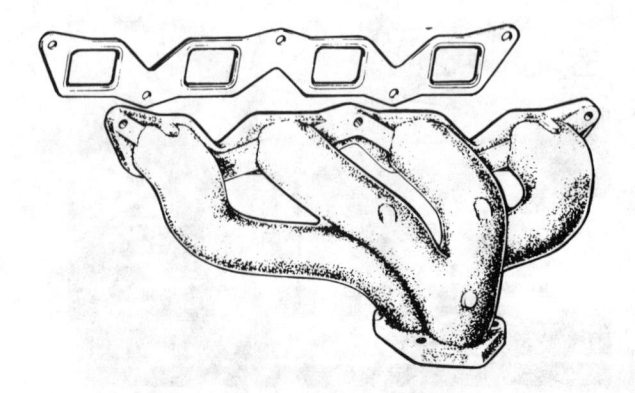

Fig. 3.63 Exhaust manifold (18R-G)

99

Fig. 3.64 Inlet manifold (20R)

1 Hose union
2 Manifold
3 Gasket
4 Gasket
5 Cover plate

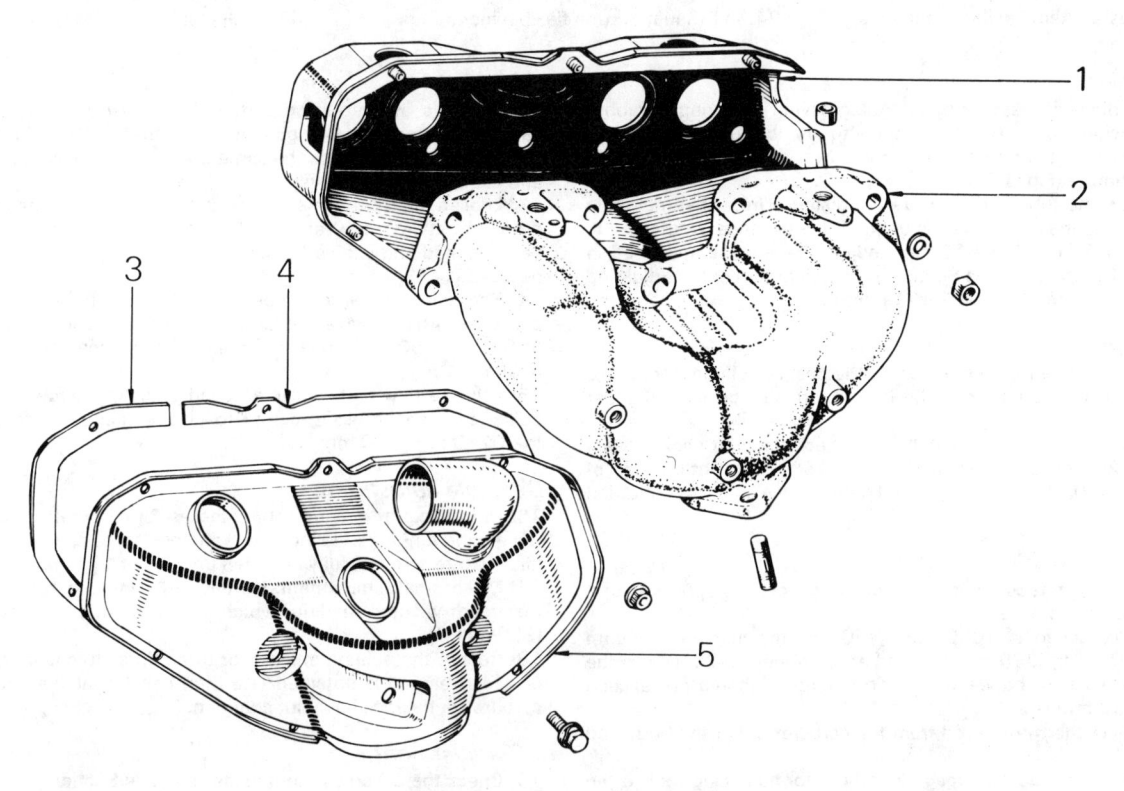

Fig. 3.65 Exhaust manifold (20R)

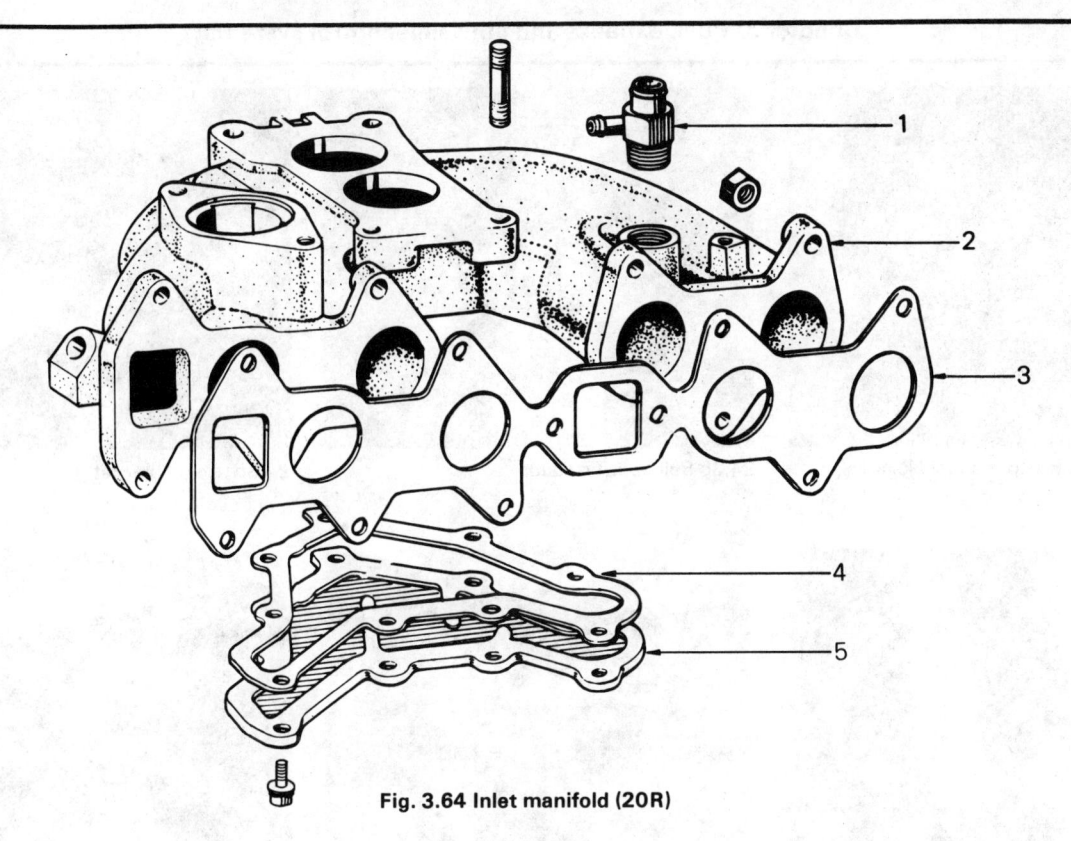

1 Inner heat insulator
2 Manifold
3 Gasket
4 Gasket
5 Outer heat insulator

Chapter 3 Fuel, exhaust and emission control systems

24.3a Exhaust pipe-to-manifold joint

24.3b Bell housing clamp ...

24.3c ... and bracket

24.5a Exhaust system flexible mountings

24.5b Exhaust system flexible mountings

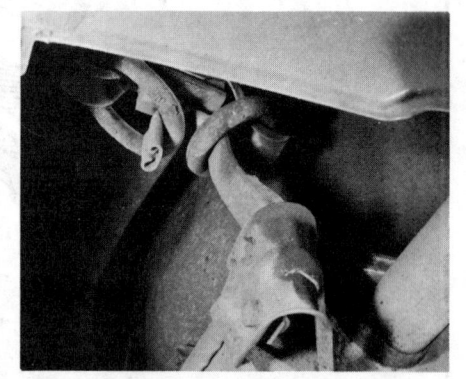

24.5c Exhaust system flexible mountings

3 Using a piece of glass, or metal tubing having an O-ring or rubber washer attached to it, insert the tube to the bottom of the float chamber and push the ring down until it is in contact with the bottom of the jet chamber (Fig. 3.56).

4 Withdraw the tube and measure the distance from the underside of the washer to the surface of the petrol. If the distance is outside the limits 0.63 to 0.71 in (16 - 18 mm), adjust the level as described in paragraph 34 of the previous Section. A complete turn of the adjusting screw will change the level 0.07 in (1.8 mm).

Choke control

5 Ensure that when the choke control is pushed fully in, the choke valves are fully closed and that the choke return springs are still under tension.

6 Pull the choke control out fully and check that the choke levers move from 25° on one side of centre to 25° on the other side of centre (Fig. 3.57). Adjust the inner and outer cables until this condition is achieved.

Carburettor balancing

7 This operation requires a vacuum gauge or a proprietary carburettor balancing device.

8 At an idle speed of 1000 rpm ± 50 rpm the manifold vacuum should be 13 in Hg (330 mm Hg) and at speeds up to 2000 rpm the difference in vacuum between the two manifolds should not exceed 0.39 in Hg (10 mm Hg).

9 Disconnect the throttle rod from the carburettors at the body end (Fig. 3.58).

10 Connect the vacuum gauges, or carburettor balancing tool to the intake manifolds of No 1 and No 4 cylinders and with all accessories switched off, run the engine until the coolant temperature has reached 180°F (80°C).

11 Loosen the rear idle speed adjusting screw until it is clear of the throttle lever (Fig. 3.59) and turn the front idle speed screw until the engine speed is 1800 rpm. Open the throttle, then release it again and check that the engine returns to a speed of 1800 rpm if not, re-adjust and re-check.

12 Turn the synchronizing screw (Fig. 3.60) until the difference in vacuum between the two gauges is less than 0.39 in Hg (10 mm Hg). Check that the reading is the same after engine speed has been raised and allowed to drop again.

13 Loosen the front idle speed adjusting screw until the engine speed drops to between 950 and 1050 rpm. Re-check after the throttle has been opened and allowed to close again, making further adjustment if necessary.

14 Screw in the rear idle speed screw enough to increase the engine speed and then unscrew it until the engine speed is again between 950 and 1050 rpm. Check the speed after opening the throttle as before.

15 After setting both the idle speed adjusting screws as described, adjust the synchronizing screw to give a vacuum difference of less than 0.39 in Hg (10 mm Hg).

Best idle adjustment

16 With vacuum gauges connected as for carburettor balancing and the engine running at idling speed and normal temperature screw each idle mixture screw in fully and then unscrew it $1\frac{1}{2}$ turns.

17 Make small adjustments to each idle mixture screw to give the maximum vacuum reading, which should be in excess of 13.0 in Hg (330 mm Hg).

18 Repeat the adjustment two or three times to ensure that the best possible reading is obtained and then check that the idle speed and carburettor balance have not changed.

CO adjustment

19 Check the CO adjustment as described in Section 13 Paragraphs 7 and 8.

23 Inlet and exhaust manifolds

1 On the 18R/18R-C engines, the inlet and exhaust manifolds are bolted together on the left-hand side of the cylinder head. They are shown in Fig. 3.61.

2 On the 18R-G and 20R engines, the inlet manifold is on the right-

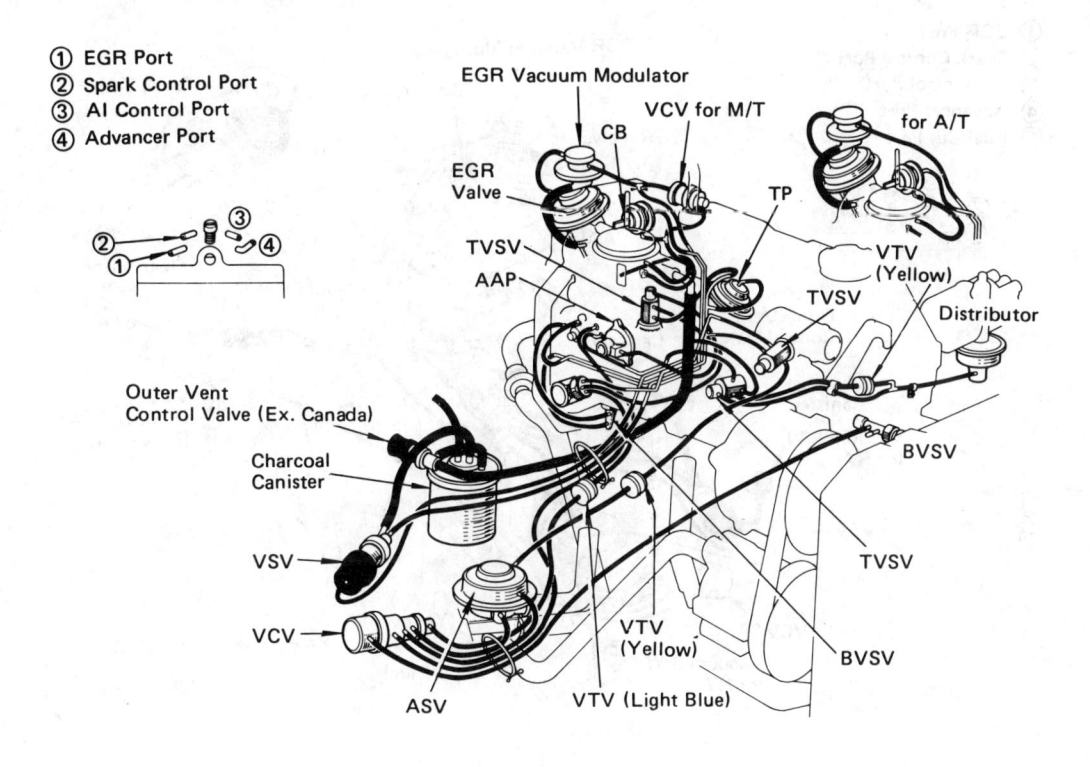

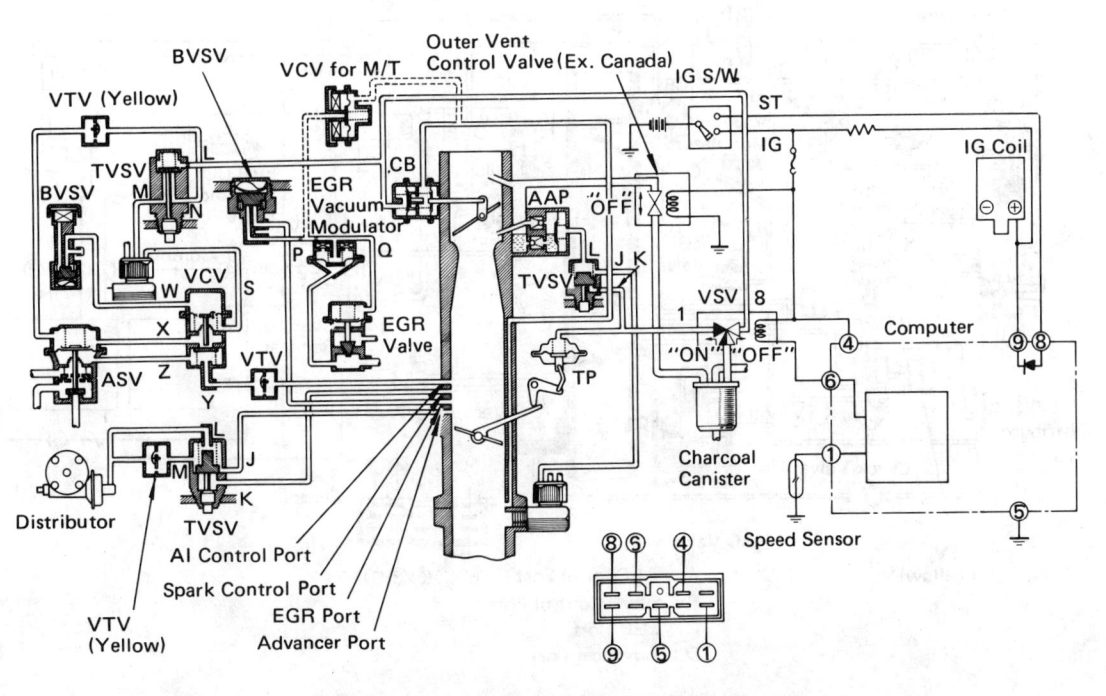

Fig. 3.65A Emission control system – Federal and Canada

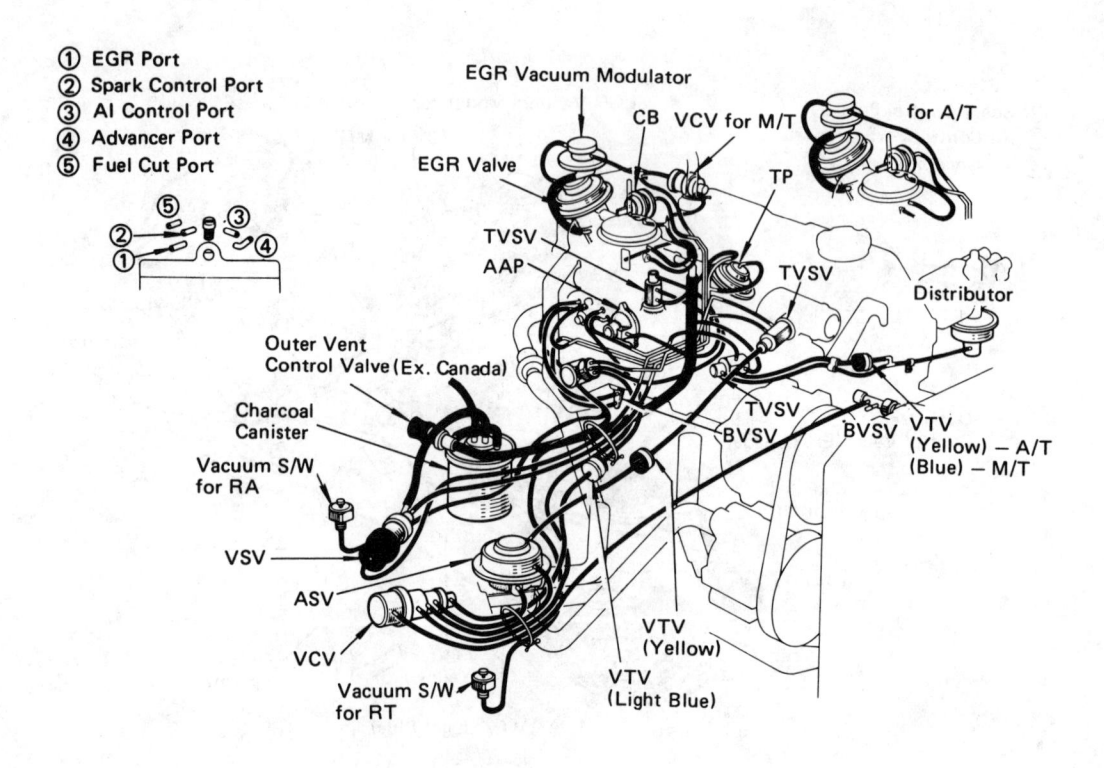

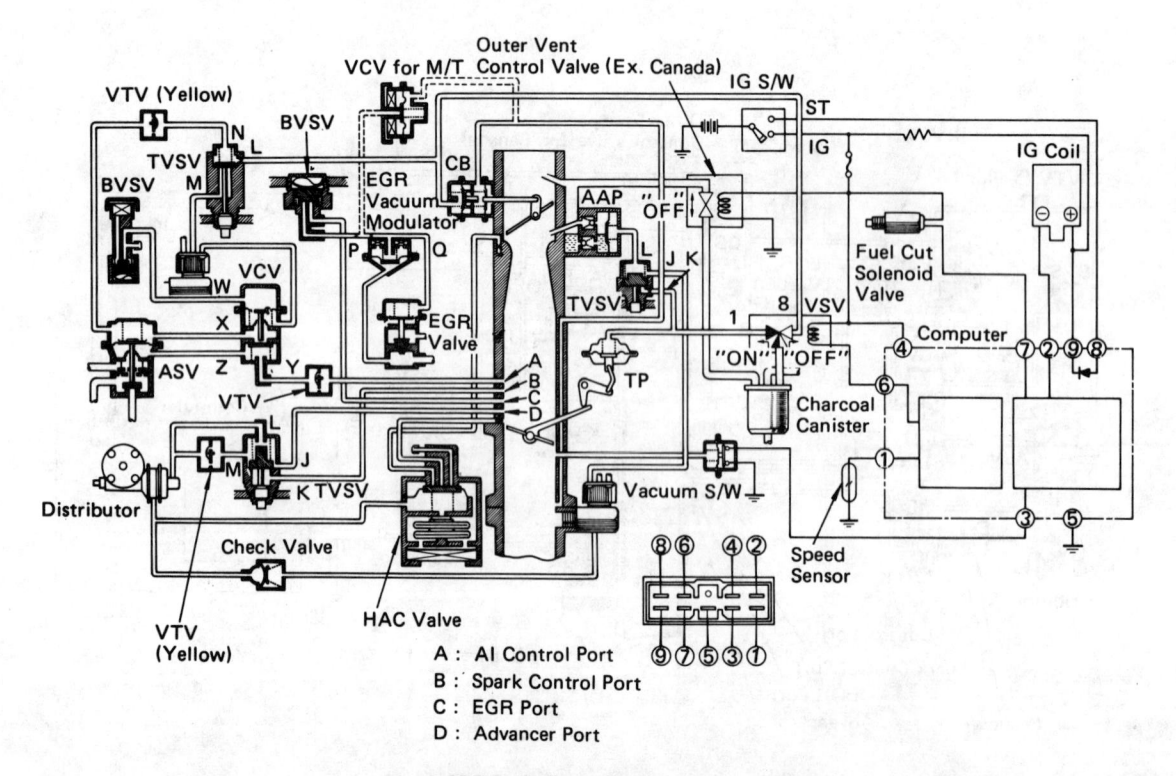

Fig. 3.65B Emission control system – California

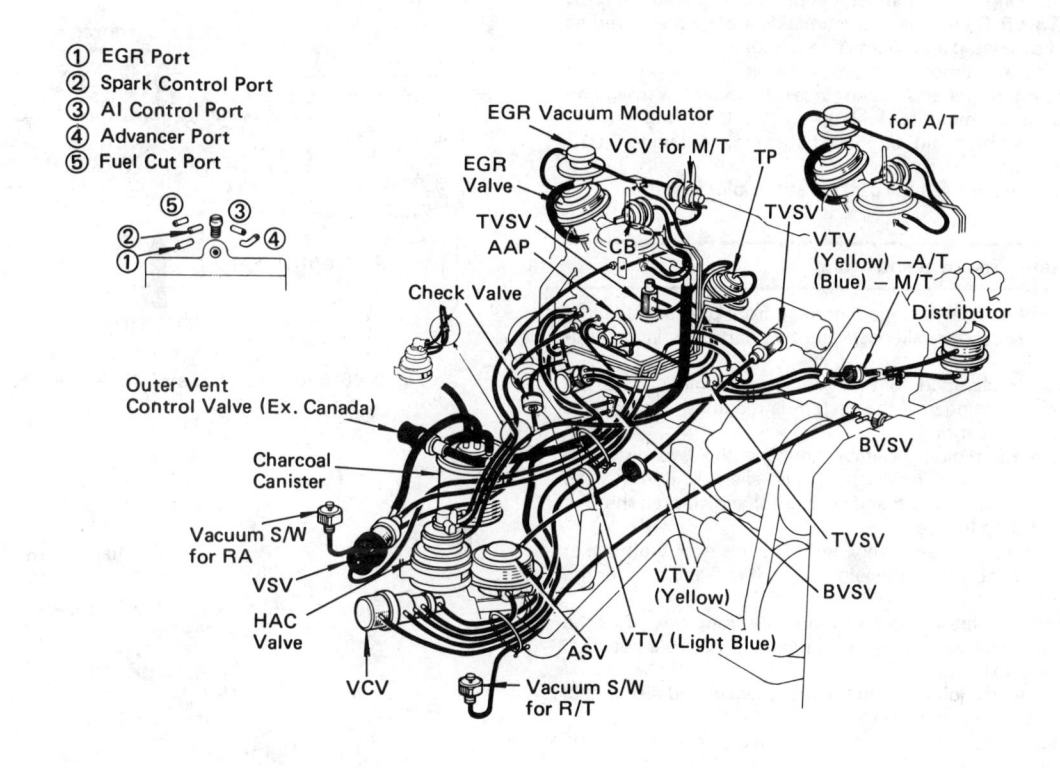

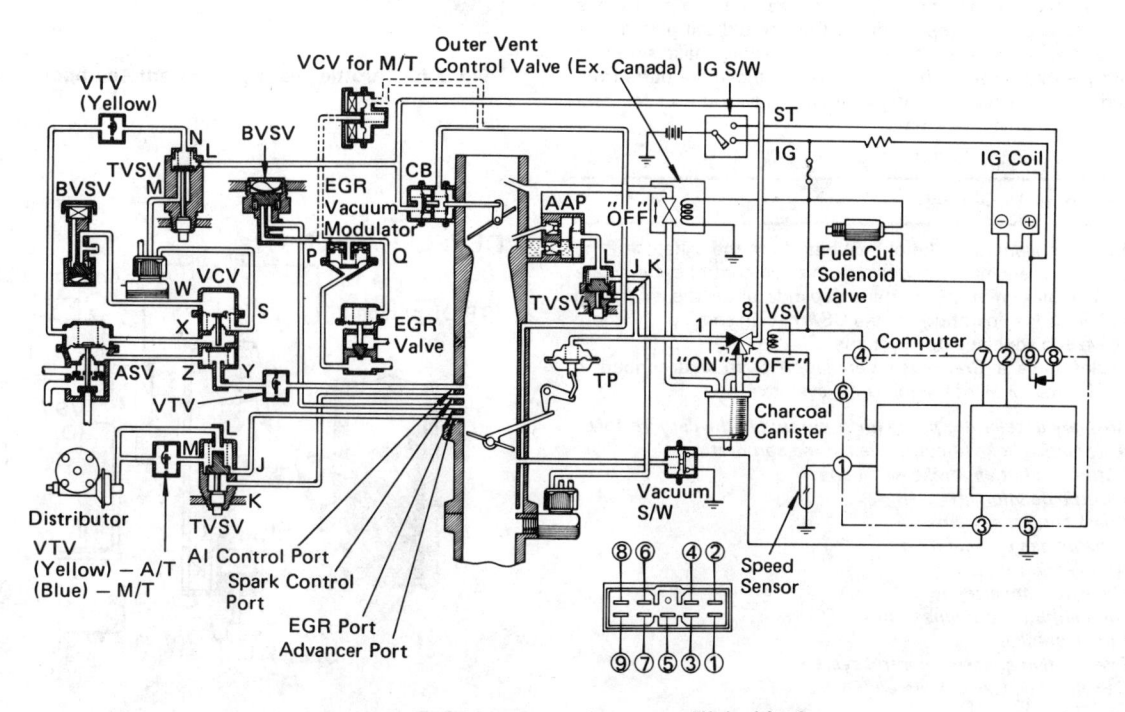

Fig. 3.65C Emission control system – High altitude

hand side of the head, and the exhaust manifold is on the left-hand side. They are shown in Figs. 3.62 and 3.63.

3 There are no special points to note when removing manifolds except that the cooling system must be partially drained on 20R engines. On the 18R/R-C engines, the manifolds are best removed as an assembly, then separated afterwards if required.

4 When manifolds are removed, it is worthwhile checking them for cracking and warping on the attachment faces. An overall warpage on the cylinder head attachment face of up to 0.008 in (0.2 mm) is acceptable for the inlet manifold; on the exhaust manifold up to 0.012 in (0.3 mm) is acceptable.

5 When refitting, always use new gaskets at the joint faces.

24 Exhaust system – removal and fitting

1 The exhaust system is in two main parts, a front pipe and silencer and a tail pipe with secondary silencer. The two parts are flange connected.

2 The front pipe is clamped to the exhaust manifold and to the bellhousing and the remainder of the system is mounted flexibly by a system of hooks and rubber rings.

3 To remove the front pipe assembly, remove the two nuts and washers from the flange joint behind the main silencer. Remove the three nuts and washers from the manifold (photo) and slacken the bolt on the bellhousing clamp (photos).

4 Knock the bellhousing clamp back until it is clear of the bracket then separate the assembly from the joints at each end and recover the gaskets.

5 To remove the tail pipe assembly, chock the front wheels of the car and jack the body to give as great a clearance as possible between the body and the rear axle tube.

6 Disconnect the flange joint behind the first silencer and disconnect the three flexible mountings (photos).

7 When the assembly is free, manoeuvre it over the rear axle and withdraw it from beneath the car.

8 Refitting is the reversal of removal, but the following points should be noted.

9 Ensure that all joint faces are clean and use new gaskets.

10 Fit the system with all nuts finger tight until the system has been aligned and is well clear of any obstruction then tighten the joints working rearwards from the manifold joint.

11 When renewing the one piece tail pipe assembly a two piece assembly can be used as a replacement. Cut the old tail pipe at the punch mark on the pipe. If the mark is no longer visible, make sure that you have left plenty of overlap before cutting. Remove the burrs from the pipe end, coat the pipe end with exhaust sealer and fit the rear silencer using an exhaust pipe clamp.

25 Emission control – general

1 To reduce atmospheric pollution resulting from the exhaust of the internal combustion engine, a number of emission control systems are in use. The complexity of the system depends upon the operating territory, with vehicles operating the the USA having some of the more comprehensive and sophisticated systems.

2 All vehicles have a *Crankcase Ventilation System* described in Chapter 1 and some, or all of the following systems:

(a) *Auxiliary accelerator pump (incorporated in the carburettor)*
(b) *Automatic choke (incorporated in the carburettor)*
(c) *Automatic hot air intake air cleaner.*
(d) *Throttle positioner*
(e) *Spark control system*
(f) *Exhaust gas recirculation*
(g) *Air injection system*
(h) *Catalytic converter*
(i) *High altitude compensation*
(j) *Choke breaker*
(k) *Evaporative emission control system*
(l) *Deceleration fuel cut system*

3 The principles and purpose of the various systems are given in the following Sections. Checking and test procedures are also given, but it is emphasised that the satisfactory completion of these will not necessarily ensure that the vehicle complies with the emission control

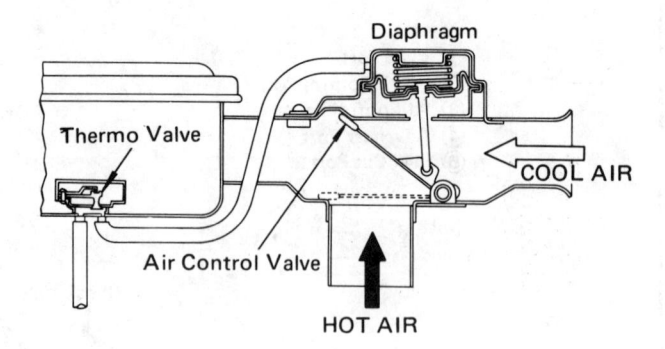

Fig. 3.66 Automatic hot air intake (HAI) system

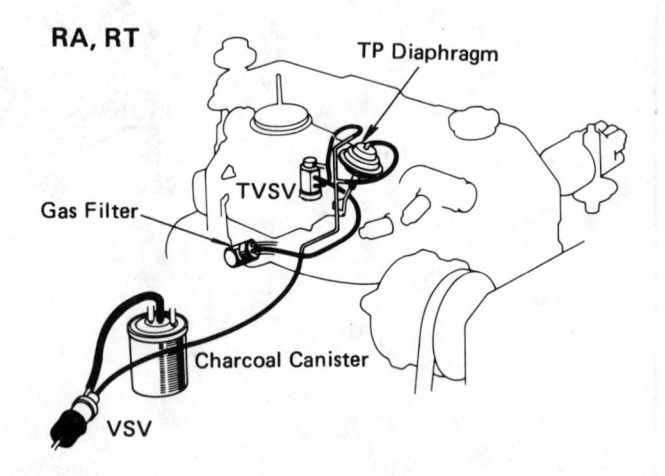

Fig. 3.67 Throttle positioner system – component layout

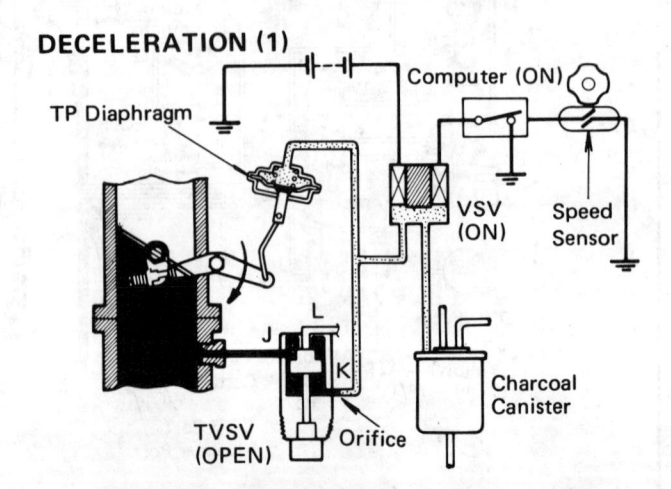

Fig. 3.68 Throttle positioner operation – decelerating from above 7 mph

Chapter 3 Fuel, exhaust and emission control systems

DECELERATION (2)

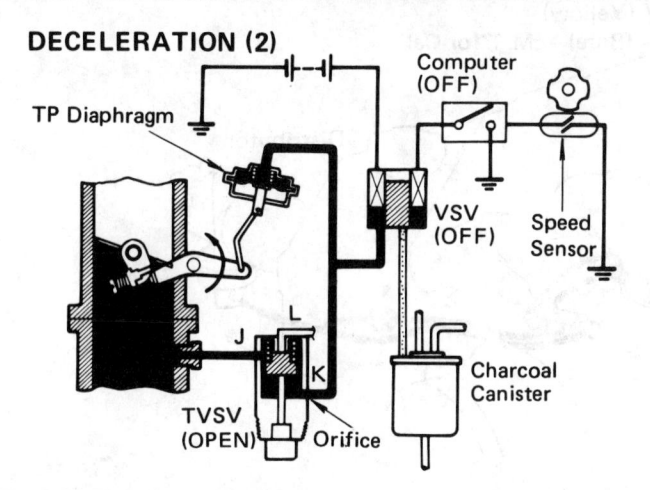

Fig. 3.69 Throttle positioner operation – decelerating from below 7 mph

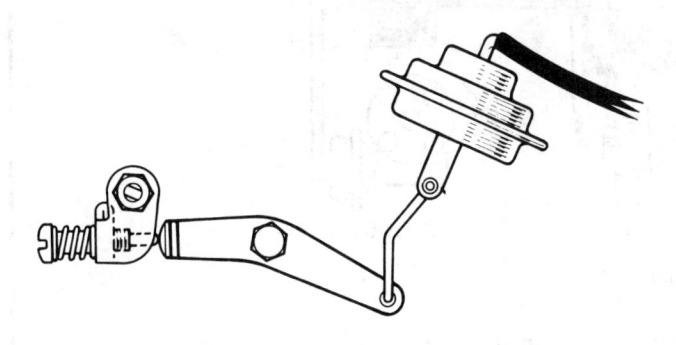

Fig. 3.70 Throttle positioner – set position

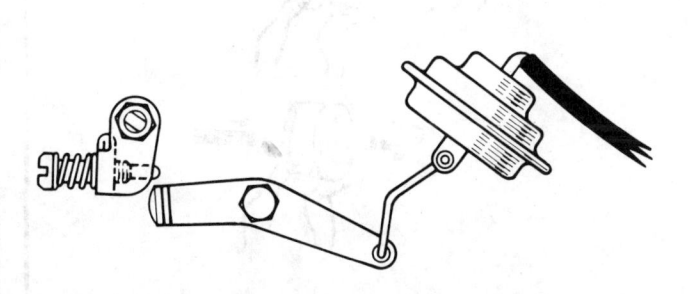

Fig. 3.71 Throttle positioner – released position

legislation of a particular territory and the vehicle should be submitted to an approved testing station.

26 Automatic hot air intake (HAI) system

1 This system leads a supply of hot air to the carburettor in cold weather, to improve driveability and to prevent the carburettor from icing up in extremely cold weather.
2 The system is incorporated in the air cleaner and consists of a duct to take hot air from the exhaust manifold through a diaphragm operated flap valve to the engine air intake.
3 The diaphragm is spring loaded to close the hot air intake and is connected via a thermo-valve to the suction of the inlet manifold.
4 Below a temperature of 100°F (38°C) the thermo-valve is closed, inlet manifold suction is applied to the diaphragm which is pulled upwards against its return spring pressure and opens the hot air intake. As the flap valve opens the hot air intake, it closes the normal cold air intake.
5 Above 106°F (41°C) the thermo-valve opens and atmospheric pressure is applied to the diaphragm. The diaphragm spring pushes downwards moving the flap valve to close the hot air intake and open the cold air intake.
6 To check the system, remove the cover of the air cleaner and blow air onto the thermo-valve to close it. Check that the air control valve closes the cold air intake.
7 Fit the air cleaner, start the engine and warm it up to normal operating temperature. Check that the air control valve opens the cold air intake.
8 As a maintenance procedure check the hose connections and the condition of the hoses.

27 Throttle positioner (TP) system

1 When the car is decelerating, the throttle positioner opens the throttle slightly more than when idling. This causes the air/fuel mixture to burn more completely and reduces the emission of hydrocarbons and carbon monoxide.
2 A spring loaded diaphragm operates a lever controlling the throttle position. The diaphragm is controlled by a vacuum servo valve (VSV) and a thermostatic vacuum servo valve (TVSV).
3 Start the engine from cold and at idling speed check that the throttle positioner is set (Fig. 3.70).
4 Warm the engine up and check that it is idling at the correct speed. Check that the throttle positioner is released (Fig. 3.71) when the engine is warm and idling.
5 Disconnect the vacuum hose from the TP diaphragm. Race the engine and then release the accelerator pedal. Check that the throttle positioner has set.
6 Repeat this test and note the speed at which the TP resets. If this speed differs from 1050 rpm, adjust the TP adjusting screw until the correct speed is achieved, then reconnect the vacuum hose.
7 As a maintenance procedure, check the hose connections and the condition of the hoses.

28 Spark control (SC) system

1 The SC system is designed to reduce the emission of hydrocarbons and the oxides of nitrogen by reducing the maximum combustion temperature. This is achieved by delaying the vacuum advance under certain operating conditions.
2 The system, which consists of a thermostatic vacuum switching valve (TVSV), and a vacuum transmitting valve (VTV) is connected between two tappings of the inlet manifold and the automatic advance diaphragm of the distributor.
3 At a coolant temperature of less than 104°F (40°C) the TVSV allows vacuum to act directly onto the distributor diaphragm and the vacuum ignition timing is normal.
4 At a coolant temperature above 131°F (55°C) when the advancer port vacuum acts on the distributor, vacuum ignition timing is delayed by the VTV and by air bled from the SC port. When the throttle setting is such that both advancer port and SC port vacuum act on the distributor, vacuum ignition timing is delayed by the VTV.
5 Checking the system consists of first checking that the vacuum is correct, then doing functional checks of the TVSV, VTV and the distributor.

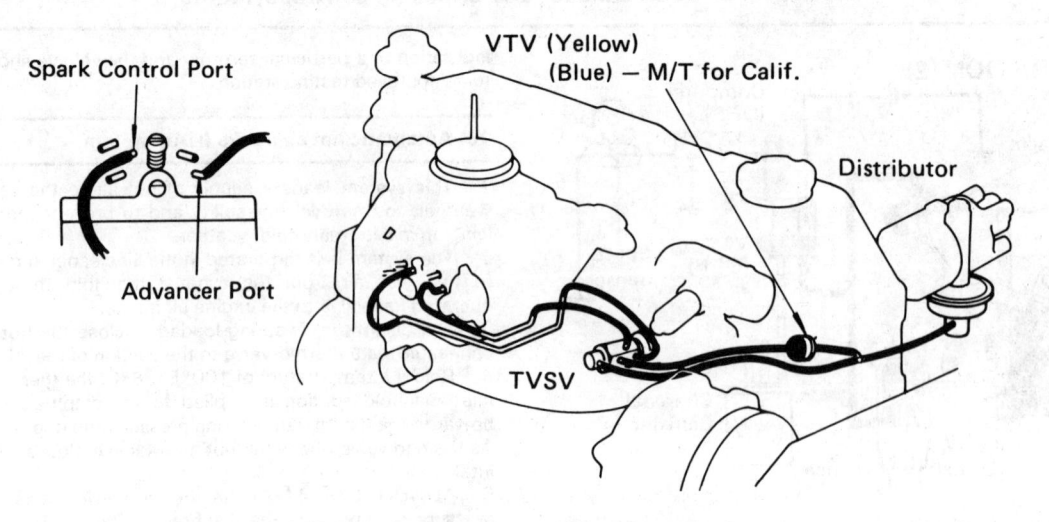

Fig. 3.72 Spark control system – component layout

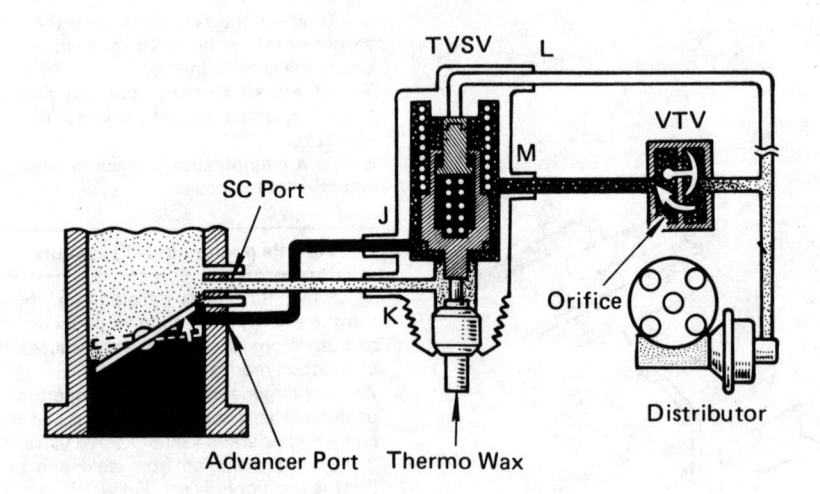

Fig. 3.73 Spark control system – schematic

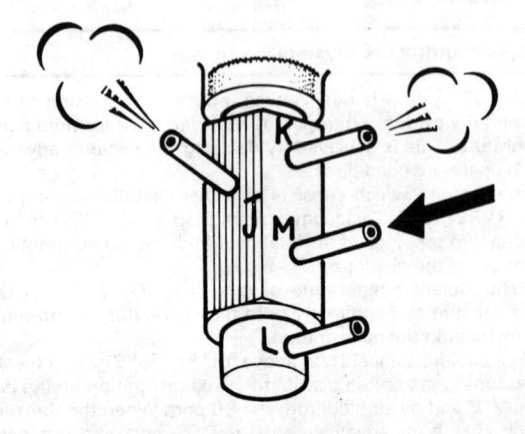

Fig. 3.74 TVSV port identification

Fig. 3.75 VTV port identification

Chapter 3 Fuel, exhaust and emission control systems

6 With the engine temperature at less than 104°F (40°C) remove the vacuum hose to the vacuum diaphragm of the distributor and fit a vacuum gauge to the hose. Start the engine, open and close the throttle valve and check that the level of vacuum changes as quickly as the engine speed changes.

7 Warm the engine up, leaving the vacuum gauge connected and then pinch the hose connected to the spark control port. Maintain an engine speed of 2000 rpm and check that a vacuum of 4 in Hg (100 mm Hg) is indicated within 2 to 4 seconds.

8 While still maintaining the engine speed at 2000 rpm, release the pinched hose and check that the vacuum falls to less than 4 in Hg. If this is the case, the system is functioning correctly and no further action is necessary except to remove the vacuum gauge and reconnect the hose to the distributor.

9 If the engine is not satisfactory, check the TVSV with the engine at a temperature of less than 104°F (40°C). Blow into port L of the valve and check that air emerges from port J and port M, but not from port K.

10 Warm the engine up, so that its temperature is above 131°F (55°C). Blow into port M and check that air emerges from ports J and K, but not from port L.

11 If these checks are not satisfactory, fit a new TVSV.

12 Check the VTV by first blowing into port B. Air should emerge from port A with little restriction. Blow into port A and a considerable resistance to flow should be felt, with only a small flow of air emerging from port B.

13 If these checks are not satisfactory, fit a new VTV.

14 If no other fault is apparent, check the operation of the vacuum advance mechanism of the distributor. Remove the distributor cap, rotor and dust cover. Disconnect the vacuum pipe from the distributor diaphragm and connect a piece of rubber tubing to the diaphragm. Suck the other end of the tube to check that the diaphragm does not allow air to leak through it and also check that when the tube is sucked, the contact breaker baseplate moves.

15 If the diaphragm is faulty, fit a new one, referring to Chapter 4 if necessary. If satisfactory, reinstall the dust cover, rotor and distributor cap and re-connect the vacuum pipe.

29 Exhaust gas recirculation (EGR) system

1 When the engine is hot and under certain throttle conditions, part of the exhaust gas is directed back into the inlet manifold to lower the temperature of combustion and so reduce the emission of oxides of nitrogen.

2 The main component of the system is the EGR valve, which is a diaphragm valve forming part of a control system consisting of a bi-metallic vacuum switching valve (BVSV), a vacuum control valve (VCV) and a vacuum modulator.

3 The vacuum modulator has a filter which should be removed and

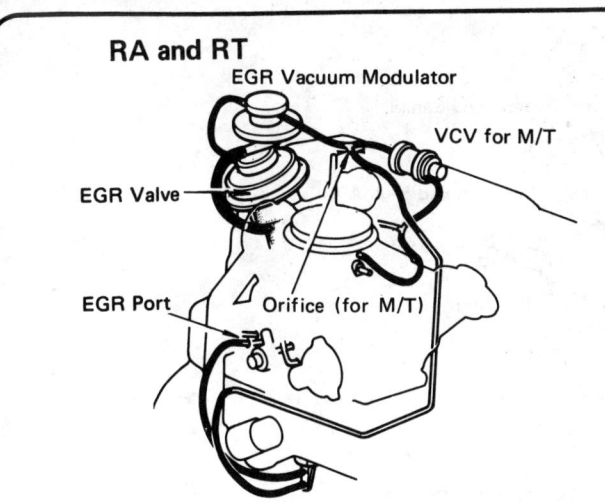

Fig. 3.76 Exhaust gas recirculation system – component layout

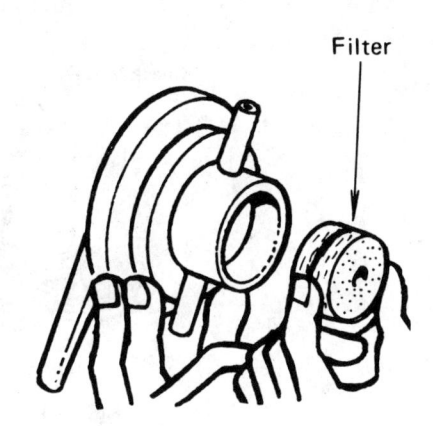

Fig. 3.77 EGR vacuum modulator filter

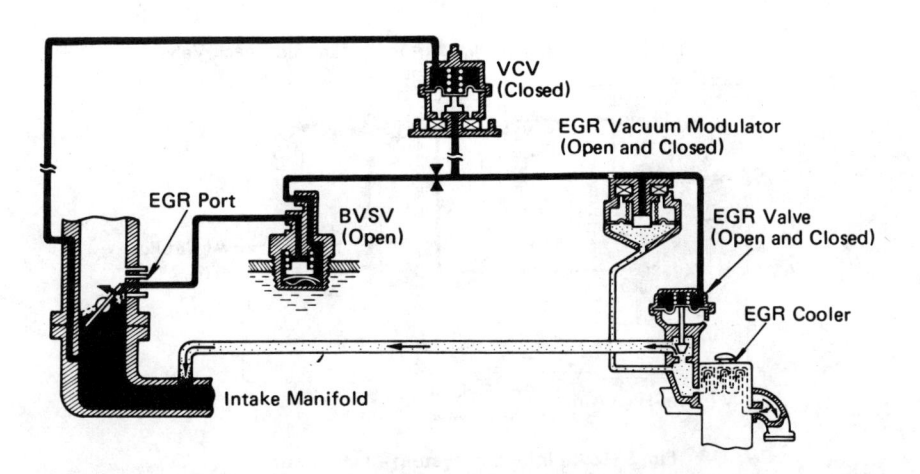

Fig. 3.78 Exhaust gas recirculation system – schematic

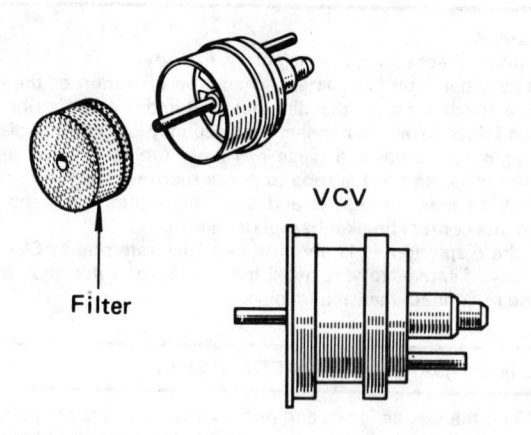

Fig. 3.79 Vacuum control valve filter

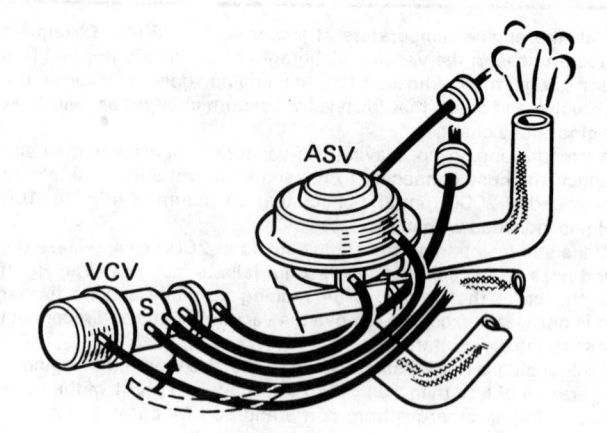

Fig. 3.80 Testing the ASV

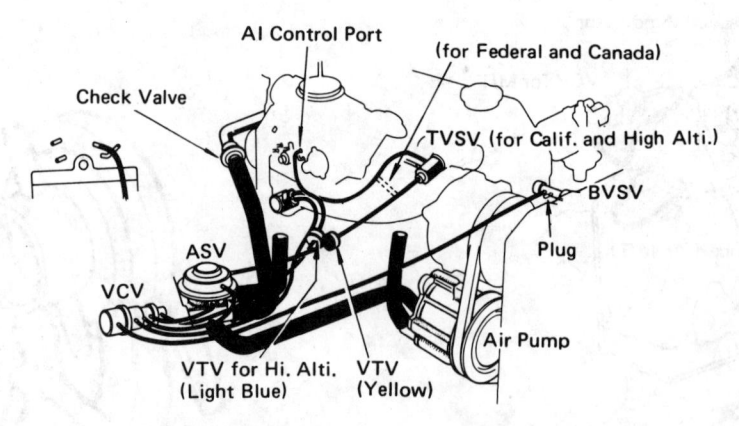

Fig. 3.81 Air injection system – component layout

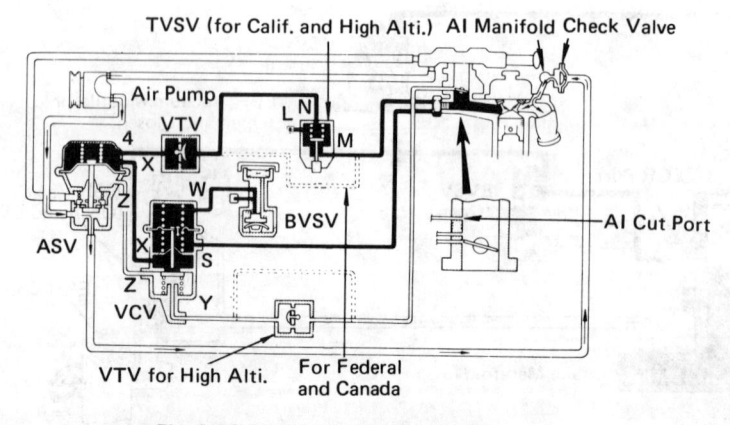

Fig. 3.82 Air injection system – schematic

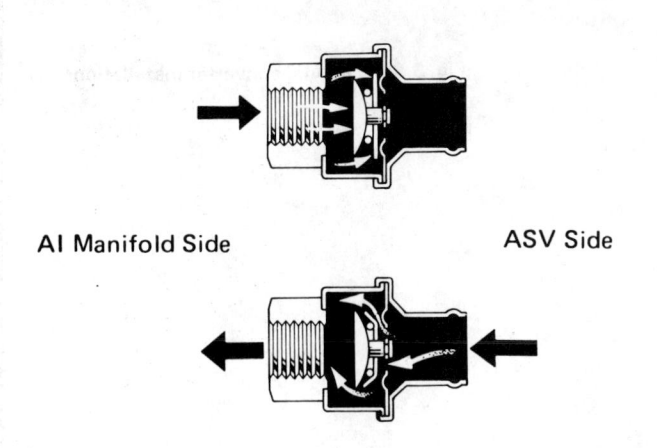

Chapter 3 Fuel, exhaust and emission control systems

Fig. 3.83 Check valve (Al Manifold Side / ASV Side)

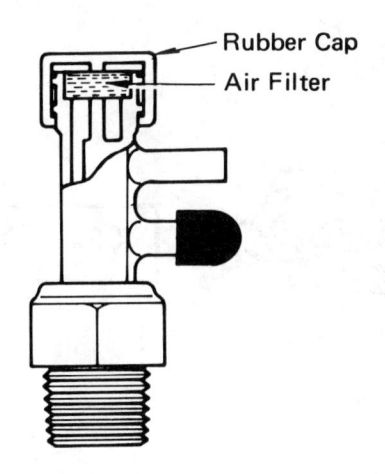

Fig. 3.84 Bi-metallic vacuum switching valve (Rubber Cap, Air Filter)

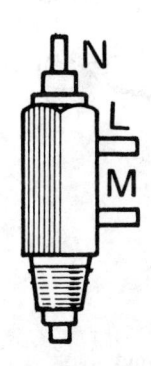

Fig. 3.85 TVSV port identification (N, L, M)

cleaned as a maintenance operation.

4 The operation of the system should be tested in the following way, after first making a visual examination of all the hoses to see that none of them are damaged, or disconnected.

5 Disconnect the vacuum hose from the EGR port of the carburettor and connect it to the inlet manifold.

6 Disconnect the vacuum hose from the EGR valve and connect a vacuum gauge to the hose, the pinch the vacuum hose joining the VCV to the carburettor and clamp it so that it remains pinched and the VCV stays closed.

7 When the engine is cold, start it and check that it will idle when cold. If it fails to start, or will not idle, the EGR valve is not seating properly and a new valve should be fitted.

8 Check the BVSV with the engine cold, when the valve should be closed and with the engine running at 2000 rpm, the vacuum gauge should indicate zero.

9 Check the opening of the BVSV when the engine warms up by leaving the engine running until the temperature of the coolant reaches 110°F (44°C). When the valve opens the gauge should indicate a low vacuum. This check will also confirm that the diaphragm of the vacuum modulator is satisfactory.

10 Check the operation of the VCV by releasing the pinched hose and with the engine running at 2000 rpm a low vacuum reading should be indicated.

11 Check the functioning of the EGR valve by pinching the vacuum hose between the VCV and the carburettor, disconnecting the vacuum gauge and applying vacuum direct to the EGR valve when the engine is idling. If the engine dies, the valve is functioning and no further tests are necessary. Release the pinched hose and reconnect the vacuum hoses to their proper places.

12 Individual components of the system may be tested as follows:

Bi-metallic vacuum switching valve

13 Drain the cooling system, then unscrew and remove the BVSV from the inlet manifold

14 With the valve at a temperature below 86°F (30°C), blow into one port of the valve and check that the valve is closed.

15 Suspend the valve in a pan of water and heat the water to a temperature of 111°F (44°C), before repeating the check of blowing into the inlet port. In this case the valve should be open and if so can be re-fitted.

16 Apply liquid sealer to the threads of the valve before refitting it, connect the two vacuum hoses to its ports and re-fill the cooling system.

ECR vacuum modulator

17 With the engine stopped, disconnect the two hoses from the modulator, cover one of the modulator ports with a finger and blow into the other one. The modulator should be open and allow air to pass out easily through the filter.

18 With the engine running at 2000 rpm, repeat the test of the previous paragraph. The valve should now be closed and there should be strong resistance to air flow.

Vacuum control valve

19 Disconnect the hoses from the VCV and remove the valve.

20 Remove the VCV cover, take out the filter and clean it by blowing air through it. Re-fit the filter with the thicker layer towards the cover and re-fit the hoses to the valve, ensuring that the filter end of the valve is connected to the EGR valve.

30 Air injection (AI) system

1 The air injection system injects compressed air into the exhaust ports to burn the hydrocarbons and carbon monoxide in the exhaust gases.

2 In addition to the principal component of the system, which is an air compressor driven from the crankshaft pulley there is a control system consisting of a thermostatic vacuum switching valve (TVSV), a bi-metallic vacuum switching valve (BVSV), a vacuum control valve (VCV) and an air switching valve (ASV).

3 Normal maintenance consists of keeping the compressor drivebelt correctly tensioned (Chapter 2, Section 10) and inspecting the hoses

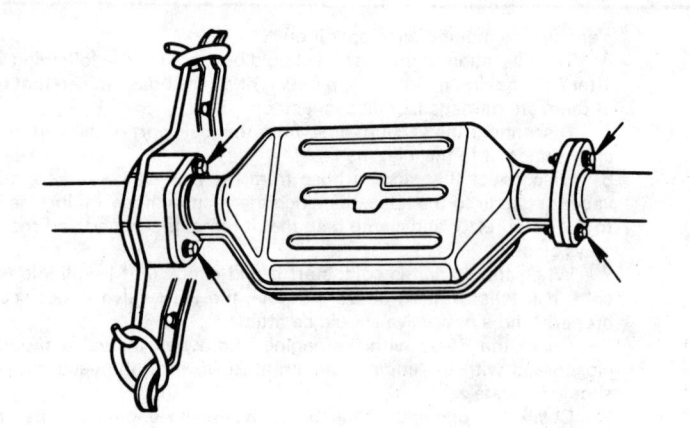

Fig. 3.86 Catalytic converter installation

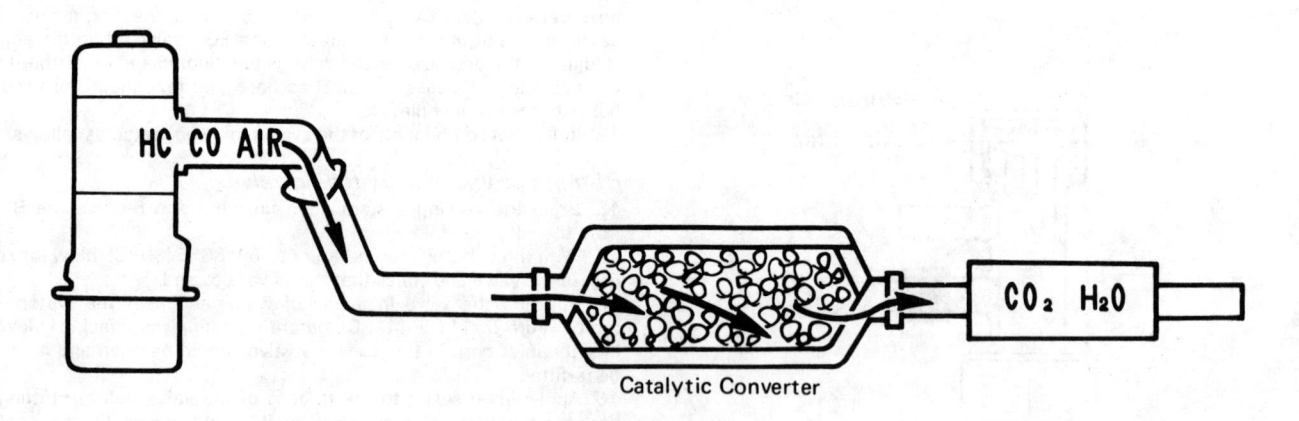

Fig. 3.87 Catalytic converter – schematic

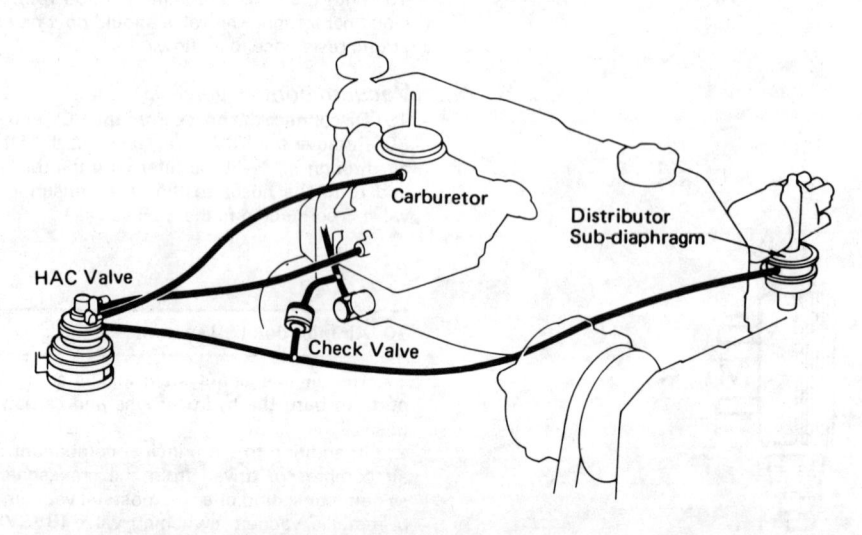

Fig. 3.88 High altitude compensation – component layout

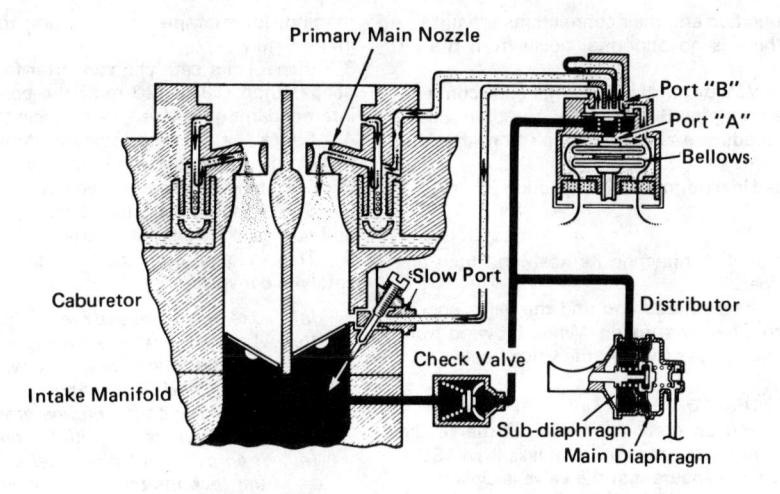

Fig. 3.89 High altitude compensation – schematic

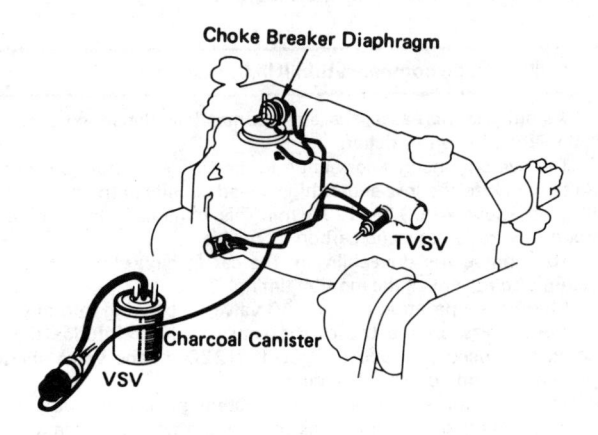

Fig. 3.90 Choke breaker – component layout

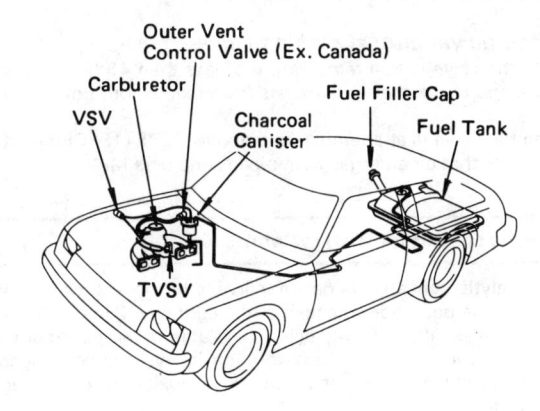

Fig. 3.92 Choke breaker – schematic

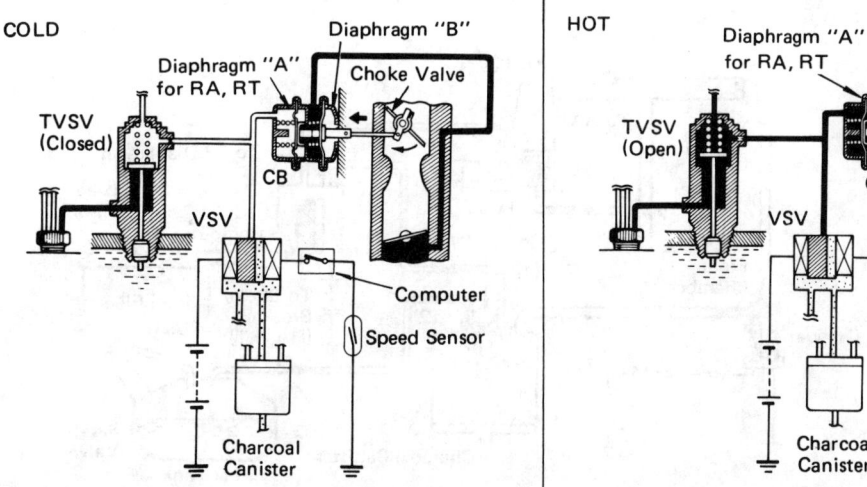

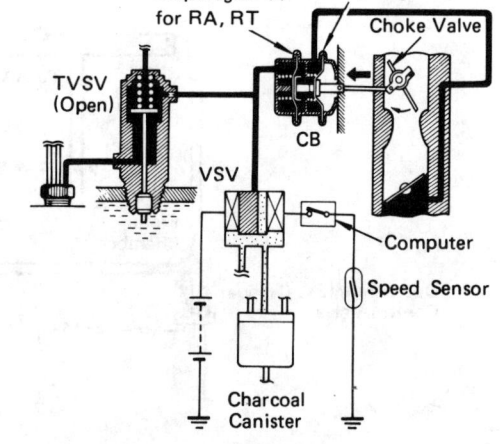

Fig. 3.91 Fuel evaporative system – component layout

112   Chapter 3 Fuel, exhaust and emission control systems

to see that they are in good condition and their connections are tight.
4 Periodically check that there is no abnormal noise from the air compressor.
5 Check the operation of the VCV by starting the engine, disconnecting the vacuum sensing hose from pipe S of the VCV and then reconnecting it. This action should produce a single pull of air from the ASV bypass hose (Fig. 3.80).
6 Check the VTV as described in paragraph 12 of Section 31.

Check valve
7 Remove the check valve by disconnecting its hose and then unscrewing the valve from the pipe.
8 Blow into the valve from the threaded end and the valve should close, making it impossible to blow air through. When blowing from the hose connection end, air should pass through the valve freely.

Bi-metallic vacuum switching valve
9 Drain the engine coolant and unscrew the BVSV. Immerse the threaded part of the valve in water at a temperature of less than 195·F (90°C), blow into the valve port and ensure that the valve is closed.
10 Suspend the threaded part of the valve in light oil and heat the oil to above 220°F (104°C). Blow into the port as before and the valve should be open.
11 Before refitting the valve, remove its top rubber cap, remove and inspect the filter, fitting a new one if the old one is clogged or damaged.
12 Apply sealing compound to the threads of the BVSV, screw it back into the manifold and re-fill the cooling system.

Thermostatic vacuum switching valve
13 When the valve is at a temperature of less than 45°F (7°C), blow into pipe L and check that air emerges from pipe N, but not from pipe M.
  When the valve is at a temperature above 59°F (15°C) repeat the test and check that air emerges from pipe N and pipe M.

31 Catalytic converter (CCO) system

1 The catalytic converter is only applicable to models for California and high altitude operation. It consists of a container filled with coated alumina granules, the coating being of platinum or palladium and when the hot exhaust gases pass through it, hydrocarbons become oxidised to produce water, and carbon monoxide is converted to carbon dioxide.
2 The catalytic converter only operates effectively at high temperatures, but because the temperature will rise excessively if a large volume of unburnt gas passes through it, a sensor and warning light are provided. If the warning light comes on during normal running, for example after starting the engine, investigate the cause immediately.
3 Inspect the catalytic converter for dents and damage. If a dent is deeper than 0.8 in (20 mm) the converter is likely to have suffered internal damage and require replacing.
4 Check the heat insulator for damage and ensure that there is sufficient clearance between the converter and the heat insulator for them not to knock against each other.
5 Shake the converter. If it rattles excessively it should be renewed and new gaskets should be fitted at each end.
6 The following precautions should be taken on systems with catalytic converters:

  (a) *Use only unleaded fuel*
  (b) *Avoid excessive fast idling*
  (c) *Avoid running the engine with a defective spark plug, or with a plug lead disconnected*
  (d) *Avoid prolonged engine braking and avoid overrunning with the engine off, or with the fuel tank empty*
  (e) *If an external tachometer is fitted for engine testing, connect the tachometer (+) terminal to the ignition coil (–) terminal. Do not connect the tachometer (+) terminal to the distributor*
  (f) *When working underneath the car after the engine has been running, take great care not to touch the converter. During running, the converter internal temperature is of the order of 1600°F (870°C)*
  (g) *Keep the converter casing and heatshield free of dried grass, or any other combustible material*

32 High altitude compensation (HAC) system

1 As altitude increases, the air becomes less dense, so that the air/fuel mixture becomes richer.
2 The system ensures correction to the air/fuel mixture by supplying additional air to the low and/or high speed circuits of the carburettor at altitudes above 4000 ft (1220 m). This minimises the emission of unburnt hydrocarbons and carbon monoxide.
3 To improve the driveability of the car at high altitudes, the HAC system also advances the ignition timing.
4 Check the operation of the HAC valve by blowing into any one of the three ports on the top of the valve. At low altitudes the valve should be closed, but above 4000 ft (1220 m) the valve should be open and air should pass through it.
5 To check the operation of the system at high altitude, run the engine at idling speed and check the ignition timing, which should be about 13° BTDC. Remove the vacuum hose from the distributor sub-diaphragm (the pipe on the side of the diaphragm unit) and the ignition timing should change to about 8° BTDC.
6 To check the operation of the system at low altitude, check that

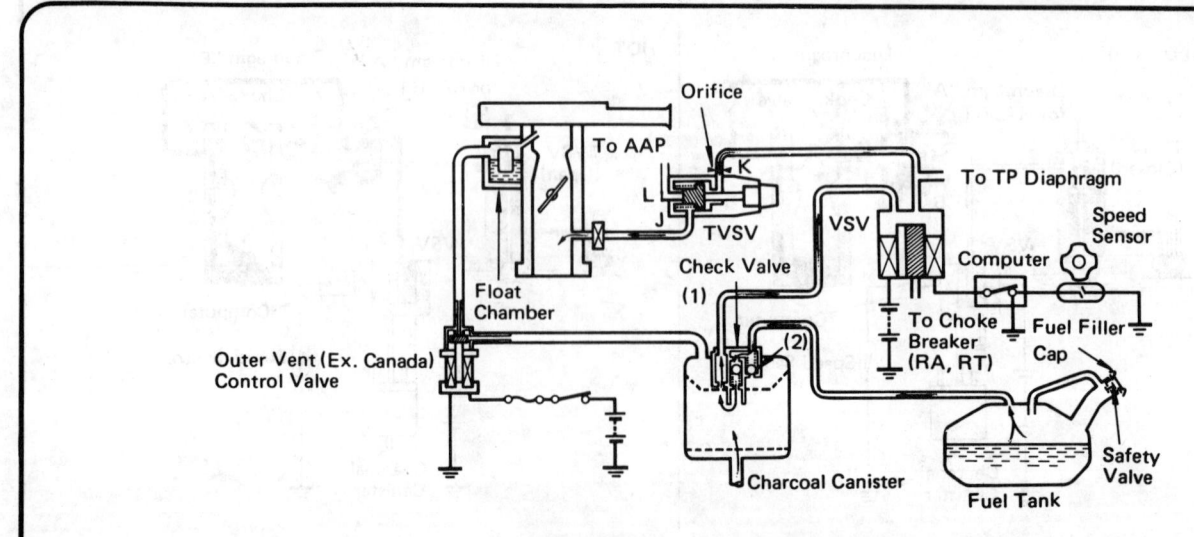

Fig. 3.93 Fuel evaporative system – schematic

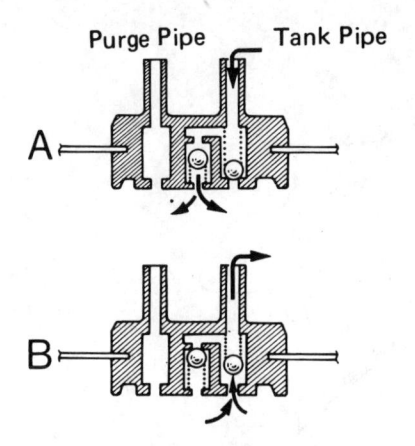

Fig. 3.94 Check valve operation

A Fuel tank at normal pressure
B Fuel tank interior under vacuum

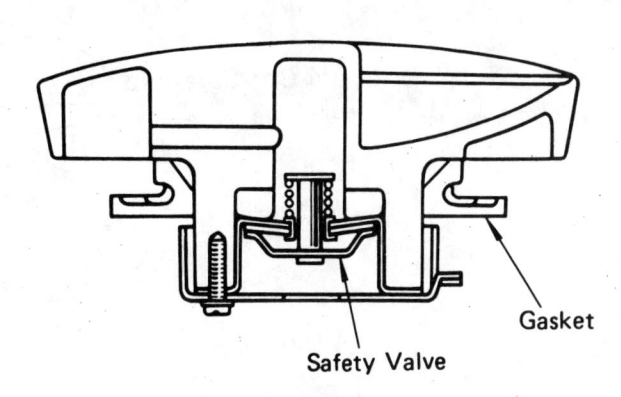

Fig. 3.95 Fuel tank safety cap

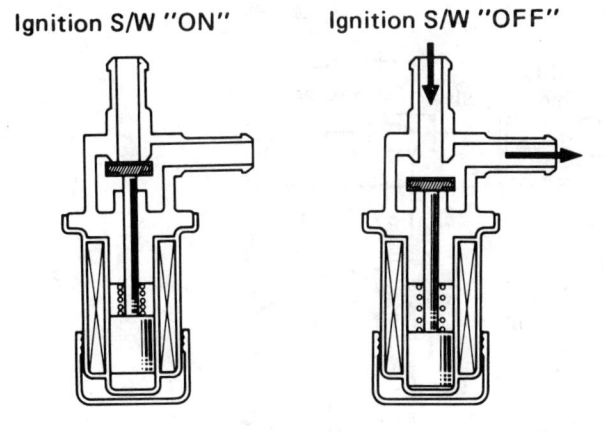

Fig. 3.96 Outer vent control valve operation (not applicable to cars for Canada)

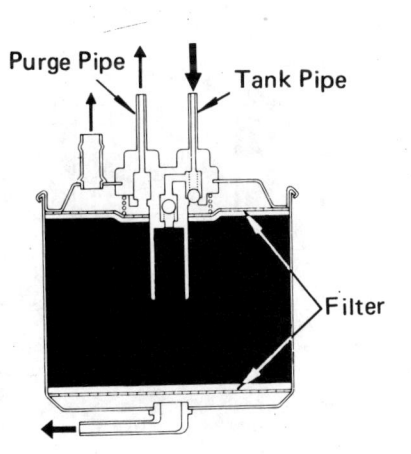

Fig. 3.97 Filter canister connections

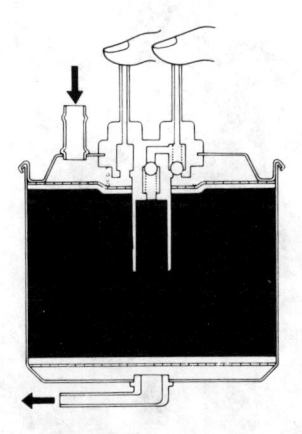

Fig. 3.98 Cleaning the canister filter

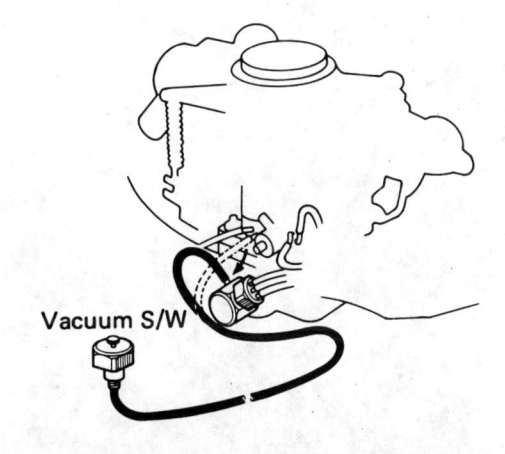

Fig. 3.99 Testing the operation of the fuel cut system

114

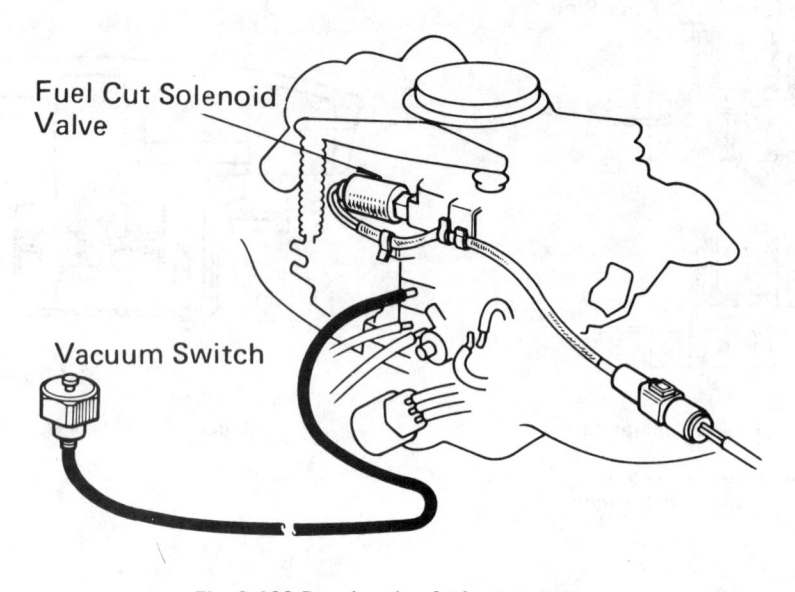

Fig. 3.100 Deceleration fuel cut system – component layout

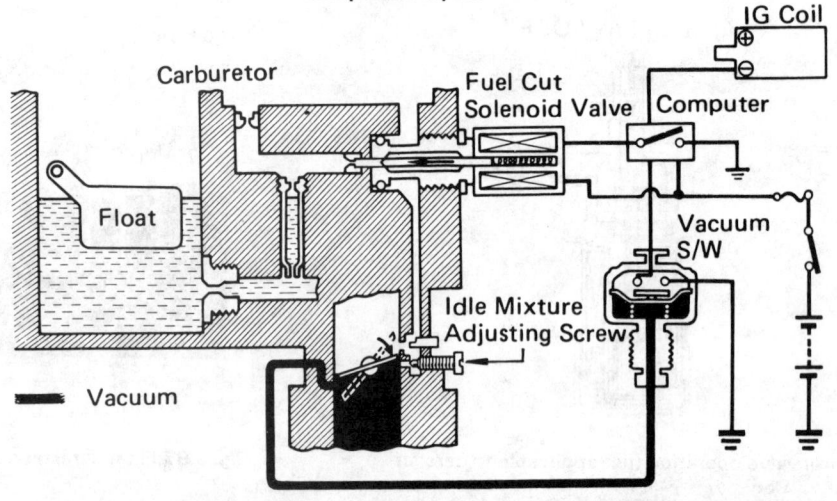

Fig. 3.101 Deceleration fuel cut system – schematic

36.1a Accelerator linkage at engine compartment bulkhead

36.1b Throttle rod connection at carburettor

Chapter 3 Fuel, exhaust and emission control systems

the ignition timing of the engine when idling is about 8° BTDC. Pinch the vacuum hose between the HAC valve and the T connector and the timing should change to about 13° BTDC.
7 To test the operation of the check valve, blow air into the white pipe to see if the valve opens and into the black pipe to see if it closes.

33 Choke breaker (CB) system

1 This system opens the choke slightly so that there is always sufficient air to prevent an excessively rich mixture. It also opens the choke further when the engine has warmed up and the car is not moving slowly.
2 The choke breaker is part of the carburettor and is described in the relevant carburettor Section. The system is controlled by a thermostatic vacuum switching valve (TVSV) and by an electrically operated vacuum switching valve (VSV). This latter valve is switched by a speed sensor.
3 Check the operation of the system by starting the engine and disconnecting each of the two diaphragm hoses in turn. When the hose is disconnected the choke link should move and then return to its former position when the hose is reconnected.
4 Check the operation of the TVSV by running the engine until it is warm and then pinching the hose from the TVSV to the carburettor. This should cause the choke linkage to move.

34 Fuel evaporative emission control (EVAP) system

1 To reduce the emission of hydrocarbon vapour from the fuel tank and the float chamber, the vapour is routed through a charcoal canister. At vehicle speeds above 11 mph, the vacuum switching valve operates and vapour stored in the canister is drawn into the inlet manifold. The vapour is then burned in the combustion chambers as a controlled fuel/air mixture.
2 Inspect the condition and security of the hoses regularly. Also check that the fuel tank cap gasket is undamaged and the safety valve free.
3 Check the canister for a stuck check valve, or a clogged filter by blowing air into the tank pipe (Fig. 3.97). Air should flow without resistance out of the other two pipes.
4 Clean the canister by blowing air at 40 psi into the pipe to the outer vent, while closing the two smaller pipes (Fig. 3.98). For vehicles which do not have an outer vent, blow air into the purge pipe. Blowing air through the canister should not cause any activated carbon to come out. Do not attempt to wash the canister.

35 Deceleration fuel cut system

1 This system is incorporated in the carburettor and is designed to overcome problems created by unburned fuel during prolonged deceleration. The system cuts off part of the fuel in the slow circuit and prevents overheating or afterburn in the exhaust system which might otherwise result from excess fuel.
2 When decelerating from above 2100 rpm with a high manifold vacuum, the solenoid valve closes the slow circuit in the carburettor so that no fuel from the float chamber reaches it.
3 Check the system by connecting the vacuum switch and the inlet manifold with a piece of vacuum tubing (Fig. 3.99), starting the engine and checking engine operation below and above 2400 rpm. Below 2400 rpm the engine should run normally and between 2400 and 2800 rpm it should misfire slightly.
4 With the engine idling, unplug the wiring connector to the solenoid valve, which should make the engine run rough. If the foregoing tests are satisfactory, the system is operating properly.
5 The tests should be carried out as quickly as possible on cars fitted with catalytic converters, so that the converter does not become overheated.

Solenoid valve
6 The solenoid valve can be checked by removing it from the engine and connecting its two terminals to a battery. The valve should click as the battery is connected and disconnected.
7 Check that the O-ring is undamaged and fit a new one if necessary.

Vacuum switch
8 With the engine stopped, check for continuity between the switch terminal and the switch body after disconnecting the switch lead from the terminals. The switch should be closed, giving continuity.
9 Start the engine and repeat the check, when there should be an open circuit.
10 On completing the test re-connect the switch lead to the switch.

36 Accelerator linkage

1 The linkage is of balljoint and rod type. No lubrication is required as nylon bushes and cups are incorporated (photo).
2 If the rod lengths are altered by releasing their locknuts, make sure there is a slight amount of free movement at the accelerator pedal and that the carburettor throttle lever is back against its stop with the pedal released. Check that full throttle can be obtained at the carburettor when the accelerator pedal is depressed to the carpet.

37 Fault diagnosis – fuel system

Symptom	Cause
Excessive fuel consumption	Air filter choked
	Leakage from pump, carburettor or fuel lines or fuel tank
	Float chamber flooding
	Distributor condenser faulty
	Distributor weights or vacuum capsule faulty
	Mixture too rich
	Contact breaker gap too wide
	Incorrect valve clearances
	Incorrect spark plug gaps
	Tyres under inflated
	Dragging brakes
	Choke sticking closed
Fuel starvation or mixture weakness	Clogged fuel line filter
	Float chamber needle valve clogged
	Faulty fuel pump valves
	Fuel pump diaphragm split
	Fuel pipe unions loose
	Fuel pump cover leaking
	Inlet maniold gasket or carburettor flange gasket leaking
	Incorrect adjustment of carburettor

38 Fault diagnosis – emission control system

Symptom or circuit	Symptom	Cause
Crankcase ventilation	Oil fume seepage from engine	Stuck or clogged PCV valve Split or collapsed hoses
Fuel Evaporative Emission Control	Fuel odour	Stuck filler cap valve Choked canister
	Vapour will not be drawn into manifold	Collapsed or split hoses Vacuum switching valve defective Speed sensor defective
	Rough running engine	Defective non-return valve
Transmission Controlled Spark	System operates at incorrect speed or temperature	Defective speed sensor Defective vacuum switch valve Defective thermostatic vacuum switching valve
Air Injection System	Fume emission from exhaust pipe	Slack or broken air pump drivebelt. Split or broken hoses. Clogged air filter. Defective air pump
Throttle Positioner System	System fails to operate during deceleration System fails to turn off during acceleration	Adjust linkage, check vacuum hose to diaphragm unit Defective diaphragm Defective vacuum switching valve Defective sensor
Exhaust Gas Recirculation System	Erratic idling Reduced power	Faulty valve Faulty valve
Choke opener or choke breaker	Choke remains closed too long	Defective choke opener diaphragm Defective vacuum switch valve Defective thermosensors or speed sensor Blocked vacuum transmitting valve
High Altitude Compensation System	Poor engine response at high altitude	Defective HAC valve Defective check valve
Spark control system	System operates at incorrect temperature of throttle opening	Defective thermostatic vacuum switch valve Defective vacuum transmitting valve

Chapter 4 Ignition system

For modifications and information applicable to later USA models, refer to Supplement at end of manual

Contents

Condenser (capacitor) – removal, testing and refitting	4
Contact breaker points – removal and refitting	3
Contact breaker points and damping spring – adjustment	2
Distributor – dismantling, inspection and reassembly	7
Distributor – removal and refitting	5
General description	1
Ignition system – fault symptoms	11
Ignition timing – checking and adjustment	6
Spark plugs and leads	9
Transistorised ignition system – fault finding	10
Transistorised ignition system – igniter	8

Specifications

System type .... Coil with contact breaker distributor. Transistorised ignition on some 20R models

Engine firing order .... 1-3-4-2 (No 1 cylinder nearest radiator)

Ignition timing:
- 2T-B .... 12° BTDC at 700 – 800 rpm
- 18R .... 7° BTDC at 650 rpm
- 18R-G .... 5° BTDC at 1000 rpm
- 20R .... 8° BTDC at 850 rpm (transmission in N)

Distributor
Contact breaker points gap:
- 2T-B .... 0.016 to 0.020 in (0.4 to 0.5 mm)
- 18R, 18R-G, 20R .... 0.018 in (0.45 mm)

Dwell angle .... 50° to 54°
Damping spring gap .... 0.004 to 0.016 in (0.1 to 0.4 mm)
Condenser capacity (not applicable to electronic ignition) .... 0.20 to 0.24 mFd

Rotor rotation:
- 2T-B, 18R, 18R-G .... Clockwise
- 20R .... Anti-clockwise

Spark plugs
Type:
- 2T-B .... Denso W16 EP or W16 EX-U / NGK BP5 ES-L
- 18R .... Denso W20 EPR or W20 EP / NGK BPR 6ES or BP 6ES
- 18R-G .... Denso W20 EXR / NGK BPR 6EZ
- 20R .... Denso W16 EP / NGK BP5 ES-L

Electrode gap .... 0.030 in (0.8 mm)

Ignition coil:

	2T-B	18R, 18R-G	20R	Electronic 20R
Primary resistance	3.3 ohms	about 1.4 ohms	1.3 to 1.5 ohms	1.35 to 1.65 ohms
Secondary resistance	8.5 K ohms	about 8.5 K ohms	6.5 to 10.5 K ohms	12.8 to 15.2 K ohms
External resistance		1.3 to 1.7 ohms	1.3 to 1.7 ohms	1.3 to 1.7 ohms

Torque wrench setting | lbf ft | Nm
Spark plugs .... | 15 | 21

Chapter 4 Ignition system

1 General description

Two different types of ignition system are used on models covered by this manual. Most models have a conventional contact breaker type, but some 20R models have a fully transistorised system which uses an ignition signal generating mechanism instead of distributor contact points.

Conventional type

The system has two main components, the ignition coil which consists of a low voltage and a high voltage winding, which are coupled electromagnetically, and a distributor, which is a mechanically operated switch which produces a spark at the required moment and routes it to the appropriate cylinder. A ballast resistor is fitted to the coil (primary circuit). During low speed operation when low primary circuit current flow is needed, the temperature of the resistor rises so increasing resistance. The current flow is thereby reduced to prolong contact points life within the distributor. During high speed operation increased primary current flow is required and the temperature of the ballast resistor falls to permit this to occur. During operation of the starter motor, the ballast resistor is by-passed to allow full battery voltage to the ignition circuit.

Ignition timing is controlled automatically by a mechanical governor, which produces the timing variations required at different engine speeds. A vacuum control, which is operated by the depression in the induction manifold, varies the ignition timing according to engine load. One some engines there is a dual diaphragm vacuum control, to provide the additional facility of retarding the ignition during idling to comply with anti-pollution legislation (See Chapter 3, Section 28).

Fully transistorised ignition system

The fully transistorised ignition system uses an ignition signal generating mechanism instead of contact breaker points which eliminates the problems associated with points. The system gives improved starting and low speed performance and because all switching is done electronically, there are no arcing problems.

As the signal rotor rotates, the air gap between the signal rotor and the pole piece on which the coil is wound changes. This differing air gap causes a change in the magnetic flux through the coil and consequently a variation in the induced voltage. The variation of the voltage generated with rotor position in shown in Figs. 4.3 and 4.4, from which it is seen that the generated voltage is zero when a rotor tooth is immediately opposite the coil pole piece and is highest when a tooth is approaching the pole piece.

With the distributor rotor in the position shown in Fig. 4.3, the voltage in the pick-up coil will be high, current will not flow from the battery to the pick-up coil and the igniter will be switched ON, energising the primary winding of the ignition coil. When the rotor has a tooth

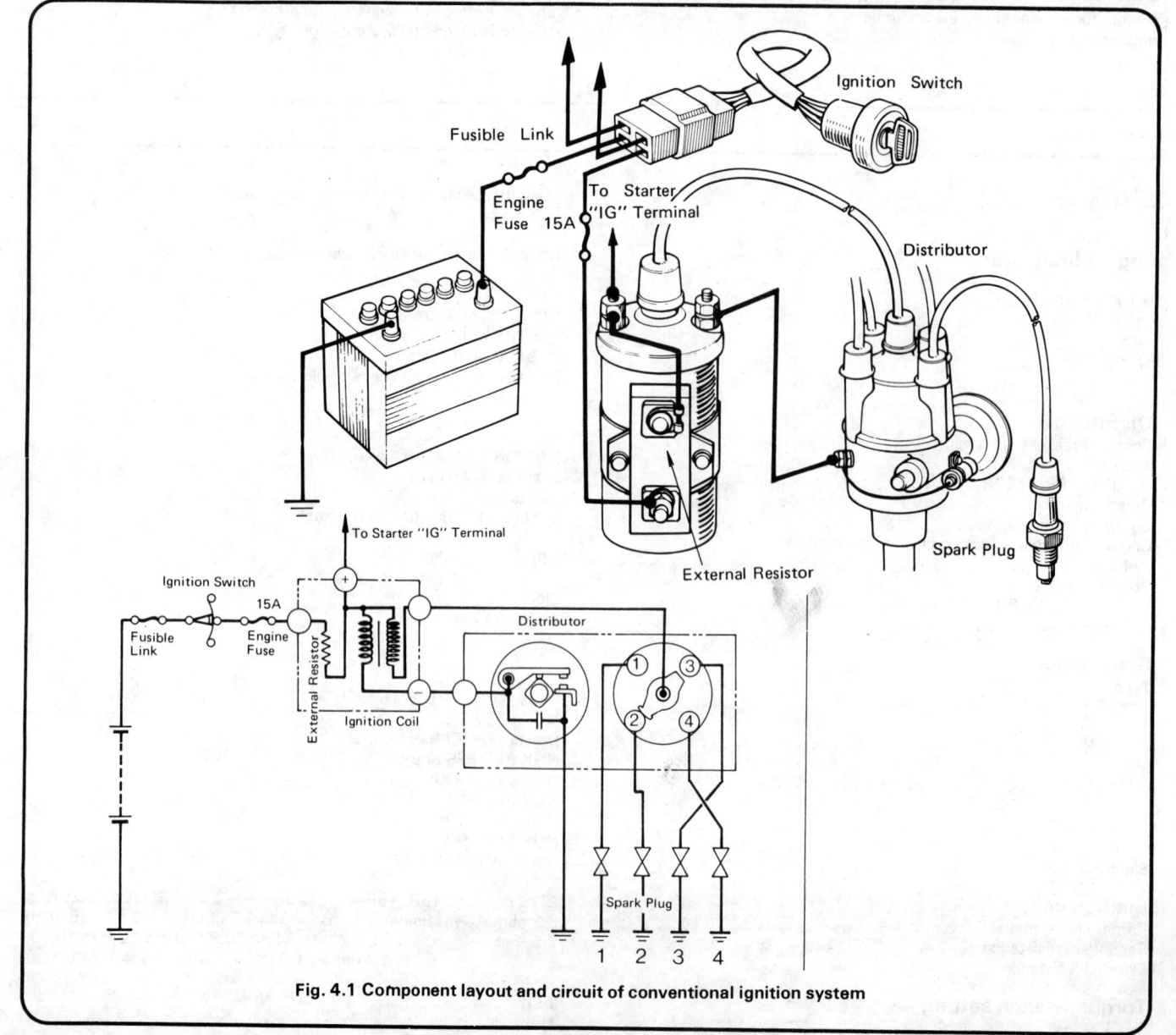

Fig. 4.1 Component layout and circuit of conventional ignition system

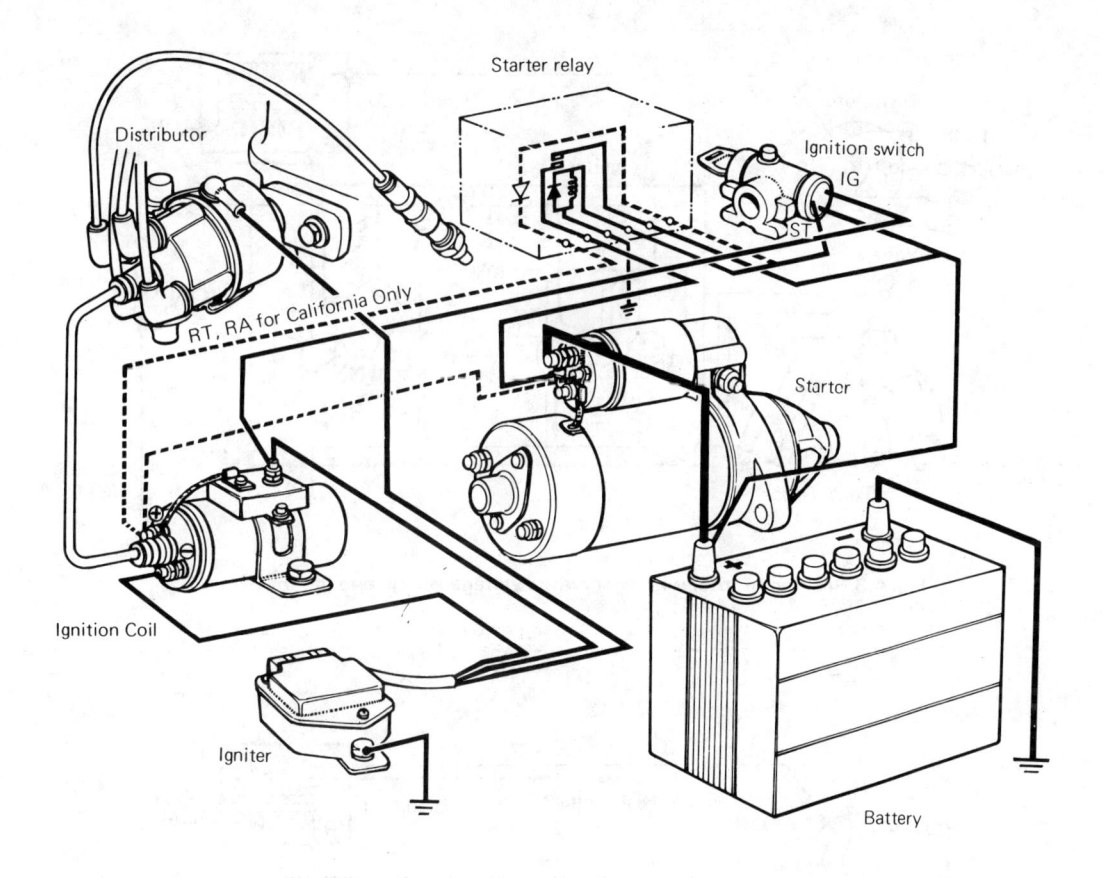

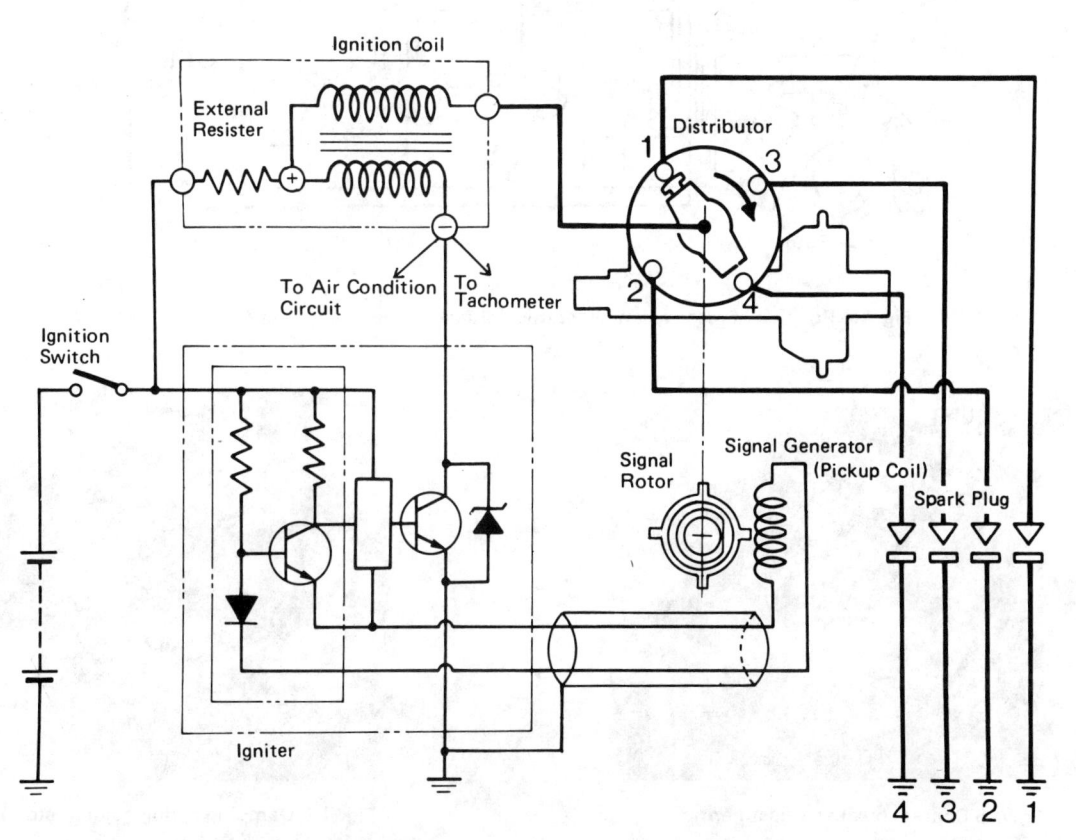

Fig. 4.2 Component layout and circuit of electronic ignition system

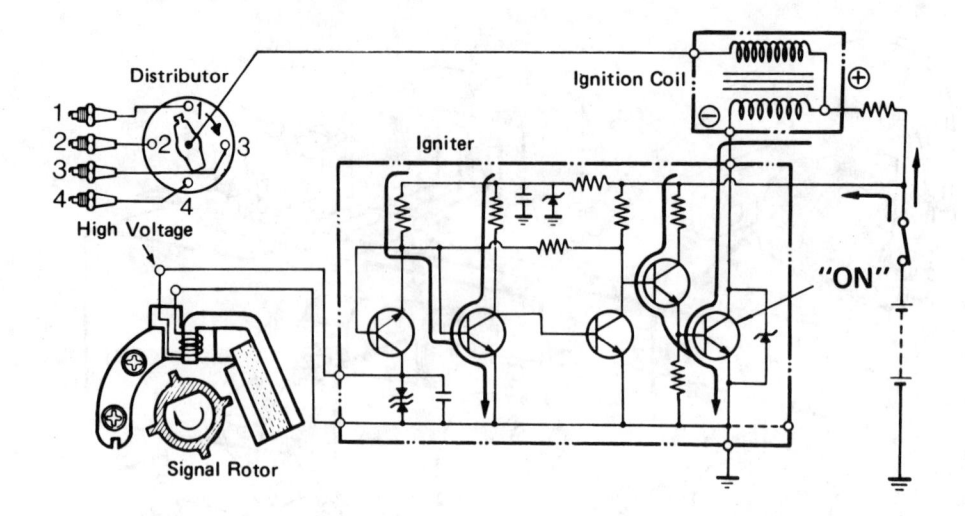

Fig. 4.3 Position of rotor when generated voltage is high and igniter ON

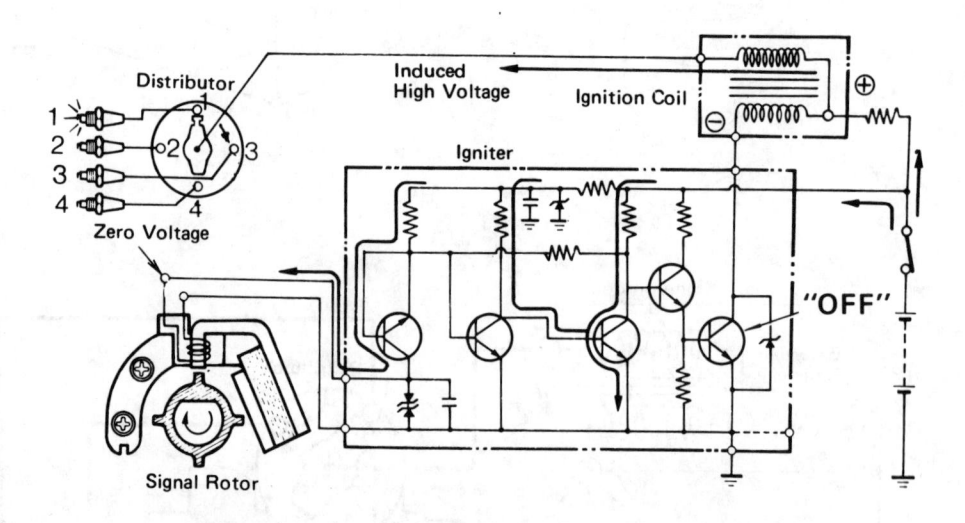

Fig. 4.4 Position of rotor when generated voltage is zero and igniter OFF

Fig. 4.5 Contact breaker adjustment

Fig. 4.6 Damping spring adjustment

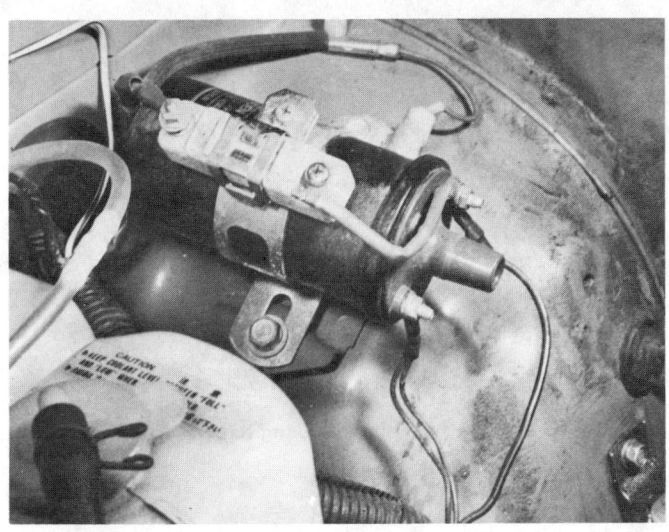

Chapter 4 Ignition system

1.2 Ignition coil and ballast resistor

opposite the pole piece, as shown in Fig. 4.4, the voltage induced in the pick-up coil will be zero, current will flow from the battery to the pick-up coil and the igniter will switch OFF. Switching the igniter OFF has the same effect as when the points of a conventional distributor open. The current through the primary winding of the ignition coil is interrupted, a very high voltage is induced in the secondary winding and a spark is produced at the spark plug.

The igniter is a separate component of the circuit, which senses the signal from the distributor rotor. It switches the primary winding of the ignition coil at the correct moment to produce a spark when required and also controls the length of time that the primary winding is energised. This latter function is the equivalent of the *dwell angle* of a conventional distributor.

Observe the following precautions when a car is fitted with an electronic ignition system:

The electronic components are very sensitive to battery polarity and the battery terminals must never be connected wrongly, even for an instant.

Electronic components can be damaged by the high voltages of short duration which are produced when current carrying circuits are broken. Do not disconnect either of the battery terminals when the engine is running. Do not disconnect the wire from the alternator B terminal to use it to produce sparks as a means of checking that the alternator is charging.

When using any piece of test equipment which utilises the pulses from contact breaker terminals, such as pulse type tachometers make sure that the pulse pick-up wire is connected to the negative terminal of the coil and not to the contact breaker terminal.

When washing the car, take care not to get the coil or igniter wet.

The ignition coil on the fully transistorised ignition system is different from that of a conventional ignition system and the two must not be interchanged.

2 Contact breaker points and damping spring – adjustment

Note: Wherever possible, adjustment of the contact breaker points should be by the Dwell Angle method to obtain the most accurate setting. For this, a proprietary Dwell Angle meter should be used in accordance with the manufacturer's instructions. Adjustment of the dwell angle is the same procedure as adjustment of the points gap.

1 Pull off the HT leads from the spark plugs and mark them 1 to 4 for easy identification.
2 Spring back the distributor cover securing clips and lift the cover to one side. Withdraw the rotor and the dustproof cover.
3 Using a spanner on the crankshaft pulley securing bolt, rotate the engine in its normal direction of rotation until the heel of the movable contact breaker arm is on one of the 4 high points of the cam. Removal of the spark plugs will make turning the engine easier.
4 Examine the contact faces of the (now open) points and, if they are pitted or burnt, they must be removed and dressed as described in the next Section.
5 If the points are in good order, check the gap by inserting a feeler gauge of the specified thickness. Insert the feeler blade in a vertical position and if the gap is correct, it will just fall by its own weight. If the gap is incorrect, adjust the fixed contact arm by loosening the retaining screw and moving it, as necessary, by means of a screwdriver blade inserted in the cut-out in the contact arm.
6 When the gap is correct, tighten the contact arm screw and remove the feeler gauge.
7 Now rotate the crankshaft, if necessary, so that the rubbing block of the damping spring is towards one of the lowest points of the cam.
8 Using a feeler gauge, check the gap between the cam and the rubbing block. If adjustment is necessary to obtain the gap specified, loosen the damping spring retaining screw and reposition the spring as necessary. Recheck the gap after tightening the screw.
9 Refit the dust cover, rotor arm and distributor cap, and reconnect the HT leads. Ensure that the ring spanner has been removed from the crankshaft pulley bolt.

3 Contact breaker points – removal and refitting

1 Carry out operations 1 and 2 of the preceding Section.
2 Detach the spring retaining clip from the top of the contact breaker arm pivot post.
3 Unscrew the nuts on the LT terminal on the outside of the distributor body just enough to enable the contact arm lead and spade terminal to be withdrawn, then lift the movable arm from the baseplate.
4 Remove the securing screws and lift the fixed contact breaker arm from the baseplate.
5 Examine the points. After a period of operation, one contact face should have a pip and the other a crater caused by arcing. This is a normal condition which should be removed by dressing the faces squarely on an oilstone.
6 Excessive pitting of the contact points may be caused by operation with an incorrect gap, the voltage regulator setting too high, faulty or wrong type of condenser, loose distributor baseplate or battery terminals.
7 Where contact breaker points are so badly worn or the pitting so deep that excessive rubbing would be required to eliminate it, then they should be renewed.
8 Wipe the faces of the points with methylated spirit before fitting, smear the high points of the cam with petroleum jelly. Apply a drop of engine oil to the pivot points of the contacts and mechanism and to the lubrication pad on top of the cam.
9 Refit the rotor, cap and HT leads.

4 Condenser (capacitor) – removal, testing and refitting

1 The condenser ensures that with the contact breaker points open, the sparking between them is not excessive, as this would cause severe pitting. The condenser is fitted in parallel and its failure will automatically cause failure of the ignition system as the points will be prevented from interrupting the low tension circuit. It is not used on the semi-transistorised ignition system.
2 Testing for an unserviceable condenser may be effected by switching on the ignition and separating the contact points by hand. If this action is accompanied by a blue flash then condenser failure is indicated. Difficult starting, missing of the engine after several miles running or badly pitted points are other indications of a faulty condenser.
3 The surest test is by substitution of a new unit.
4 To remove the condenser, unscrew its retaining screw and detach its lead from the LT terminal on the distributor body.
5 Refitting is the reverse of the removal procedure.

5 Distributor – removal and refitting

1 Remove No 1 spark plug and place a finger over the hole to feel the compression being generated as the engine is rotated by means of the crankshaft pulley bolt.

2   As soon as compression is felt this will indicate that No 1 piston is rising on its compression stroke. Continue turning the engine until the timing mark (not TDC mark) on the crankshaft pulley is in line with the pointer on the timing cover (photo).

3   Remove the distributor cap, and mark the rim of the distributor body at a point opposite to the centre line of the contact end of the rotor arm. This is equivalent to alignment with No 1 contact in the distributor cap. Also mark the installed position of the distributor body for reference.

4   Disconnect the LT wire from the terminal on the distributor body.

5   Disconnect the vacuum tube from the distributor advance diaphragm unit.

6   Unbolt the distributor clamp plate and withdraw the distributor.

7   The distributor cap may be withdrawn if the HT leads are first disconnected from the spark plugs and coil centre socket.

8   If the engine is turned while the distributor is removed, reset the crankshaft pulley timing mark opposite the timing cover pointer (No 1 piston on compression stroke) as described in paragraph 1, before fitting the distributor.

2T-B engines

9   Set the rotor arm so that it is pointing to the No 1 contact in the distributor cap (if it was fitted). Check that the drive slots of the distributor shaft and oil pump shaft are in alignment. The oil pump shaft can be turned if necessary with a screwdriver.

10   Now turn the rotor 30° in a clockwise direction.

11   Hold the distributor with the octane selector (vernier adjuster) parallel to the engine centre line and install it into the cylinder block. As the drivegears mesh, the rotor will turn anti-clockwise and take up its position in alignment with No 1 contact in the distributor cap.

12   Turn the distributor body until the contact points are just opening and tighten the clamp bolts. Check the ignition timing as described in the next Section.

18R engines

13   Hold the distributor over its cylinder block recess so that the octane selector (vernier adjuster) is at 30° to the engine centre line as shown, and the rotor arm is pointing to No 1 contact in the distributor cap (as if fitted). Now turn the rotor arm 30° in an anti-clockwise direction.

14   Check that the driveshaft slots in the oil pump and distributor are in alignment, if not, turn the oil pump shaft with a screwdriver.

15   Install the distributor and as the gears mesh, the rotor will turn to take up its No 1 cap contact alignment. Turn the distributor until the contact points are just about to open and tighten the clamp screw.

16   Check the ignition timing as described in the next Section.

18R-G engines

17   Set the engine so that the two front cam lobes are as shown in Fig. 4.14.

18   Hold the distributor above its recess so that the octane selector (vernier adjuster) is at 20° to the engine centre line. Set the rotor arm to point vertically.

19   Install the distributor and as the gears mesh, the rotor will turn so that it aligns with No 1 contact in the distributor cap (if it was fitted).

20   Turn the distributor body until the contact points are just about to open and tighten the clamp screw. Check the ignition timing as described in the next Section.

20R engines (mechanical contact breaker)

21   On 20R engines, fit the distributor body with the octane selector (vernier adjuster) pointing upwards and the rotor pointing towards the cap retaining clip spring. As the distributor is pushed fully home, the rotor will turn anti-clockwise and should then be in the position shown in Fig. 4.18. Turn the distributor clockwise to close the points then slowly anti-clockwise until the points **just** open; tighten the clamp plate bolt in this position.

22   Refit the distributor cap, leads and vacuum pipe, then check the timing as described in Section 6.

20R engines (transistorized ignition)

23   Before installing the distributor, set No 1 piston to 8° BTDC. This will be correct when the rocker arms for No 1 cylinder valves have clearance (valves closed) but those on No 4 cylinder have not (valves open) and the crankshaft pulley timing notch is in alignment with the pointer.

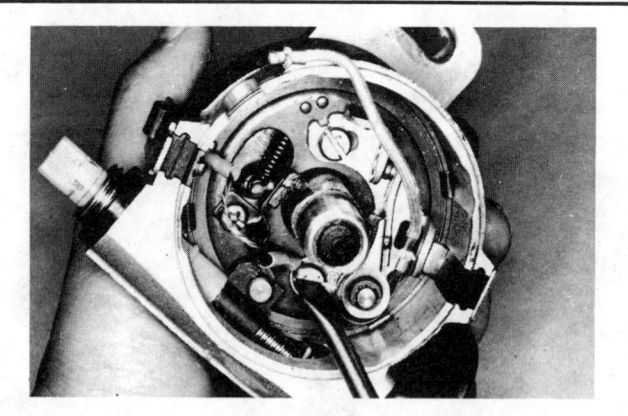

Fig. 4.7 Points securing screws

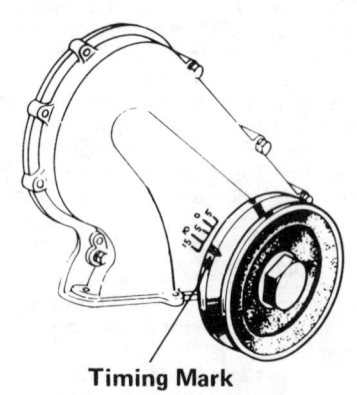

Fig. 4.8 Timing marks (2T–B)

Fig. 4.9 Distributor installed position 2T-B

Measuring plug gap. A feeler gauge of the correct size (see ignition system specifications) should have a slight 'drag' when slid between the electrodes. Adjust gap if necessary

Adjusting plug gap. The plug gap is adjusted by bending the earth electrode inwards, or outwards, as necessary until the correct clearance is obtained. Note the use of the correct tool

Normal. Grey-brown deposits, lightly coated core nose. Gap increasing by around 0.001 in (0.025 mm) per 1000 miles (1600 km). Plugs ideally suited to engine, and engine in good condition

Carbon fouling. Dry, black, sooty deposits. Will cause weak spark and eventually misfire. Fault: over-rich fuel mixture. Check: carburettor mixture settings, float level and jet sizes; choke operation and cleanliness of air filter. Plugs can be re-used after cleaning

Oil fouling. Wet, oily deposits. Will cause weak spark and eventually misfire. Fault: worn bores/piston rings or valve guides; sometimes occurs (temporarily) during running-in period. Plugs can be re-used after thorough cleaning

Overheating. Electrodes have glazed appearance, core nose very white – few deposits. Fault: plug overheating. Check: plug value, ignition timing, fuel octane rating (too low) and fuel mixture (too weak). Discard plugs and cure fault immediately

Electrode damage. Electrodes burned away; core nose has burned, glazed appearance. Fault: pre-ignition. Check: as for 'Overheating' but may be more severe. Discard plugs and remedy fault before piston or valve damage occurs

Split core nose (may appear initially as a crack). Damage is self-evident, but cracks will only show after cleaning. Fault: pre-ignition or wrong gap-setting technique. Check: ignition timing, cooling system, fuel octane rating (too low) and fuel mixture (too weak). Discard plugs, rectify fault immediately

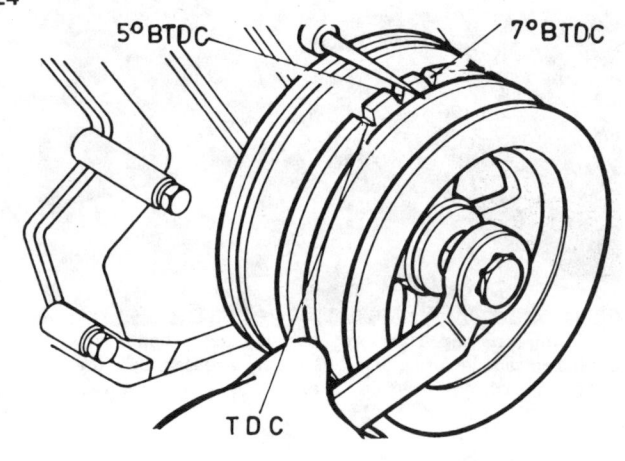

Fig. 4.10 Timing marks (18R and 18R-G)

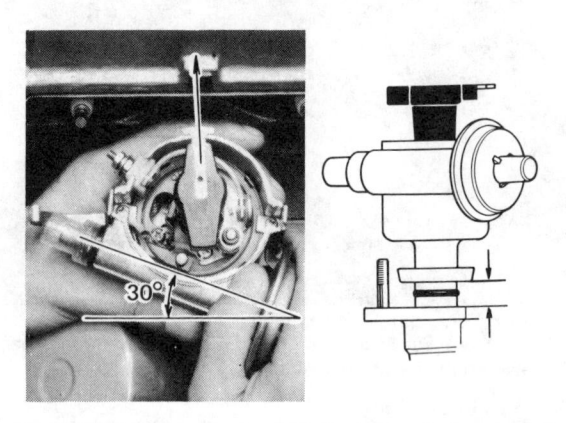

Fig. 4.11 Position of rotor and diaphragm before inserting distributor (18R)

Fig. 4.12 Position of oil pump slot before inserting distributor (18R)

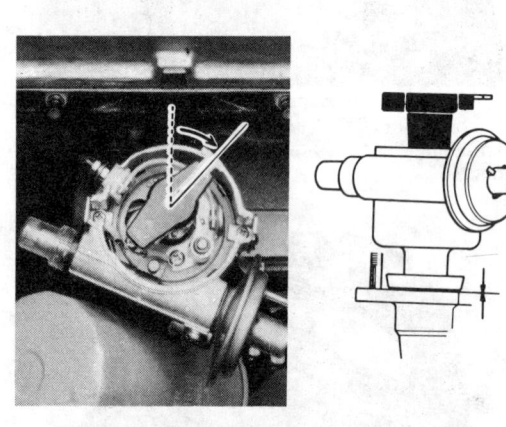

Fig. 4.13 Position of rotor and diaphragm with distributor fitted (18R)

Fig. 4.14 Position of cams when inserting distributor (18R-G)

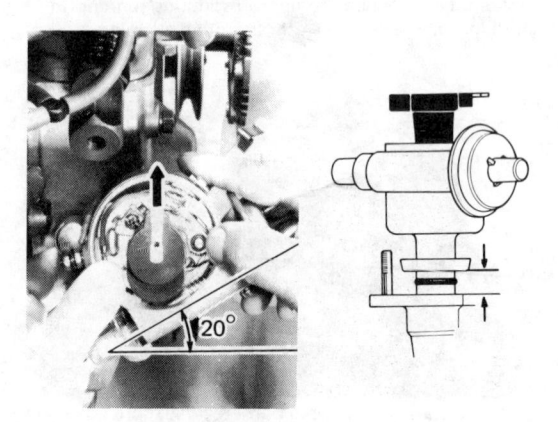

Fig. 4.15 Position of rotor and diaphragm before inserting distributor (18R-G)

Chapter 4 Ignition system

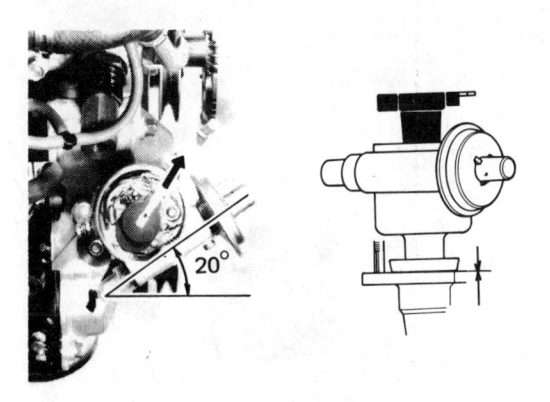

Fig. 4.16 Position of rotor and diaphragm with distributor fitted (18R-G)

Fig. 4.17 Timing mark (20R)

Fig. 4.18 Position of diaphragm with distributor fitted (20R) (Line shows position of rotor before inserting distributor)

24 Hold the distributor over its recess so that the octane selector (vernier adjuster) is pointing vertically upward and the rotor arm is aligned with the upper distributor cap clip.
25 Install the distributor and the rotor will turn as the gears mesh so that the rotor is aligned with the No 1 contact in the distributor cap (if it were fitted).
26 Turn the distributor body so that the high point of the rotor is square to the signal generator pick-up coil as shown in Fig. 4.28.
27 Check the ignition timing with a stroboscope as described in the next Section.

6 Ignition timing – checking and adjustment

1 If the ignition timing point has been lost completely, it will be necessary to check that the distributor is fitted in the correct position with regard to the crankshaft. This is dealt with in Section 5.
2 To check the static ignition timing, first check the distributor points gap and/or dwell angle as described previously (Section 2).
3 Rotate the cranshaft to get No 1 piston at the correct timing point (see Specifications) as described in Paragraphs 1 and 2 of Section 5. Remove the distributor cap.
4 Connect a 12V bulb of 5 watts maximum rating between an earth point on the engine and the terminal on the distributor body.
5 Switch on the ignition. If the timing is correctly set, the bulb should have just illuminated at this point, but it will be necessary to rotate the crankshaft anti-clockwise slightly until the bulb extinguishes, then rotate it clockwise again until it illuminates.
6 Check the point at which the bulb **just** illuminates; this should be as the timing mark aligns with the pointer. If the bulb illuminates before the mark on the flywheel is reached, the ignition is too far advanced and the distributor must be turned clockwise a little to correct it. If the bulb illuminates after the mark on the flywheel reaches the point, the ignition is too far retarded and the distributor must be turned anti-clockwise a little. *Note: (1)–This adjustment can be done using the octane selector (vernier adjuster) to advance or retard the ignition, but if this method is used the datum point will be altered. For most practical purposes this does not matter. (2) Rotor rotation on the 20R engine is anti-clockwise so these instructions must be reversed.*
7 After any adjustment, tighten the distributor clamp bolt and recheck the setting. Refit the distributor cap.
8 Having checked the static setting, the timing should now be checked dynamically with the engine running at idle speed using a stroboscopic lamp in accordance with the manufacturer's instructions. It may be necessary to reposition the distributor slightly now, or to alter the position of the octane selector (vernier adjuster). No variation in timing should be obtained whether the distributor vacuum line is connected or not at idle speed.
9 With the engine idling and the vacuum line disconnected, check that the timing point varies with throttle opening. This checks that the centrifugal advance is operating.
10 With the engine idling, detach the distributor vacuum hose at the carburettor and apply suction (by mouth) to the distributor diaphragm; the timing should now advance.
11 It may be necessary to make a minor adjustment to prevent 'pinking' from the engine when labouring or when a low octane fuel is being used. In the event, retard the ignition slightly by means of the octane selector (vernier adjuster).

7 Distributor – dismantling, inspection and reassembly

Conventional distributor
1 Remove the distributor cap, rotor and dustproof cover (photo).
2 Remove the vacuum adjuster (octane selector) cap (photo).
3 Pull off the point cover, where applicable.
4 Detach the points wire from the terminal on the distributor body; also detach the condenser wire, where applicable. Lift out the terminal insulator.
5 Remove the contact breaker points assembly and the damping spring.
6 Where applicable, remove the condensor from its attachment point.
7 Remove the screw (models without a condenser) and detach the vacuum advance unit.

126

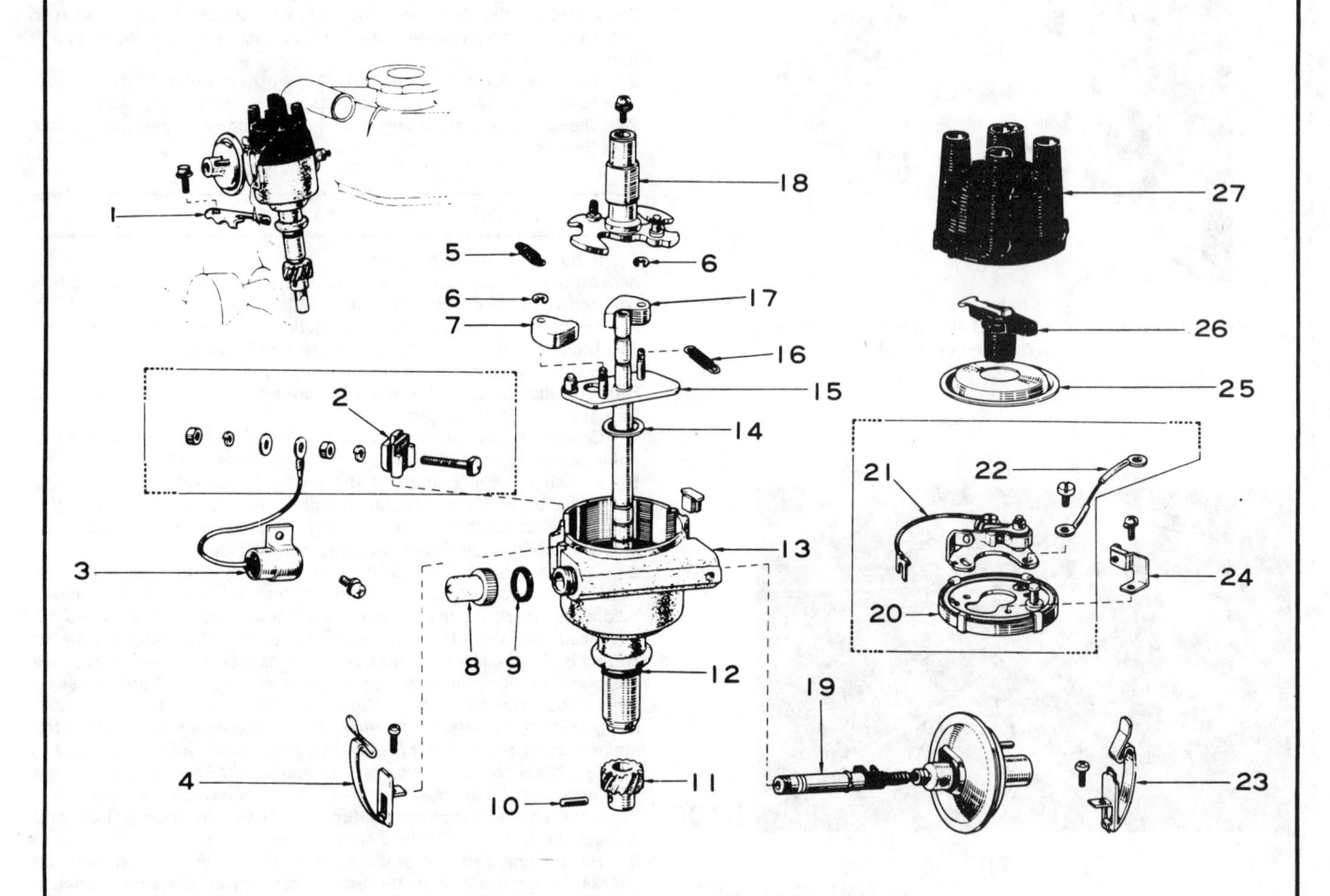

Fig. 4.19 Distributor (2T–B, 18R and 18R–G) – exploded view

1	Clamp	8	Octane adjuster cap
2	Terminal assembly	9	O-ring
3	Condenser	10	Pin
4	Cap retaining clip	11	Gear
5	Governor spring	12	O-ring
6	Spring clip	13	Distributor housing
7	Governor weight	14	Steel washer
15	Governor shaft and plate		assembly
16	Governor spring	22	Earth wire
17	Governor weight	23	Cap retaining clip
18	Cam	24	Damper spring
19	Vacuum advance unit	25	Dust cover
20	Contact breaker plate	26	Rotor
21	Contact breaker points	27	Cap

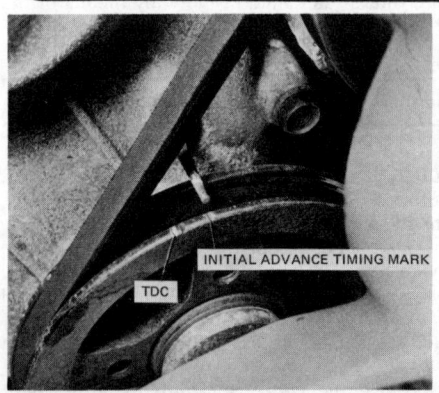

5.2 Timing mark (18R)

7.1 Distributor with cap, rotor and dustproof cover removed

7.2 Octane selector (vernier adjuster) with cap removed

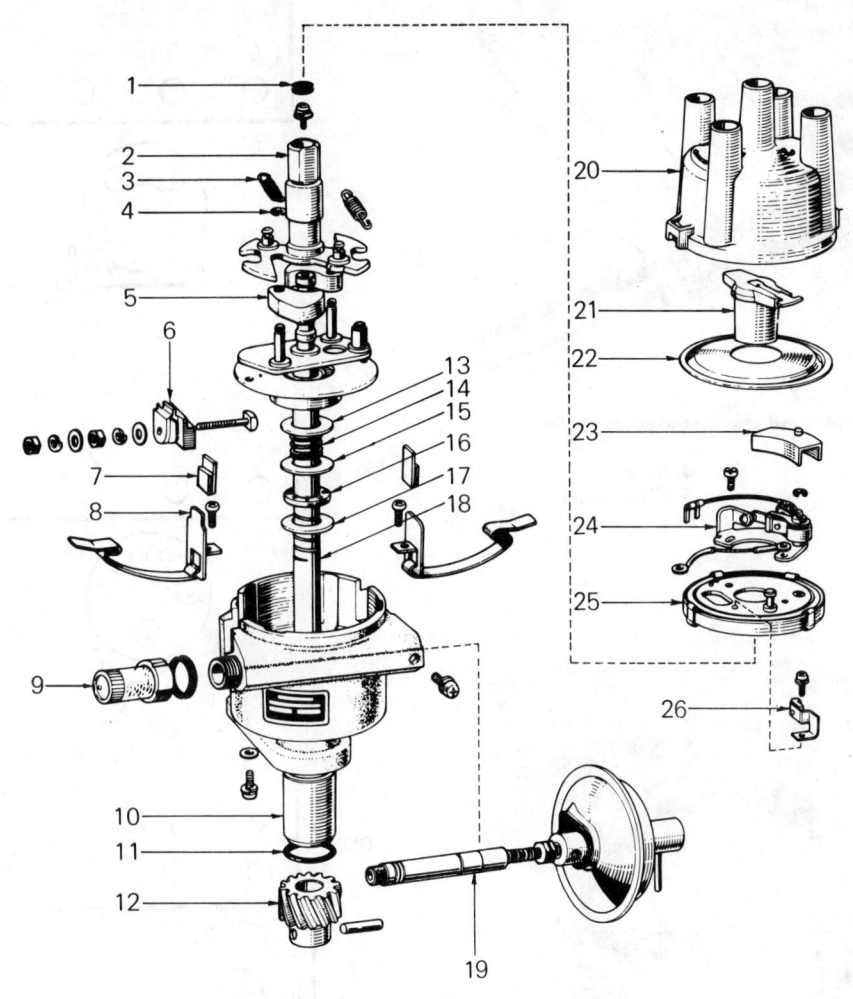

Fig. 4.20 Distributor (20R) – conventional ignition system

1 Grease pad	8 Cap retaining clip	15 Washer	21 Rotor
2 Cam	9 Cap	16 Bearing	22 Dustproof cover
3 Governor spring	10 Distributor housing	17 Washer	23 Point cover
4 Spring clip	11 O-ring	18 Governor shaft	24 Contact breaker points
5 Governor weight	12 Gear	19 Vacuum advance unit	25 Contact breaker plate
6 Terminal insulator	13 Washer	20 Distributor cap	26 Damping spring
7 Rubber plate	14 Spring		

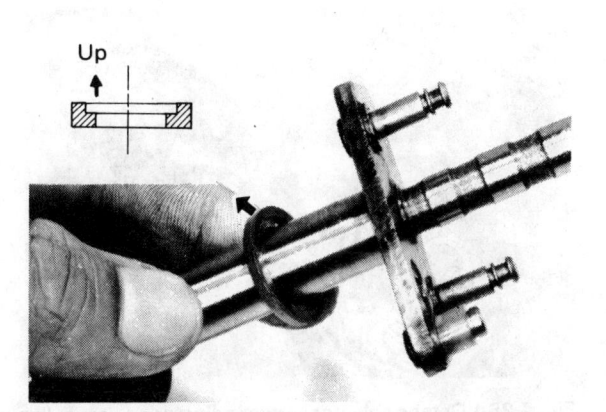

Fig. 4.21 Spindle and washer (2T-B, 18R)

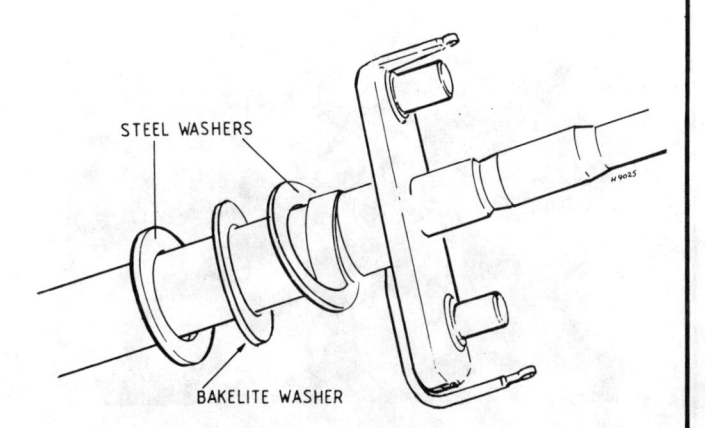

Fig. 4.22 Spindle and washers (18R-G)

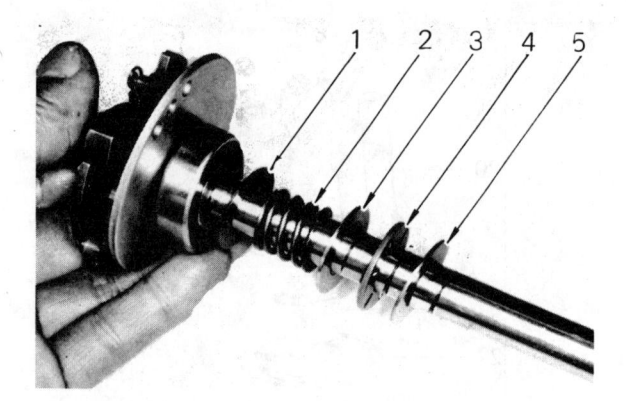

Fig. 4.23 Spindle and associated parts (20R)

1 Blue washer
2 Spring
3 Thin washer
4 Bearing
5 Thick washer

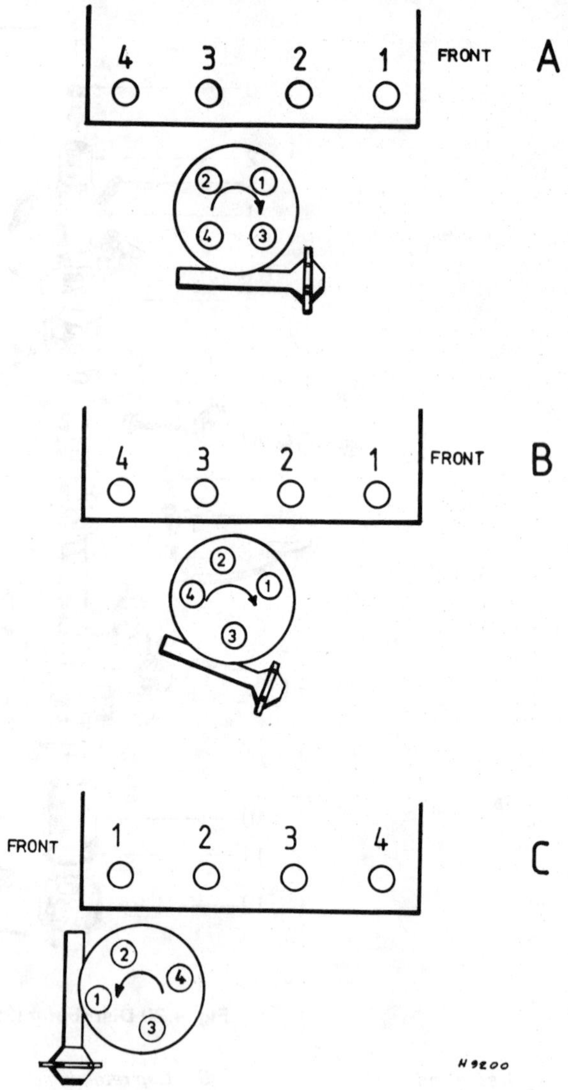

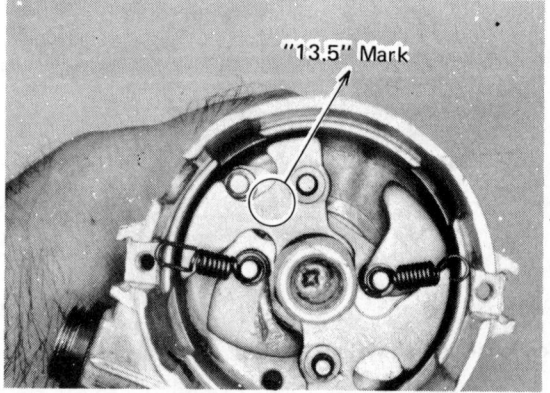

Fig. 4.25 Mark alignment (18R, 18R–G)

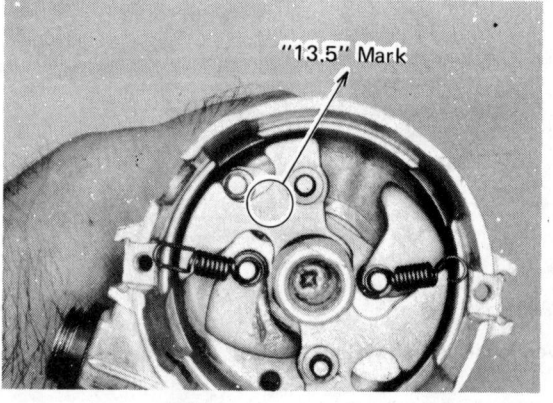

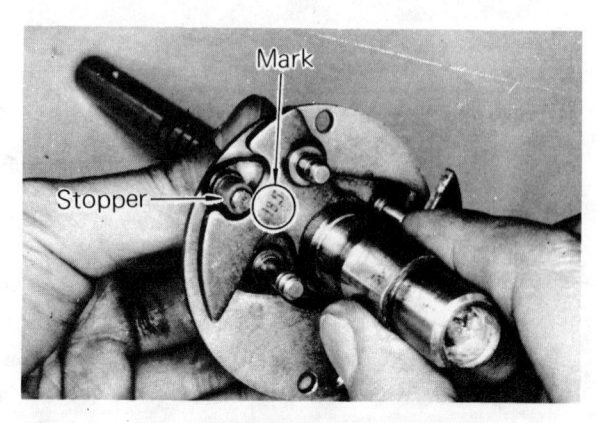

"13.5" Mark

Fig. 4.24 HT lead connection diagrams

A 2T–B
B 18–R
C 20R

Fig. 4.26 Mark alignment (20R)

Mark
Stopper

Fig. 4.26A Octane selector vernier adjuster on datum line

Chapter 4 Ignition system

8 Pull out the two rubber plates, and unscrew the hold-down clips and earth wire screws.
9 Lift out the breaker plate, and detach the mechanical advance springs. The springs are not identical, so check which way round they are fitted.
10 Remove the E-ring, and take out the governor weights and bearings.
11 Remove the grease stopper and screw from the top of the spindle, and remove the cam.
12 Carefully drill out the pin retaining the gear; detach the gear from the spindle.
13 Remove the screws from the base of the distributor body and pull out the spindle (and bearing, where applicable).
14 Remove the remaining parts from the spindle.
15 Check for wear between the spindle and body on the 18R distributor. If evident, it is recommended that a new distributor body is obtained. Examine all the other parts for wear and damage (including the bearing on 20R distributors), renewing parts as necessary. Suck on the diaphragm unit connection and check that the spindle moves. Examine the cap for cracks or tracking; if evident renew the cap.
16 Assembly of the distributor is basically the reverse of the dismantling procedure, but note the following:

 (a) *Lightly lubricate the rubbing surfaces of the spindle and body or bearing with a general purpose grease.*
 (b) *Ensure that the spindle and washer(s) (18R, 18R-G) or spindle, bearings and washers (20R) are assembled as shown in Figs. 4.21, 4.22 and 4.23.*
 (c) *When fitting the weights, align the 13·5 mark with the stopper – see Figs. 4.25 and 4.26.*
 (d) *Set the damping spring gap to that given in the Specifications. The gap is measured between the end of the rubbing block and one of the flats of the cam, and is adjusted by means of the spring attachment screw.*
 (e) *After setting the points gap, set the octane selector (vernier adjuster) to the standard (datum) line.*
 (f) *Do not forget to peen over the ends of the gear pin if this was renewed.*

Electronic ignition system – distributor

17 Remove the distributor cap, rotor and dustproof cover, then peel off the dust sealing ring.
18 Remove the two fixing screws and take off the signal generator assembly.
19 Remove the screw securing the cable connector to the distributor body and take the connector off.
20 Remove the vacuum advance unit cap and pull out the vacuum advance assembly.
21 Pull out the two rubber plates, remove the two holding down screws and clips and lift out the contact breaker plate.
22 The remainder of the operations are as in paragraph 9 onwards.

8 Transistorised ignition system – igniter

1 This unit needs to be kept clean and dry, but requires no routine attention.
2 If an internal fault occurs and the unit ceases functioning (See Section 10), the unit should be repaired by a Toyota dealer, or exchanged for a new unit.

9 Spark plugs and leads

1 The correct functioning of the spark plugs is vital for the correct running and efficiency of the engine. The plugs fitted as standard are listed in Specifications.
2 At the intervals stated in Routine Maintenance the plugs should be removed, examined, cleaned and, if worn excessively, renewed. The condition of the spark plug will also tell much about the overall condition of the engine.
3 If the insulator nose of the spark plug is clean and white, with no deposits, this is indicative of a weak mixture, or too hot a plug. (A hot plug transfers heat away from the electrode slowly – a cold plug transfer it away quickly).
4 If the insulator nose is covered with hard black looking deposits, then this is indicative that the mixture is too rich. Should the plug be black and oily, then it is likely that the engine is fairly worn, as well as the mixture being too rich.
5 If the insulator nose is covered with light tan to greyish brown deposits, then the mixture is correct and it is likely that the engine is in good condition.
6 If there are any traces of long brown tapering stains on the outside of the white portion of the plug, then the plug will have to be renewed, as this shows that there is a faulty joint between the plug body and the insulator, and compression is being allowed to leak away.
7 Plugs should be cleaned by a sand blasting machine, which will free them from carbon more thoroughly than cleaning by hand. The machine will also test the condition of the plugs under compression. Any plug that fails to spark at the recommended pressure should be renewed.
8 The spark plug gap is of considerable importance, as, if it is too large or too small the size of the spark and its efficiency will be seriously impaired. The spark plug gap should be set to the gap given in Specifications.
9 To set it, measure the gap with a feeler gauge, and then bend open, or close the outer plug electrode until the correct gap is achieved. The centre electrode should never be bent as this may crack the insulation and cause plug failure.
10 The HT leads to the coil and sparking plugs are of internal resistance, carbon core type. They are used in the interest of eliminating interference caused by the ignition system and have a resistance not exceeding 25 kohms. They are much more easily damaged than copper cored cable and they should be pulled from the spark plug terminals by gripping the metal end fitting at the end of the cable. Occasionally wipe the external surfaces of the leads free from oil and dirt using a fuel moistened cloth.
11 Always check the connection of the HT leads to the spark plugs is in the correct firing order sequence 1 – 3 – 4 – 2.

10 Transistorized ignition system – fault finding

Failures of the ignition system will either be due to faults in the HT or LT circuits. Initial checks should be made by observing the security of spark plug terminals, switch terminals, coil and battery connection. More detailed investigation and the explanation and remedial action in respect of symptoms of ignition malfunction in a 'conventional' system are described in the next Section. If a fault develops in a transistorized ignition system, the following checks should be carried out:

 (a) *Check that a spark is present by pulling the high tension lead from the coil to the distributor out of the distributor and holding its end close to an earthed part of the engine while the engine is cranked. Because of possible damage to the system resulting from high voltages of short duration (see precautions to be taken in Section 1) do this checking for a spark as quickly as possible. If a spark is present look for a distributor cap fault*
 (b) *Check the wiring and wiring connectors for damage or bad connections*
 (c) *Check the ignition coil using an ohmeter. The resistances measured between the ponts indicated in Fig. 4.27 should be as follows:*

Item	Terminals	Valve
Primary winding resistance	c and e	1·35 to 1·65 ohms
Secondary winding resistance	c and d	12·8 to 15·2 kohms
Resistor resistance	a and b	1·3 to 1·7 ohms
Insulation resistance	c and f	Infinity

  If the coil is defective it must be replaced by a transistorized ignition system coil and not by one from a conventional ignition system
 (d) *Check the air gap between the rotor and the coil pole piece, using a feeler gauge. If the gap is outside the limits 0·008 to 0·016 in (0·2 to 0·4 mm), make the necessary adjustment*
 (e) *Check the resistance of the signal generator with an ohmeter. Fit a new coil if the measurement obtained is outside the limits 130 to 190 ohms*
 (f) *With the ignition switch turned ON, check that the voltage*

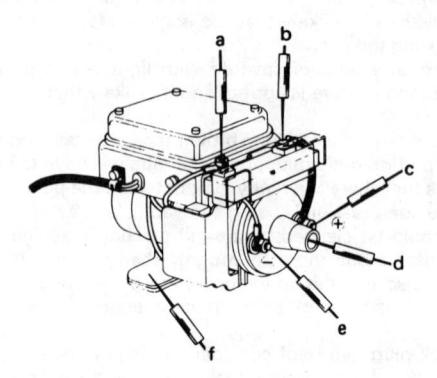

Fig. 4.27 Resistance check points (transistorised ignition)

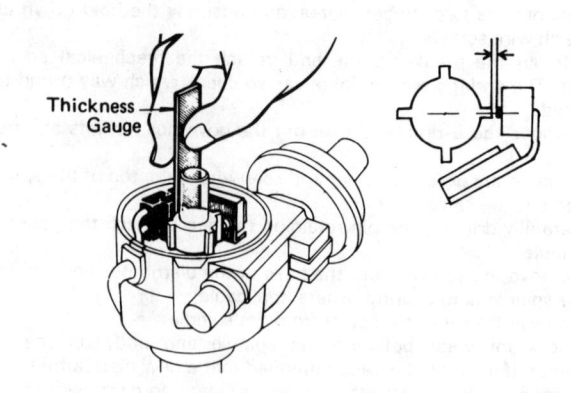

Fig. 4.28 Checking the air gap (transistorised ignition)

between the ignition coil (−) terminal and the resistor terminal is 12 volts

(g) Disconnect the wiring connector from the distributor and with an ohmeter set to its lowest range (not more than 10 ohms full scale deflection), connect it between the igniter terminals. It is very important that the (+) and (−) terminals of the ohmeter are connected as shown, otherwise the igniter may be damaged. With the ohmeter thus connected, turn the ignition switch ON and measure the voltage between the ignition coil (−) terminal and the resistor terminal as in test (f). On this occasion, the voltage should be almost zero

11 Ignition system – fault symptoms

Engine fails to start

1 If the engine fails to start and the car was running normally when it was last used, first check there is fuel in the tank. If the engine turns over normally on the starter motor and the battery is evidently well charged, then the fault may be in either the high or low tension circuits. First check the HT circuit. Note: If the battery is known to be fully charged; the ignition light comes on, and the starter motor fails to turn the engine check the tightness of the leads on the battery terminals and also the secureness of the earth lead to its connection to the body. It is quite common for the leads to have worked loose, even if they look and feel secure. If one of the battery posts gets very hot when trying to work the starter motor this is a sure indication of a faulty connection to that terminal.

2 One of the commonest reasons for bad starting is wet or damp spark plug leads and distributor. Remove the distributor cap. If condensation is visible internally, dry the cap with a rag and also wipe over the leads. Refit the cap.

3 If the engine still fails to start, check that current is reaching the plugs, by disconnecting each plug lead in turn at the spark plug end, and hold the end of the cable about $\frac{3}{16}$th inch (4 mm) away from the cylinder block. Spin the engine on the starter motor.

4 Sparking between the end of the cable and the block should be fairly strong with a regular blue spark. (Hold the lead with rubber gloves to avoid electric shocks). If current is reaching the plugs, then remove them, and clean and regap them. The engine should now start.

5 If there is no spark at the plug leads take off the HT lead from the centre of the distributor cap and hold it to the block as before. Spin the engine on the starter once more. A rapid succession of blue sparks between the end of the lead and the block indicate that the coil is in order and that the distributor cap is cracked, the rotor arm faulty, or the carbon brush in the top of the distributor cap is not making good contact with the spring on the rotor arm. Possibly the points are in bad condition. Clean and reset them as described in Section 2.

6 If there are no sparks from the end of the lead from the coil, check the connections at the coil end of the lead. If it is in order start checking the low tension circuit.

7 Use a 12V voltmeter or a 12V bulb and two lengths of wire. With the ignition switch on and the points open test between the low tension wire to the coil (it is marked +) and earth. No reading indicates a break in the supply from the ignition switch. Check the connections at the switch to see if any are loose. Refit them and the engine should run. A reading shows a faulty coil or condenser, or broken lead between the coil and the distributor.

8 Take the condenser wire off the points assembly and with the points open, test between the moving point and earth. If there now is a reading, then the fault is in the condenser. Fit a new one and the fault is cleared.

9 With no reading from the moving point to earth, take a reading between earth and the (−) terminal of the coil. A reading here shows a broken wire which will need to be renewed between the coil and distributor. No reading confirms that the coil has failed and must be renewed, after which the engine will run once more. Remember to refit the condenser wire to the points assembly. For these tests it is sufficient to separate the points with a piece of dry paper while testing with the points open.

Engine misfires

10 If the engine misfires regularly run it at a fast idling speed. Pull off each of the plug caps in turn and listen to the note of the engine. Hold the plug cap in a dry cloth or with a rubber glove as additional protection against a shock from the HT supply.

11 No difference in engine running will be noticed when the lead from the defective circuit is removed. Removing the lead from one of the good cylinders will accentuate the misfire.

12 Remove the plug lead from the end of the defective plug and hold it about $\frac{3}{16}$in (4 mm) away from the block. Restart the engine. If the sparking is fairly strong and regular the fault must lie in the spark plug.

13 The plug may be loose, the insulation may be cracked, or the points may have burnt away giving too wide a gap for the spark to jump. Worse still, one of the points may have broken off. Either renew the plug, or clean it, reset the gap, and then test it.

14 If there is no spark at the end of the plug lead, or if it is weak and intermittent, check the ignition lead from the distributor to the plug. If the insulation is cracked or perished, renew the lead. Check the connections at the distributor cap.

15 If there is still no spark, examine the distributor cap carefully for tracking. This can be recognised by a very thin black line running between two or more contacts, or between a contact and some other part of the distributor. These lines are paths which now conduct electricity across the cap thus letting it run to earth. The only answer is a new distributor cap.

16 Apart from the ignition timing being incorrect, other causes of misfiring have already been dealt with under the Section dealing with the failure of the engine to start. These are:

(a) The coil may be faulty giving an intermittent misfire.
(b) There may be a damaged wire or loose connection in the low tension circuit.
(c) The condenser may be faulty.
(d) There may be a mechanical fault in the distributor (broken driving spindle or contact breaker spring).

17 If the ignition timing is too far retarded, it should be noted that the engine will tend to overheat, and there will be a quite a noticeable drop in power. If the engine is overheating and the power is down, and the ignition timing is correct, then the carburettor should be checked, as it is likely that this is where the fault lies.

Chapter 5 Clutch

Contents

Clutch – adjustment	2
Clutch – inspection and renovation	9
Clutch – refitting	11
Clutch – removal	8
Clutch pedal – removal and refitting	12
Clutch release bearing – renewal	10
Fault diagnosis – clutch	13
General description	1
Hydraulic system – bleeding	7
Master cylinder – dismantling and reassembly	4
Master cylinder – removal and refitting	3
Operating cylinder – dismantling and reassembly	6
Operating cylinder – removal and refitting	5

Specifications

Type .... Single dry plate, diaphragm spring, hydraulically operated

Driven plate
Diameter:
- 2T-B engine .... 7·5 in
- 18R, 20R .... 8·8 in
- 18R-G .... 8·8 in heavy duty
- Minimum plate thickness above rivets .... 0·01 in (0·3 mm)
- Maximum friction plate run-out .... 0·03 in (0·8 mm)

Clutch pedal
- Height from floor (with carpet removed) .... 6·48 to 6·87 in (164·5 to 174·5 mm)
- Free-play .... 0·04 to 0·20 in (1·0 to 5·0 mm)

Clutch fluid type .... SAE 1703e

Torque wrench settings

	lbf ft	Nm
Clutch cover-to-flywheel	15	21
Clutch bellhousing-to-engine:		
2T-B	45	62
18R, 20R	50	69
Master cylinder reservoir bolt	20	28

1 General description

All vehicles fitted with a manual gearbox, are equipped with a single dry plate diaphragm spring type clutch. Operation is by means of a pendant foot pedal and hydraulic circuit. The clutch comprises a pressure plate assembly which contains the pressure plate, diaphragm spring and fulcrum rings. The assembly is bolted by means of its cover to the rear face of the flywheel.

The driven plate (friction disc) is free to slide along the gearbox input shaft and it is held in place between the flywheel and pressure plate faces by the pressure exerted by the diaphragm spring. The friction lining material is riveted to the driven plate which incorporates a rubber cushioned hub designed to absorb transmission rotational shocks and to assist in ensuring smooth take offs. The circular diaphragm spring assembly is mounted on shouldered pins and held in place in the cover by two fulcrum rings. The spring itself is held in place by three spring steel clips.

Depressing the clutch pedal pushes the release bearing, mounted on its hub retainer, forward to bear against the fingers of the diaphragm spring. This action causes the diaphragm spring outer edge to deflect and so move the pressure plate rearwards to disengage the pressure plate from the driven plate.

When the clutch pedal is released, the diaphragm spring forces the pressure plate into contact with the friction linings of the driven plate and at the same time pushes the driven plate fractionally forward on its splines to ensure full engagement with the flywheel. The driven plate is now firmly sandwiched between the pressure plate and the flywheel and so the drive is taken up.

2 Clutch adjustment

1 Two adjustments are required to ensure that the clutch operates

Chapter 5 Clutch

properly. When the pedal is released, there must be free-play between the master cylinder pushrod and the master cylinder piston, so that no hydraulic pressure is applied to the clutch operating cylinder and the clutch spring can exert its full force. When the pedal is depressed, the stroke of the master cylinder piston must be sufficient to cause the clutch operating lever to travel far enough to disengage the clutch completely.

2 If the pedal free-play is insufficient, there will be excessive and unnecessary wear to the clutch thrust race and the fingers of the diaphragm spring. Insufficient pedal movement will result in the clutch not disengaging completely, making gear changing difficult.

3 Adjust the pedal height by releasing the locknut of the pedal stop and screwing the pedal stop in or out until the distance from the mid-point of the pedal rubber to the floor (Fig. 5.2) is the pedal height given in Specifications.

4 When the pedal height is correct, lock the stop in position with the locknut and adjust the pedal play.

5 Pedal play is the distance the pedal moves before the pushrod makes contact with the piston of the master cylinder. If the brake pedal is pushed by hand, this point of contact can be felt. Adjust the pedal play by releasing the locknut of the pushrod and screwing the rod in or out of the pushrod clevis yoke until the specified free-play is obtained. Screw the locknut against the yoke when the adjustment has been completed.

3 Master cylinder – removal and refitting

1 Remove the split cotter from the clevis pin connecting the master cylinder yoke to the foot pedal.

2 Remove the bolt from the banjo connection of the hydraulic pipe to the master cylinder and recover the two washers on LHD cars. On RHD cars disconnect the hydraulic pipe union from the master cylinder. After disconnecting the pipe, seal its end to prevent the entry of dirt and loss of fluid.

3 Plug the hole in the master cylinder and wipe up any fluid which has leaked out. Remove the two nuts securing the master cylinder to the bulkhead and remove the master cylinder.

4 Refitting is the reversal of removal. On LHD cars use new washers on the banjo joint.

5 On completion of refitting, check the pedal height and free-play and bleed the hydraulic system as described in Section 7.

4 Master cylinder – dismantling and reassembly

1 Remove the reservoir cap and pour out the fluid.

2 Remove any temporary plug which has been fitted to the cylinder outlet and operate the pushrod two or three times to eject any fluid from the cylinder bore.

3 Unscrew and remove the reservoir securing bolt inside the reservoir. Remove the washer under the bolt and the reservoir.

4 Remove the circlip from the end of the cylinder bore and pull out the pushrod assembly.

5 Extract the piston and seal assembly from the cylinder.

6 Discard all the hydraulic seals and the cylinder boot and wash all the components in clean hydraulic fluid. Inspect all the components for wear, damage and corrosion and discard any unserviceable parts.

7 Using a service kit, rebuild the master cylinder as a reverse of removal. Fit all rubber seals using the fingers only, so as not to damage them and note that the lips of all the seals face the reservoir end of the cylinder.

8 Smear rubber grease on to the areas arrowed in Fig. 5.8 and reassemble the master cylinder assembly.

9 Tighten the reservoir securing bolt to the specified torque.

5 Operating cylinder – removal and refitting

1 Remove the two bolts securing the operating cylinder to the bellhousing (photo) and separate the cylinder from the housing.

2 Fit a spanner to the hydraulic pipe union with the cylinder and while holding the union still so as not to twist the flexible pipe, turn the cylinder to screw it off the pipe union. Take care to recover the washer from the joint.

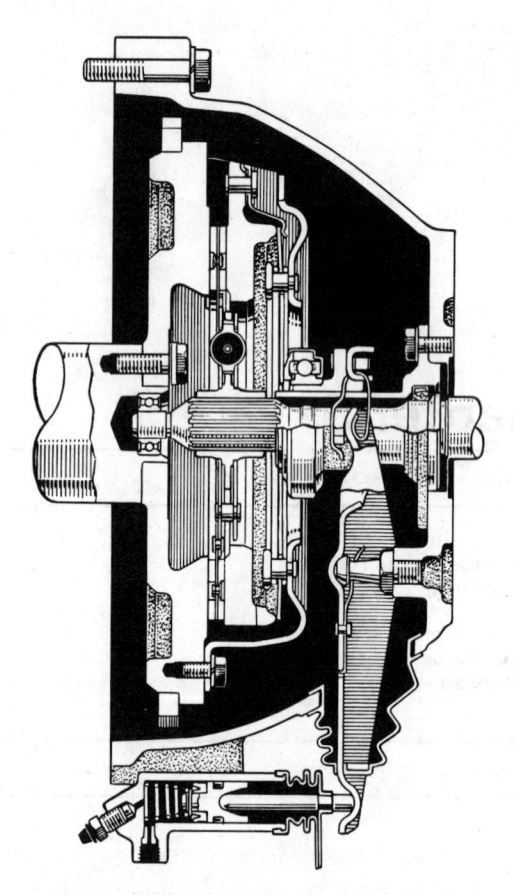

Fig. 5.1 Clutch assembly – sectional view

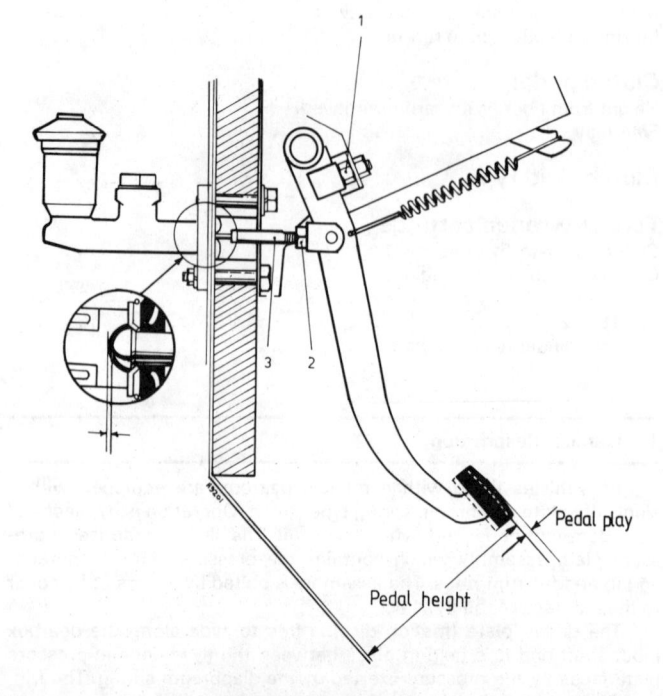

Fig. 5.2 Clutch pedal adjustment

1 Pedal stop locknut
2 Pushrod locknut
3 Pushrod

Chapter 5 Clutch

5.1 Operating cylinder

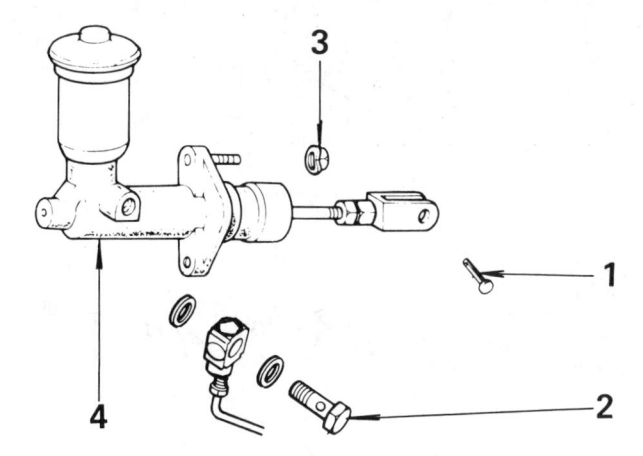

Fig. 5.3 Master cylinder (LHD)

1 Pushrod pin
2 Union bolt
3 Securing nut
4 Master cylinder

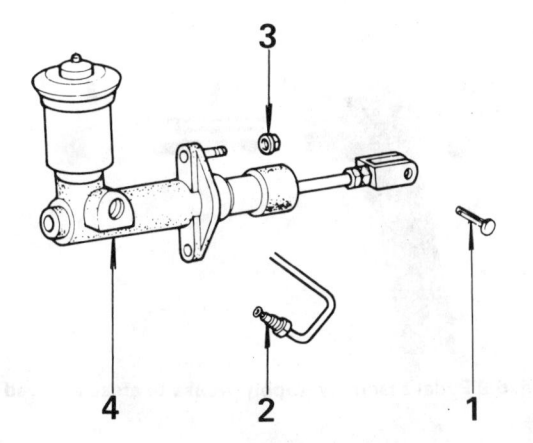

Fig. 5.4 Master cylinder (RHD)

1 Pushrod pin
2 Pipe union nut
3 Securing nut
4 Master cylinder

3 Before fitting the union to the cylinder when reassembling, make sure that the joint washer has been fitted and again keep the union still and screw the cylinder on to it.
4 After fitting the cylinder to the bellhousing and bleeding the hydraulic system, check the clutch pedal adjustment (See Section 2).

6 Operating cylinder – dismantling and reassembly

1 Depress the pushrod two or three times to eject any hydraulic fluid and then pull out the pushrod assembly complete with rubber boot.
2 Eject the piston assembly by applying air pressure at the fluid inlet or by tapping the end of the cylinder on a block of wood.
3 Wash components in clean hydraulic fluid and discard all rubber seals. If there are any 'bright' wear areas on the piston or cylinder bore surfaces, renew the complete unit.
4 If the components are in good condition, obtain a repair kit which will contain all the necessary seals and other items requiring renewal.
5 Reassembly is the reverse of the dismantling procedure, but manipulate the seals into position using the fingers only to avoid damaging them. Dip each component in clean hydraulic fluid before inserting it into the cylinder.

7 Hydraulic system – bleeding

1 Gather together a clean glass jar, a length of rubber or plastic tubing which fits tightly over the bleed nipple on the operating cylinder, a tin of hydraulic brake fluid and someone to help.
2 Check that the master cylinder is full. If it is not, fill it and cover the bottom two inches of the jar with hydraulic fluid.
3 Remove the rubber dust cap from the bleed nipple on the operating cylinder and, with a suitable spanner, open the bleed nipple approximately three quarters of a turn.
4 Place one end of the tube securely over the nipple and insert the other end into the jar so that its open end will remain submerged in the fluid.
5 Have your assistant depress the clutch pedal to the limit of its travel and then remove his foot, so that the pedal can return to its normal position without being obstructed.
6 Repeat this operation until no more air can be seen being expelled from the end of the tube submerged in the jar. Keep the reservoir well topped-up with fluid to prevent air being drawn into the system again.
7 With the pedal fully depressed, tighten the bleed nipple.
8 Refit the dust cap. Always use new hydraulic fluid for topping-up the reservoir, which has been stored in an air tight tin and has not been shaken during the preceding 24 hours. Always discard fluid which has been bled from the system or retain it for bleed jar purposes only.

8 Clutch – removal

1 Access to the clutch assembly and release mechanism may be gained either by removing the gearbox leaving the engine in position, or by removing the power plant as a unit and then separating the engine and gearbox.
2 With the gearbox removed from the engine, put mating marks on the clutch cover and on the flywheel to ensure that the clutch will be refitted in exactly the same position.
3 Unscrew the clutch cover securing bolts one turn at a time in diagonal pairs to avoid distorting the clutch cover mounting flange.
4 When all the bolts have been removed, insert a rod or screwdriver through the centre hole of the driven plate so that the plate does not drop out when the clutch cover is removed. Then remove the cover assembly and driven plate.

9 Clutch – inspection and renovation

1 Examine the clutch plate friction linings for wear and loose rivets and the disc for rim distortion, cracks, perished torsion rubbers and worn splines. The surface of the friction linings may be highly glazed, but as long as the clutch material pattern can be clearly seen this is satisfactory. If the amount of friction lining remaining is less than the minimum given in the Specifications, a new clutch plate should be obtained.

Fig. 5.5 Master cylinder – exploded view

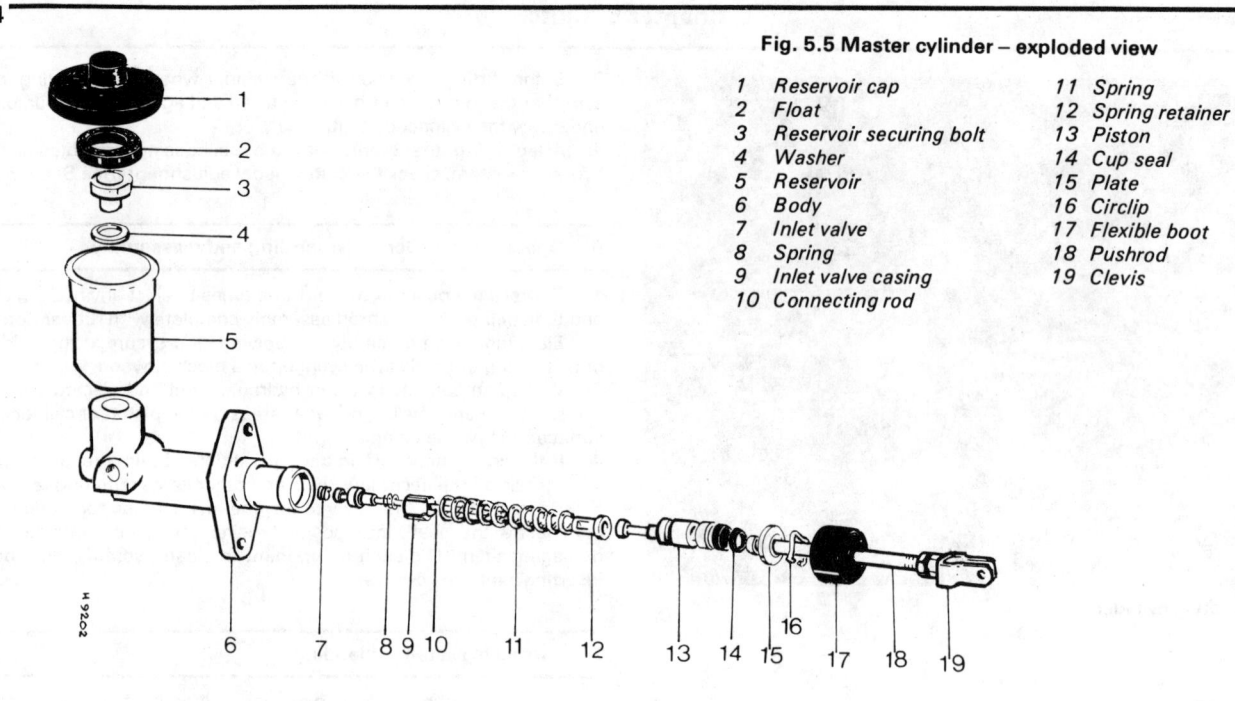

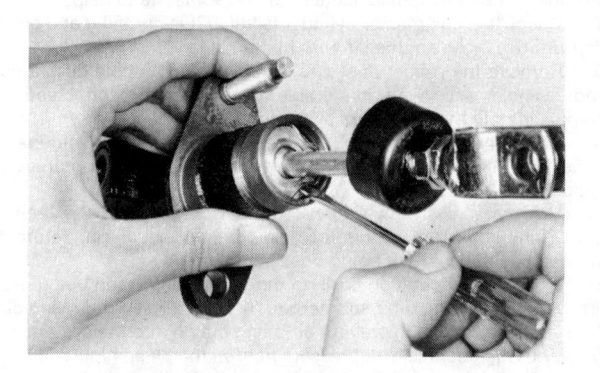

1. Reservoir cap
2. Float
3. Reservoir securing bolt
4. Washer
5. Reservoir
6. Body
7. Inlet valve
8. Spring
9. Inlet valve casing
10. Connecting rod
11. Spring
12. Spring retainer
13. Piston
14. Cup seal
15. Plate
16. Circlip
17. Flexible boot
18. Pushrod
19. Clevis

Fig. 5.6 Removing the circlip

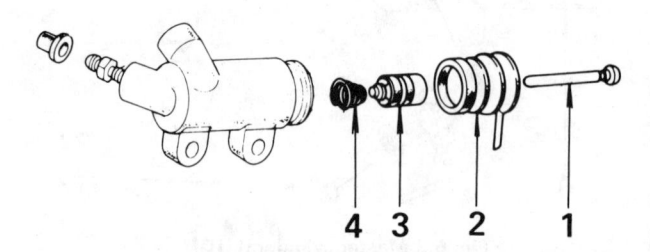

Fig. 5.7 Operating cylinder – exploded view

1. Pushrod
2. Boot
3. Piston
4. Spring

Fig. 5.8 Piston assembly – apply rubber grease to areas arrowed

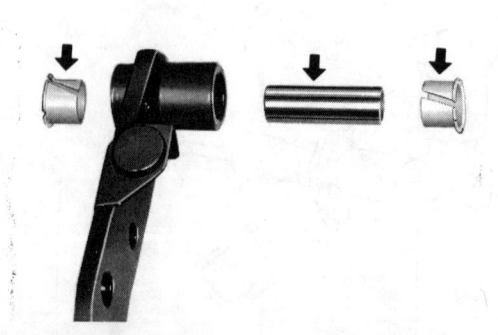

Fig. 5.9 Pedal assembly – apply grease to areas arrowed

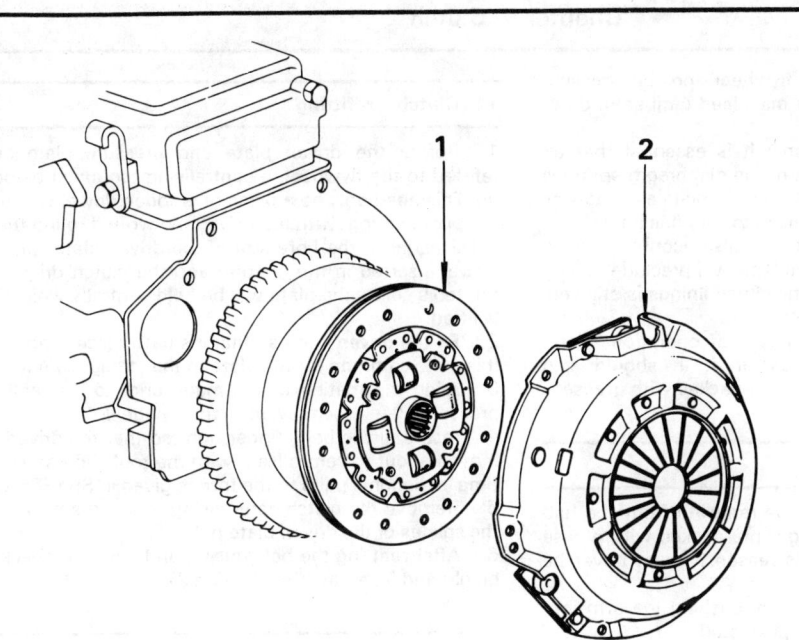

Fig. 5.10 Clutch components

1 Driven plate
2 Pressure plate assembly

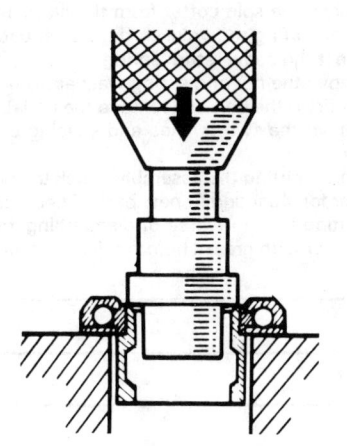

Fig. 5.11 Pressing the hub from the release bearing

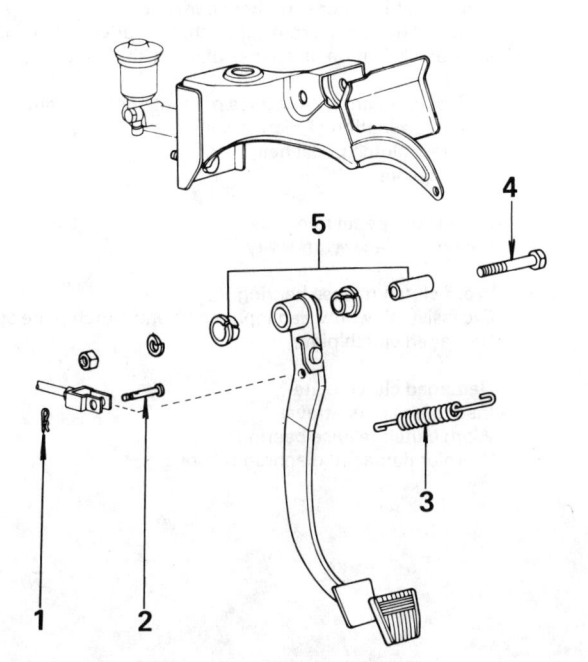

Fig. 5.12 Fitting the release bearing

Fig. 5.13 Clutch pedal components

1 Pin
2 Pushrod pin
3 Spring
4 Pedal shaft
5 Bushes, collar and pedal

2 Check the machined surfaces of the flywheel and the pressure plate. If either are grooved they should be machined until smooth, or renewed.
3 If the pressure plate is cracked or split it is essential that an exchange unit is fitted, also if the pressure of the diaphragm spring is suspect. It is not practical to dismantle the pressure plate assembly as it will have been accurately set up and balanced to very fine limits.
4 If a new clutch plate is being fitted it is a false economy not to renew the release bearing at the same time. This will preclude having to renew it at a later date when wear on the clutch linings is still very small.
5 Check the release bearing for smoothness of operation. There should be no harshness and no slackness in it. It should spin reasonably freely bearing in mind it has been pre-packed with grease.

10 Clutch release bearing – renewal

1 The clutch release bearing should not be removed from the hub unless it needs to be renewed. The bearing is pre-packed with grease to give permanent lubrication and for this reason it must never be cleaned with solvent.
2 Remove the bearing and fork assembly by pushing the arm into the bellhousing so that the fork slides off its pivot stud.
3 Detach the two clips which secure the hub and bearing assembly to the release fork.
4 Using a mandrel of suitable diameter, press or drive the hub out from the centre of the bearing. Press or drive the new bearing onto the hub until the upper face of the bearing is flush with the end of the hub with the chamfered end of the bearing uppermost.
5 When refitting the hub and bearing assembly to the fork arm use new spring clips and grease the pivot pin before refitting the arm to it.
6 Check the condition of the dust excluding bellows on the clutch arm and fit a new one if necessary.
7 Before refitting the clutch assembly, check the condition of the input shaft spigot bearing in the end of the crankshaft. If there is excessive play in the bearing, extract it and fit a new one (See Chapter 1).

11 Clutch – refitting

1 Before the driven plate and pressure plate assembly can be refitted to the flywheel, a centralising tool must be obtained, or made up. This need only be a piece of wooden dowel which is a snug fit into the pilot bearing, with adhesive tape wound round the dowel to form a collar equal to the bore size of the driven plate splined hub. With the dowel inserted in the bearing and the clutch driven plate located on the tools collar, the plate will be held centrally while the pressure plate is fitted.
2 Fit the driven plate so that it's flatter side is against the flywheel – it is quite common for the plate to the stamped *Flywheel* on one side.
3 Align the mating marks made prior to dismantling and offer the pressure plate assembly up to the flywheel.
4 Insert all the bolts finger tight so that the driven plate can still be moved about to centralise it with the tool. Tighten the bolts a turn at a time in diagonal pairs to the torque given in Specifications.
5 Remove the clutch centralising tool and smear a little grease onto the splines of the driven plate hub.
6 After refitting the bellhousing and gearbox, check the clutch pedal height and free play (See Section 2).

12 Clutch pedal – removal and refitting

1 Remove the split cotter from the clevis pin connecting the master cylinder operating rod yoke to the clutch pedal and withdraw the pin. Disconnect the pedal return spring.
2 Remove the nut and spring washer from the end of the pedal pivot bolt, withdraw the bolt and remove the pedal.
3 Remove the two bushes and spacing collar from the bore of the pedal arm.
4 Before refitting the assembly check the pedal rubber, pedal bushes and collar for wear and fit new parts if necessary.
5 Refitting is the reverse of dismantling, but smear the bushes and spacer collar with grease before refitting them.

13 Fault diagnosis – clutch

Symptoms	Reason/s
Judder when taking up drive	Loose engine or gearbox mountings Friction surface contaminated with oil Driven plate torsional rubbers perished Worn splines on gearbox input shaft or driven plate hub Lack of lubrication on splines of input shaft and hub
Clutch spin – failure to disengage so that gears cannot be meshed	Lack of lubrication on splines of input shaft and hub Oil contamination of friction surface Incorrect clutch pedal height Worn splines
Clutch slip – increase in engine speed does not result in a corresponding increase in road speed	Insufficient pedal free-play Friction surface worn away
Noise when clutch pedal is depressed	Worn clutch release bearing Excessive play between input shaft and clutch plate splines Damaged clutch plate
Noise when clutch pedal is released	Damaged clutch plate Insufficient pedal travel Worn clutch release bearing Worn or damaged diaphragm spring

Chapter 6 Transmission

For modifications and information applicable to later USA models, refer to Supplement at end of manual

Contents

Part A – Manual gearbox

Fault diagnosis – manual gearbox	23
Gearbox – removal and installation	2
Gearbox (Type P51) – dismantling	16
Gearbox (Type T50) – dismantling	3
Gearbox (Type W50) – dismantling	9
Gearbox (Type P51) – reassembly	22
Gearbox (Type T50) – reassembly	8
Gearbox (Type W50) – reassembly	15
Gear case (P51) – overhaul	21
Gear case (T50) – overhaul	7
Gear case (W50) – overhaul	14
General description	1
Input shaft (P51) – overhaul	17
Input shaft (T50) – overhaul	6
Input shaft (W50) – overhaul	11
Laygear (P51) – overhaul	18
Laygear (T50) – overhaul	5
Laygear (W50) – overhaul	12
Mainshaft (T50) – overhaul	4
Mainshaft (W50) – overhaul	10
Reverse idler gear (P51) – overhaul	19
Reverse idler gear (W50) – overhaul	13
Synchro-hub (P51) – overhaul	20

Part B – Automatic transmission

Automatic transmission – general description and precautions	24
Automatic transmission – removal and refitting	30
Extension housing oil seal – renewal	29
Fault diagnosis – automatic transmission	31
Maintenance	25
Neutral start switch – adjustment	28
Selector linkage – adjustment	26
Throttle cable (downshift) – adjustment	27

Specifications

Part A – Manual gearbox

Type .... Five forward speeds and reverse. Synchromesh on all forward speeds. Remote-type floor-mounted lever

Application:
- Type T50 .... 1600 ST models
- Type W50 .... All models except GT
- Type P51 .... GT twin-cam models

Gear ratios

	T50	W50	P51
1st	3·587 : 1	3·287 : 1	3·525 : 1
2nd	2·022 : 1	2·043 : 1	2·054 : 1
3rd	1·384 : 1	1·394 : 1	1·396 : 1
4th	1·000 : 1	1·000 : 1	1·000 : 1
5th	0·861 : 1	0·853 : 1	0·858 : 1
Reverse	3·484 : 1	4·039 : 1	3·755 : 1

Oil capacity
- Type T50 .... 2·8 pints (1·6 US qt) 1·5 litres
- Type W50 .... 4·6 pints (2·7 US qt) 2·6 litres
- Type P51 .... 4·6 pints (2·7 US qt) 2·6 litres

Torque wrench settings	lbf ft	Nm
Type T50		
Casing half bolts	20	28
Mainshaft rear nut	50	68
Extension housing bolts	20	28
Clutch bellhousing-to-engine	45	62
Reverse idler shaft lockbolt	13	18
Restrictor pin plugs	30	41
Clutch bellhousing bolts	30	41
Gearchange lever turret bolts	40	55
Type W50		
Extension housing bolts	35	48
Restrictor pin plugs	30	41
Detent ball plugs	18	25
Gearchange lever turret bolts	14	19
Front bearing retainer bolts	6·0	8·0
Clutch bellhousing-to-gearcase bolts	50	69
Clutch bellhousing-to-engine bolts	50	69
Type P51		
As Type W50 except for:		
Reverse idler shaft nut	40	55
Countershaft rear bearing screws	50	69
Detent ball plugs	20	28
Speedometer driven gear lockplate bolt	11	15

Part B – Automatic transmission

Type	A40 torque converter, bandless type with three forward speeds and reverse
Fluid capacity:	
Drain and refill	4·2 pints (2·5 US qt/2·4 litres)
Dry refill	11 pints (6·7 US qt/6·3 litres)
Fluid type	ATF type F
Gear ratios:	
L1	2·450 : 1
L2	1·450 : 1
D	1·000 : 1
Reverse	2·222 : 1

Torque wrench settings	lbf ft	Nm
Oil sump bolts	4·0	6·0
Oil screen bolts	4·0	6·0
Drain plug	12	16
Extension housing-to-main casing	30	41
Torque converter housing-to-engine	50	69
Driveplate-to-crankshaft:		
2T, 18R engine	45	62
20R engine	60	83
Driveplate-to-torque converter	20	28
Oil cooler hose unions-to-transmission	20	28

PART A — MANUAL GEARBOX

1 General description

Three types of manual gearbox may be fitted, according to the model and engine combination. All have five forward speeds, each with synchromesh, controlled by a floor-mounted gearlever. The gear ratios vary according to the type of gearbox specified.

2 Gearbox – removal and installation

1 The gearbox may be removed from the car leaving the engine in position. Where it is required to remove the engine and gearbox together as a combined assembly, refer to Chapter 1.
2 The gearbox removal operations apply to all types of unit.
3 With the car over an inspection pit or with the rear raised sufficiently high to enable the gearbox to pass out beneath the car disconnect the battery.
4 Working inside the car, remove the centre console and flexible gaiter. With the gearchange lever in neutral, unbolt and withdraw the lever (photos).
5 Working under the car, drain the gearbox and disconnect the reversing lamp switch wires (photo).
6 Disconnect the exhaust bracket from the transmission casing and uncouple the exhaust pipe from the manifold.
7 Unbolt, disconnect and remove the starter motor.
8 Unbolt and remove the reinforcement brackets from the junction of the clutch bellhousing and the engine crankcase lower flange.
9 Disconnect the clutch operating cylinder. There is no need to uncouple the hydraulic pipe, simply unbolt the cylinder and tie it up out of the way.
10 Disconnect the speedometer drive cable from the transmission casing, also the earthing leads (photo).
11 Remove the propeller shaft as described in Chapter 7.
12 Place a jack undr the gearbox and just take its weight (a trolley

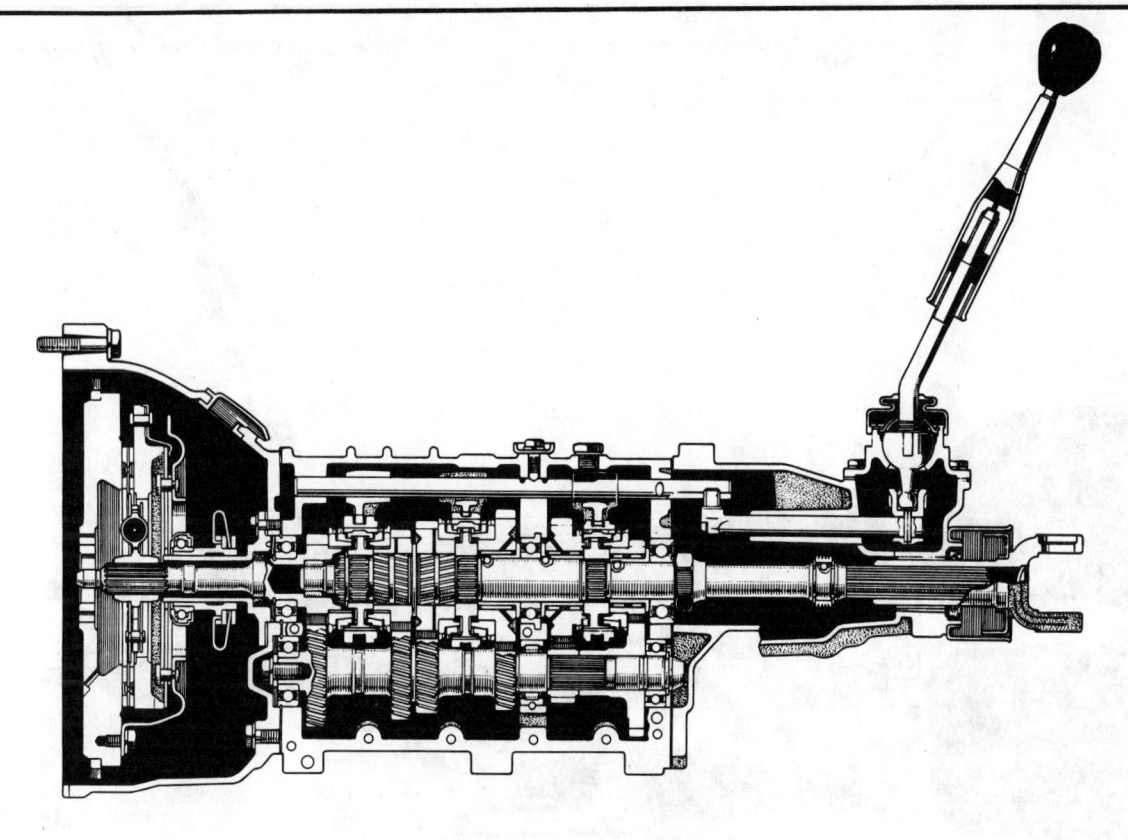

Fig. 6.1 Cutaway view of Type T50 gearbox

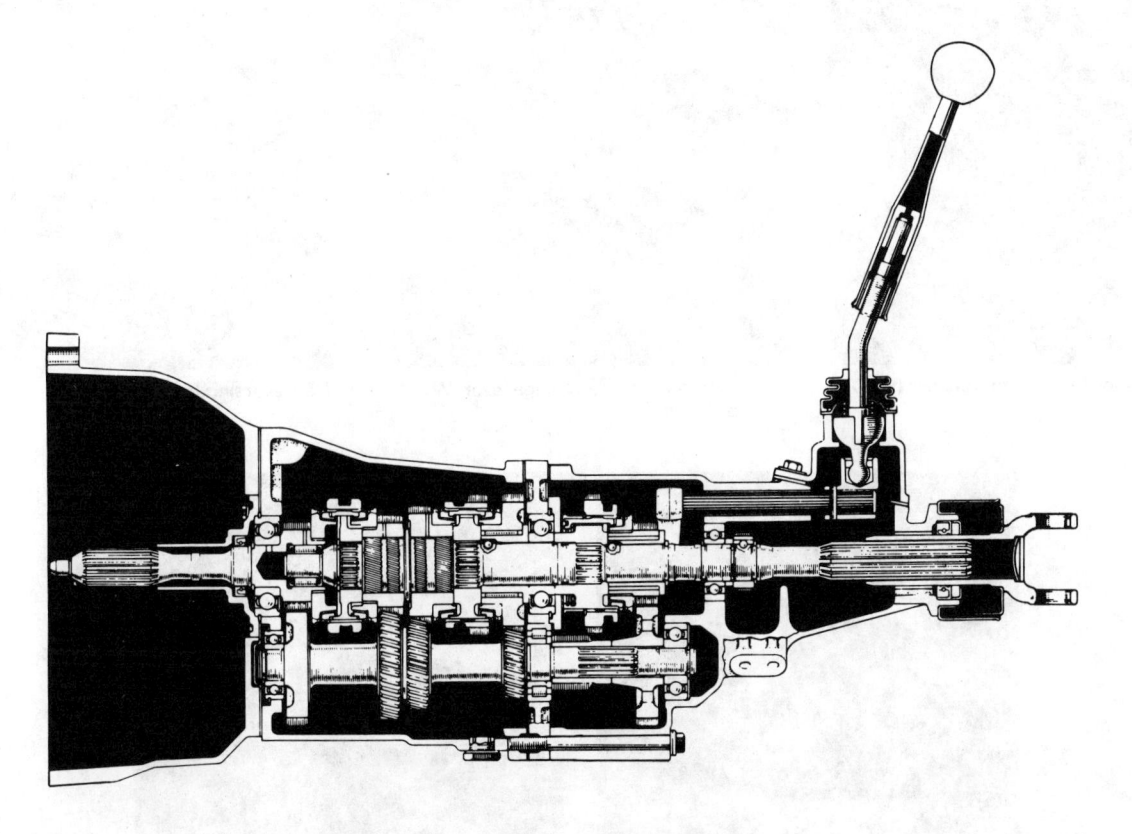

Fig. 6.2 Cutaway view of Type W50 gearbox

Fig. 6.3 Cutaway view of Type P51 gearbox

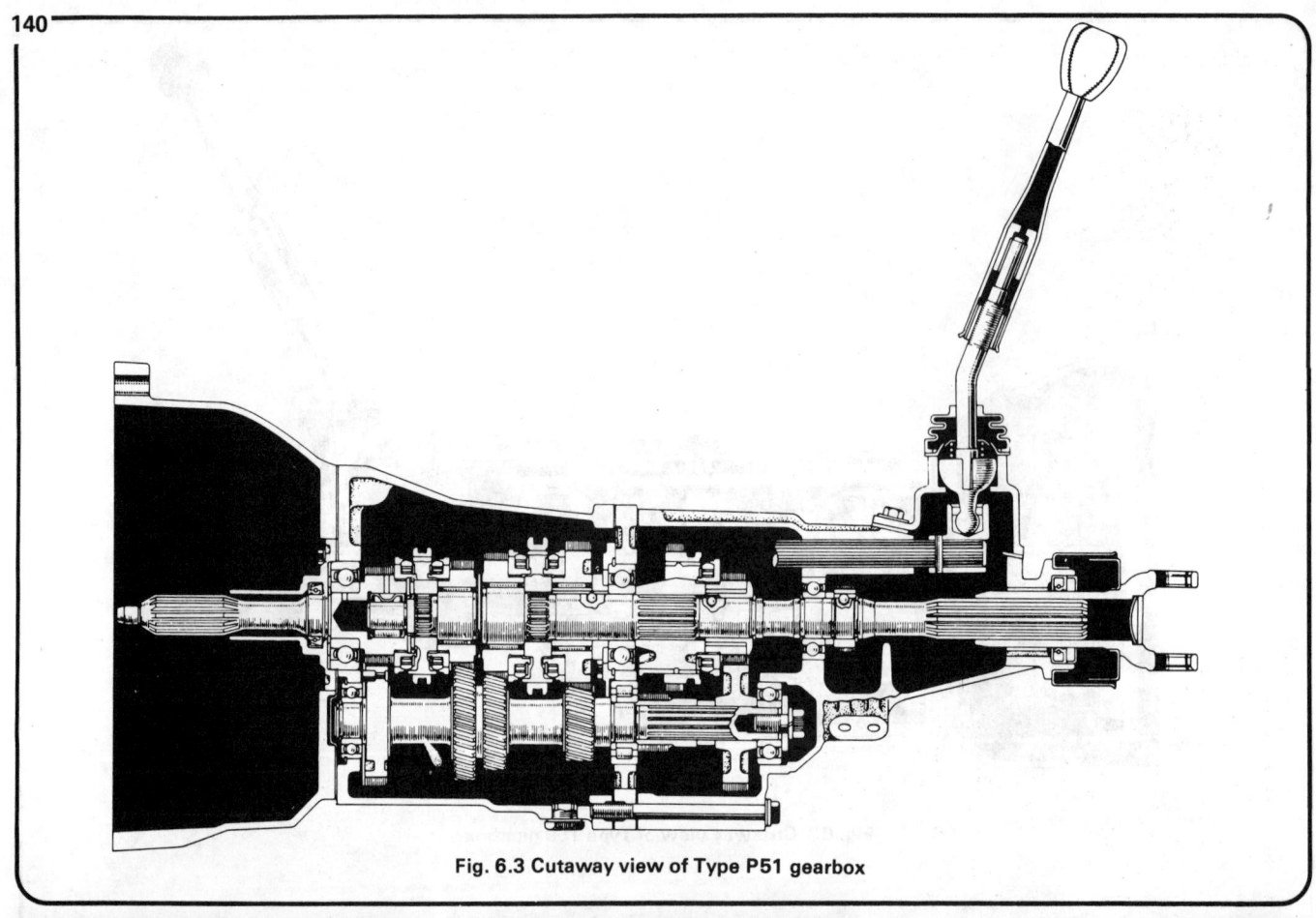

2.4a Gearchange lever gaiter retainer (W50)

2.4b Withdrawing gearchange lever (W50)

2.5 Reversing lamp switch (W50)

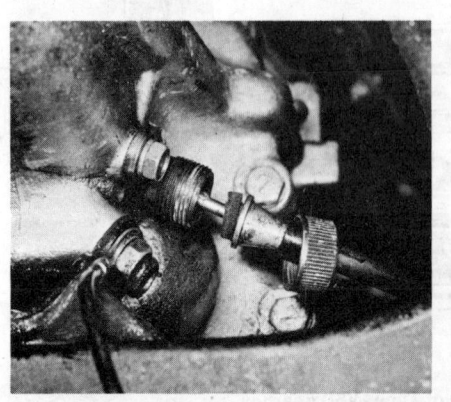

2.10 Withdrawing speedometer cable from W50 transmission

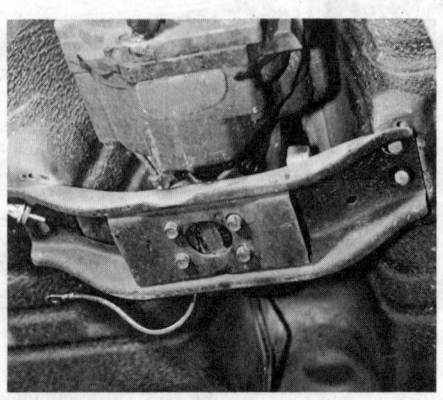

2.12 Gearbox rear crossmember and mounting (W50)

Chapter 6 Transmission

Fig. 6.4 Gearbox external components (T50)

1 Clutch release mechanism
2 Bellhousing
3 Mainshaft convex washer
4 Layshaft convex and plain washers
5 Speedometer driven gear
6 Restrictor pins
7 Extension housing
8 Reversing lamp switch
9 Detent balls and springs

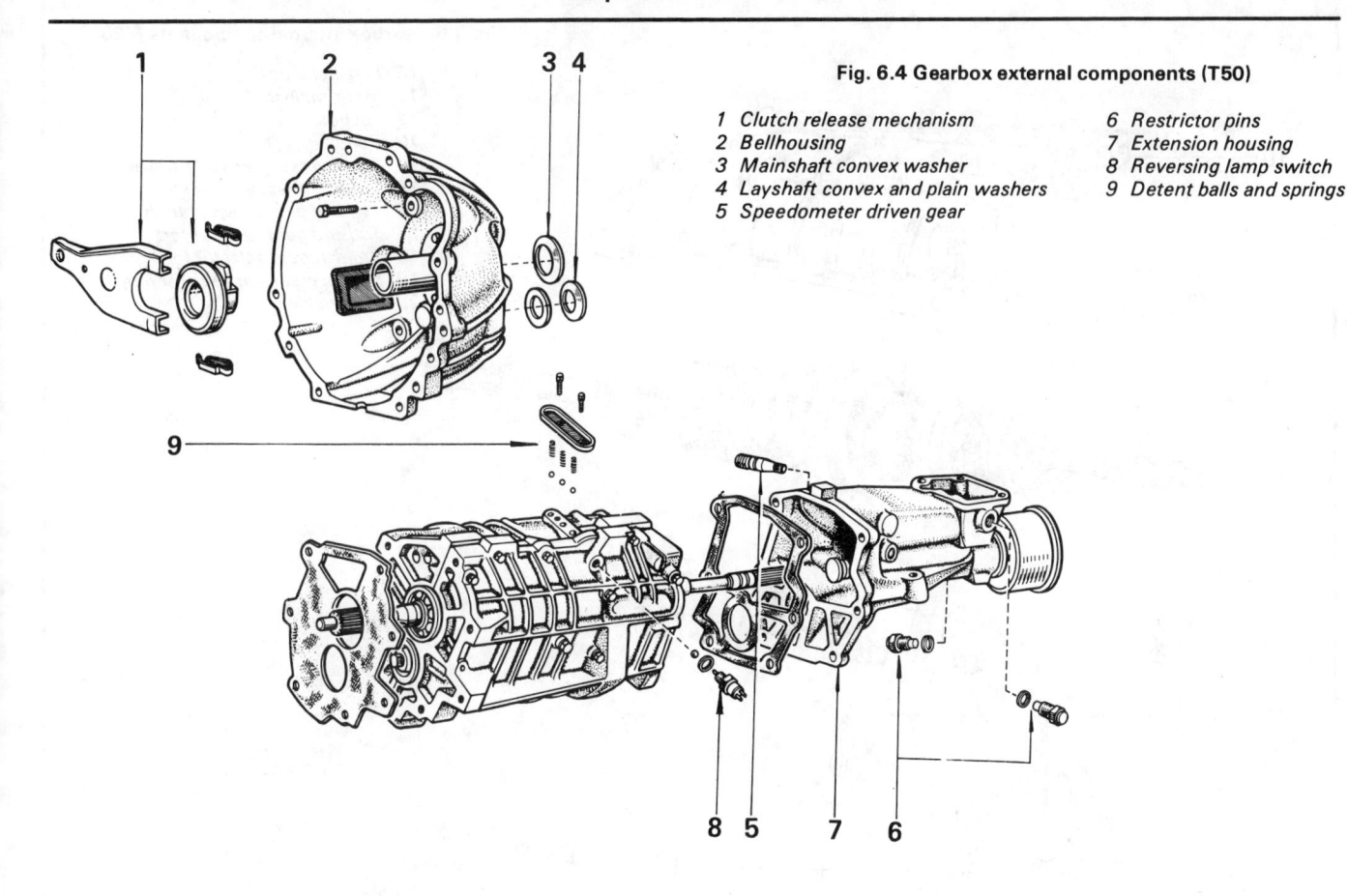

jack is recommended). Unbolt the rear crossmember from the bodyframe and then unbolt and remove the crossmember from the gearbox rear mounting (photo).
13 Unscrew and remove the bolts which hold the clutch bellhousing to the engine.
14 Lower the jack carefully until the gearbox can be withdrawn to the rear. Take care that the engine does not damage components on the engine compartment rear bulkhead during this operation as the cylinder head inclines rearward. If necessary, fit a block of wood between the engine sump and the crossmember or between the rear face of the cylinder block and the bulkhead to prevent the engine tilting excessively. Watch for strain on the radiator top hose and throttle linkage. Release these connections if necessary (which will require the cooling system be drained).
15 When withdrawing the gearbox, support its weight while the input shaft is still engaged with the splines of the clutch driven plate.
16 Installation is a reversal of removal but if the clutch has been dismantled make sure that this has been centralised as described in Chapter 5.
17 Refill the gearbox with lubricant on completion of installation.

3 Gearbox (Type T50) – dismantling

1 With the gearbox removed from the car, clean off all external dirt with paraffin or a water soluble solvent.
2 From inside the clutch bellhousing, remove the clutch release mechanism (Chapter 5).
3 Unbolt and withdraw the clutch bellhousing.
4 Unscrew and remove the speedometer driven gear.
5 Unscrew and remove the reversing lamp switch and the two restrictor pins.
6 Unbolt and withdraw the extension housing.
7 Place a cloth over the detent ball cover plate, unbolt the plate and remove it carefully so that the detent balls and springs do not fly out. Extract the springs and balls by turning the gearbox upside down or using a pencil magnet.

8 Extract the bolts which hold both halves of the gearcase together and tap the left-hand half off with a soft faced hammer. Watch out for the locking ball as the two halves separate.
9 Lift out the mainshaft/input shaft assembly.
10 Lift out the layshaft assembly.
11 Drive out the roll-pins which secure the selector forks to the selector shafts, withdraw the shafts noting their location in the casing. Lift off the forks as the shafts are withdrawn.
12 Extract the interlock plungers from the holes which connect the selector shaft holes in the gear case.

4 Mainshaft (T50) – overhaul

1 Before dismantling the mainshaft, check the endfloat between the gearwheels with a feeler blade. If the clearances exceed the following tolerances, the gearwheels are worn and will have to be renewed.

1st, 2nd, 5th	0·020 in (0·5 mm)
3rd, Reverse	0·024 in (0·6 mm)

2 Commence dismantling by extracting the circlips and sliding off the speedometer drivegear. Extract the gear locking ball.
3 Relieve the staking on the mainshaft nut, grip the shaft in a vice fitted with jaw protectors and then unscrew the nut.
4 Withdraw the gears, synchronizer units and other components, keeping them in their originally fitted order. Remove reverse gear by supporting it and pressing the mainshaft from it. Remove 1st gear in a similar manner.
5 From the front end of the mainshaft, extract the circlip, support the rear face of the 3rd gear and press the mainshaft from the gear.
6 Check all components for wear or damage.
7 If the synchronizer assemblies are worn, there has been a history of noisy gear changing, or if the synchromesh has been weak, the complete assembly should be renewed.
8 The clearance between the selector forks and their respective

Fig. 6.5 Gearbox internal components (T50)

10 Left-hand casing
11 Input/mainshaft
12 Laygear
13 Rollpin
14 5th/reverse gear selector shaft
15 3rd/4th gear selector shaft
16 1st/2nd gear selector shaft
17 1st/2nd gear selector fork
18 3rd/4th gear selector fork
19 5th/reverse gear selector fork
20 Interlock pin

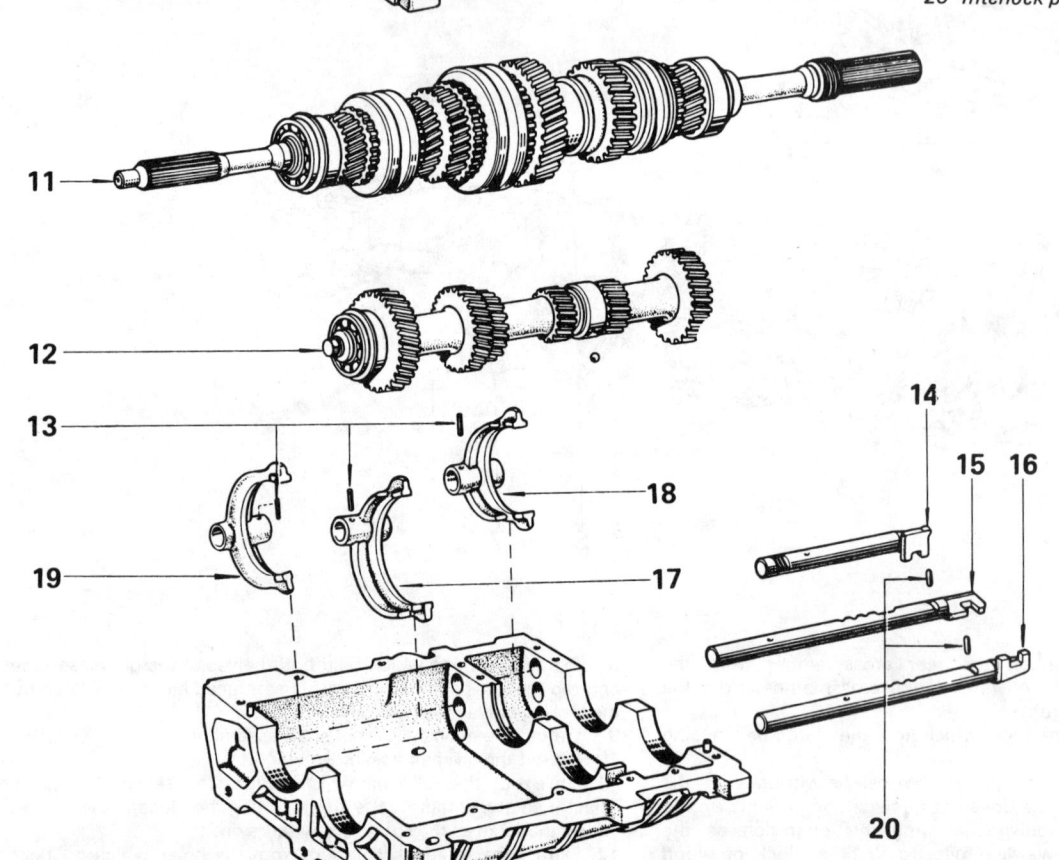

Fig. 6.6 Checking gear endfloat (T50)

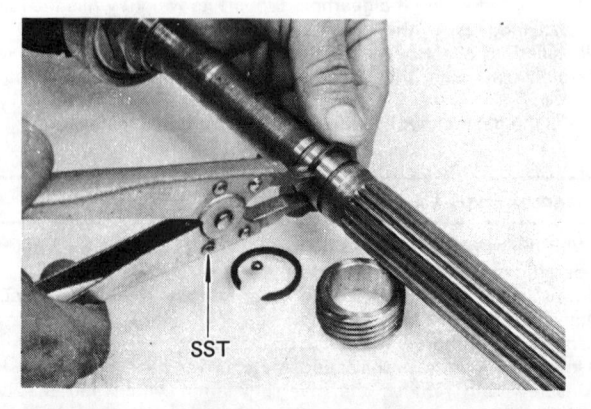

Fig. 6.7 Extracting speedometer drivegear circlip (T50)

SST

143

Fig. 6.8 Relieving mainshaft nut staking (T50)

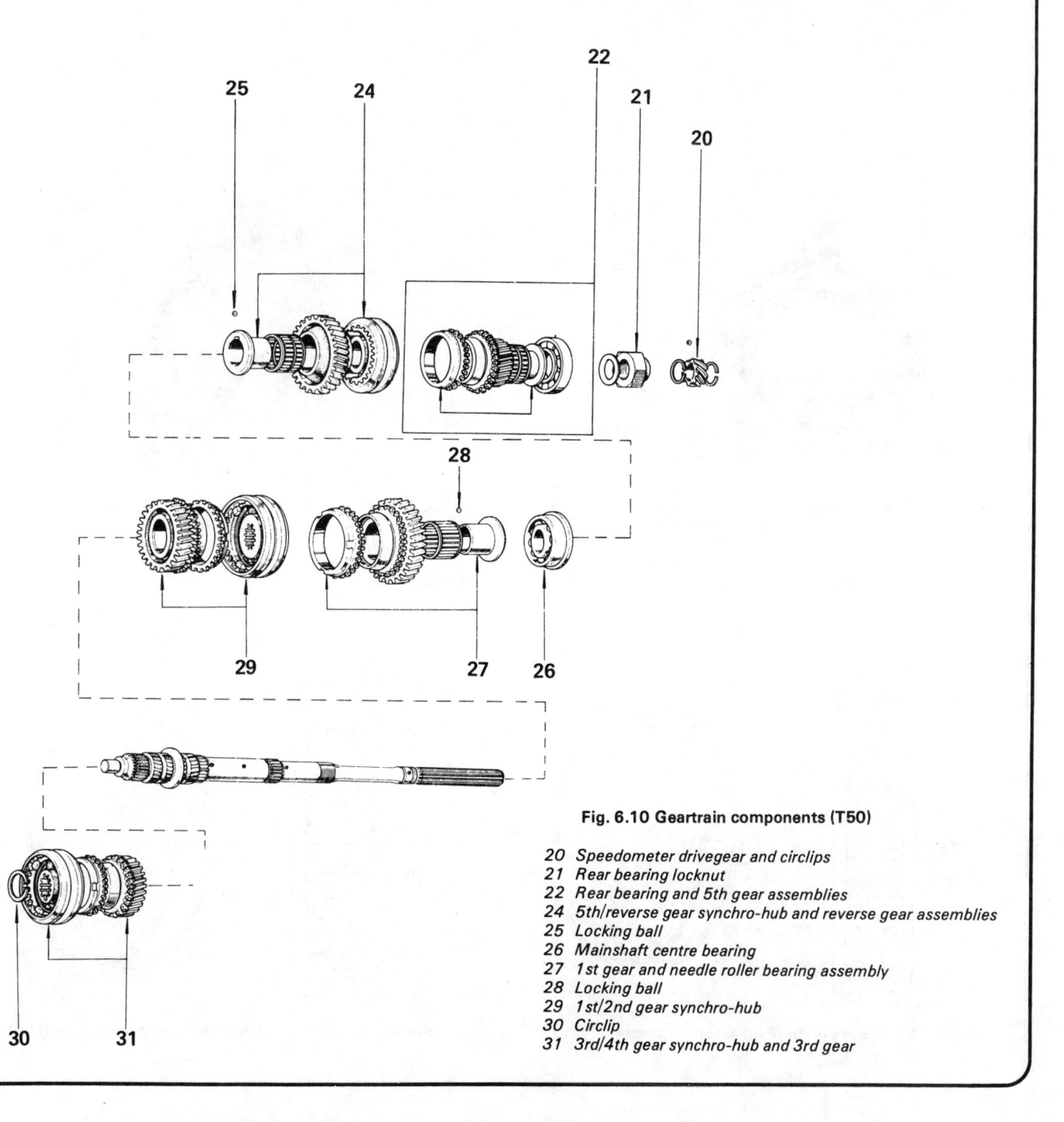

Fig. 6.9 Pressing mainshaft from reverse gear (T50)

Fig. 6.10 Geartrain components (T50)

20 Speedometer drivegear and circlips
21 Rear bearing locknut
22 Rear bearing and 5th gear assemblies
24 5th/reverse gear synchro-hub and reverse gear assemblies
25 Locking ball
26 Mainshaft centre bearing
27 1st gear and needle roller bearing assembly
28 Locking ball
29 1st/2nd gear synchro-hub
30 Circlip
31 3rd/4th gear synchro-hub and 3rd gear

Fig. 6.11 Pressing mainshaft from 1st gear (T50)

Fig. 6.12 Pressing mainshaft from 3rd gear (T50)

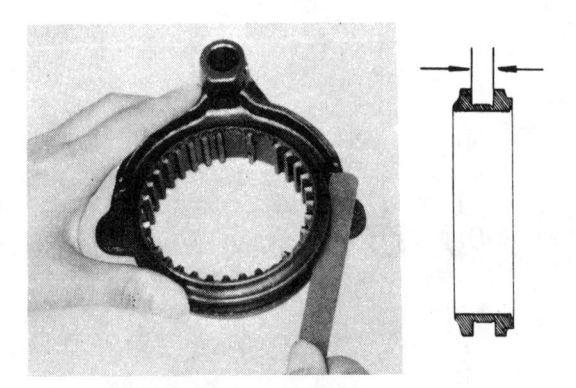

Fig. 6.13 Checking selector fork-to-synchro sleeve groove clearance

Fig. 6.14 Checking synchro baulk ring-to-gear clearance

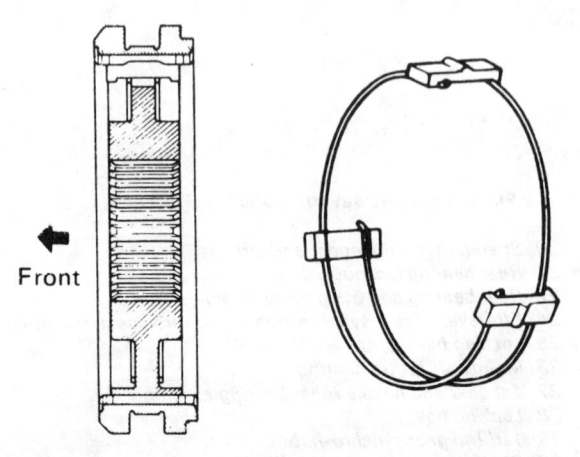

Fig. 6.15 1st/2nd synchro-hub assembly diagram (T50)

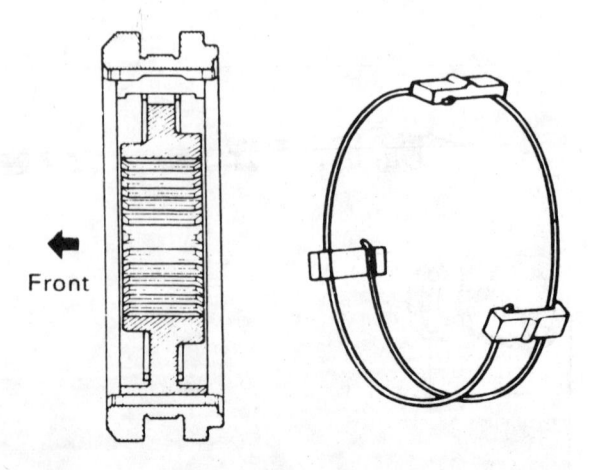

Fig. 6.16 3rd/4th synchro-hub assembly diagram (T50)

Chapter 6 Transmission

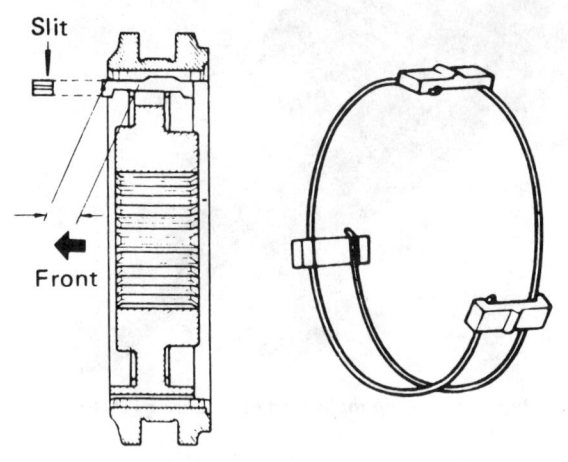

Fig. 6.17 5th/reverse synchro-hub assembly diagram (T50)

Fig. 6.18 1st/2nd synchro bush and locking ball (T50)

Fig. 6.19 Fitting centre bearing and reverse gear to mainshaft (T50)

synchro sleeve groove should not exceed 0·04 in (1·0 mm). If it does, renew the components as necessary.

9 The clearance between the synchro rings and their respective gears should not exceed 0·03 in (0·8 mm) also press each ring to its gear cone at the same time turning the ring. The ring should stick to the cone proving effective braking action.

10 If the synchronising units have been dismantled, reassemble them in accordance with the diagrams, making sure that the sliding keys engage in the sleeve and hub slots, also that the two spring rings are correctly assembled.

11 Reassemble the mainshaft by reversing the dismantling sequence but observe the following points.

12 Make sure that the bush groove aligns properly with the locking ball as the 1st/2nd gear synchro-hub is installed.

13 Fit the centre bearing and reverse gear so that the latter slides over its locking ball.

14 After installing the reverse/5th gear synchro-hub slide on 5th gear engaging it with the locking ball.

15 Fit the mainshaft rear bearing so that its ball shroud is towards the rear.

16 If the old mainshaft locknut is to be used again, insert a shim under it so that the nut staking position will be altered. Tighten the nut to the specified torque wrench setting.

17 Fit the 3rd/4th gear synchro-hub to the front end of the mainshaft and then fit the thickest circlip which will fit in the shaft groove yet provide the minimum clearance between the circlip and the bearing. Circlips are available in the following thicknesses:

0·077 – 0·079 in	(1·95 to 2·00 mm)
0·079 – 0·081 in	(2·00 to 2·05 mm)
0·081 – 0·083 in	(2·05 – 2·10 mm)
0·083 – 0·085 in	(2·10 – 2·15 mm)
0·085 – 0·087 in	(2·15 – 2·20 mm)

18 Check the gear endfloat (see paragraph 1), stake the nut and fit the speedometer drivegear and the two circlips.

5 Laygear (T50) – overhaul

1 Inspect the laygear carefully for wear or damage.
2 To dismantle, remove the nut from the end of the shaft and extract the circlip from the outer diameter of the front bearing. Using a suitable puller, draw the bearing from the shaft.
3 Extract the circlip from the rear end of the layshaft and withdraw the bearing with a puller.
4 Support the rear face of 5th gear and press the layshaft from it.
5 Remove reverse gear and the centre bearing.
6 Reassembly is a reversal of removal, but note the following points. The centre bearing is fitted with the larger diameter end of the roller cage towards the front (nut) end of the shaft. When fitting the rear bearing circlip, select the thickest one which will fit and provide the minimum clearance between the circlip and the bearing. Circlips are available in the following thicknesses:

0·079 in	(2·00 mm)
0·071 in	(1·80 mm)
0·069 in	(1·75 mm)

6 Input shaft (T50) – overhaul

1 Check the gear teeth and bearing for wear or damage.
2 If the bearing is to be renewed, extract the circlip and while supporting the bearing, press the input shaft from it.
3 When the new bearing has been installed, select a circlip which will have the minimum clearance between it and the bearing yet will fit fully into its groove. Circlips are available in the following thicknesses:

0·077 – 0·079 in	1·95 – 2·00 mm
0·079 – 0·081 in	2·00 – 2·05 mm
0·081 – 0·083 in	2·05 – 2·10 mm
0·083 – 0·085 in	2·10 – 2·15 mm
0·085 – 0·087 in	2·15 – 2·20 mm

7 Gearcase (T50) – overhaul

1 Check the casing for cracks particularly around the bolt holes.

146

Fig. 6.20 Fitting 5th gear to mainshaft (T50)

Fig. 6.21 Fitting mainshaft rear bearing (T50)

Fig. 6.22 Fitting 3rd/4th gear synchro-hub to mainshaft (T50)

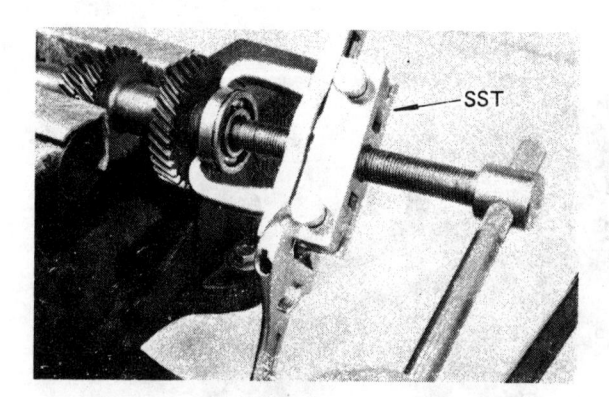

Fig. 6.23 Removing layshaft front bearing (T50)

Fig. 6.24 Pressing layshaft from 5th gear (T50)

Fig. 6.25 Removing reverse gear and centre bearing from layshaft (T50)

Chapter 6 Transmission

2 Renew the oil seals in the bellhousing and the extension housing as a matter of course.
3 Working inside the gearcase, measure the reverse idler gear endfloat. This should not exceed 0·04 in (1·0 mm). The thrust washers can be renewed if the reverse idler shaft lockbolt is removed. The idler gear centre bush can also be renewed but when pressing in the new bush, make sure that the oil holes align.
4 If the reverse restrictor pin is worn or does not slide easily it can be removed and renewed by unscrewing the blanking plug with an Allen key and then driving out the roll-pin with a punch. When refitting the plug apply sealant to its threads.
5 The gearchange selector levers and restrictor pins can be renewed by driving out the roll-pin or unscrewing the restrictor pin nuts.

8 Gearbox (Type T50) – reassembly

1 Make sure that the casing halves are clean inside and out.
2 Insert the interlock pins between the selector shaft holes in the casing, retaining them in position with thick grease.
3 Insert the selector shafts with their detent grooves upward and pick up the selector forks with the shaft as each one passes through into position. Commence with the 3rd/4th gear selector shaft, then 1st/2nd gear selector shaft and finally fit the reverse/5th gear selector shaft.
4 Pin the selector forks to their selector shafts and then set the shafts in the neutral position (cut outs in shaft dogs aligned).
5 Apply grease to the needle roller bearing in the end of the input shaft and fit the input shaft to the front end of the mainshaft making sure that the shift keys align with the slots in the synchroniser ring.
6 Set the synchro-sleeves in the neutral positions and fit the mainshaft/input shaft and layshaft assemblies to the gear case.
7 Fit the layshaft bearing locking ball into the groove in the gearcase web.
8 Smear the mating surfaces of the gear casings with jointing compound and bring them together.
9 Smear the threads of the bolts with jointing compound, insert and tighten them progressively in the sequence shown to the specified torque wrench settings.
10 Grip the end of the input shaft and check that it has a slight side-to-side movement, this is correct.
11 Smear the extension housing mating flange and the gearbox casing flange with jointing compound and place a new gasket in position.
12 Pass the extension housing over the output shaft taking care not to damage the extension housing oil seal. Engage the end of the remote control shift rod with the 3rd/4th gear selector shaft dog. Bolt up the extension housing to the specified torque. Fit the restrictor pins.
13 Stick the cone shaped washers into position on the rear face of the clutch bellhousing using heavy grease. Also apply some grease to the

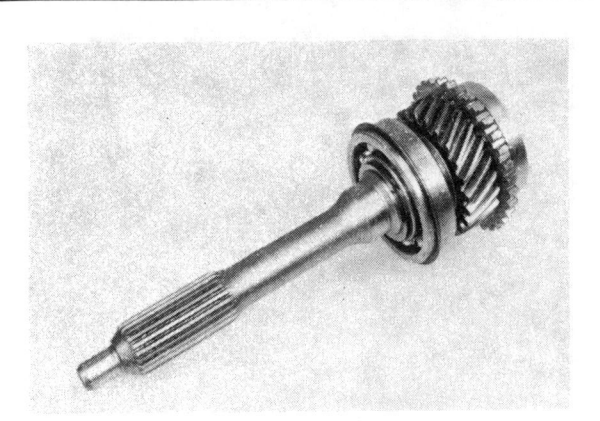

Fig. 6.26 Input shaft (T50)

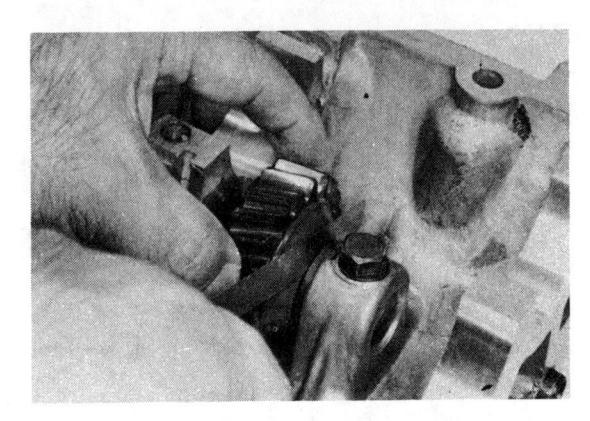

Fig. 6.27 Checking reverse idler gear endfloat (T50)

Fig. 6.28 Unscrewing reverse idler restrictor pin plug and driving out roll-pin (T50)

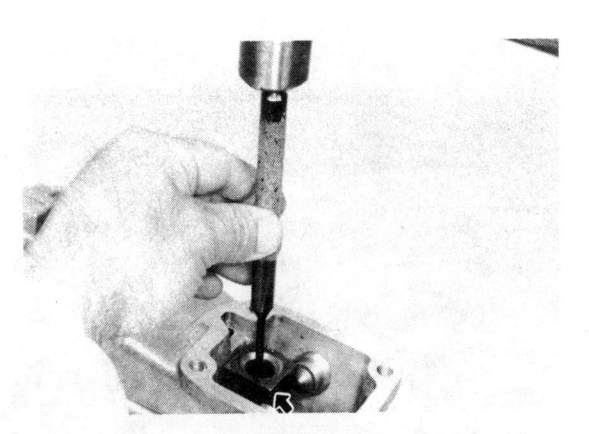

Fig. 6.29 Driving out the selector lever roll-pin (T50)

148

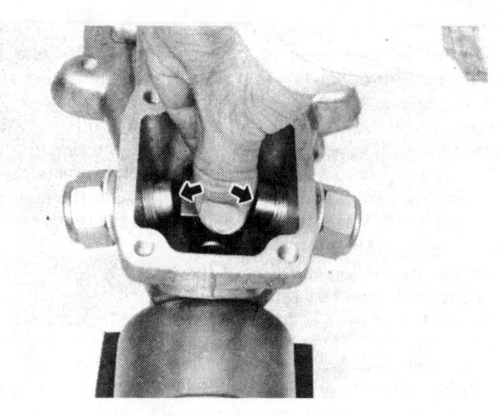

Fig. 6.30 Selector rod restrictor pins (T50)

Fig. 6.31 Selector shaft location (T50)

Top – 5th/reverse
Centre – 3rd/4th
Bottom – 1st/2nd

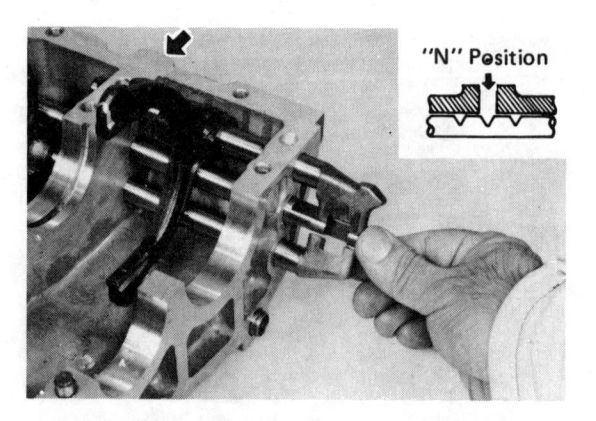

Fig. 6.32 Selector shafts installed (T50)

"N" Position

Fig. 6.33 Laygear assembly and locking ball installed (T50)

Fig. 6.34 Gearcase bolt tightening sequence (T50)

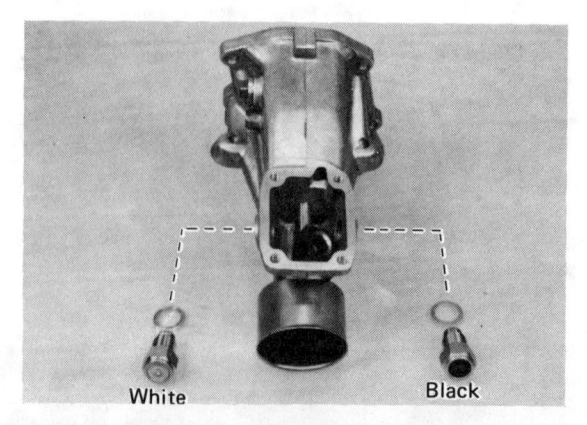

Fig. 6.35 Restrictor pin identification (T50)

White   Black

149

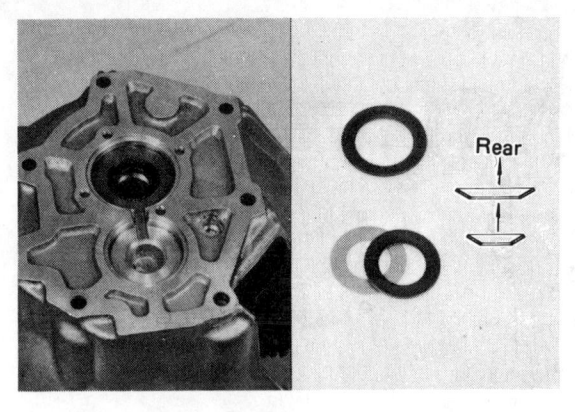

Fig. 6.36 Clutch bellhousing convex washers (T50)

Fig. 6.37 Applying grease to the bellhousing oil seal (T50)

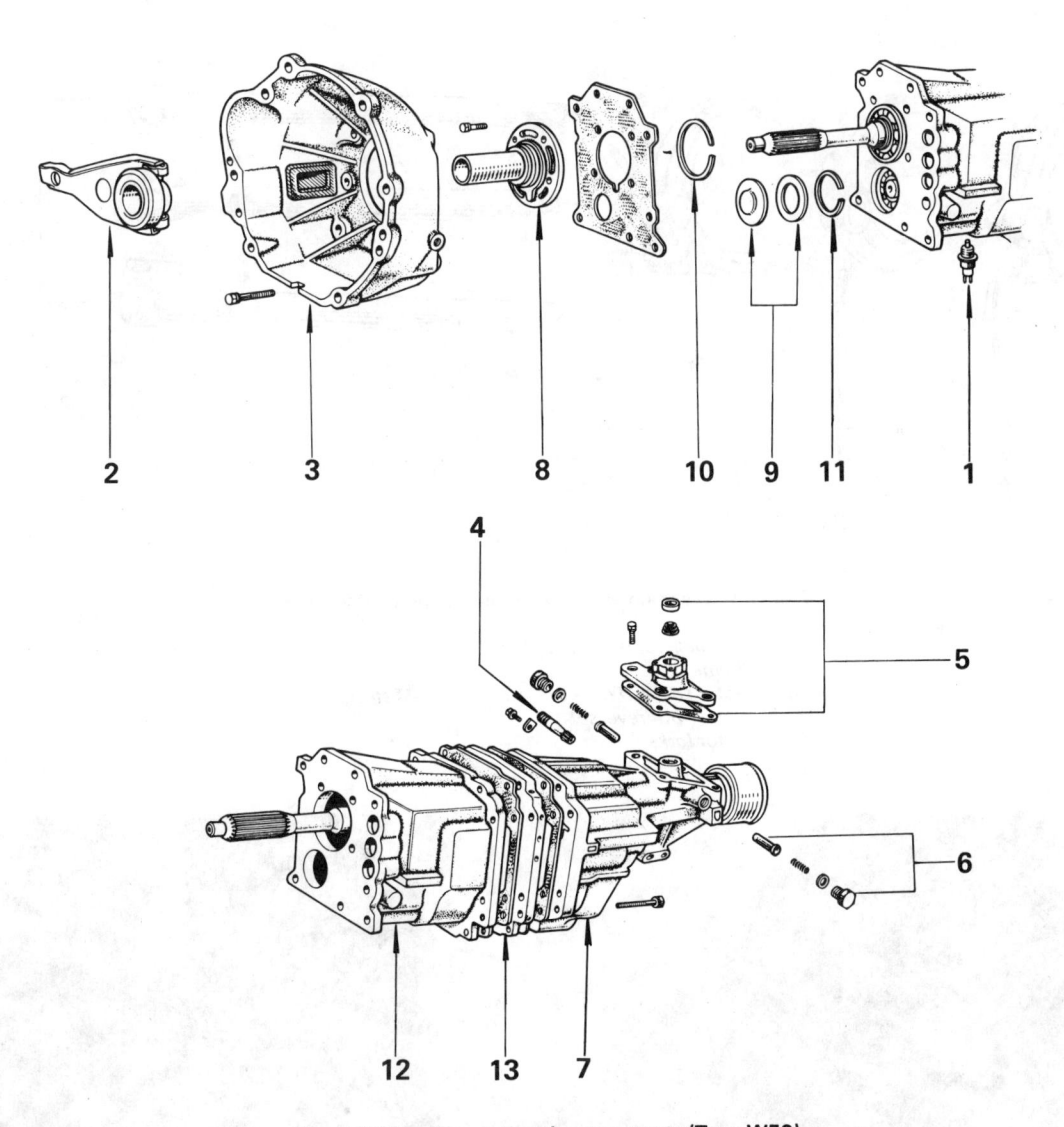

Fig. 6.38 Gearbox external components (Type W50)

1 Reversing lamp switch
2 Clutch release assembly
3 Clutch bellhousing
4 Speedometer driven gear
5 Gearchange lever turret
6 Restrictor pins, springs and plugs
7 Extension housing
8 Front bearing retainer
9 Layshaft spacer and cover
10 Circlip
11 Circlip
12 Gearbox casing
13 Intermediate plate

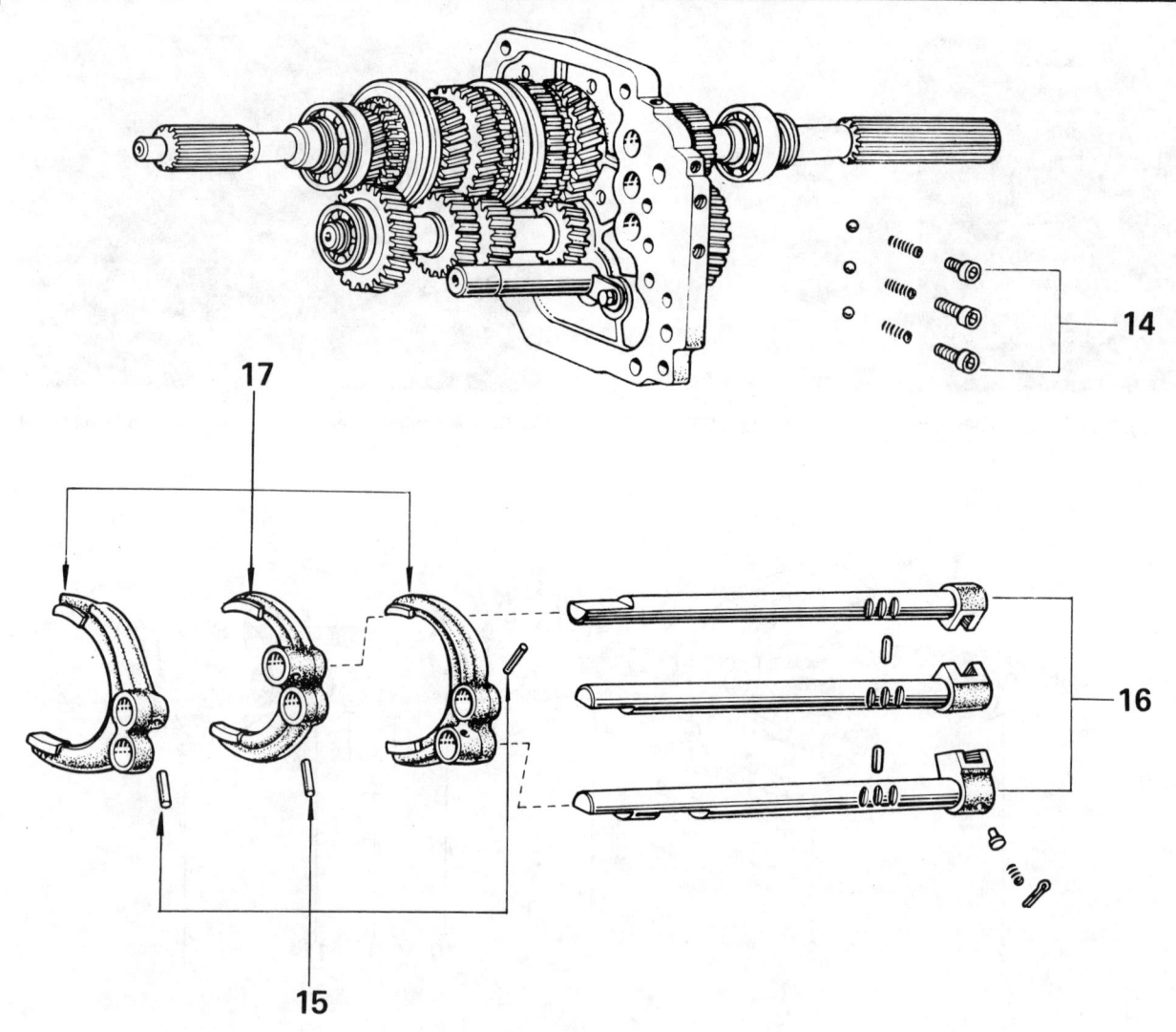

Fig. 6.39 Geartrain and selector components (W50)

14 Detent balls, springs and plugs
15 Roll-pins
16 Selector shafts (Top – 3rd/4th. Centre – 1st/2nd. Bottom – 5th/reverse)
17 Selector forks

Fig. 6.40 Clutch bellhousing bolt tightening sequence (T50)

Fig. 6.41 Front bearing retainer and layshaft bearing cover (W50)

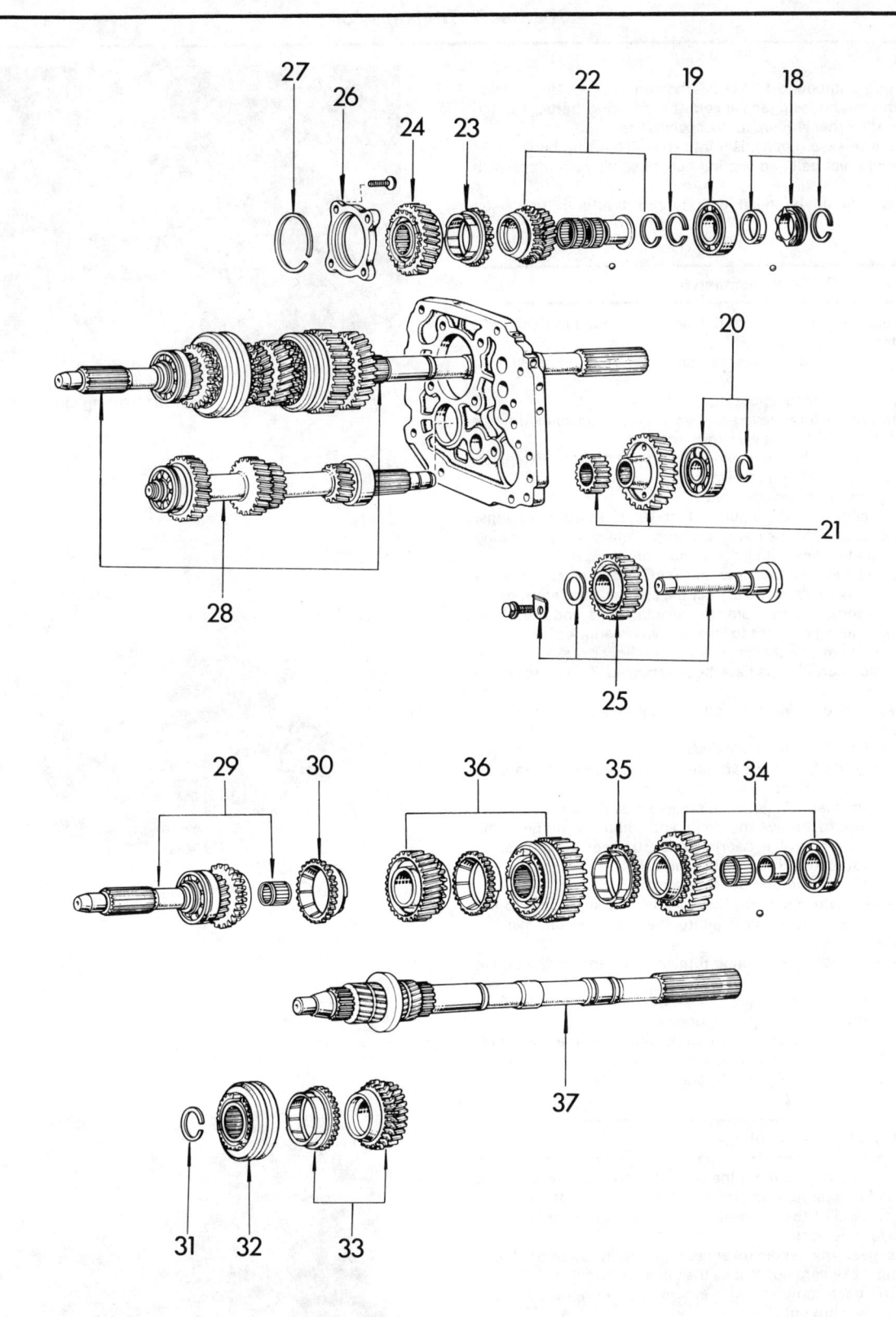

Fig. 6.42 Geartrain components (W50)

18 Speedometer drivegear	25 Reverse idler gear	32 3rd/4th gear synchro-hub
19 Mainshaft rear bearing	26 Bearing retainer	33 Synchro-ring and 3rd gear
20 Layshaft rear bearing	27 Circlip	34 Mainshaft bearing and 1st gear
21 5th gear and reverse gear	28 Input shaft, mainshaft and layshaft assemblies	35 Synchro-ring
22 5th gear	29 Input shaft and needle roller bearing	36 1st/2nd gear synchro-hub and 2nd gear
23 Synchro-ring	30 Synchro-ring	37 Mainshaft
24 5th/reverse gear synchro-hub	31 Circlip	

oil seal lips.

14 Apply jointing compound to the gearbox and bellhousing mating surfaces. Fit the bellhousing and insert the securing bolts. Tighten progressively in the order shown, to the specified torque.

15 Fit the clutch release mechanism into the bellhousing having first smeared the lever pivot ball and bearing hub inner sliding surface with grease.

16 The gearbox should be filled with lubricant after it has been installed in the car.

9 Gearbox (Type W50) – dismantling

1 With the gearbox removed from the car as described in Section 2, clean off all external dirt.
2 Remove the clutch release mechanism from inside the bellhousing.
3 Unbolt and remove the bellhousing.
4 Unscrew and remove the reversing lamp switch, the speedometer driven gear and the restrictor pins, springs and plugs.
5 Unbolt and remove the gearchange lever retainer, the extension housing and the front bearing retainer.
6 Remove the layshaft cover and spacer.
7 Extract the circlips from the input shaft and layshaft front bearings.
8 Remove the gear casing from the intermediate plate by tapping the intermediate plate away with a soft metal drift.
9 Using a socket wrench, unscrew and remove the plugs from the edge of the intermediate plate and extract the springs and detent balls.
10 Drive out the tension pins from the selector forks and withdraw the selector shafts taking care not to lose the two interlock pins.
11 Remove the speedometer drivegear and its spacer from the mainshaft after the shaft circlips have been removed. Take care not to lose the locking ball.
12 Using a two legged puller, draw off the bearing from the rear end of the mainshaft.
13 Remove the circlip from the mainshaft.
14 From the rear end of the layshaft, remove the circlip and draw off the bearing.
15 Withdraw the layshaft 5th gear and reverse gear.
16 From the mainshaft, remove the circlip and withdraw 5th gear, the synchronizer ring, needle roller bearing and 5th gear inner bearing track, taking care not to lose the track locking ball.
17 Remove the reverse gear and 5th gear synchro unit.
18 Slacken the bolt which secures the reverse idler shaft stop to the intermediate plate, withdraw the shaft to the rear and remove the reverse idler gear and spacer.
19 Unbolt the mainshaft rear bearing retainer and remove the circlip from the bearing.
20 From the rear end of the layshaft, push the bearing outer track to the rear and withdraw the bearing components.
21 Remove the input shaft and synchronizer ring from the mainshaft and then withdraw the mainshaft assembly from the intermediate plate, at the same time removing the layshaft assembly.

10 Mainshaft (W50) – overhaul

1 Extract the circlip and remove the 3rd/4th gear synchro unit, the synchro-ring and 3rd gear from the front end of the mainshaft.
2 From the rear end of the mainshaft, draw off the bearing. A press will be required for this operation.
3 Remove 1st gear, the needle roller bearing, bearing inner track and synchronizer ring. Take care not to lose the inner track locking ball.
4 Press off 2nd gear complete with synchronizer ring, reverse gear and 1st/2nd gear synchro unit.
5 Clean all components thoroughly and examine for worn or chipped teeth and grooving or scoring of the shaft. The gears should have a running clearance between their internal bores and the shaft of between 0·008 and 0·0020 in for 1st and 5th gears and between 0·0014 and 0·0039 in for 2nd and 3rd gears.
6 Check the synchronizer units as described in Section 4, paragraphs 7, 8 9 and 10. Reassemble the units in accordance with the diagrams.
7 Commence reassembly of the mainshaft by installing the 3rd/4th gear synchronizer ring to 3rd gear and then fitting them to the shaft.
8 Fit the 3rd/4th gear synchronizer unit, positioning it tightly against

Fig. 6.43 Removing the front bearing circlips (W50)

Fig. 6.44 Unscrewing detent plugs (W50)

Fig. 6.45 Removing a selector fork roll-pin (W50)

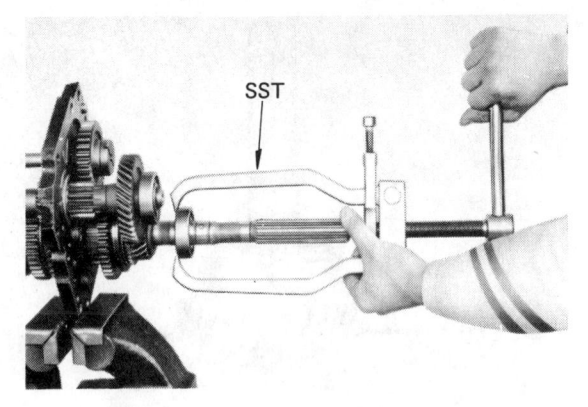

Fig. 6.46 Extracting mainshaft rear bearing (W50)

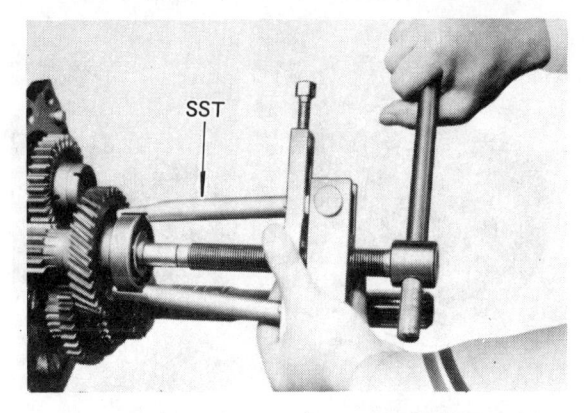

Fig. 6.47 Removing layshaft rear bearing (W50)

Fig. 6.48 Removing 5th and reverse gears from the layshaft (W50)

Fig. 6.49 Removing circlip and 5th gear from the mainshaft (W50)

Fig. 6.50 Extracting mainshaft rear bearing retainer screws (W50)

Fig. 6.51 Withdrawing mainshaft and layshaft gear trains (W50)

154

Fig. 6.52 Removing 3rd/4th gear synchro unit (W50)

Fig. 6.53 Pressing off the mainshaft rear bearing (W50)

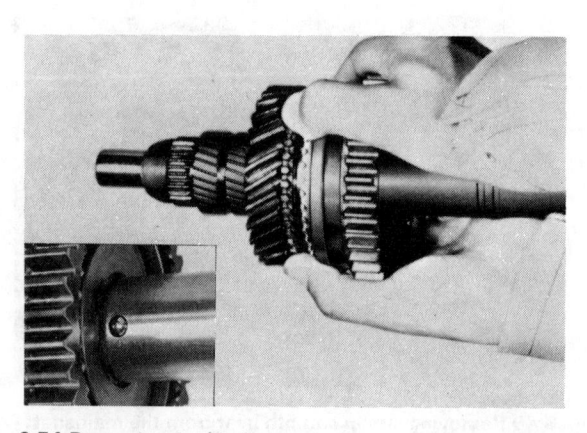

Fig. 6.54 Removing 1st/2nd gear synchro-unit and lockball (W50)

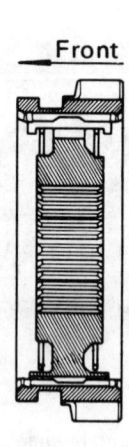

Fig. 6.55 Correct assembly of 1st/2nd gear synchro-hub (W50)

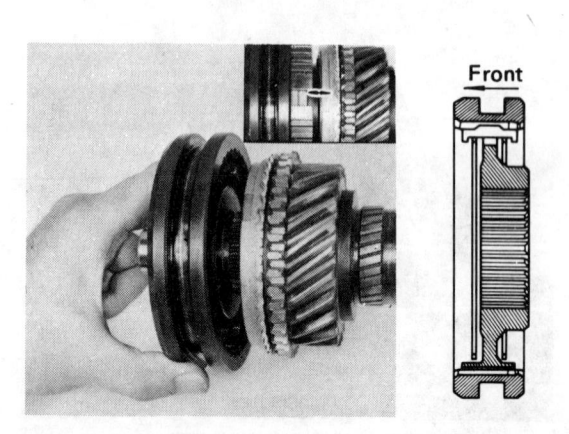

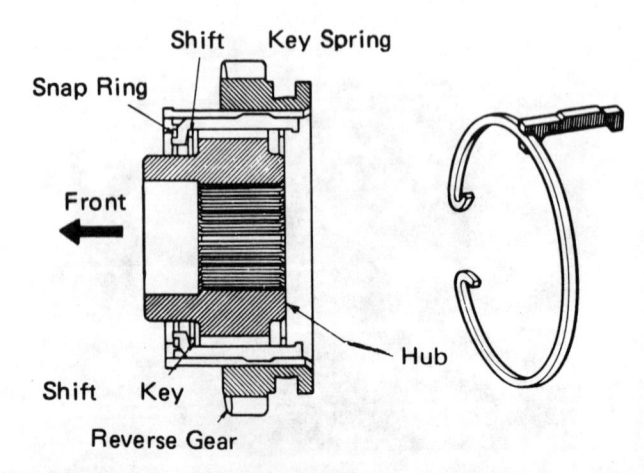

Fig. 6.56 Correct assembly of 3rd/4th gear synchro-hub (W50)

Fig. 6.57 Correct assembly of 5th/reverse gear synchro-hub (W50)

155

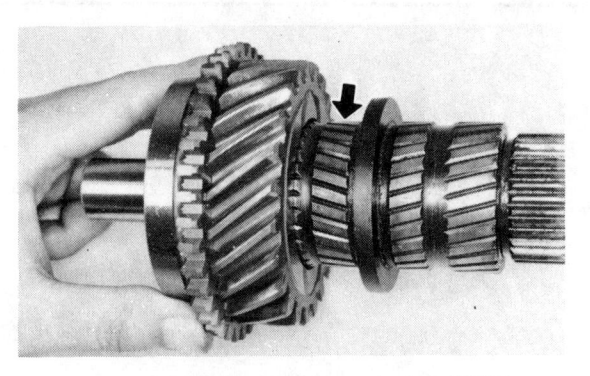

Fig. 6.58 Fitting 3rd gear to mainshaft (W50)

Fig. 6.59 The circlip thickness controls the 3rd gear endfloat (W50)

Fig. 6.60 Checking 3rd gear endfloat (W50)

Fig. 6.61 Fitting 2nd gear to mainshaft (W50)

Fig. 6.62 Fitting 1st/2nd gear synchro unit (W50)

Fig. 6.63 Fitting locking ball and 1st gear components (W50)

Fig. 6.64 Checking 2nd gear endfloat (W50)

Fig. 6.65 Checking 1st gear endfloat (W50)

Fig. 6.66 Removing remote control selector rod pin (W50)

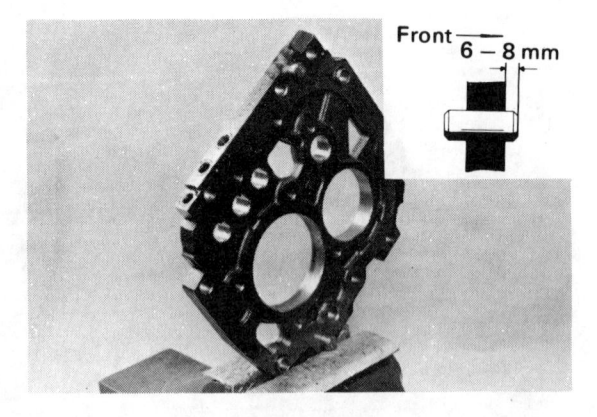

Fig. 6.67 Intermediate plate dowel pin fitting (W50)

the mainshaft shoulder. Secure it with a circlip to give a minimum clearance between circlip and synchro unit yet will still fit fully into the groove. Six circlip thicknesses are available:

0·079 – 0·081 in	(2·00 – 2·05 mm)
0·081 – 0·083 in	(2·05 – 2·10 mm)
0·083 – 0·085 in	(2·10 – 2·15 mm)
0·085 – 0·087 in	(2·15 – 2·20 mm)
0·087 – 0·089 in	(2·20 – 2·25 mm)
0·089 – 0·091 in	(2·25 – 2·30 mm)

9 Check the 3rd gear endfloat, this should not exceed 0·012 in (0·30 mm).
10 Fit 2nd gear and the 1st/2nd gear synchro-hub.
11 Fit the locking ball and 1st gear.
12 Press on the mainshaft bearing and then check the 2nd gear endfloat. This again should not exceed 0·012 in (0·30 mm).
13 Check the 1st gear has the same endfloat.

11 Input shaft (W50) – overhaul

1 Inspect the splines and gear teeth for wear or damage.
2 If the bearing is worn, it must be pressed off after removing the retaining circlip.
3 After the new bearing is installed, fit a circlip which will give the minimum clearance between the bearing and the bearing yet fit snugly in its groove. Circlips are available in six thicknesses:

0·081 – 0·083 in	(2·05 – 2·10 mm)
0·083 – 0·085 in	(2·10 – 2·15 mm)
0·085 – 0·087 in	(2·15 – 2·20 mm)
0·087 – 0·089 in	(2·20 – 2·25 mm)
0·089 – 0·091 in	(2·25 – 2·30 mm)
0·091 – 0·093 in	(2·30 – 2·35 mm)

12 Laygear (W50) – overhaul

1 Check for wear or damage.
2 The bearings can be drawn-off for renewal after extracting the circlips.

13 Reverse idler gear (W50) – overhaul

1 The gear bush can be pressed out for renewal but when installing the new one, make sure that the oil holes are in alignment.

14 Gear case (W50) – overhaul

1 Inspect the case for cracks, particularly around the bolt holes.
2 Renew the front and rear oil seals as a matter of routine.
3 If necessary, the remote control lever can be removed after driving out the securing pin.

Fig. 6.68 Installing the gear trains meshed together (W50)

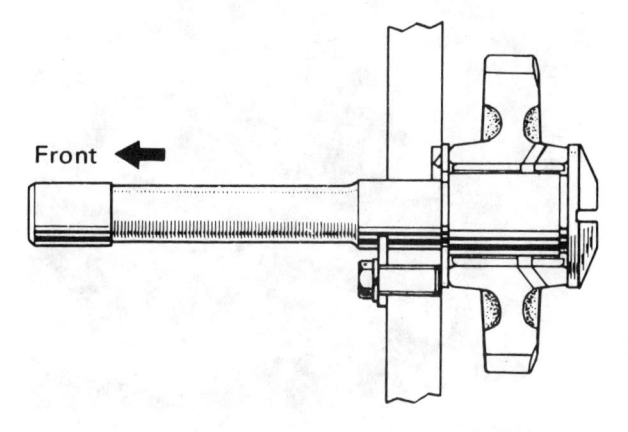

Fig. 6.69 Reverse idler gear and spacer (W50)

Fig. 6.70 Checking reverse idler gear endfloat (W50)

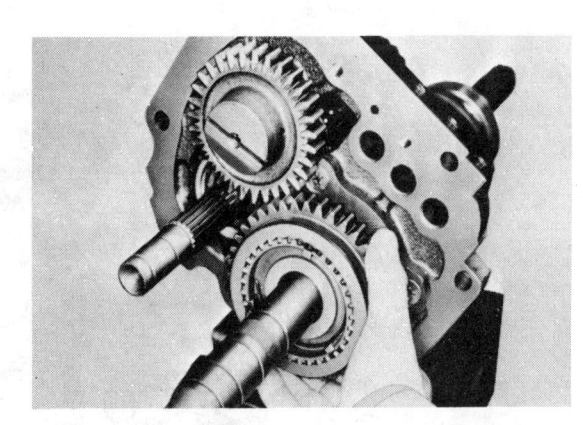

Fig. 6.71 Fitting 5th/reverse synchro-unit (W50)

Fig. 6.72 Checking 5th gear endfloat (W50)

Fig. 6.73 Fitting speedometer drivegear and spacer to mainshaft (W50)

Fig. 6.74 Installing selector shafts and forks (W50)

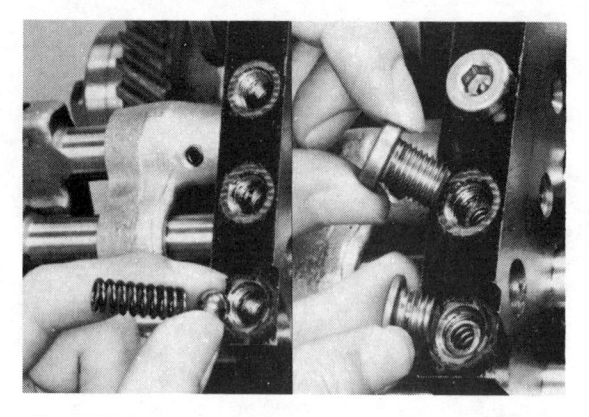

Fig. 6.75 Fitting detent balls, springs and plugs (W50)

Fig. 6.76 Inserting selector fork roll-pins (W50)

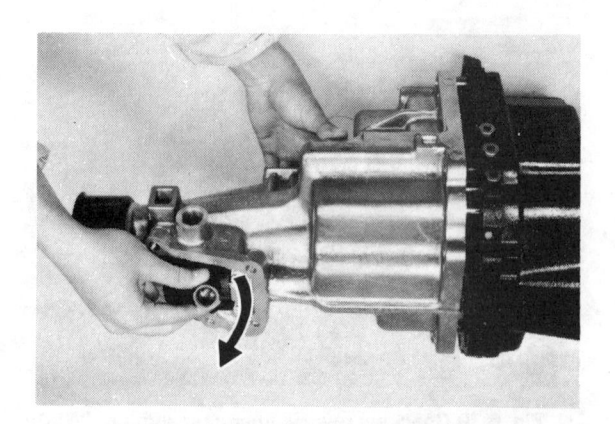

Fig. 6.77 Installing extension housing (W50)

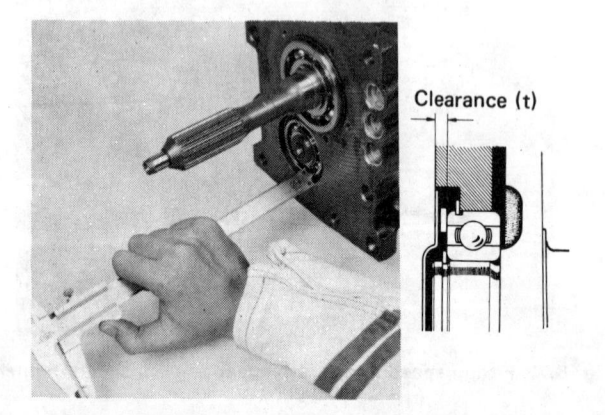

Fig. 6.78 Measuring layshaft bearing depth (W50)

Fig. 6.79 Fitting layshaft bearing spacer and cover (W50)

Fig. 6.80 Front bearing retainer (W50)

Fig. 6.81 Removing circlip and 5th gear from mainshaft (P51)

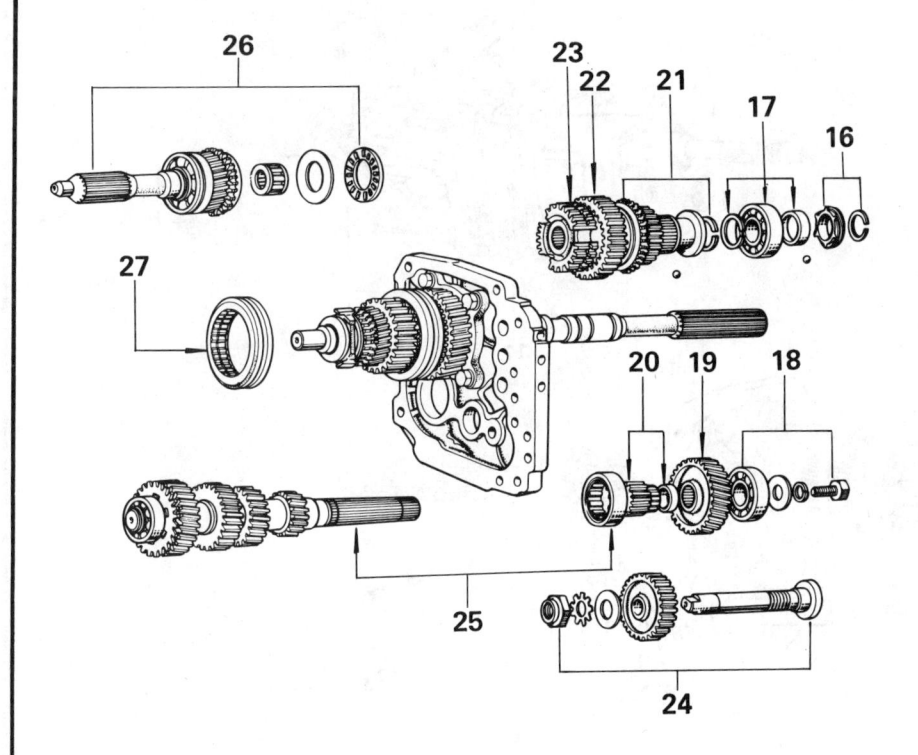

Fig. 6.82 Gear train components (P51)

16 Speedometer drivegear
17 Mainshaft rear bearing
18 Layshaft rear bearing
19 Layshaft 5th gear
20 Layshaft reverse gear
21 5th gear
22 Reverse gear
23 5th/reverse synchro-hub
24 Reverse idler gear
25 Layshaft and centre bearing
26 Input shaft
27 3rd/4th gear synchro-hub
28 1st gear, intermediate plate, bearing and retainer
29 1st/2nd gear synchro-sleeve
30 2nd gear, needle roller at 1st/2nd synchro-hub
31 3rd gear, needle roller at 3rd/4th gear synchro-hub
32 Mainshaft
33 Mainshaft centre bearing

Fig. 6.83 Gearbox external components (P51)

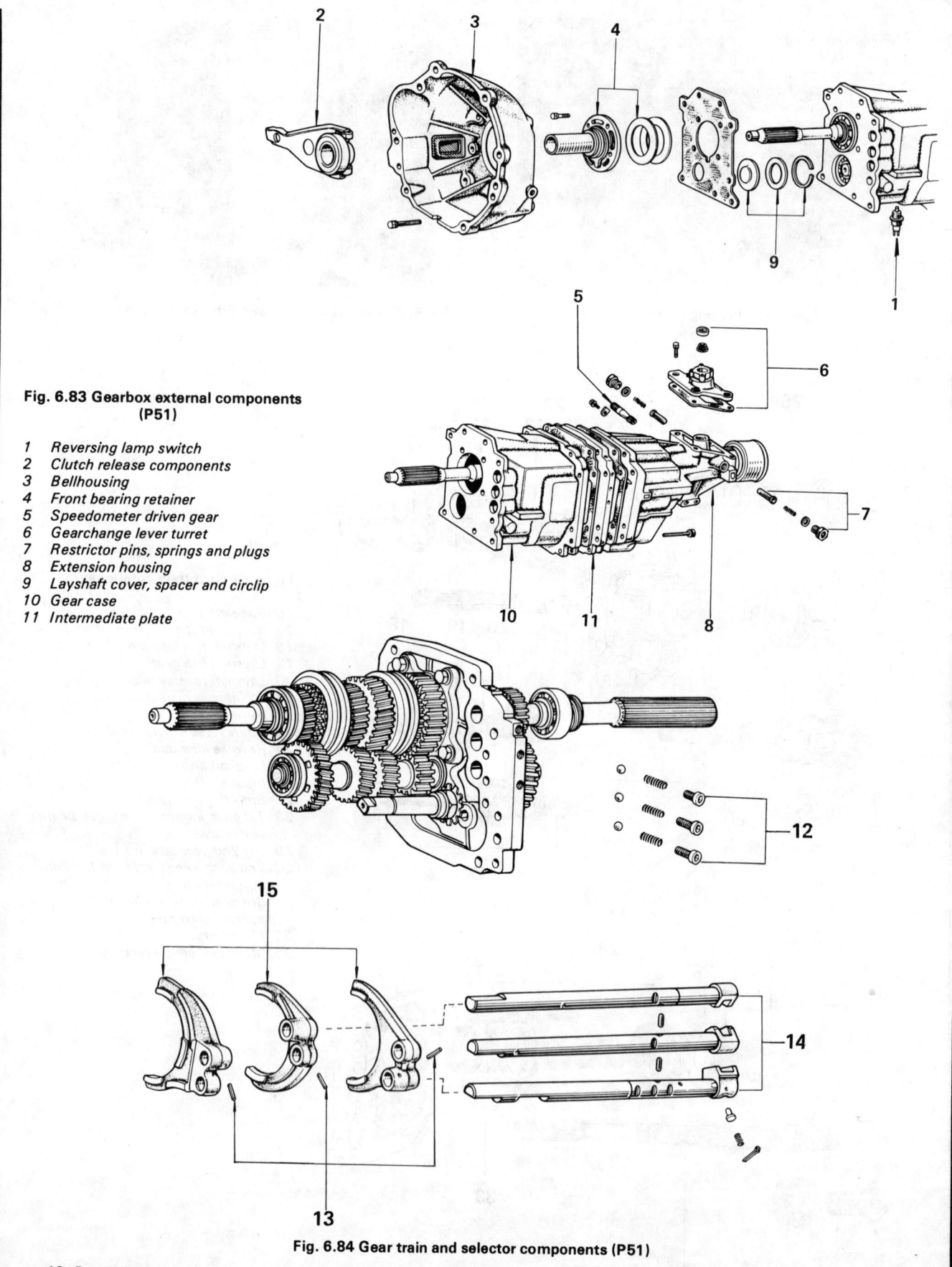

1. Reversing lamp switch
2. Clutch release components
3. Bellhousing
4. Front bearing retainer
5. Speedometer driven gear
6. Gearchange lever turret
7. Restrictor pins, springs and plugs
8. Extension housing
9. Layshaft cover, spacer and circlip
10. Gear case
11. Intermediate plate

Fig. 6.84 Gear train and selector components (P51)

12. Detent balls, springs and plugs
13. Roll-pins
14. Selector shafts (Top – 1st/2nd. Centre 3rd/4th. Bottom – 5th/Reverse)
15. Selector forks

15 Gearbox (Type W50) – reassembly

1 Grip the intermediate plate in a soft-jawed vice and check that the dowel pin projects between 0.24 and 0.32 in (6.0 and 8.0 mm) from the front face of the plate.
2 Apply grease to the needle bearing in the recess at the end of the input shaft and fit the input shaft to the front end of the mainshaft.
3 Mesh the teeth of the mainshaft and laygear assemblies and install them simultaneously to the intermediate plate.
4 Fit the retaining circlip to the mainshaft bearing.
5 Fit the bearing retainer to the intermediate plate.
6 Install the reverse idler gear and spacer noting that the oil hole in the gear must face towards the rear. Lock the idler shaft with the stop plate and bolt and check that the endfloat is betwen 0.006 and 0.012 in (0.15 and 0.30 mm).
7 Fit the reverse/5th gear synchro hub.
8 Fit the locking ball into the mainshaft recess. Use a dab of thick grease to retain it.
9 To the mainshaft, fit 5th gear, synchronizer ring, needle roller bearing and inner track (all assembled together) until the assembly rests against the face of the synchronizer unit.
10 Secure the assembly to the mainshaft by selecting a circlip from the selection available to give the minimum clearance.
11 Check 5th gear endfloat, this should be between 0.004, and 0.010 in with a maximum clearance of 0.012 in.
12 Fit the layshaft reverse gear, 5th gear and then drive on the bearing using a piece of tubing as a drift.
13 Fit a circlip to the layshaft and another to the mainshaft. Select the circlips from the four available thicknesses:

0.055 – 0.057 in	(1.40 – 1.45 mm)
0.063 – 0.065 in	(1.60 – 1.65 mm)
0.071 – 0.073 in	(1.80 – 1.85 mm)
0.079 – 0.089 in	(2.00 – 2.05 mm)

14 Drive the rear bearing onto the mainshaft, again using a piece of tubing as a drift and making sure that it rests against the inner track of the bearing.
15 To the mainshaft, fit the spacer, locking ball and speedometer drive gear and secure them with a circlip.
16 Locate the selector forks in their respective synchro-hub grooves, ensuring that the bosses of the forks face the correct way as shown.
17 Insert the 1st/2nd gear selector shaft and the 5th/reverse gear selector shaft. The gears should be in neutral when assembling each shaft and the interlock pins correctly inserted.
18 Insert the 3rd/4th gear selector shaft.
19 Insert the detent balls and springs into their holes in the edge of the intermediate plate. Tighten the socket screws to the specified torque and in order to prevent oil leaks, ensure that their threads are coated with jointing compound.
20 Secure the selector forks to the selector shafts by driving in the roll-pins.
21 Connect the gear case to the intermediate plate using a new gasket.
22 Fit the mainshaft and layshaft bearing circlips.
23 Fit the extension housing using a new gakset. Turn the remote control rod during the operation so that the rod dog connects with the selector rods.
24 Tighten the extension housing bolts to the specified torque.
25 Install the restrictor pins and springs noting that the longer spring goes on the right-hand side looking towards the bellhousing. Tighten the plugs.
26 Fit the oil baffle and gearchange lever turret.
27 Push the layshaft fully to the rear and using a depth gauge, measure the distance between the front face of the layshaft front bearing and the front face of the transmission casing. Select a spacer to match this dimension from the thicknesses available.

Bearing depth measured	Spacer thickness to use
0.113 – 0.118 in	0.081 – 0.085 in
(2.87 – 2.99 mm)	(2.05 – 2.15 mm)
0.120 – 0.122 in	0.087 – 0.091 mm
(3.00 – 3.09 mm)	(2.20 – 2.30 mm)
0.122 – 0.126 in	0.093 – 0.096 in
(3.10 – 3.19 mm)	(2.35 – 2.45 mm)
0.126 – 0.131 in	0.098 – 0.102 in
(3.20 – 3.32 mm)	(2.50 – 2.60 mm)

The use of the specified spacer will provide the bearing with the correct clearance. Fit the spacer and its cover.
28 Using a new gasket, bolt on the front bearing retainer making sure that the oil groove and hole are in alignment.
29 Fit the bellhousing, tightening the bolts to specified torque.
30 Fill the gearbox with oil after it has been installed in the car.

16 Gearbox (Type P51) – dismantling

1 With the gearbox removed from the car as described in Section 2, clean away external dirt.
2 From the interior of the bellhousing remove the clutch release mechanism.
3 Unbolt and remove the bellhousing.
4 Unbolt and remove the front bearing retainer, retrieving any shims.
5 Remove the speedometer driven gear, the reversing lamp switch, the gearchange lever turret and the restrict plugs, pins and springs.
6 Unbolt and withdraw the extension housing.
7 Remove the cover, spacer and circlip from the front end of the layshaft.
8 Drive off the gear case from the intermediate plate.
9 Grip the intermediate plate in a soft-jawed vice.
10 Unscrew the plugs from the detent holes and extract the springs and balls.
11 Drive out the roll-pins which hold the selector forks to the selector rods, withdraw the rods and take out the forks.
12 Extract the circlips and remove the speedometer drivegear from the mainshaft. Retrieve the locking ball.
13 Pull off the mainshaft rear bearing with a two legged extractor.
14 Remove the bolt and laygear bearing. To prevent the layshaft rotating, lock-up the gear train by engaging two sets of gears simultaneously.
15 Remove the circlip and pull 5th gear from the front end of the mainshaft.
16 Withdraw the locking ball and reverse/5th gear synchro-hub.
17 Relieve the staking on the reverse idler gear shaft and bend back the locktab. Unscrew the nut and withdraw reverse idler gear and shaft.
18 Remove the laygear.
19 Remove the input shaft from the mainshaft.
20 Support 1st gear and press the mainshaft out of the intermediate plate and 1st gear.
21 Withdraw 1st/2nd gear synchro sleeve, the locking ball from the shaft and press the mainshaft out of 1st/2nd gear synchro-hub and 2nd gear.
22 Press off 3rd/4th gear synchro-hub and 3rd gear.
23 Unbolt and remove the mainshaft centre bearing.

17 Input shaft (P51) – overhaul

1 Check the splines and 4th gear teeth for wear or damage.
2 If the bearing is worn and must be renewed, extract the circlip and use a puller to draw the bearing from the shaft.
3 Press on a new bearing, applying pressure to the inner track only.
4 Select a circlip which will give minimum clearance between the circlip and the bearing inner track. Circlips are available in six different thicknesses:

0.081 – 0.083 in	(2.05 – 2.10 mm)
0.083 – 0.085 in	(2.10 – 2.15 mm)
0.085 – 0.087 in	(2.15 – 2.20 mm)
0.087 – 0.089 in	(2.20 – 2.25 mm)
0.089 – 0.091 in	(2.25 – 2.30 mm)
0.091 – 0.093 in	(2.30 – 2.35 mm)

18 Laygear (P51) – overhaul

1 Check for worn or damaged gear teeth.
2 If the front bearing is worn, extract the circlip and draw off the bearing with its spacer using a two-legged extractor.

Fig. 6.85 Removing 5th/reverse gear synchro-hub (P51)

Fig. 6.86 Relieving reverse idler gear shaft nut staking (P51)

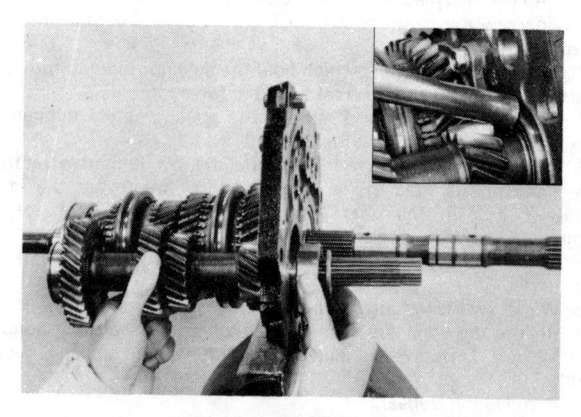

Fig. 6.87 Removing the laygear assembly (P51)

Fig. 6.88 Pulling the input shaft from the mainshaft (P51)

Fig. 6.89 Pressing mainshaft from 1st gear and intermediate plate (P51)

Fig. 6.90 Pressing mainshaft from 1st/2nd gear synchro-hub and 2nd gear (P51)

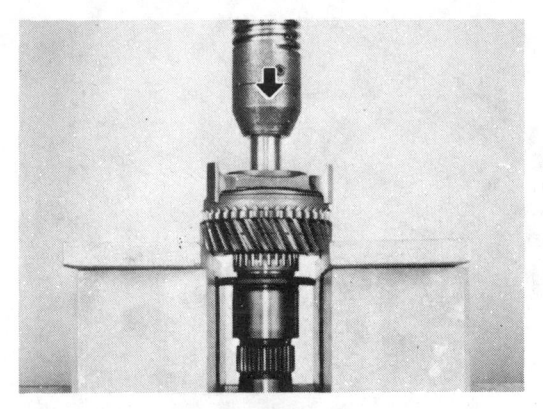

Fig. 6.91 Pressing mainshaft from 3rd/4th gear synchro-hub and 3rd gear (P51)

Fig. 6.92 Removing mainshaft centre bearing retainer (P51)

Fig. 6.93 Laygear anti-lash plate (P51)

Fig. 6.94 Laygear anti-lash plate alignment holes (P51)

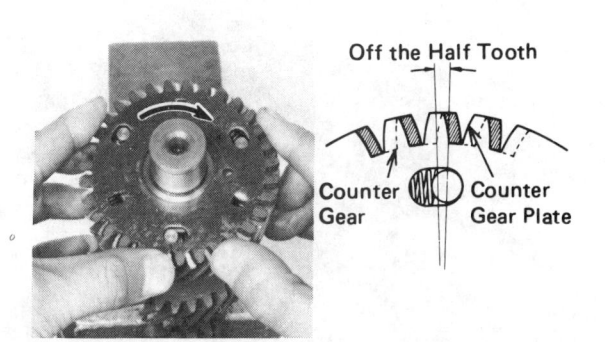

Fig. 6.95 Turning laygear anti-lash plate out of alignment with gear teeth (P51)

Fig. 6.96 Checking synchro-ring outside diameter (P51)

Fig. 6.97 Extracting a synchro-unit circlip (P51)

Fig. 6.98 Dismantling gearchange remote control rod (P51)

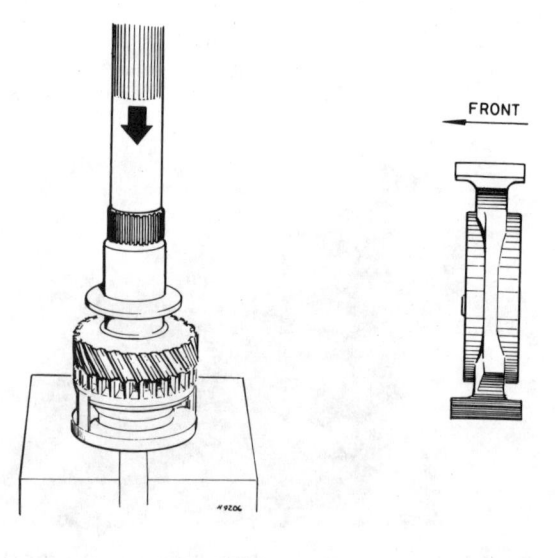

Fig. 6.99 Pressing mainshaft into 3rd/4th gear synchro-hub (P51)

Fig. 6.100 Checking 3rd gear endfloat (P51)

Fig. 6.101 Fitting needle roller and 2nd gear assembly to mainshaft (P51)

Fig. 6.102 Checking 2nd gear endfloat (P51)

Chapter 6 Transmission

Fig. 6.103 Fitting 1st/2nd gear synchro-sleeve (P51)

Fig. 6.104 Fitting 1st gear assembly (P51)

Fig. 6.105 Fitting 1st gear thrust washer (P51)

3   Remove the anti-lash plate and springs. This device helps to absorb gear backlash.
4   When reassembling, align the hole in the plate with the mark on the gear. Turn the plate through half a tooth as shown so that its teeth are out of alignment with those on the gear.
5   Press on the spacer and new bearing.
6   The roller bearing on the layshaft can only be removed and a new one installed using a press.

19 Reverse idler gear (P51) – overhaul

1   Check the gear teeth for wear and the shaft for grooving.
2   If the gear bush is worn, it may be renewed if the old one is pressed out. When installing the new one make sure that the oil holes are in alignment.

20 Synchro-hub (P51) – overhaul

1   If there has been a history of noisy gearchanging or if the synchromesh has been weak, then the particular assembly should be renewed.
2   Check each unit for wear. The selector fork to groove clearance should not exceed 0·04 in (1·0 mm). The synchro-ring outside diameters should not be less than 3·39 – 3·41 in (86·2 – 86·6 mm) for the 1st/2nd gear unit and 3·00 – 3·01 in (76·1 – 76·5 mm) for the other units.
3   To dismantle a synchro-hub extract the circlip and take out the ring, thrust blocks, anchor blocks and brake bands. Keep the components in their originally fitted order and the correct way round.

21 Gear case (P51) – overhaul

1   Check the casing for cracks, particularly around the bolt holes.
2   Renew the front and rear oil seals as a matter of routine.
3   The gear lever remote control rod can be removed if the securing pin is driven out. Renew any worn components.

22 Gearbox (Type P51) – reassembly

1   Commence reassembly by fitting the mainshaft centre bearing to the intermediate plate.
2   Fit the needle roller bearing and 3rd gear to the mainshaft.
3   Support the 3rd/4th gear synchro-hub and press the shaft into it.
4   At this point check the 3rd gear endfloat. This should not exceed 0·012 in (0·30 mm).
5   Fit the needle roller bearing and 2nd gear.
6   Press the 1st/2nd gear synchro-hub onto the mainshaft.
7   Check the 2nd gear endfloat with a feeler blade. This should not exceed 0·012 in (0·30 mm).
8   Fit the 1st/2nd gear synchro sleeve.
9   Fit the inner track, needle roller cage and 1st gear, making sure that the locking ball is in position in the shaft.
10 Fit 1st gear thrust washer.
11 Support the intermediate plate and press the assembled mainshaft into it by applying pressure to the end of the shaft.
12 With the corner of the intermediate plate gripped in a soft-jawed vice, fit the 3rd/4th gear synchro sleeve.
13 Fit the input shaft complete with needle bearings and thrust plate to the end of the mainshaft.
14 Fit the laygear to the intermediate plate, then fit the centre bearing.
15 Fit reverse idler gear and spacer, then tighten the nut and lock it. Always use a new nut and lockwasher.
16 Check the reverse idler gear endfloat which should not exceed 0·012 in (0·30 mm).
17 Fit reverse/5th gear synchro-hub.
18 Fit reverse gear.
19 Fit the locking ball, 5th gear, the needle roller bearing and inner race.
20 Fit a circlip which will give the minimum clearance between the circlip and the thrust plate yet fit fully into the shaft groove. Circlips are available in the following thicknesses:

Fig. 6.106 Fitting 3rd/4th gear synchro-sleeve (P51)

Fig. 6.107 Insert the input shaft into the mainshaft (P51)

Fig. 6.108 Fitting laygear and centre bearing (P51)

Fig. 6.109 Fitting reverse idler gear and spacer (P51)

Fig. 6.110 Fitting 5th/reverse gear synchro-hub (P51)

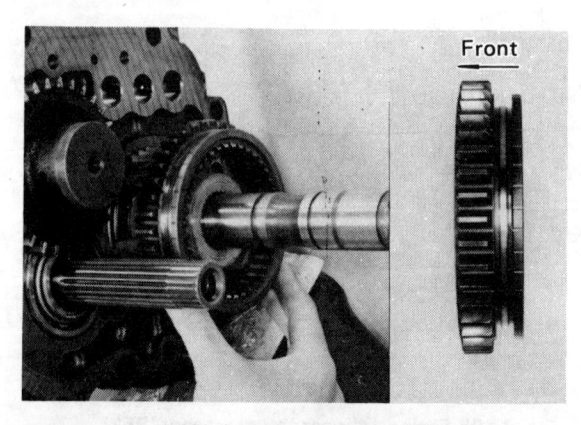

Fig. 6.111 Fitting reverse gear (P51)

Fig. 6.112 Fitting 5th gear assembly to mainshaft (P51)

Fig. 6.113 5th gear endfloat is controlled by the thickness of the circlip (P51)

Fig. 6.114 Checking 5th gear endfloat (P51)

Fig. 6.115 Checking 1st gear endfloat (P51)

Fig. 6.116 Checking 2nd gear endfloat (P51)

Fig. 6.117 Fitting reverse gear to layshaft (P51)

Fig. 6.118 Fitting 3rd/4th gear selector shaft and forks (P51)

Fig. 6.119 Inserting an interlock pin (P51)

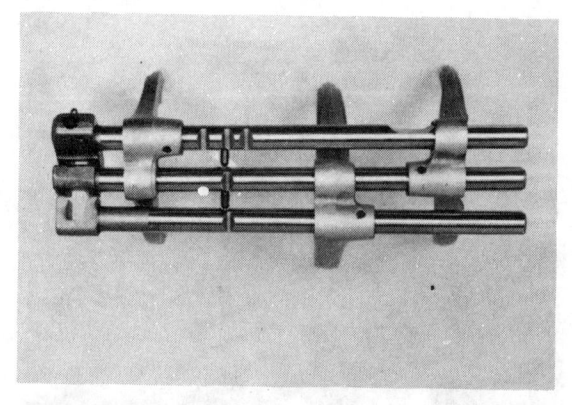

Fig. 6.120 Selector shaft and shift fork identification (P51)
(Top – 5th/reverse. Centre 3rd/4th. Bottom 1st/2nd)

Fig. 6.121 Selector shaft location in intermediate plate (P51)

Fig. 6.122 Layshaft front bearing circlip (P51)

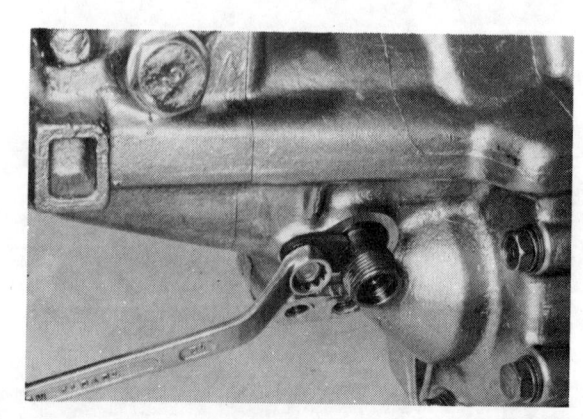

Fig. 6.123 Fitting the speedometer driven gear retaining plate (P51)

Chapter 6 Transmission

Fig. 6.124 Fitting the gearchange lever turret (P51)

Fig. 6.125 Fitting spring and ball seat into gearchange lever turret (P51)

Fig. 6.126 Checking input shaft bearing projection (P51)

0·074 – 0·076 in	1·89 – 1·94 mm
0·077 – 0·079 in	1·95 – 2·00 mm
0·079 – 0·081 in	2·01 – 2·06 mm
0·082 – 0·084 in	2·07 – 2·12 mm
0·084 – 0·086 in	2·13 – 2·18 mm
0·086 – 0·088 in	2·19 – 2·24 mm
0·089 – 0·089 in	2·25 – 2·30 mm
0·091 – 0·093 in	2·31 – 2·36 mm
0·093 – 0·095 in	2·37 – 2·42 mm

21 Check the 5th gear endfloat. This should not exceed 0·012 in (0·30 mm).
22 Check 1st gear thrust clearance, this should not exceed 0·012 in (0·30 mm).
23 Check the 2nd gear endfloat, this should not exceed 0·012 in (0·30 mm).
24 To the layshaft fit reverse gear, spacer and 5th gear.
25 Fit the layshaft rear bearing, lock-up two sets of gears and tighten and lock the retaining nut.
26 Fit the circlip and tap the rear bearing onto the mainshaft.
27 Fit the locking ball, the speedometer drivegear and the circlip to the mainshaft.
28 Locate the selector forks into the synchro sleeve grooves and pass the 3rd/4th gear selector shaft through the forks and intermediate plate.
29 Insert the interlock pins between the shaft and the other two holes. Retain them in position with thick grease.
30 Install the 1st/2nd and 5th/reverse gear selector rods. Fix the forks to the shafts with new roll-pins.
31 Fit the detent balls, springs and renew in the plugs having applied sealant to the threads.
32 Using a new gasket, connect the gearcase to the intermediate plate.
33 Fit the circlip to the layshaft front bearing.
34 Using a new gasket, fit the extension housing to the gear case at the same time turning the remote control rod to engage it with the selector lever dog cut-outs.
35 Tighten the extension housing bolts to specified torque.
36 Fit the restrictor pins and their springs. The longer spring goes on the right-hand side when looking towards the bellhousing. Tighten the plugs to the specified torque.
37 Fit the speedometer driven gear.
38 Fit the oil baffle and the gearchange lever turret.
39 Fit the spring and ball seat into the turret so that the larger diameter of the spring enters first.
40 Select and fit the laygear spacer as described in paragraph 27 of Section 15.
41 Apply pressure to the input shaft in a rearward direction and using a depth gauge, measure the projection of the input shaft bearing from the front face of the transmission casing.
42 Fit the appropriate number of shims in accordance with the following table:

Bearing projection	Number of shims
0·196 to 0·212 in (4·98 to 5·38 mm)	0
0·180 to 0·196 in (4·57 to 4·97 mm)	1
0·164 to 0·180 in (4·16 to 4·56 mm)	2
Below 0·163 in (4·15 mm)	3

43 Fit the selected shims, then the input shaft bearing retainer with a new gasket. Make sure that the retainer oil groove and hole are in alignment and apply grease to the oil seal lips.
44 Fit the clutch bellhousing and screw in the reversing lamp switch.
45 Temporarily fit the gear selector lever and check all gear selections. Fill the gearbox with oil after it has been installed in the car.

23 Fault diagnosis – manual gearbox

Symptom	Cause
Ineffective synchromesh on one or more gears	Worn baulk rings or sliding keys
Jumps out of one or more gears	Weak detent springs Worn selector forks Worn engagement.dogs Worn synchro-hubs
Whining, roughness, vibration allied to other faults	Bearing failure and/or overall wear
Noisy and difficult gear engagement	Clutch not operating correctly
Sloppy and impositive gear selection	Overall wear in selector mechanism

PART B – AUTOMATIC TRANSMISSION

24 Automatic transmission – general description and precautions

The Toyota A40 automatic transmission is a bandless type, and therefore has no internal adjustments. The only external adjustments are to the throttle cable and shift linkage.

Internally, the transmission has a three-element torque converter, two clutches and three multi-disc brakes which actuate the planetary gears, and an oil pump which supplies pressure for actuation of the clutches and brakes.

In the event of breakdown, the vehicle must not be towed in excess of 30 mph (48 km/h), or further than 50 miles (80 km), unless the propeller shaft is disconnected. Failure to observe this requirement may cause damage to the transmission due to lack of lubrication as the fluid pump will not be working.

Due to the complexities of dismantling and reassembly of automtic transmission units, the operations described in this Chapter are limited to maintenance, adjustment, and removal and refitting.

25 Maintenance

Fluid level checking

1 With the engine running at idle speed and the handbrake applied, briefly select each gear in turn and finally select *PARK*.
2 If the transmission fluid is cold, withdraw the dipstick, wipe it, reinsert it and withdraw it again. The fluid level should be within the cold range. If the vehicle has travelled at least 5 miles (8 km) the fluid level should be within the hot range of the dipstick when the same checking procedure is followed. Top-up with fluid of the specified type.

Other maintenance

3 Keep the external surfaces of the transmission unit clean and free from mud and grease to prevent overheating. Check the fluid cooler (where applicable) and make sure that the connecting pipes are secure and in good condition.
4 Periodically, check the condition of the transmission fluid. If it looks black or smells as though it is burnt, it requires renewal, although this may be indicative of an internal fault if the fluid has been changed regularly.
5 At the time interval given in the Routine Maintenance Section, the

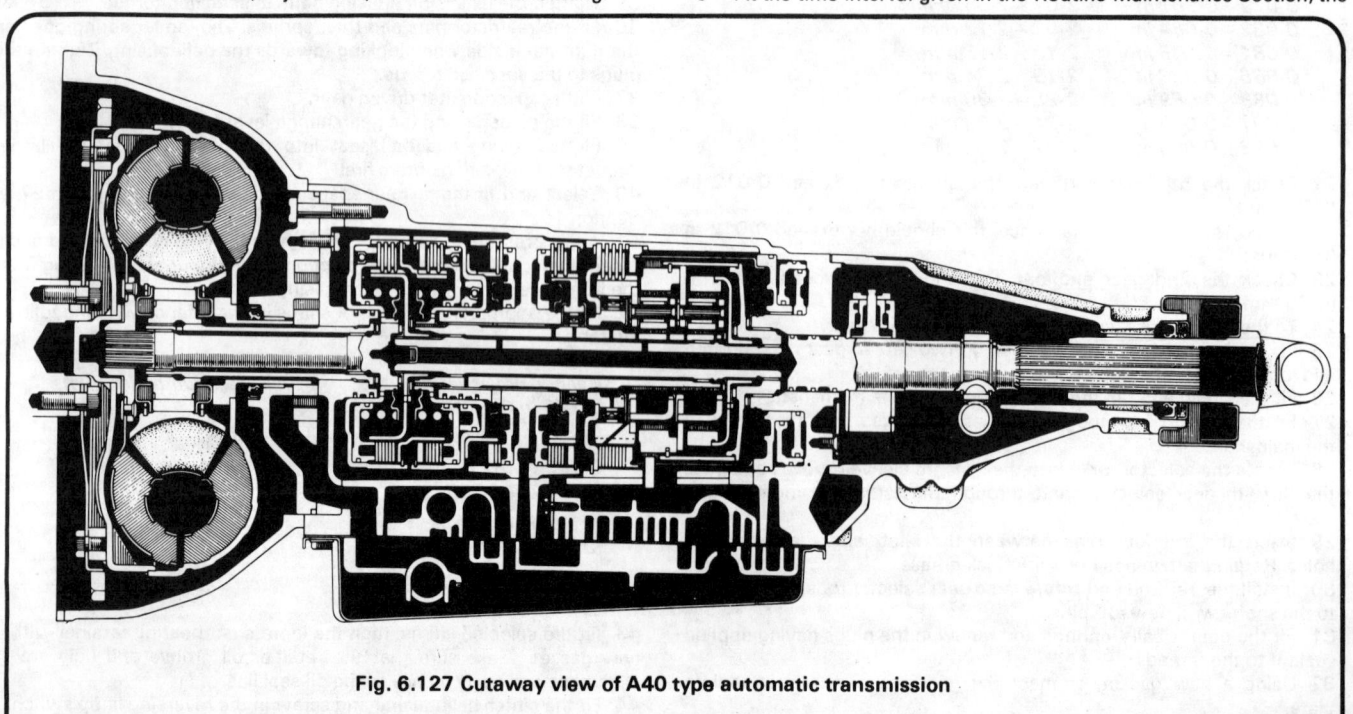

Fig. 6.127 Cutaway view of A40 type automatic transmission

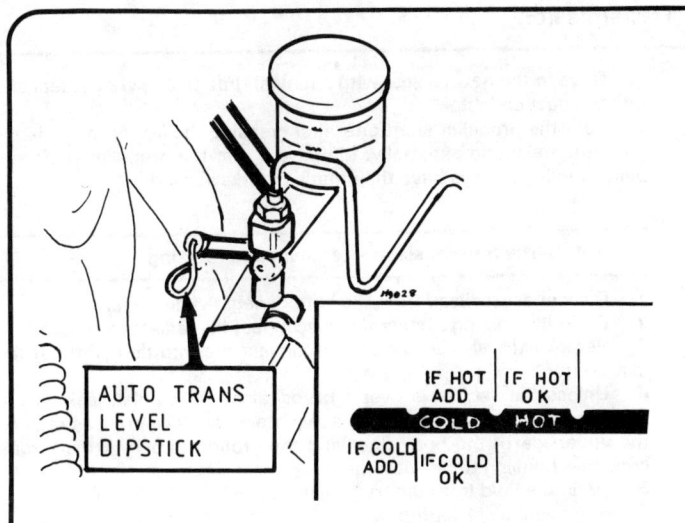

Fig. 6.128 Automatic transmission fluid dipstick markings

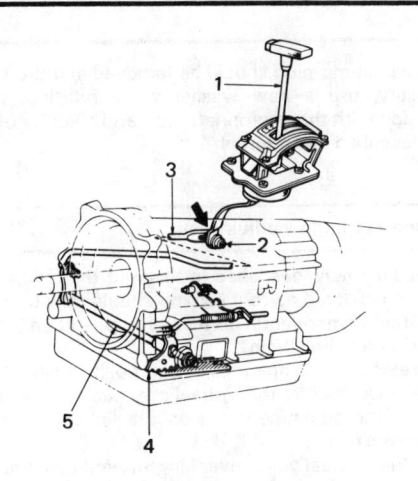

Fig. 6.129 Selector linkage

1 Control lever
2 Adjuster nut
3 Control rod
4 Valve lever
5 Valve shaft

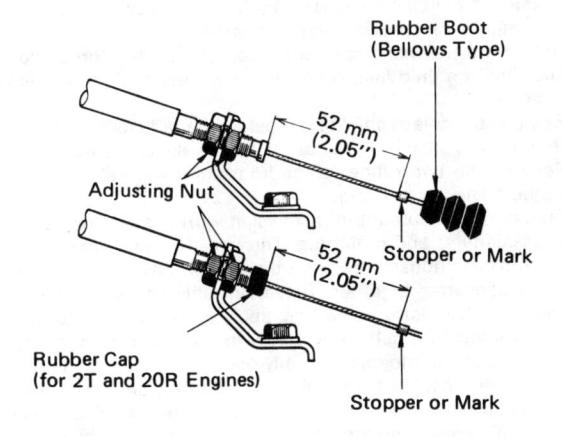

Fig. 6.130 Downshift cable (convoluted type boot)

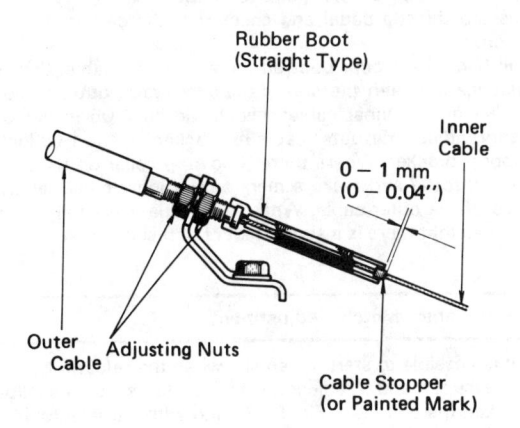

Fig. 6.131 Downshift cable (straight sleeve boot)

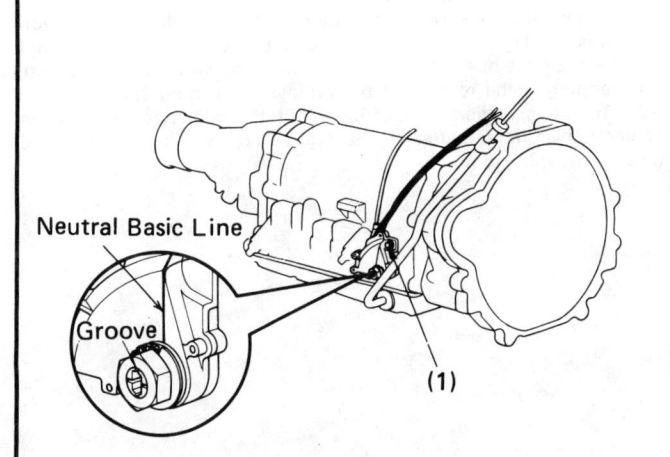

Fig. 6.132 Neutral start switch adjustment diagram

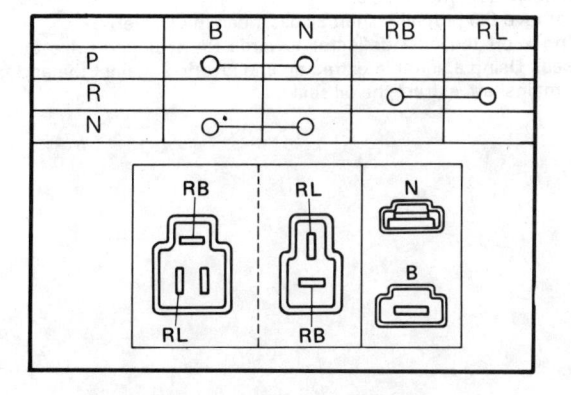

Fig. 6.133 Neutral start switch connections

transmission sump plug should be removed and the contents drained. If necessary, use a new washer when refitting, then top-up the transmission with the appropriate type and quantity of fluid. Check the level as described in paragraph 1.

26 Selector linkage – adjustment

1 The adjustment described in this and the following Sections are not to be considered as routine, and should only be carried out when wear in the components or incorrect operation of the automatic transmission requires them.
2 The selector lever operates in a six position gate, through a control rod and a cross-shaft to the hydraulic manual valve lever.
3 Loosen the adjustment nut on the linkage, and check that the linkage moves freely.
4 Push the manual valve lever fully forward then back three notches; this is the Neutral position. Now, with the selector held in Neutral by an assistant, tighten the linkage adjustment nut. Check the operation of the lever throughout the full range of travel.

27 Throttle cable (downshift) – adjustment

1 Remove the air cleaner (refer to Chapter 3 if necessary), then fully depress the throttle pedal and check that the carburettor throttle is fully open.
2 Pull back the rubber boot on the end of the cable, then measure the distance between the end of the accelerator outer cable and the stop collar on the inner cable; this should be 2·05 in (52·0 mm). If necessary, adjust the outer cable by slackening the two locknuts on the support bracket. Where there is no stop collar on the inner cable, close the throttle and make a mark on the inner cable relative to the end face of the outer cable. With the throttle valve fully open, adjust the cable so that there is a small amount of reserve travel.

28 Neutral start switch – adjustment

1 If it is possible to start the engine when the selector lever is in any position except *P* and *N*, adjust the neutral start switch as follows:
2 Slacken the bolt (1) in Fig. 6.132 and with the selector lever set to *N* align the switch so that the groove in its shaft is along the *neutral basic line* scribed on the quadrant.
3 Tighten the bolt with the switch in the above position.
4 If the switch is thought to be faulty, check for continuity between the terminals as shown in Fig. 6.133. Fit a new switch if the existing one is shown to be faulty.

29 Extension housing oil seal – renewal

1 Renewal of the oil seal may be carried out with the transmission unit in position in the vehicle.
2 Remove the propeller shaft as described in Chapter 7.
3 Knock off the dust deflector towards the rear and prise out the dust seal. Using a suitable extractor, and levering against the end face of the mainshaft, extract the oil seal.

4 Drive in the new oil seal with a tubular drift, fit a new dust seal and refit the dust deflector.
5 Refit the propeller shaft after first greasing the front sliding sleeve both internally and externally. Make sure that the propeller shaft and pinion driving flanges have their mating marks aligned.

30 Automatic transmission – removal and refitting

1 Disconnect the lead from the battery negative terminal.
2 Drain the cooling system and disconnect the radiator top hose.
3 Remove the air cleaner and disconnect the throttle control at the carburettor.
4 Unless the vehicle is over a pit or raised on a hoist, jack-up the front and rear so that there is an adequate working clearance between the underside of the body floor and the ground to permit the torque converter housing to be withdrawn.
5 Drain the fluid from the transmission unit.
6 Remove the starter motor.
7 Disconnect the propeller shaft from the rear axle (see Chapter 7) and withdraw it from the transmission rear extension housing.
8 Disconnect the speed selector linkage at the transmission unit.
9 Disconnect the exhaust downpipe from the manifold and remove the support bracket from the transmission unit.
10 Disconnect the fluid cooler pipes from the transmission and plug them. Remove the pipe supports from the transmission.
11 Disconnect the speedometer drive cable.
12 Unbolt the two reinforcement brackets from the torque converter housing. Pull the fluid filler tube from the transmission and retain the O-ring seals.
13 Remove the splash shield from below the radiator.
14 Remove the support plate for the handbrake equaliser.
15 Remove the two rubber plugs from the lower half of the front of the torque converter housing.
16 Support the automatic transmission with a jack, then remove the rear crossmember and mounting. Through the open lower half of the torque converter housing, remove the six bolts which secure the drive plate and converter together. These can only be removed in turn by rotating the drive plate. To do this, apply a ring spanner to the crankshaft pulley securing bolt. Now screw in two guide pins (easily made from two old bolts) into diametrically opposite bolt holes in the front of the drive plate, then rotate the engine until they are horizontal. These pins will act as pivot points during removal of the transmission unit.
17 Place a jack under the engine sump (use a block of wood to protect it), and remove the bolts which secure the torque converter housing to the engine.
18 Lower both jacks progressively until the transmission unit will clear the lower edge of the engine rear bulkhead. Insert two levers between the engine rear plate and the temporary pivot pins, and prise the transmission unit from the engine. Catch the fluid which will run from the torque converter during this operation. *On no account should levers be placed between the drive plate and the torque converter as damage or distortion will result.*
19 The torque converter can now be pulled forward to remove it from the housing. The drive plate can be unbolted from the crankshaft flange if the plate has to be renewed because of worn starter ring gear.
20 Refitting is the reverse of the removal procedure, but tighten all bolts to the specified torque; carry out the adjustments described earlier in this Chapter, after first having refilled the unit with the correct type and quantity of fluid.

31 Fault diagnosis – automatic transmission

Symptom	Cause
Delayed upshifts	Throttle cable adjustment incorrect Internal transmission fault
Downshifts from 3rd to 2nd, then back to 3rd	Throttle cable adjustment incorrect Internal transmission fault
Slip on upshifts	Shift linkage incorrectly adjusted Throttle cable incorrectly adjusted Internal transmission fault
Slow upshifts or failure to upshift	Shift linkage incorrectly adjusted Internal transmission fault
Slip, squawk or shudder on take-off	Shift linkage incorrectly adjusted Internal transmission fault
Transmission noise (possibly increasing with engine speed)	Broken connector driveplate. (The driveplate can be checked by removing the inspection covers from the converter housing and the engine being slowly turned) Internal transmission fault
Harsh engagement or shifting	Throttle cable adjustment incorrect Internal transmission fault
No engagement after downshift	Throttle cable adjustment incorrect Internal transmission fault
No kickdown	Throttle cable adjustment incorrect
No downshift, or downshift at incorrect speed	Internal transmission fault
No engine braking in 2nd gear	Internal transmission fault
Automatic 2nd to 3rd shift in 2nd range	Internal transmission fault
Coast downshift occurs at incorrect speed	Throttle cable adjustment incorrect Internal transmission fault
Vehicle does not hold in **P**	Shift linkage incorrectly adjusted Internal transmission fault

Chapter 7 Propeller shaft

Contents

Centre bearing – removal and installation	3	Propeller shaft – removal and installation	2
Fault diagnosis – propeller shaft	6	Universal joints – inspection, dismantling and	
Front sliding yoke – inspection	5	reassembly	4
General description	1		

Specifications

Type .... Tubular, with universal joints and splined sliding yoke at gearbox end. Some models have a rubber mounted centre bearing and a third universal joint.

Universal joints
Type .... Hooke's joint with sealed for life needle roller bearings
Available circlip thicknesses .... 0.0935 to 0.0955 in (2.375 to 2.425 mm)
0.0955 to 0.0974 in (2.425 to 2.475 mm)
0.0974 to 0.0994 in (2.475 to 2.525 mm)
0.0994 to 0.1014 in (2.525 to 2.575 mm)
Spider endfloat .... Less than 0.002 in (0.05 mm)

Torque wrench settings   lbf ft   Nm
Centre joint self-sealing nut:
  First stage .... 123 to 144   170 to 200
  Second stage .... 19 to 25   25 to 35
Universal joint flanges .... 15 to 28   20 to 40
Centre bearing bracket-to-body .... 22 to 32   30 to 45

1 General description

The drive from the gearbox or automatic transmission is transmitted to the rear axle by a tubular propeller shaft having Hooke's joints to take up the axial misalignment between the two ends. A sliding coupling at the gearbox end permits a small amount of longitudinal movement.

The shaft may be in one or two sections, the two section shaft having a rubber mounted bearing at the mid-point between the gearbox and the rear axle. Each shaft is connected by a universal joint, comprising a four way trunnion, or "spider", each leg of which runs in a needle roller bearing, prepacked with grease and located in the coupling flange yokes. The universal joints are replaceable by means of a repair kit. The bearings are of the sealed type and do not require subsequent lubrication.

2 Propeller shaft – removal and refitting

1 Jack-up the rear of the car to minimise the quantity of oil which runs out when the gearbox end of the shaft is removed and to provide adequate working clearance.
2 Mark the two flanges at the rear axle end, so that the coupling will be reassembled in exactly the same relative position. Then remove the four bolts from the flange (photo).
3 Unscrew and remove the two securing bolts from the centre bearing (photo), if fitted. Support the shaft while separating the two flanges of the shaft-to-rear axle joint and then lower the free end of the shaft to the ground.
4 Place a container under the rear of the gearbox to catch any oil which runs out and then pull the shaft out of the gearbox (photo). When removing the sleeve yoke, take care not to damage the oil seal at the rear of the gearbox.
5 Before refitting the propeller shaft, clean the outside of the sleeve yoke very carefully and smear it with oil before inserting it.
6 Insert the yoke carefully, taking care not to damage the oil seal.
7 Line up the mating marks on the two parts of the rear flange joint, insert the bolts, then screw on and tighten the nuts.
8 Insert the centre support bearing height spacer between the body and the bearing and fit the bearing with the bolts finger tight.
9 Align the bearing so that it is at right angles to the propeller shaft and its centre is offset 1 mm towards the rear end (Fig. 7.5), then tighten the fixing bolts.

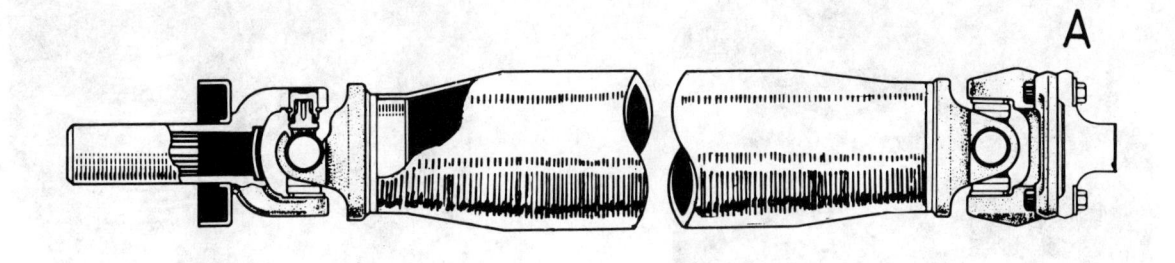

Fig. 7.1 Propeller shafts – sectional views

A One piece type
B Centre bearing type

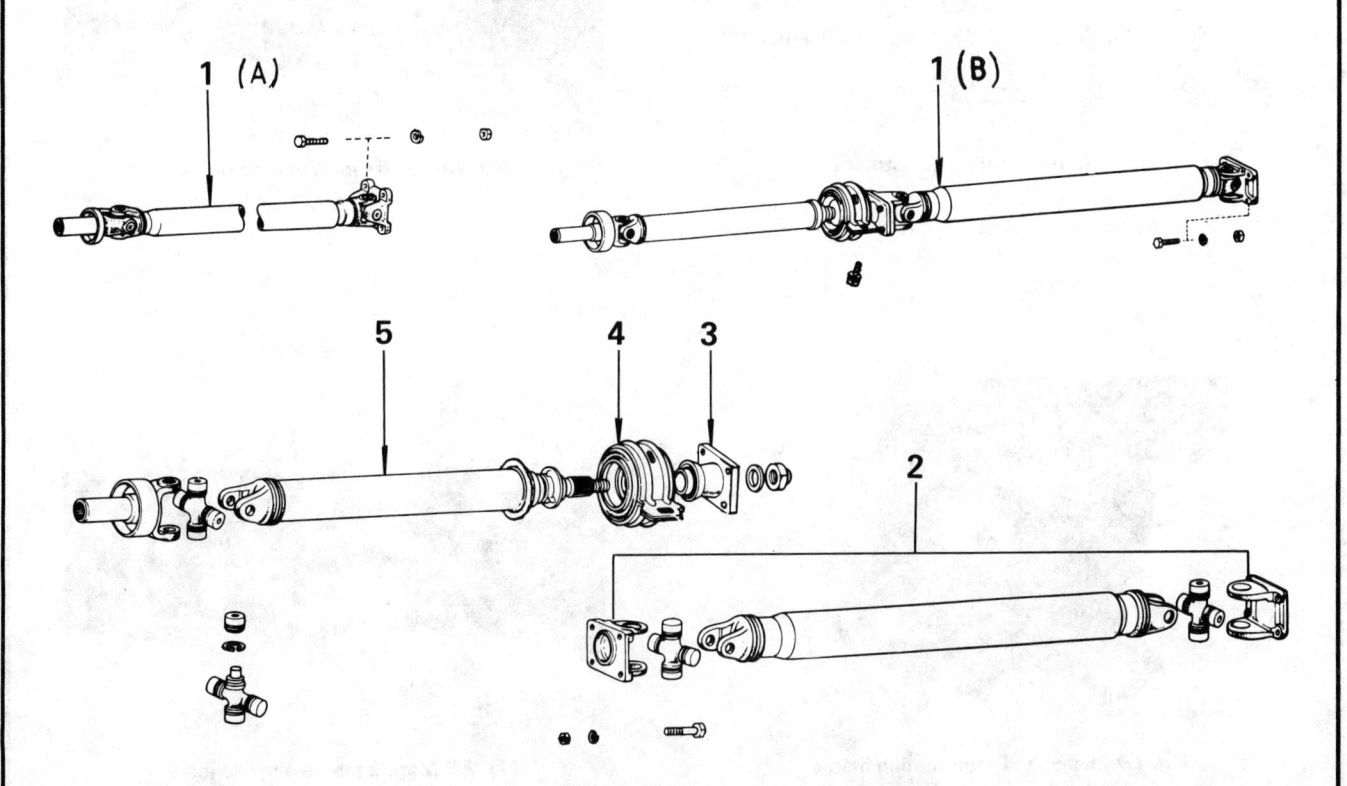

Fig. 7.2 Propeller shafts – component parts

1(A) One piece type	2 Rear shaft components	4 Centre bearing assembly
1(B) Centre bearing type	3 Rear shaft coupling flange	5 Front shaft

Fig. 7.3 Mating marks on rear axle flange

Fig. 7.4 Centre bearing height spacer

Bracket Center Line

Front

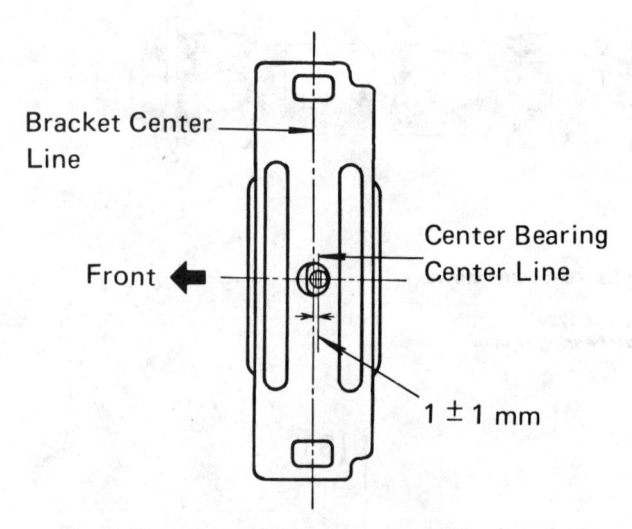

Center Bearing Center Line

1 ± 1 mm

Fig. 7.5 Centre bearing adjustment

Fig. 7.6 Centre joint mating marks

Fig. 7.7 Flange and shaft mating marks

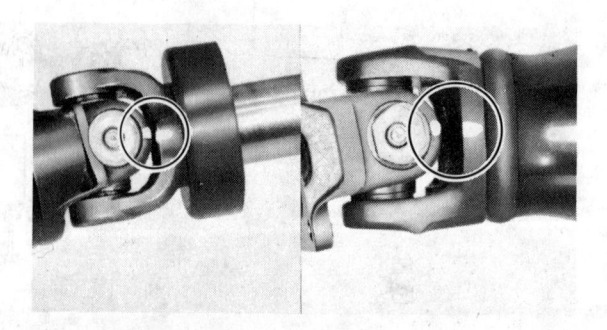

Fig. 7.8 Mating marks on yoke joints

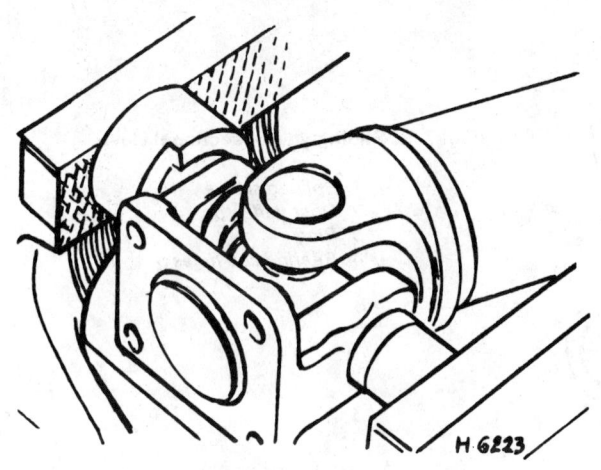

Fig. 7.9 Pressing out a bearing

2.2 Rear coupling flange bolts

Fig. 7.10 Separating the bearing and shaft

2.3 Centre bearing

Fig. 7.11 Fitting the circlips

2.4 Removing the sliding yoke from the rear of the gearbox

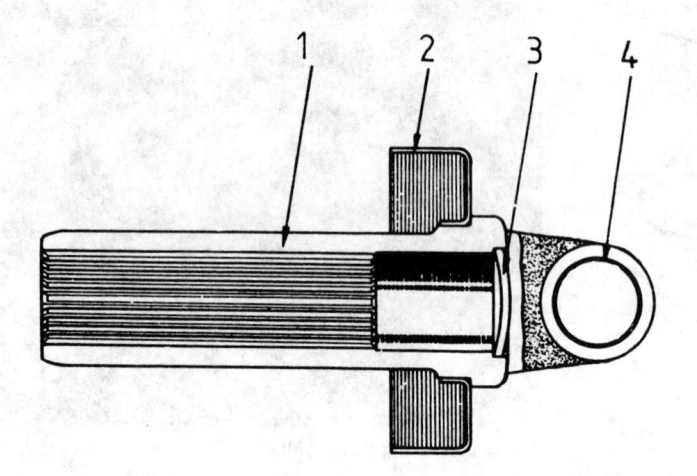

Fig. 7.12 Sliding yoke – sectional view

1 Splined sleeve
2 Dust deflector
3 Plug
4 Bearing cup holes

3 Centre bearing – removal and installation

1 Remove the propeller shaft assembly from the car as described previously.
2 Put mating marks on the two flanges of the propeller shaft centre joint, then unbolt and separate the flanges.
3 Put mating marks on the shaft at the rear of the front section and on the flange splined to it. Unscrew and remove the securing nut and washer, then draw off the flange.
4 Remove the centre bearing, inspect it for wear and renew it as an assembly if it is not reusable.
5 When reassembling (which is the reverse of removal) take care to align the mating marks on the spindle and its flange and those on the two flanges of the centre joint. Use a new self-locking nut to attach the flange to the spindle and tighten the nut to the specified torque (Stage 1), then slacken the nut and retighten it to the specified torque (Stage 2). Finally stake the nut.
6 Position centre bearing as described in paragraph 9 of Section 2.

4 Universal joints – inspection, dismantling and reassembly

1 Preliminary inspection of the universal joints can be carried out with the propeller shaft still in position on the car.
2 Grasp each side of a joint and use a twisting action to see if there is any play in the joint. Also check to see whether there is any lateral slackness. If there is felt to be slackness, or if there is a "clunk" when letting in the clutch and taking up the drive, renew the joint.
3 Remove the propeller shaft assembly from the car (see Section 2). Put mating marks on the shaft and flange, or yokes.
4 Lightly tap the end of the bearing to give a slight clearance between the bearing and its circlip, then remove the circlip. Repeat the operation on the bearing diametrically opposite.
5 Using a suitable sized mandrel against one bearing and a cup, or socket against the opposite one, mount the assembly in a vice and tighten until a bearing has been pushed about 0.2 in (5 mm) out of the yoke.
6 Remove the mandrel and cup, grip the exposed bearing end in the vice and tap the shaft clear of the bearing. Remove the other bearings in the same manner.
7 Inspect the holes in the yoke for elongation. If they are worn so that the bearings do not fit properly, a new propeller shaft must be fitted.
8 Grease both the spider and the bearing bores before refitting and then proceed as follows.
9 With the spider held in position in the yoke, press in one bearing until the exposed width of circlip groove is about 0.10 in (2.5 mm). Press in the opposite bearing to about the same extent. Measure the gap on each side and adjust the bearings so that the gaps are equal.
10 Fit a circlip of the same thickness as the gap, it being important that the same thickness of circlip is fitted on each side of the spider.
11 Check that the spider moves freely and smoothly and that its axial play is less than 0.002 in (0.05 mm).

5 Front sliding yoke – inspection

1 When the propeller shaft is checked for wear, also examine the condition of the sliding yoke assembly.
2 Check for wear in the splines and for looseness and distortion of the dust deflector. Check that the outer surface is free from dirt, damage or corrosion, otherwise the gearbox rear oil seal will wear excessively and cease to be effective.
3 Check that there is no leakage of oil from the plug at the rear end of the sleeve yoke. If there is evidence of a leak, fit a new plug after removing the old one, cleaning the plug recess and coating it with jointing compound.

6 Fault diagnosis – propeller shaft

Symptom	Cause
Vibration	Worn universal joints
	Worn or loose centre bearing
	Bent propeller shaft
	Worn extension housing bush
	Loose drive flange bolts
	Propeller shaft out of balance
Knocking during starting, gear shifting or deceleration	Worn joints or splines
	Loose drive flange bolts

Chapter 8 Rear axle

Contents

Axleshaft, bearings and oil seals – removal and installation	2	General description	1
Fault diagnosis – rear axle	6	Pinion oil seal – renewal	3
Final drive unit – dismantling and reassembly	4	Rear axle – removal and installation	5

Specifications

Rear axle type .... Hypoid semi-floating, with limited slip differential on GT models

Final drive ratio
Except GT .... 4.100 : 1
GT .... 3.909 : 1

Lubricant
Type .... Hypoid gear oil API GL-S
Viscosity:
  Above -10°F (-23°C) .... SAE 90
  Below -10°F (-23°C) .... SAE 80W or 85W
Capacity:
  Detachable final drive unit .... 2.4 Imp pints (1.3 litres)
  Integral final drive unit .... 2.2 Imp pints (1.2 litres)

Torque wrench settings

	lbf ft	Nm
Pinion coupling nut*	80 to 173	110 to 240
Final drive unit-to-axle nuts	27	38
Brake backplate/retainer plate bolts	36	50
Drain plug	30	42
Filter plug	23	32

For suspension torque wrench settings see Chapter 11.

*Do not exceed tightening torque required to give pinion flange starting torque recorded prior to dismantling (see text).

1 General description

The rear axle is of the hypoid semi-floating type and the final drive assembly may be of the conventional type where the differential carrier may be removed as a unit, or it may be of the type where the differential is assembled directly into the axle casing. The differential used on GT models is of the "limited slip" type, having a slipping clutch built into it.

With a standard differential, if one wheel is on a slippery surface such as ice, snow or mud and the other wheel is able to grip, the wheel on the slippery surface spins and the other wheel cannot produce any traction. A limited slip differential enables both wheels to provide traction under these conditions and so reduces the possibility of the car becoming immobile.

Operations on the rear axle should be limited to those described in this Chapter. Dismantling and reassembly of the rear axle crown wheel and pinion is not considered to be within the capability of the home mechanic and if a fault develops, the axle or final drive assembly should be exchanged for a factory reconditioned unit.

2 Axleshafts, bearings, oil seals – removal and installation

1 The axleshafts may be withdrawn without disturbing the differential gear. They are removed in order to renew the bearings or oil seals located at the outer ends of the shafts. They will have to be partially withdrawn if the differential carrier is to be removed on conventional type rear axles.

2 Jack-up the car at the rear and support the axle casings on stands. Remove the road wheels.

3 Release the handbrake and remove the rear brake drums (see Chapter 9).

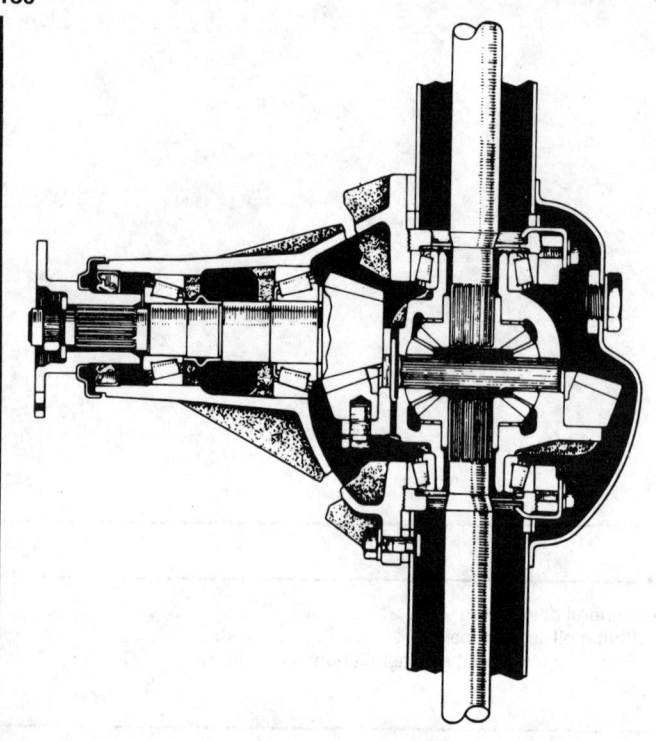

Fig. 8.1 Conventional differential (sectional view)

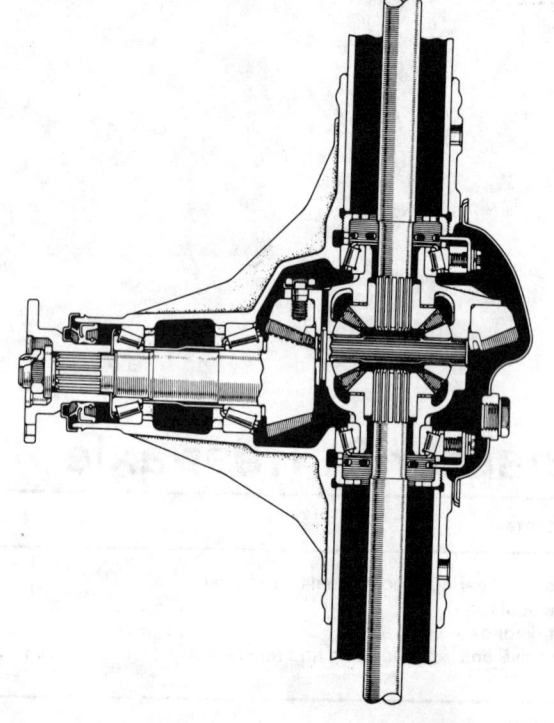

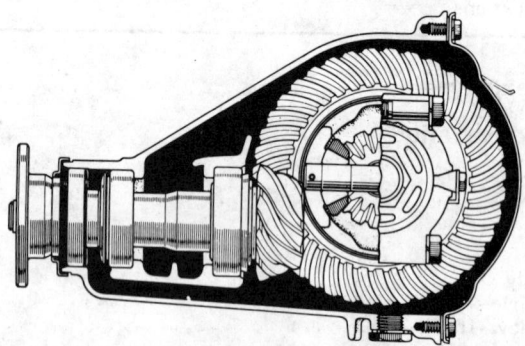

Fig. 8.2 Unitized differential (sectional view)

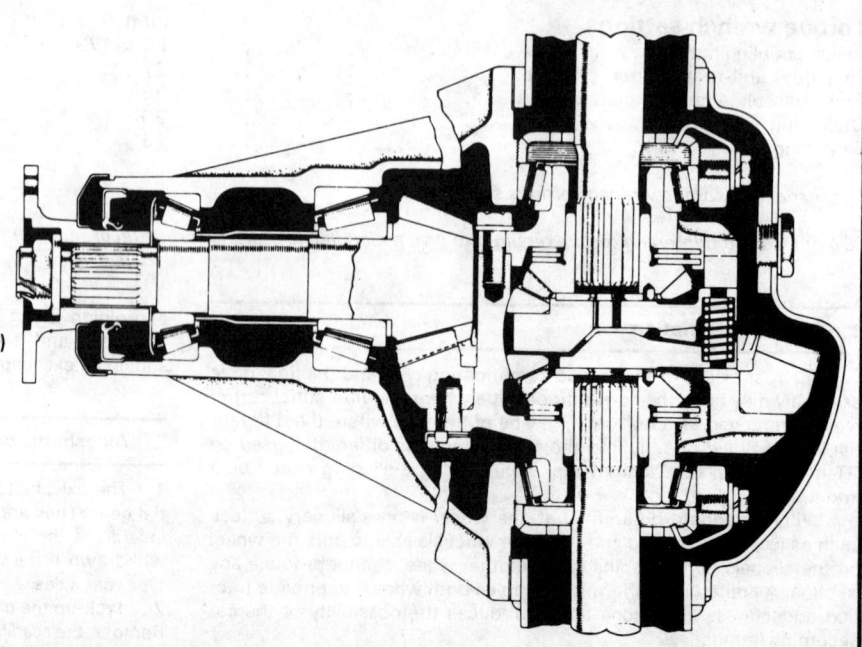

Fig. 8.3 Limited slip differential (sectional view)

Chapter 8 Rear axle

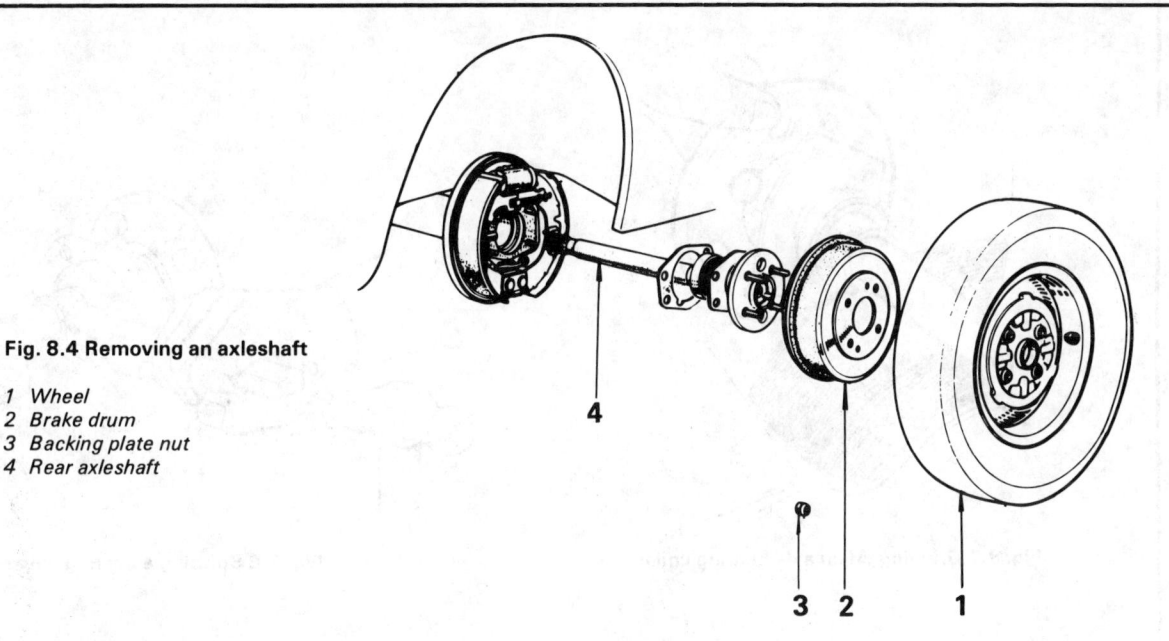

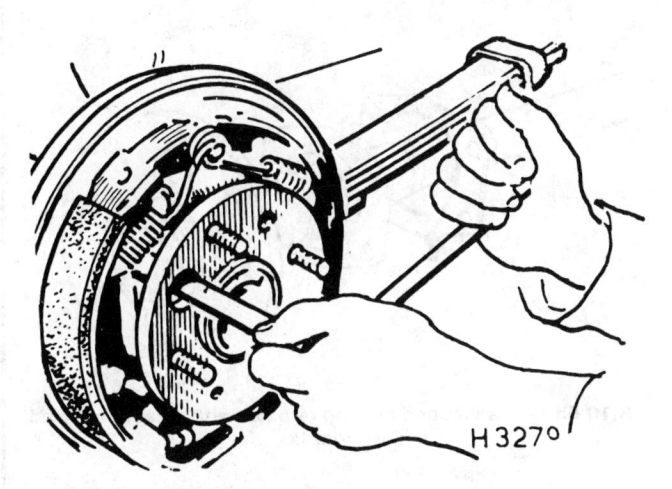

Fig. 8.4 Removing an axleshaft

1. Wheel
2. Brake drum
3. Backing plate nut
4. Rear axleshaft

Fig. 8.5 Removing the backing plate nuts

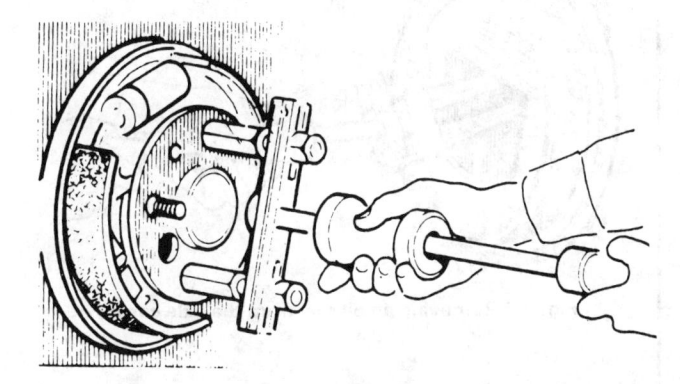

Fig. 8.6 Using a slide hammer to withdraw the axleshaft

4 Remove the bolts which secure the bearing retainer plate and the brake backplate to the flange on the end of the axle casing. These bolts are accessible through the holes in the axleshaft flanges.

5 The axleshaft, bearing and other components are now held in position by the fit of the outer race of the bearing in the axle casing. Ideally, a slide hammer should be attached to the roadwheel studs and the axleshaft extracted. Alternatively, an old roadwheel can be bolted onto the axleshaft flange and opposite points on its inner rim struck. On no account strike or prise the axleshaft flange itself. Pulling on the shaft flange will prove quite ineffective and will probably only result in dislodging the car from the axle stands.

6 With the axleshaft removed, grind or hacksaw off the bearing retaining collar. When a deep cut or groove has been made in the collar, a sharp chisel can be used to split it. In this way, no damage will be caused to the axleshaft.

7 Using a suitable press or extractor, remove the bearing from the shaft, followed by the spacer, gasket and bearing retainer plate.

8 Always renew an oil seal whenever an axleshaft is removed. The seal can be levered from the end of the axle casing and a new one tapped into position using a piece of tubing as a drift.

9 Examine the axleshaft for cracks, distortion and spline wear.

10 To reassemble the axleshaft, fit the bearing retainer plate, gasket, spacer, bearing and collar. The unchamfered face of the collar goes against the bearing. A press will be required to install the new bearing and collar and the operation will be facilitated if both components are first heated to about 300°F in an oven. Do not exceed this temperature.

11 Repack the bearing by pressing in wheel bearing grease.

12 Check the condition of the brake backplate to axle-end flange gasket and renew it if necessary by gently prising the backplate from the axle flange.

13 Stick the bearing retainer plate gasket to the outside face of the brake backplate.

14 Insert the axleshaft into the axle casing taking care not to damage the lips of the new oil seal. The splines at the inner end of the axleshaft should engage with those in the differential side gears. Now enter the bearing into the recess in the axle casing until the outer face of the bearing is flush with the axle casing.

15 Insert the backplate/retainer plate bolts and tighten to the specified torque.

16 Install the brake drum, roadwheel and lower the car.

17 Check the oil level in the rear axle.

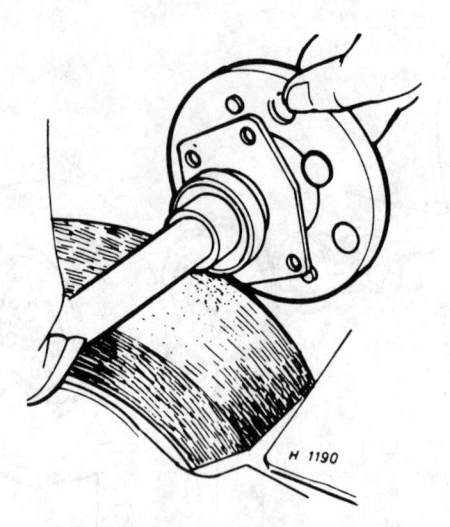

Fig. 8.7 Grinding off an axle bearing collar

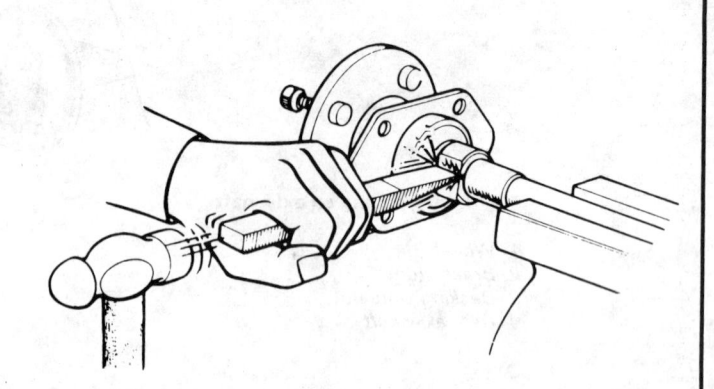

Fig. 8.8 Splitting a bearing collar

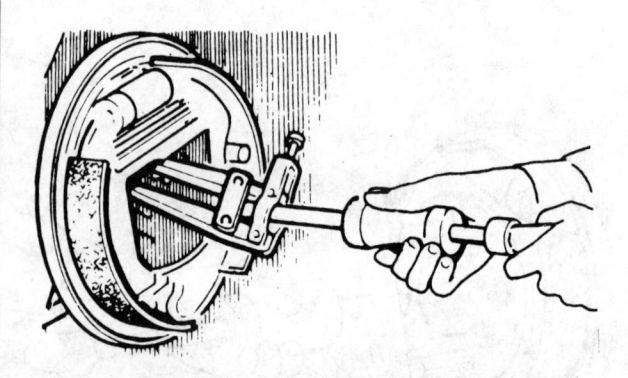

Fig. 8.9 Removing an oil seal from the axle casing

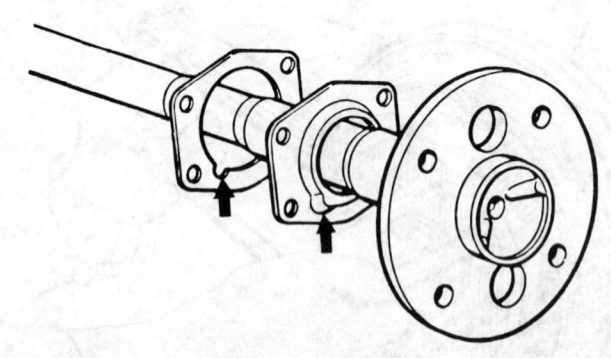

Fig. 8.10 Fit the gasket and bearing retainer with recess (arrowed) downwards

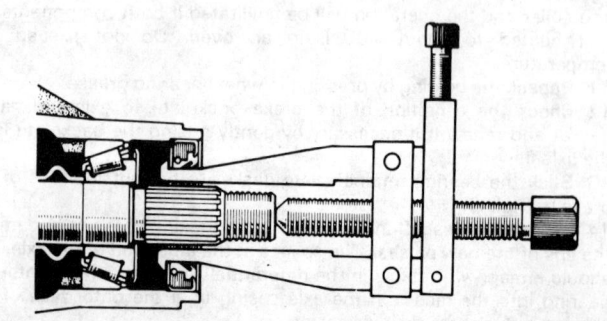

Fig. 8.11 Extracting the pinion oil seal with a two legged puller

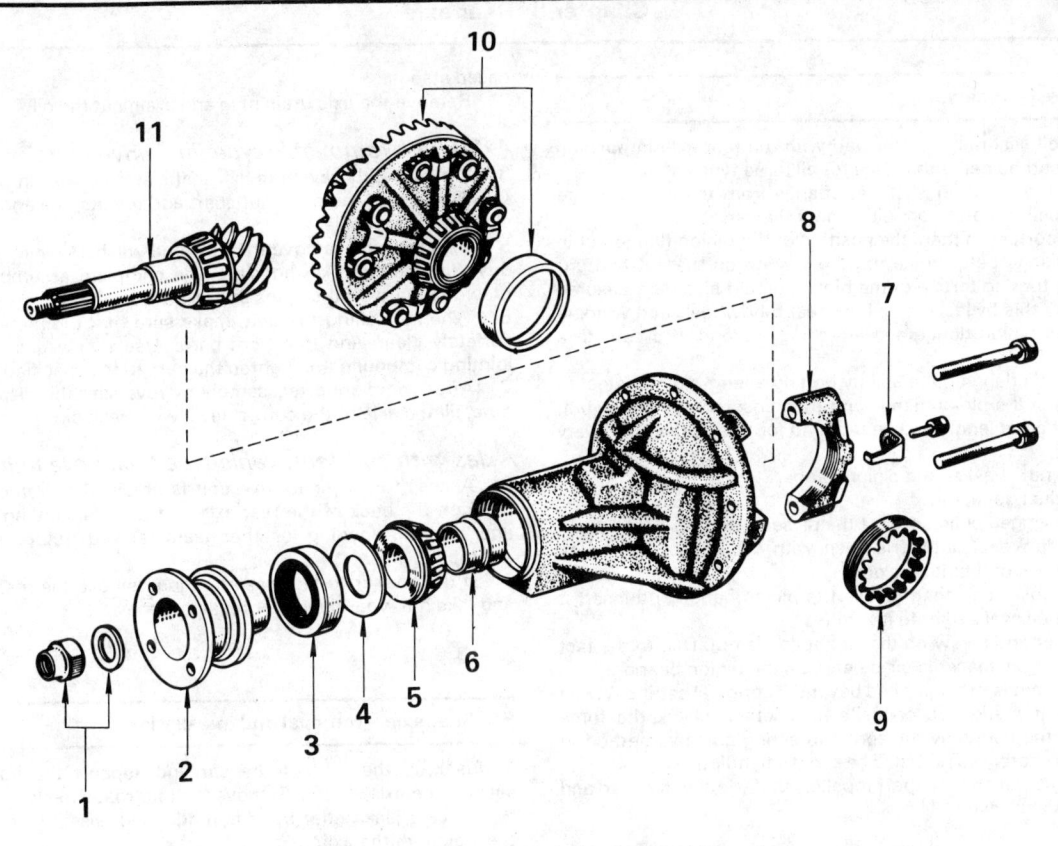

Fig. 8.12 Conventional differential (with detachable carrier) components

1 Nut & washer	4 Oil slinger	7 Lockwasher	10 Differential assembly
2 Drive flange	5 Front bearing	8 Bearing cap	11 Drive pinion
3 Oil seal	6 Bearing spacer	9 Adjusting nut	

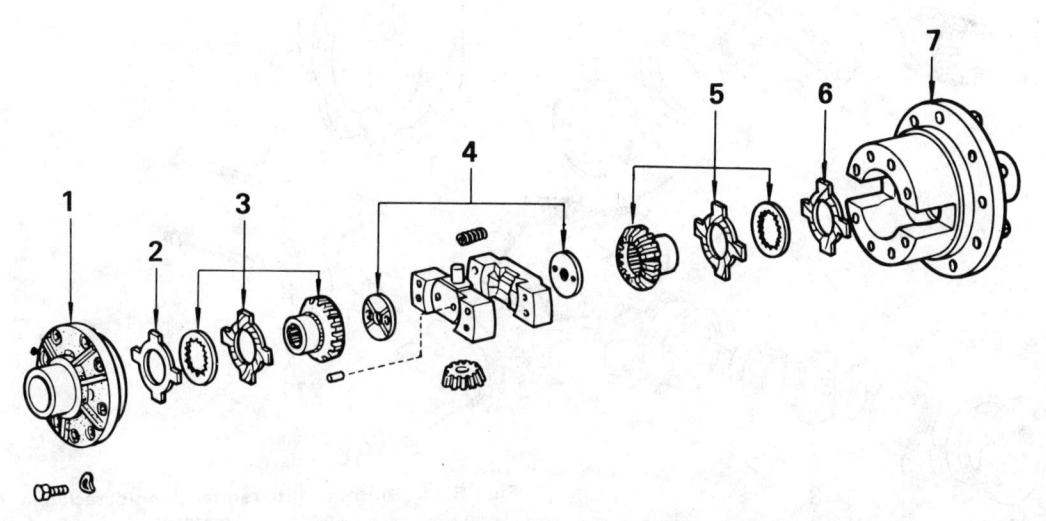

Fig. 8.13 Limited slip differential components

1 Differential case cover
2 Side gear thrust washer
3 Clutch plate, thrust washer, side gear
4 Clutch assembly, thrust washer, spring and differential pinion
5 Side gear, thrust washer, clutch plate
6 Side gear thrust washer
7 Differential case

Chapter 8 Rear axle

3 Pinion oil seal – renewal

1 The pinion oil seal may be renewed with the rear axle in position on the car, the method being the same for all three types of axle.
2 Disconnect the propeller shaft rear flange from the pinion flange and rest the propeller shaft to one side under the car.
3 It is very important to mark the position of the pinion flange nut in relation to the flange before relieving the staking on the nut and unscrewing it. The turning torque of the pinion should also be measured and recorded. Do this by jacking up both rear roadwheels and winding a cord round the pinion flange and attaching the end of a cord to a spring balance.
4 Hold the pinion flange quite still by bolting a length of flat steel to two of the holes in the pinion flange and then unscrew the pinion nut. A ring wrench of good length will be required for this as the nut is very tight.
5 Remove the nut, washer and pinion flange.
6 Detach the dust deflector.
7 Using a two-legged puller, extract the oil seal.
8 Tap in the new seal until it is flush with the end of the pinion housing. Use a piece of tubing as a drift.
9 Fit the dust cover and then grease the pinion flange and insert it carefully into the oil seal and onto its splines.
10 Fit the washer and screw on the pinion flange nut. Due to the fact that a collapsible type spacer is used between the pinion bearings, it is essential that the nut is not tightened beyond its original setting. When the nut alignment marks are opposite each other, check the force required to turn the flange by the cord and spring balance method. It should be approximately as recorded before dismantling.
11 Stake the nut, reconnect the propeller shaft, lower the car and check the rear axle oil level.

4 Final drive unit – dismantling and reassembly

1 Jack up the rear of the car and support the axle casing on firmly based stands.
2 Remove the axle drain plug and drain out the oil.

Axles with removable type final drive units

3 Partially withdraw both axleshafts as described in Section 2.
4 Disconnect the propeller shaft and rest its free end on the ground, clear of the rear axle.
5 Unscrew and remove the nuts and washers which secure the final drive unit to the axle housing. The complete assembly can then be drawn forward off the studs and removed.
6 When installing the unit, make sure that the mating surfaces are perfectly clean and free from burrs. Use a new gasket, coated with jointing compound and tighten the nuts to the specified torque.
7 After completing reassembly by reversing the dismantling procedure, fill the axle to the correct level with fresh oil.

Axles with non-removable type final drive unit

8 Access to the final drive unit is obtained by removing the cover plate on the back of the rear axle, but any dismantling of the unit will alter the pre-loading of the gears, and therefore should not be attempted.
9 If the final drive requires attention, remove the rear axle complete and take it to your dealer for repair or renewal.

5 Rear axle – removal and installation

1 Jack-up the rear of the car and support the body side frame securely on axle stands. Remove the rear roadwheels.
2 Place a jack under the differential and raise it just enough to take the weight of the axle.
3 Disconnect the brake flexible hose from the hydraulic rigid line. Plug both pipes.
4 Slacken the parking brake primary cable adjustment completely and then remove each of the rear brake drums.

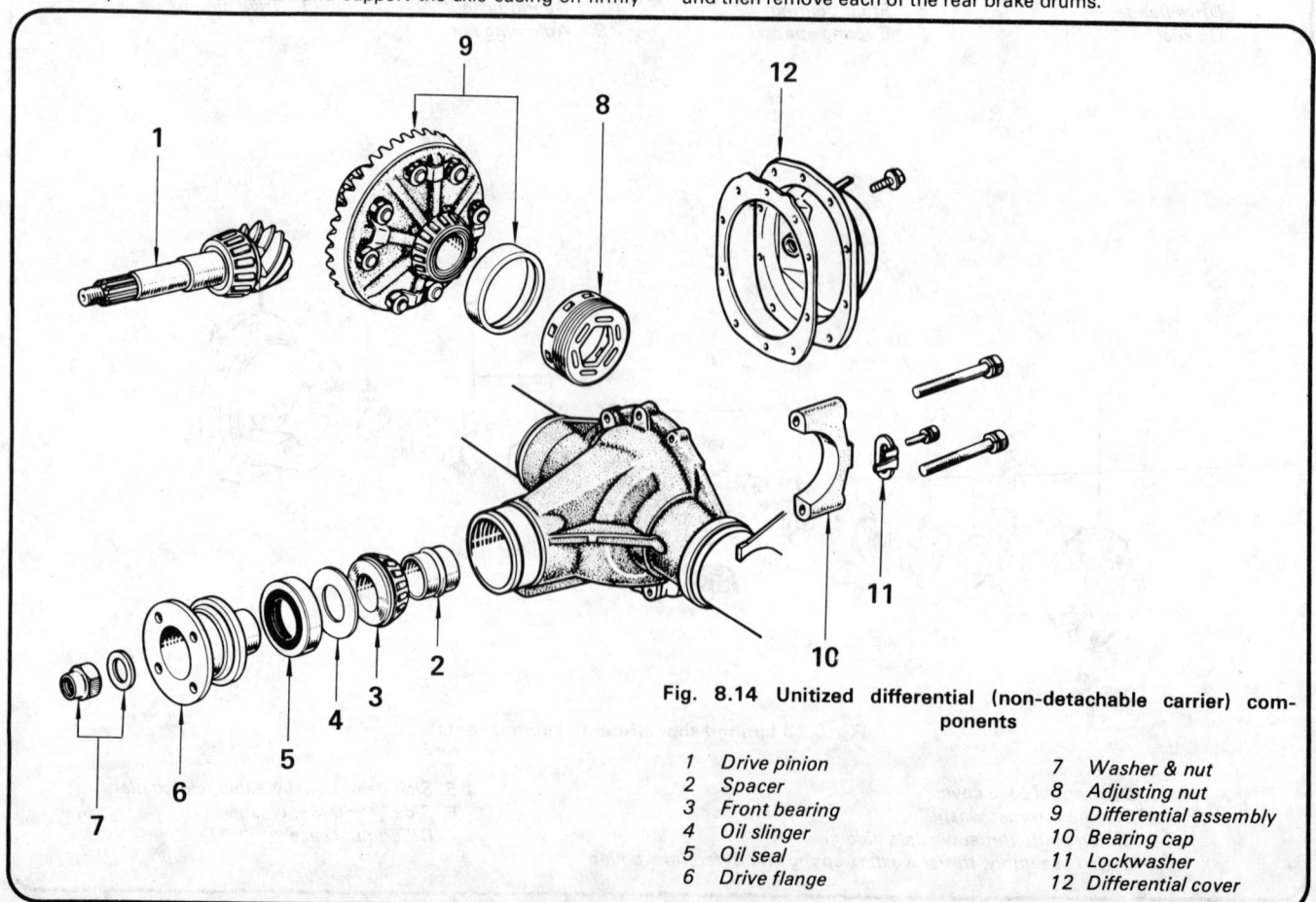

Fig. 8.14 Unitized differential (non-detachable carrier) components

1 Drive pinion
2 Spacer
3 Front bearing
4 Oil slinger
5 Oil seal
6 Drive flange
7 Washer & nut
8 Adjusting nut
9 Differential assembly
10 Bearing cap
11 Lockwasher
12 Differential cover

5 Disconnect the ends of the parking brake cables from the brake shoe levers and then withdraw the cables from the backplates.
6 Disconnect the propeller shaft rear flange and rest the shaft to one side.
7 Disconnect the rear shock absorber lower mountings.
8 Lower the jack previously placed under the differential until the coil springs and their insulators can be removed from the spring pans.
9 Disconnect the lower control arms by removing the pivot bolts from the axle mountings.
10 Disconnect the upper control arms by removing the pivot bolts from the axle mountings.
11 Disconnect the lateral control rod from the rear axle casing by removing the split-pin, nut and washer.
12 The rear axle can now be removed sideways through the wheel arch on one side of the car.
13 Installation is a reversal of removal but before tightening the control arm bolts to the specified torque, lower the car to the ground so that its weight is on the roadwheels.

6 Fault diagnosis – rear axle

Symptom	Reason/s
Vibration	Worn halfshaft bearing Loose bolts (propeller shaft-to-drive flange) Tyres require balancing Propeller shaft out of balance
Noise on turns	Worn differential gear
Noise on drive or coasting*	Worn or incorrectly adjusted ring and pinion gear
'Clunk' on acceleration or deceleration	Worn differential gear cross-shaft Worn propeller shaft universal joints Loose bolts (propeller shaft-to-drive flange)

*It must be appreciated that tyre noise, wear in the rear suspension bushes and worn or loose shock absorber mountings can all mislead the mechanic into thinking that components of the rear axle are the source of trouble.

Chapter 9 Braking system

For modifications and information applicable to later USA models, refer to Supplement at end of manual

Contents

Booster (vacuum servo unit) – description	15
Booster unit – dismantling and reassembly	17
Booster unit – removal and installation	16
Brake disc – examination and renovation	5
Brake drums – inspection and removal	7
Brake pedal – adjustment, removal and installation	13
Disc caliper – removal, servicing, installation	4
Fault diagnosis – braking system	18
Front disc pads – inspection and renewal	2
General description	1
Handbrake – adjustment	11
Handbrake cables – renewal	12
Hydraulic lines – inspection and renewal	10
Hydraulic system – bleeding	9
Master cylinder – removal, servicing and installation	8
Pressure regulating valve	14
Rear brake shoes – inspection and renewal	3
Rear brake wheel cylinders – removal, servicing and installation	6

Specifications

System type
Front disc, rear drum, dual hydraulic circuit with vacuum servo assistance. Parking brake mechanically operated on rear wheels

Front discs
Type .... Single piston sliding caliper

	Coupe	Liftback
Disc diameter	13 in (330 mm)	14 in (355 mm)
Disc thickness:		
Standard	0.39 in (10.0 mm)	0.49 in (12.5 mm)
Minimum	0.35 in (9.0 mm)	0.45 in (11.5 mm)

Disc run-out (maximum) .... 0.006 in (0.15 mm)
Pad thickness (minimum) .... 0.04 in (1.0 mm)

Rear drums
Type .... Self-adjusting, leading and trailing shoes
Drum inner diameter:
   Standard .... 9.00 in (228.6 mm)
   Maximum .... 9.08 in (230.6 mm)
Lining thickness (minimum) .... 0.04 in (1.0 mm)
Shoe clearance .... 0.024 in (0.6 mm)
Handbrake lever travel .... 3-7 notches

Brake pedal
Pedal height from floor (carpet removed) .... 6.48 to 6.87 in (164.5 to 174.5 mm)
Pedal free play .... 0.1 to 0.2 in (3 to 6 mm)

Brake booster
Booster pushrod to piston clearance:
   At idling vacuum .... 0.004 to 0.020 in (0.1 to 0.5 mm)
   At zero vacuum .... 0.024 to 0.026 in (0.60 to 0.65 mm)

Brake fluid type
SAE J 1703E

Torque wrench settings
	lbf ft	Nm
Pedal shaft bolt	35	48

Chapter 9 Braking system

Master cylinder piston stop bolt	10	14
Master cylinder outlet check valve	35	48
Fluid reservoir retaining bolt	20	28
Brake pipe union nuts	12	16
Disc brake caliper mounting bolts	50	69
Brake disc-to-hub bolts	35	48

1 General description

The braking system is of four wheel hydraulic type. Front brakes are of disc with self-adjusting drums at the rear.

The hydraulic system is of dual circuit design using a tandem master cylinder. A vacuum servo booster is used to reduce the braking effort required on the foot pedal.

The parking brake is mechanically operated and operates on the rear wheels only.

A pressure regulating valve is incorporated in the hydraulic circuit to regulate the pressure between the front and rear circuits in order to prevent the rear wheels locking in advance of the front wheels during heavy application of the brakes.

2 Front disc pads – inspection and renewal

Note: *More than one type of caliper has been fitted to these cars. Pad replacement kits normally contain parts for all caliper types – only fit the clips/springs specific to your caliper*

1 Jack-up the front of the car and remove the roadwheels.

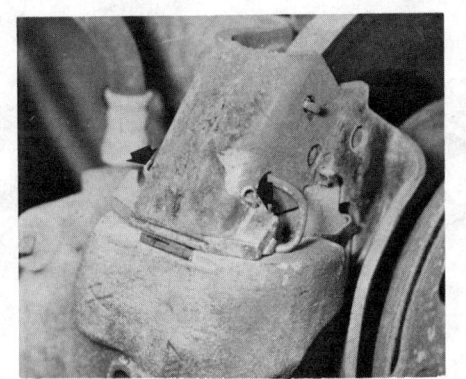

2.4a Cylinder guide clips (arrowed)

2.4b Cylinder guide partially removed

2.5a Removing the cylinder assembly

2.5b Removing a brake pad

3.5 Shoe steady pins and springs (arrowed)

3.7a Access hole to automatic adjusting lever

3.7b Upper shoe return spring and automatic adjuster

3.8 Handbrake cable attachment (arrowed)

3.9 Lower shoe return spring

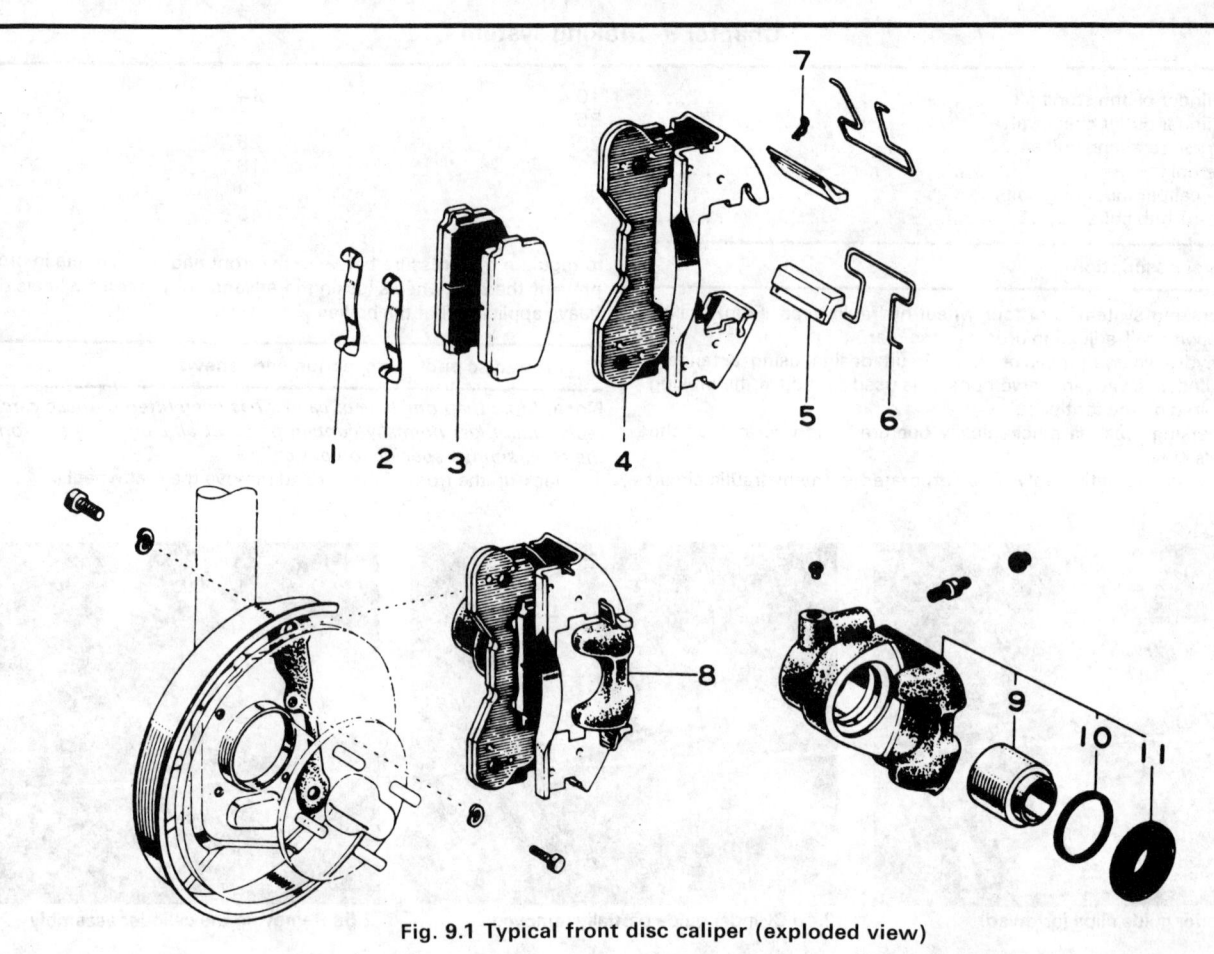

Fig. 9.1 Typical front disc caliper (exploded view)

1 and 2 Disc pad support plates	4 Cylinder mounting	7 Clip	10 Seal
3 Disc pad	5 Guide	8 Cylinder assembly	11 Dust excluding boot*
	6 Cylinder support spring	9 Piston	* On some models this is retained by a set ring

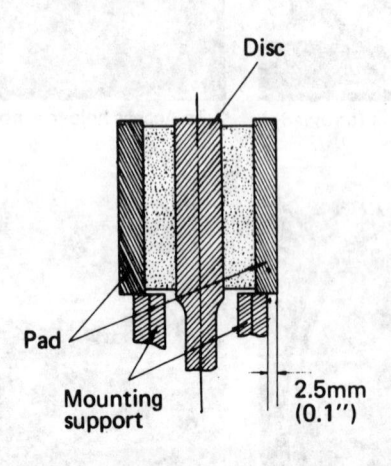

BEFORE WEAR

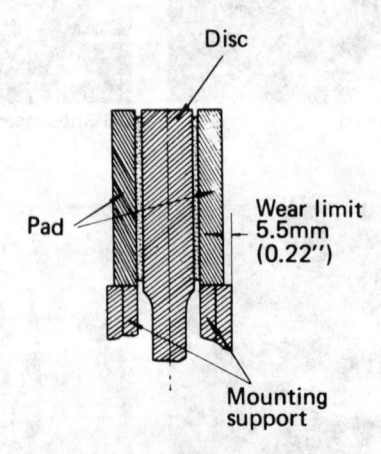

AFTER WEAR

Fig. 9.2 Brake pad wear measurement

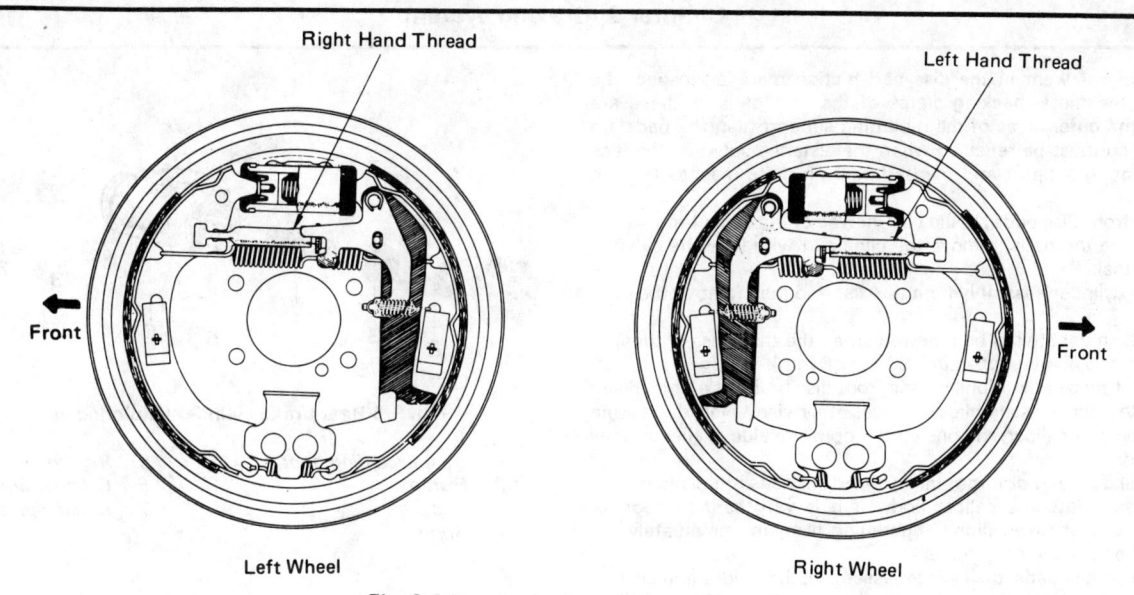

Fig. 9.3 Rear brake drum assemblies

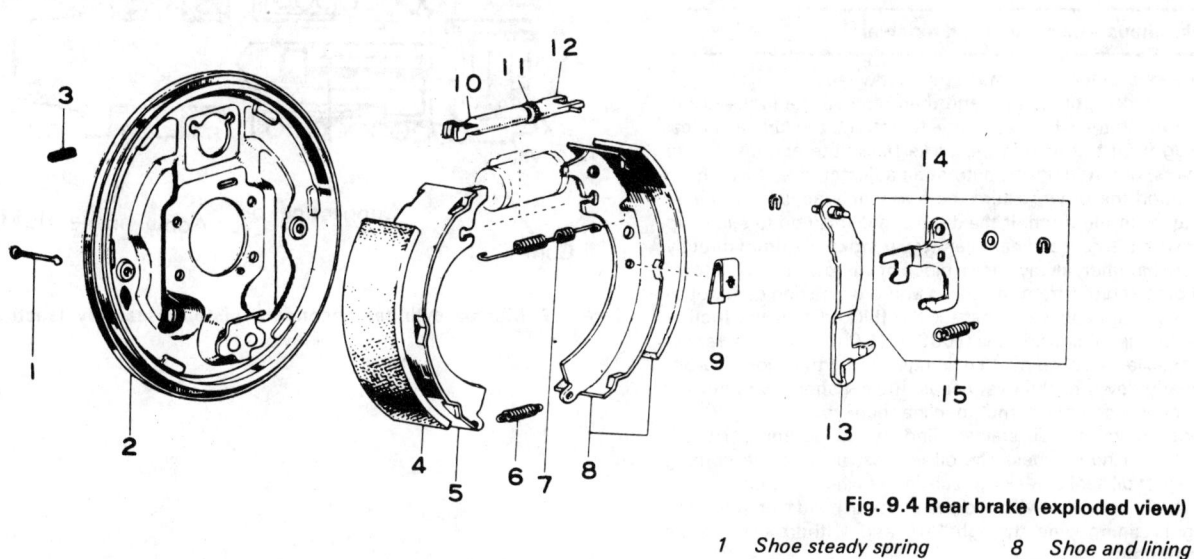

Fig. 9.4 Rear brake (exploded view)

1 Shoe steady spring retaining pin
2 Backplate
3 Plug for adjuster access hole
4 Lining
5 Shoe
6 Lower return spring
7 Upper return spring
8 Shoe and lining
9 Shoe steady spring
10 Strut
11 Knurled adjuster wheel
12 Strut
13 Parking brake shoe lever
14 Automatic adjuster lever
15 Return spring

Fig. 9.5 Releasing automatic adjuster through backplate hole

190 Chapter 9 Braking system

2 To check for wear in the disc pad friction material, inspect the position of the metal backing plates of the disc pads. If these are inboard of the outer faces of the mounting support then the pads are well worn and must be renewed when the distance between the face of the mounting support and the pad backing plate reaches 0.22 in (5.6 mm).
3 All four front disc pads should be renewed at the same time.
4 To remove the pads, remove the clips and withdraw the cylinder guides (photos).
5 Pull the cylinder assembly from the disc and then remove the pads (photo).
6 The pads must always be renewed when the thickness of the friction material has been reduced to 0.04 in (1.0 mm).
7 Carry out these operations on one front disc brake at a time, never remove both caliper assemblies at once, otherwise when depressing the piston on one caliper, the one on the opposite side of the car may be displaced.
8 Clean all dust and dirt from the exposed surfaces of the piston and then depress it into the caliper body. If it is very hard to depress, release the caliper bleed nipple, tightening it again immediately the piston reaches the end of its travel.
9 Install the new pads, the cylinder assembly, the guides and clips.
10 Depress the brake pedal several times to bring the pads into contact with the disc.
11 Repeat the operations on the opposite disc brake and then check and top-up the master cylinder front circuit reservoir. Should there by any feeling of sponginess in the foot brake, bleed the front hydraulic circuit, as described in Section 9.

3 Rear brake shoes – inspection and renewal

1 Jack-up the rear of the car, remove the roadwheel.
2 Release the parking brake fully and then remove the brake drum. To free the drum it may be necessary to retract the brake shoes. Extract the plug from the hole in the brake backplate and insert two thin screwdrivers, one to hold the automatic adjuster lever away from the star wheel and the other to turn the star wheel adjuster to move the shoes away from the drum. If the drum is still inclined to stick, tap it off carefully with a soft-faced hammer. Never strike the drum directly with an ordinary hammer, always use a block of wood as an insulator.
3 Brush all dust and dirt from the shoes and drum taking care not to inhale it as it contains asbestos. Inspect the thickness of the friction material. With bonded linings, if it is 0.039 in (1.0 mm) or less, renew the shoes as an axle set. With rivetted linings, renew the shoes if worn down to or nearly down to the rivet heads. Renew shoes with new or reconditioned ones – do not attempt to reline them yourself.
4 If the shoe linings are oil stained, find the cause and correct it before installing the new shoes. The oil is most likely to be coming from the axle-shaft oil seal or leaking seals in the wheel cylinder.
5 To remove the shoes, twist the tee-shaped heads of the shoe steady spring retaining pins through 90° and withdraw the shoe steady springs (photo).
6 Note carefully the positions of the leading and trailing ends of the shoes (the relative positions of the friction linings where they stop short before the ends of the shoes).
7 Unhook the upper shoe return spring and the spring on the automatic adjuster lever (photo).
8 Disconnect the parking brake cable from the shoe lever (photo).
9 Pull both shoes outwards to expand the lower return spring (photo) and lift the shoes from the brake backplate.
10 Remove the C washer and withdraw the parking brake cable shoe lever and automatic adjuster.
11 Lay the new shoes on the bench in their correct leading and trailing positions.
12 Fit the parking brake shoe lever and automatic adjuster securing them to the new shoe with the C washer.
13 Connect the parking brake cable to the shoe lever and locate the adjuster. Turn the star wheel on the adjuster to fully retract the adjuster. Failure to do this will prevent the fitting of the brake drum.
14 Connect the automatic adjuster lever spring.
15 Connect both shoes at their lower ends by means of the return spring and engage the shoes in the lower anchorage.
16 Engage the upper ends of the shoes with the wheel cylinder pistons taking care not to trap the rubber dust excluders.
17 Reconnect the shoe upper return spring and install the shoe

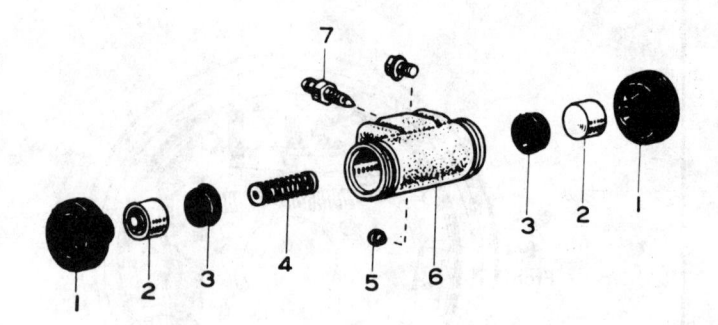

Fig. 9.6 Rear brake cylinder (exploded view)

1 Dust excluding boot
2 Piston
3 Cup seal
4 Spring
5 Rigid pipe union
6 Cylinder body
7 Bleed nipple

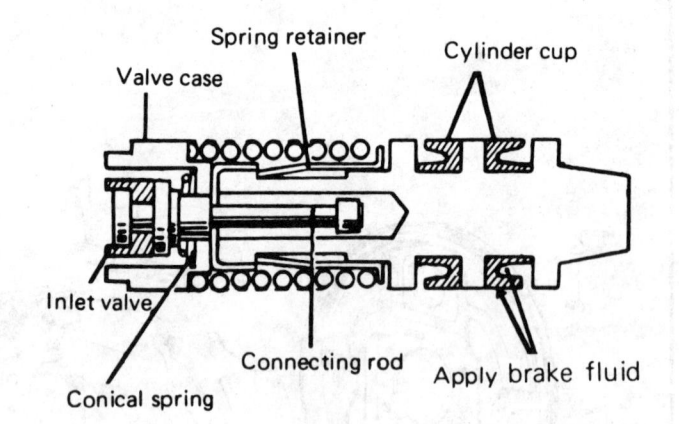

Fig. 9.7 Master cylinder secondary piston assembly (sectional view)

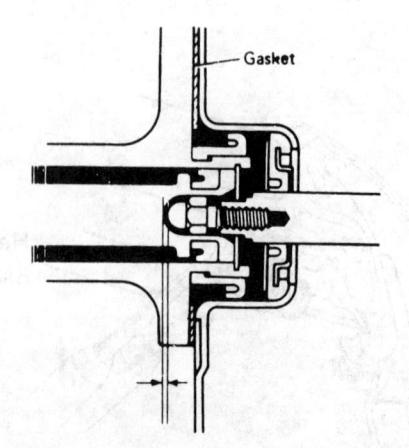

Fig. 9.8 Booster to master cylinder pushrod clearance

Chapter 9 Braking system

steady springs.
18 Check the operation of the automatic adjuster by moving the shoe lever.
19 Clean the interior of the brake drum and install it.
20 Apply the parking brake several times to adjust the lining to drum clearance.
21 Repeat the operations on the opposite brake.
22 Refit the roadwheels and lower the car.

4 Disc caliper – removal, servicing and installation

1 Remove the master cylinder reservoir cap and place a thin sheet of plastic film over the opening and then refit the cap. This will help to restrict the flow of hydraulic fluid from the circuit when the pipes are disconnected.
2 Disconnect the flexible hydraulic hose from the rigid pipe at the support bracket. Plug or cap the open ends of the pipes to prevent loss of fluid or dirt entering.
3 Unscrew the flexible pipe from the caliper.
4 Remove the clips and withdraw the cylinder guide (see Section 2).
5 Withdraw the caliper from the disc.
6 Brush or wipe the rust and dust from the caliper external surfaces.
7 Remove the rubber dust excluder.
8 Eject the piston by applying air pressure at the fluid inlet on the caliper body.
9 At this stage, inspect the condition of the piston and cylinder surfaces. If any scoring or bright wear areas are apparent, renew the caliper assembly complete.
10 Where the components are in good condition, obtain a repair kit and discard the old seal and rubber dust excluder.
11 Manipulate the new seal into the groove in the cylinder using the fingers only. Dip the piston into clean hydraulic fluid and insert it into the cylinder. Install the new dust excluder.
12 Installation is a reversal of removal but on completion, bleed the system (see Section 9).

5 Brake disc – examination and renovation

1 Whenever the caliper cylinder assembly is withdrawn to inspect the condition of the pads, it is a good plan to check the disc itself.
2 Inspect the friction surfaces of the disc for cracks or deep grooves. Light scoring is normal. Any deep grooves will have to be removed by surface grinding but if this action will reduce the thickness of the disc beyond the specified minimum thickness then the disc will have to be renewed.
3 Check the disc for distortion (run-out) by using a dial gauge or feeler blades between a fixed point and the disc. Maximum run-out is 0.006 in (0.15 mm).
4 To remove a disc, withdraw the cylinder assembly and disc pads. Unbolt and remove the cylinder assembly mounting, support springs and plates.
5 Knock off the hub grease cap, withdraw the split-pin, nut retainer, nut and thrust washer.
6 Withdraw the combined hub/disc assembly taking care to catch the bearings.
7 Unbolt the disc from the hub.
8 Reassembly and installation are reversals of removal and dismantling but tighten bolts to specified torque and adjust the front hub bearing preload, as described in Chapter 11.

6 Rear brake wheel cylinders – removal, servicing and installation

Note: Certain models may have been fitted with non-standard wheel cylinders in production, While this does not cause any safety problems in service, it is important to renew the rear wheel cylinders on both sides if one should fail.

1 New seals can be fitted without removing the cylinder from the brake backplate.
2 Remove the brake shoes, as described in Section 3.
3 Remove the rubber dust excluders from both ends of the cylinder body. To prevent loss of fluid, cover the master cylinder reservoir neck with plastic film, as described in Section 4.
4 Extract the pistons, cup seals and spring.
5 At this stage, inspect the condition of the piston and cylinder surfaces. If they appear scored or show any bright wear areas, the complete cylinder assembly must be renewed by removing it in the following way.
6 Disconnect the fluid pipe from the wheel cylinder at the inner face of the brake backplate.
7 Unscrew and remove the two wheel cylinder securing bolts and lift the cylinder from the backplate.
8 Where the wheel cylinder components are in good order, obtain a repair kit and install the new cup seals, spring, pistons and boots. Dip the components in clean hydraulic fluid before inserting them into the cylinder bore.
9 Bleed the system on completion of reassembly (Section 9).
10 If a new wheel cylinder has been installed, do not tighten the securing bolts beyond the specified limit.

7 Brake drums – inspection and renovation

1 Whenever the rear brake drums are removed to inspect the condition of the shoe linings, check the friction surfaces of the drum for cracks or deep grooves.
2 In addition to those faults, the drum may have become out-of-round (oval in shape). This can usually be detected if the brakes tend to grab when applied or when turning the roadwheels by hand, they bind at two opposite points.
3 The drums can be machined to overcome these problems provided the inside diameter does not then exceed 9.079 in (230.6 mm) otherwise the drum will have to be renewed.

8 Master cylinder – removal, servicing and installation

1 Disconnect the brake pipes from the master cylinder body. Plug their open ends to prevent entry of dirt.
2 Disconnect the leads from the fluid pressure indicator switches.
3 Unbolt and remove the master cylinder from the front face of the vacuum servo booster unit.
4 Secure the master cylinder assembly in a vice and unscrew and remove the fluid pressure indicator switches.
5 From the end of the master cylinder, extract the circlip, piston assembly and spring.
6 Unscrew and remove the stop bolt from the side of the master cylinder body and extract the second piston and spring, also the inlet valve seat.
7 Unscrew and remove the outlet plugs and extract the check valves.
8 At this stage, inspect the surfaces of the pistons and cylinder bore. If they appear scored or show any bright wear areas, renew the master cylinder complete.
9 If the components are in good condition, obtain a repair kit, discard all old seals and fit the new ones, using the fingers only to manipulate them into position. Dismantling the piston assemblies is quite straight-forward except that on the secondary piston, the lips of the return spring retainer must be prised up while at the same time compressing the spring before the components can be separated.
10 Dip all components in clean hydraulic fluid before reassembling them and tighten the plugs, switches and other components to the torque wrench settings specified in Specification.
11 Before installing the master cylinder to the booster unit, check that there will be a clearance between the end of the booster unit pushrod and the end of the master cylinder piston. This is easily carried out by measuring the projection of the booster pushrod and the depth of the recess in the master cylinder piston using a vernier gauge or a bolt, nut and feeler blades. By subtracting one measurement from the other, the clearance can be established. If necessary, adjust the legnth of the booster pushrod by releasing the locknut and turning the end screws to give a dimension of 0.004 to 0.020 in (0.1 to 0.5 mm).
12 Check the brake pedal height and free-play, as described in Section 13.
13 When installation is complete, bleed the hydraulic system (both circuits), as described in Section 9.

9 Hydraulic system – bleeding

1 Use only clean fluid (which has remained unshaken for 24 hours and has been stored in an airtight container) for topping-up the master

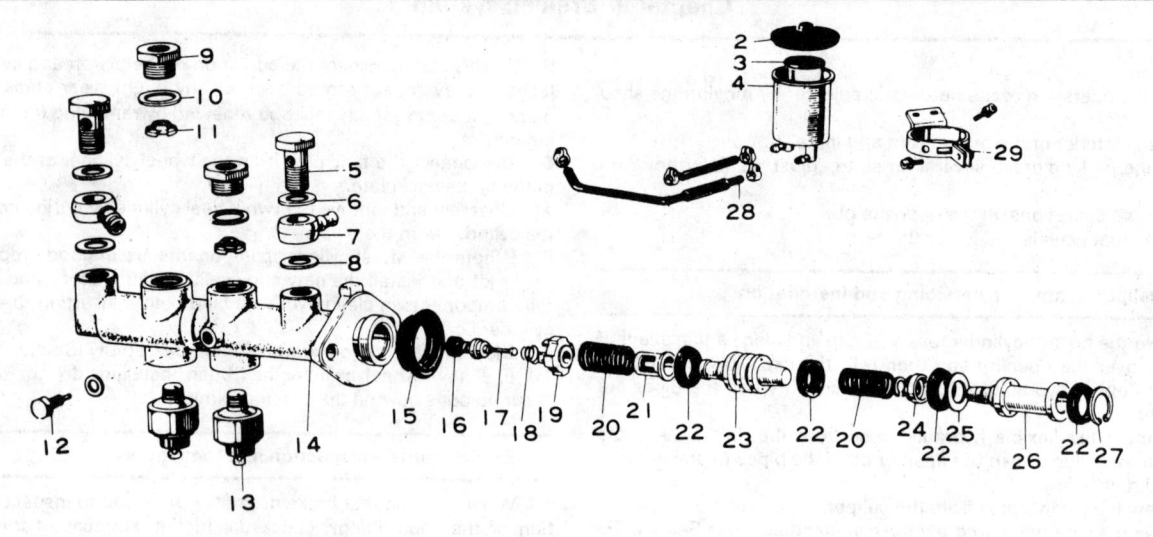

Fig. 9.9 Tandem master cylinder (exploded view)

2 Reservoir cap
3 Float
4 Reservoir
5 Union (hollow) bolt
6 Washer
7 Banjo union
8 Washer
9 Fluid outlet plug
10 Washer
11 Outlet check valve
12 Stop bolt
13 Pressure switch
14 Body
15 Dust excluding boot
16 Inlet valve
17 Inlet valve connecting rod
18 Spring
19 Case
20 Spring
21 Spring retainer
22 Cup seal
23 Secondary piston
24 Spring retainer
25 Piston stop plate
26 Primary piston
27 Circlip
28 Supply hose
29 Reservoir support clip

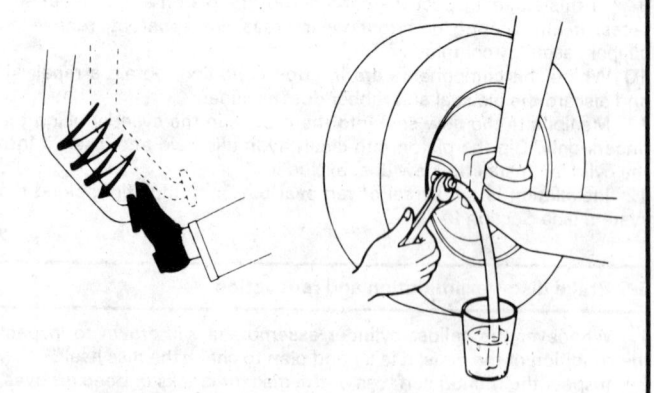

Fig. 9.10 Bleeding the hydraulic system

Fig. 9.11 Handbrake components

1 Brake lever
2 Cover
3 Boot
4 Pull rod
5 Brake equalizer
6 Brake cable (LH)
7 Brake cable (RH)

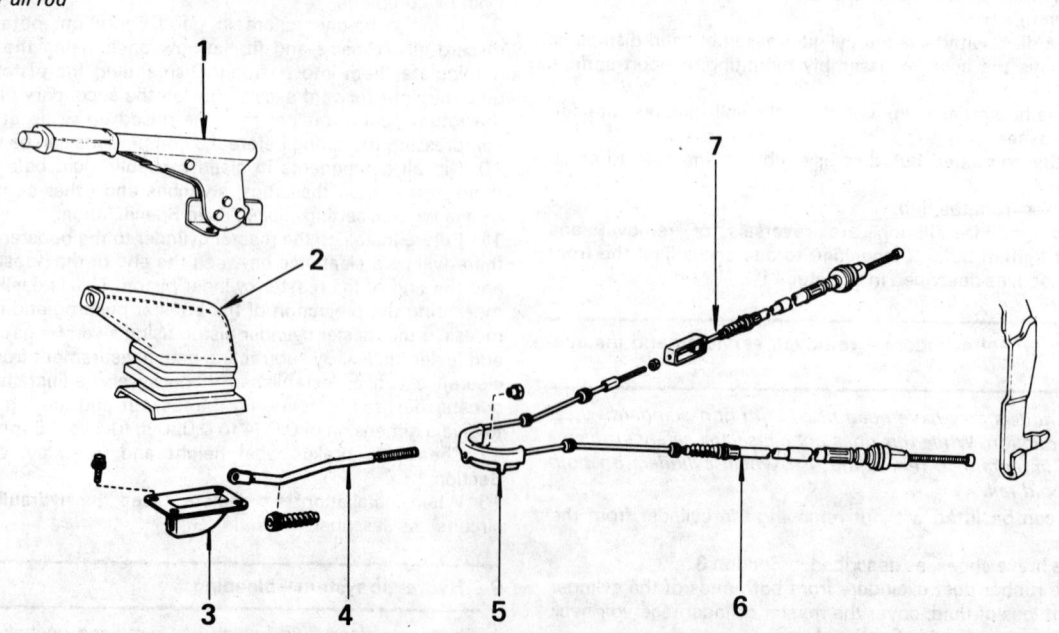

Chapter 9 Braking system

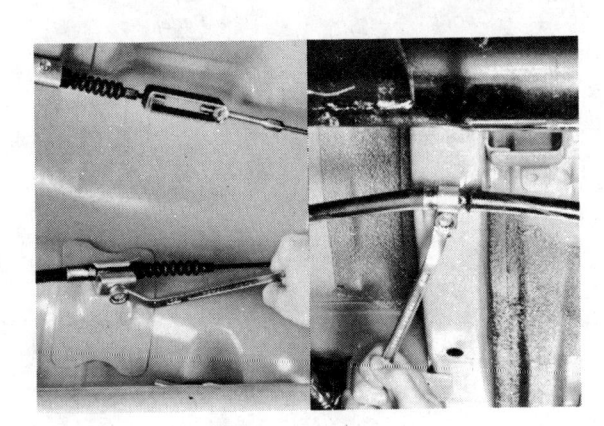

Fig. 9.12 Brake cable fixing clips

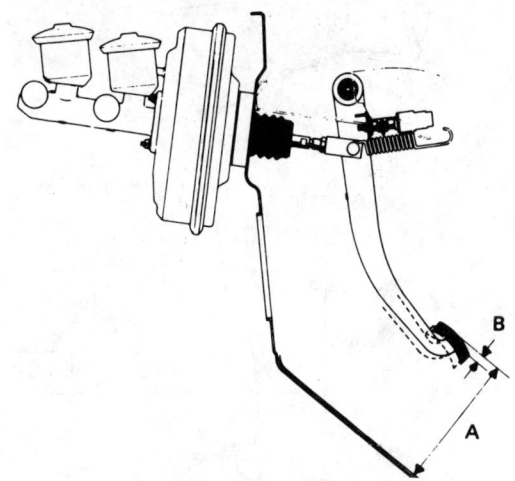

Fig. 9.13 Brake pedal adjustment

A – Pedal height
B – Pedal free play

cylinder reservoir during the following operations. Keep the reservoirs well topped-up during the whole of the bleeding operations otherwise air will be drawn into the system and the whole sequence of bleeding will have to start over again.

2 Depress the foot brake pedal several times and exhaust the vacuum effect in the servo unit.

3 Fit a rubber or plastic bleed tube to the nipple on the right rear wheel cylinder and immerse the free end of the tube in a jar containing sufficient hydraulic fluid to keep the end of the tube submerged.

4 Unscrew the bleed nipple one half turn and watch the air being expelled from the submerged end of the tube as an assistant depresses the brake pedal slowly to the full extent of its travel.

5 After each stroke of the brake pedal, allow it to fly back to its stop with the foot completely removed.

6 Continue to depress the brake pedal until air bubbles cease to emerge from the end of the tube. During these operations, it is essential to keep the end of the bleed tube covered with fluid all the time and keep the reservoir well topped-up with clean fluid.

7 Tighten the bleed nipple when the foot pedal is in the fully depressed position. Do not overtighten the nipple and use only a spanner of short length. Remove the bleed tube and refit the nipple dust cap.

8 Repeat the bleed operation on the left rear wheel cylinder and then the right-hand caliper and finally the left-hand front caliper.

9 When bleeding is complete, top-up the reservoir with clean fluid. Always discard the expelled fluid in the bleed jar unless it is desired to retain it for bleed jar tube covering purposes only.

10 Hydraulic lines – inspection and renewal

1 Periodically carefully examine all brake pipes, both rigid and flexible, for rusting, chafing and deterioration. Check the security of unions and connections.

2 First examine for signs of leakage where the pipe unions occur. Then examine the flexible hoses for signs of chafing and fraying and of course, leakage. This is only a preliminary part of the flexible hose inspection, as exterior condition does not necessarily indicate their interior condition which will be considered later in this Section.

3 The steel pipes must be examined equally carefully. They must be cleaned off and examined for any signs of dents, or damage, rust and corrosion. Rust and corrosion should be scraped off and if the depth of pitting in the pipes is significant, they will need replacement. This is particularly likely in those areas underneath the car body and along the rear axle where the pipes are exposed to the full force of road and weather conditions.

4 If any section of pipe is to be taken off, first of all remove the fluid reservoir cap and line it with a piece of polythene film to make it air tight and replace it. This will minimise the amount of fluid dripping out of the system when pipes are removed, by sealing the vent in the reservoir cap.

5 Rigid pipe removal is usually quite straightforward. The unions at each end are undone and the pipe and union pulled out and the centre sections of the pipe removed from the body clips where necessary. Underneath the car, exposed unions can sometimes be very tight. As one can use only an open ended spanner and the unions are not large, burring of the flats is not uncommon when attempting to undo them. For this reason a self-locking grip wrench (mole) is often the only way to remove a stubborn union.

6 Flexible hoses are always mounted at both ends in a rigid bracket attached to the body or a sub-assembly. To remove them it is necessary first of all to unscrew the pipe unions of the rigid pipes which go into them. Then, with a spanner on the hexagonal end of the flexible pipe union, the locknut and washer on the other side of the mounting bracket need to be removed. Here again exposure to the elements often tends to seize the locknut and in this case the use of penetrating oil is necessary. The mounting brackets, particularly on the body frame, are not very heavy gauge and care must be taken not to wrench them off. A self-grip wrench is often of use here as well. Use it on pipe unions if unable to get a ring spanner on the locknut.

7 With the flexible hose removed, examine the internal bore. If it is blown through first, it should be possible to see through it. Any specks of rubber which come out, or signs of restriction in the bore, mean that the inner lining is breaking up and the pipe must be renewed.

8 Rigid pipes which need renewal can usually be purchased at any local garage where they have the pipe, unions, and special tools to make them. All they need to know is the total length of the pipe and the type of flare used at each end with the union. This is very important as one can have a flare and a mushroom on the same pipe.

9 Replacement of pipes is a straightforward reversal of the removal procedure. If the rigid pipes have been made up it is best to get all the sets (or bends) in them before trying to install them. Also if there are any acute bends, ask your supplier to put these in for you on a tube bender. Otherwise you may kink the pipe and thereby restrict the bore area and fluid flow.

10 With the pipes replaced, remove the polythene film from the reservoir cap (paragraph 4) and bleed the system, as described in Section 9.

11 Handbrake – adjustment

1 Adjustment of the handbrake is normally quite automatic by the action of the automatic adjustment mechanism of the rear brake shoes.

2 Due to cable stretch or renewal or cables or other components, the following adjustment can be carried out when the travel of the handbrake brake lever becomes excessive. The handbrake should be fully applied when its pawl has passed over between 3 and 7 notches of the quadrant. This can be measured by counting the number of clicks.

3 To adjust, release the locknut and turn the cable adjuster until the

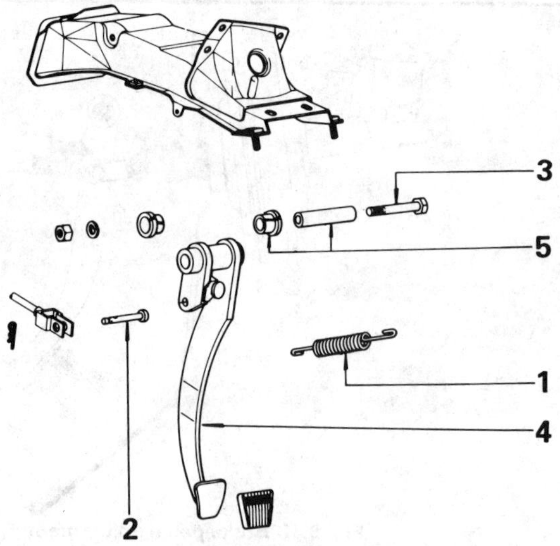

Fig. 9.14 Brake pedal assembly (exploded view)

1 Spring
2 Clevis pin
3 Pivot bolt
4 Pedal
5 Bushes and spacer

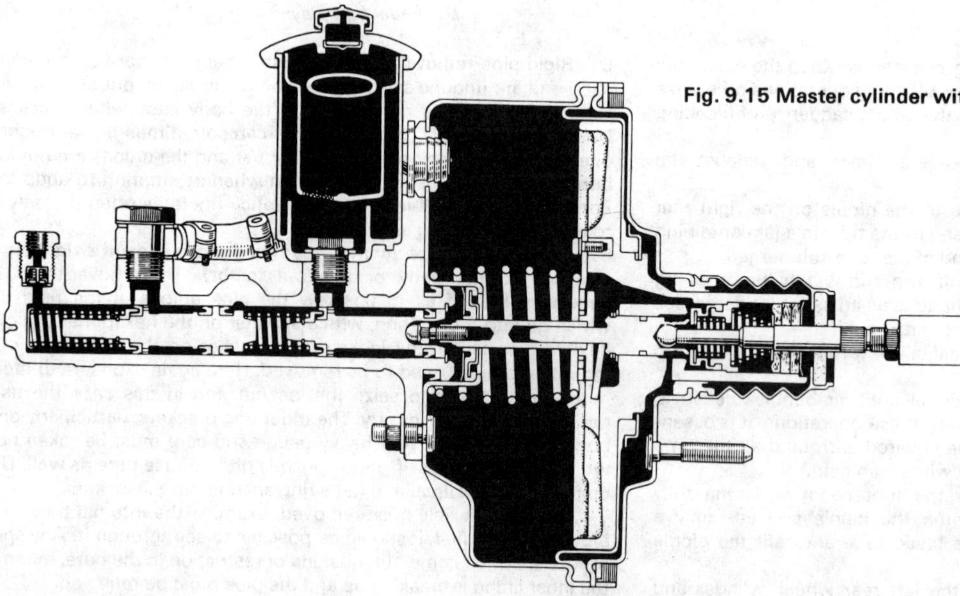

Fig. 9.15 Master cylinder with ASCO booster (sectional view)

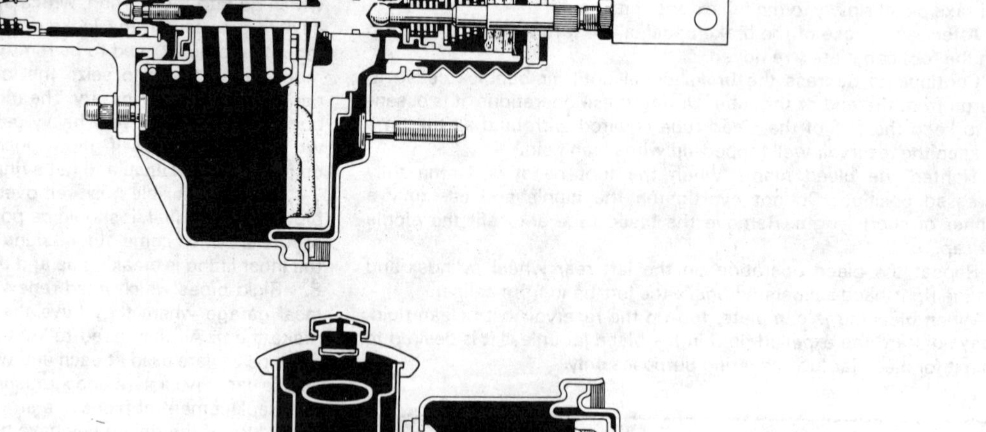

Fig. 9.16 Master cylinder with JKK booster (sectional view)

adjustment is correct (photo).
4   Check that the rear wheels rotate without binding when the handbrake is fully released.
5   Fully tighten the cable locknut.
6   A handbrake warning switch is fitted adjacent to the handbrake lever. Adjust the switch if necessary so that the warning lamp is out when the handbrake is released (ignition switched *on*).

12 Handbrake cable – renewal

1   Remove the propeller shaft, as described in Chapter 7.
2   Disconnect the brake cylinder from the handbrake lever pull rod (photo).
3   Remove the handbrake cable clamps (Fig. 9.12).
4   Remove the rear brake drums and disconnect the brake cable from the brake shoes, as described in Section 3 of this Chapter.
5   Before fitting a new cable by reversing the operations necessary for removal, apply grease to the part of the cable which is in contact with the equalizer.

13 Brake pedal – adjustment, removal and installation

1   The upper surface of the brake pedal should be 6.48 to 6.87 in (164.5 to 174.5 mm) above the floor panel with the carpet removed. If adjustment is required, disconnect the pushrod from the pedal arm and adjust the position of the stop lamp switch to give the correct dimension when the switch is pushed in fully.
2   Pedal free movement is checked by pushing the pedal with the fingers until resistance is felt and should be 0.1 to 0.2 in (3 to 6 mm). Adjust pedal play, by unscrewing the locknut of the pushrod and screwing the pushrod in, or out as required. Tighten the locknut when the adjustment is correct.
3   To remove the brake pedal, disconnect the pushrod from the pedal arm and detach the return spring.

4   Unscrew the nut from the pivot bolt and withdraw the bolt.
5   Examine the bushes and collar for wear and renew as appropriate.
6   Installation is a reversal of removal but apply grease to the collar bushes and pivot bolt.

14 Pressure regulating valve

1   This valve which is incorporated in the rear hydraulic circuit, regulates the pressure applied to the rear wheels to obviate any tendency for these wheels to lock under heavy braking application (photo).
2   The valve is mounted within the engine compartment and it cannot be repaired or adjusted. In the event of the valve leaking or if there is a tendency for the rear wheels to lock, then the valve must be renewed complete.
3   To do this, disconnect the fluid pipe unions from the valve body and unscrew and remove the valve securing bolt.
4   Installation of the new valve is a reversal of removal but on completion, bleed the hydraulic system.

15 Booster (vacuum servo unit) – description

The vacuum servo unit is designed to supplement the effort applied by the driver's foot to the brake pedal.
The unit is an independent mechanism so that in the event of its failure the normal braking effort of the master cylinder is retained. A vacuum is created in the servo unit by its connection to the engine inlet manifold and with this condition applying on one side of a diaphragm, atmospheric pressure applied on the other side of the diaphragm is harnessed to assist the foot pressure on the master cylinder. With the brake pedal released, the diaphragm is fully retracted and held against the rear shell by the return spring. The operating rod assembly is also fully retracted and a condition of vacuum exists each side of the diaphragm.

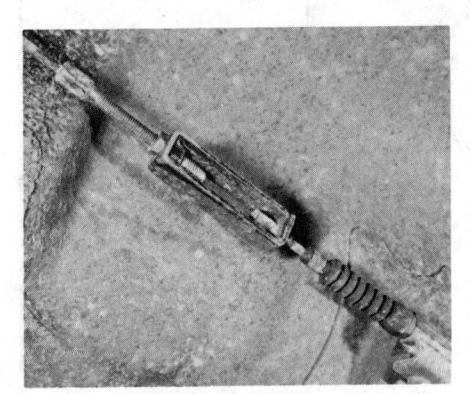

11.3 Handbrake cable adjuster

11.6 Handbrake ON warning switch

12.2 Handbrake cable equalizer

14.1 Pressure regulating valve

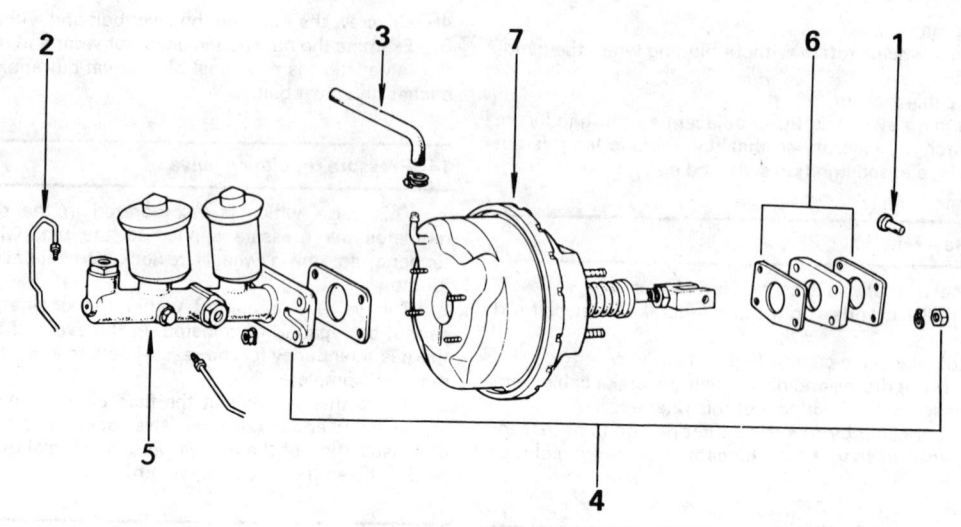

Fig. 9.17 Brake booster removal sequence

1 Clevis pin	3 Vacuum hose	5 Master cylinder	7 Brake booster
2 Brake pipes	4 Brake booster with master cylinder	6 Bracket and gasket	

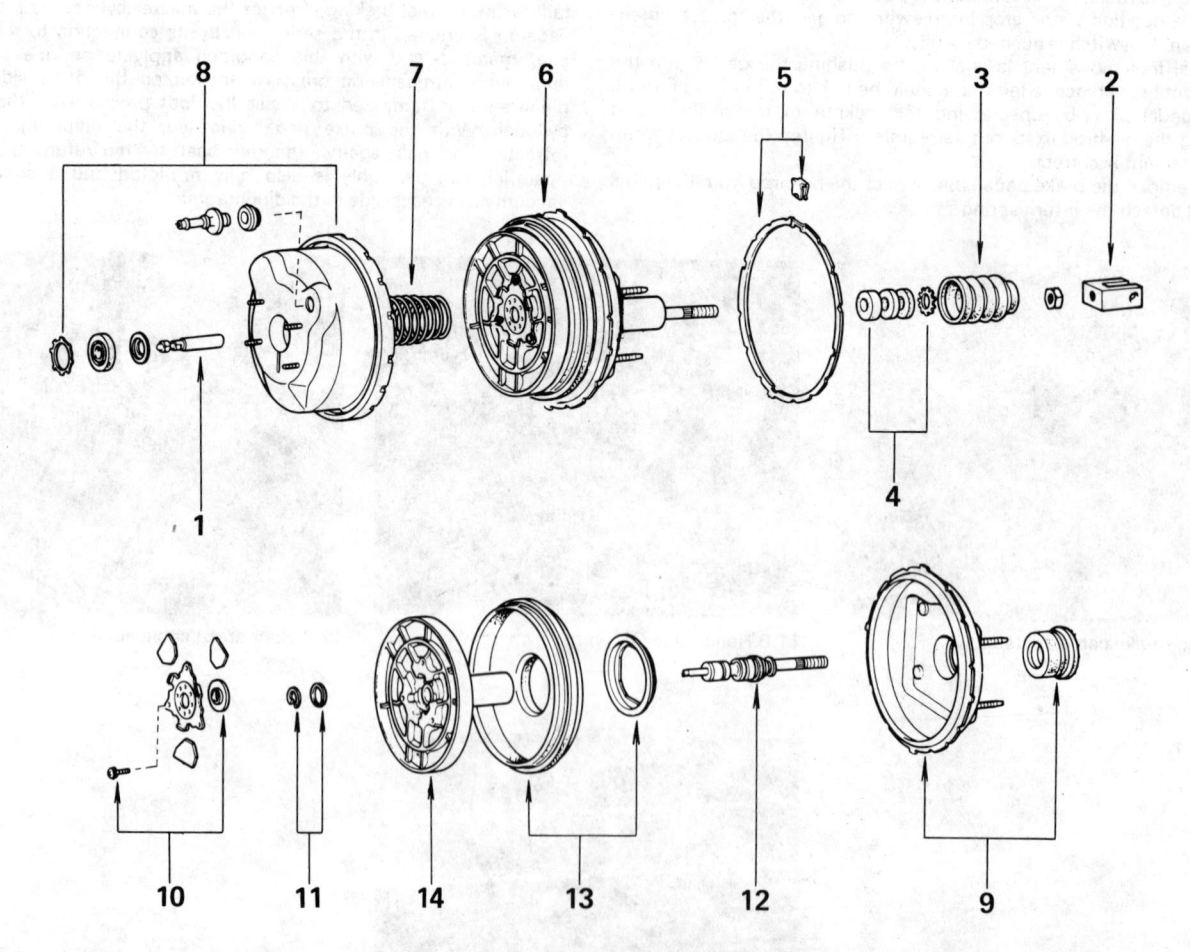

Fig. 9.18 Booster dismantling sequence (ASCO)

1 Push rod	5 Lock clip & body connector	9 Rear shell	13 Diaphragm & retainer
2 Clevis	6 Rear body with diaphragm plate	10 Reaction retainer, lever & plate	14 Diaphragm plate
3 Boot	7 Spring	11 Snap ring & plate	
4 Air filter elements	8 Front shell	12 Operating rod	

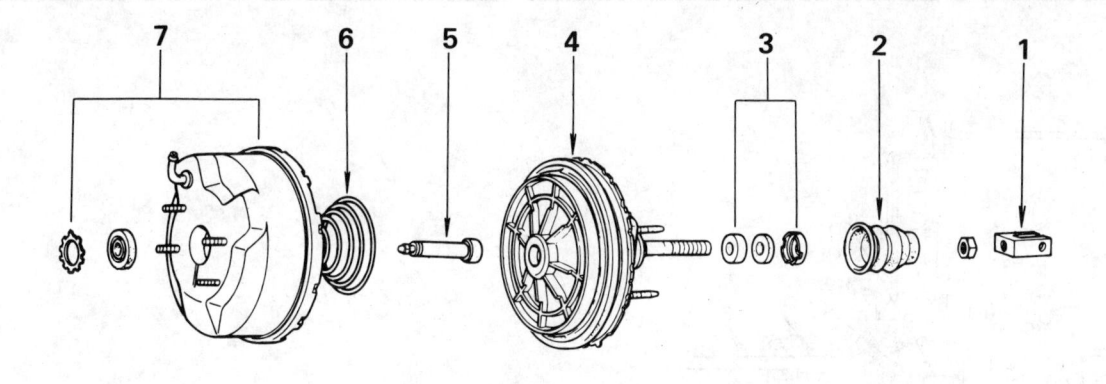

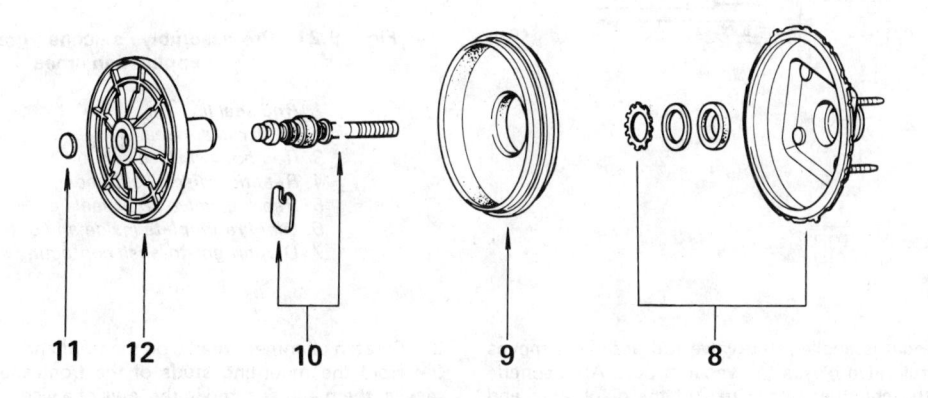

Fig. 9.19 Booster dismantling sequence (JKK)

1 Clevis	4 Rear shell with diaphragm plate	7 Front shell	10 Stopper key & operating rod
2 Boot	5 Push rod	8 Rear shell	11 Reaction disc
3 Air filter elements	6 Spring	9 Diaphragm	12 Diaphragm plate

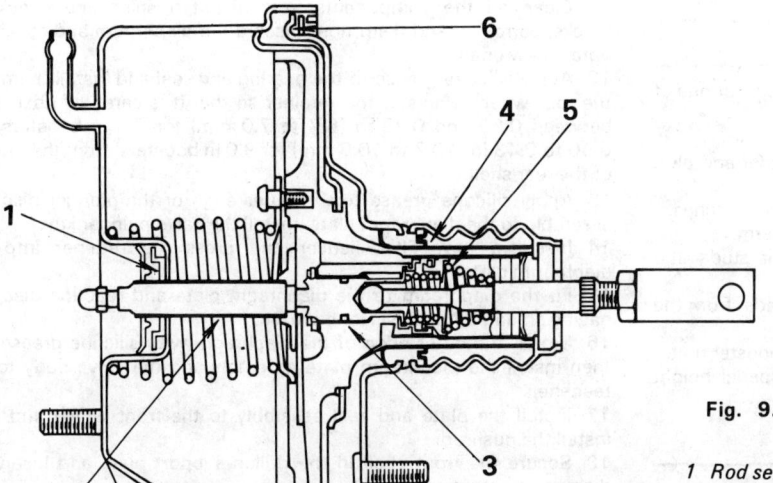

Fig. 9.20 Pre-assembly silicone greasing (ASCO) – application areas

1 Rod seal lip	4 Bearing, valve body seal lip
2 Push rod surface	5 Diaphragm plate inside & outside
3 Operating rod	6 Diaphragm-to-shell contacting surfaces

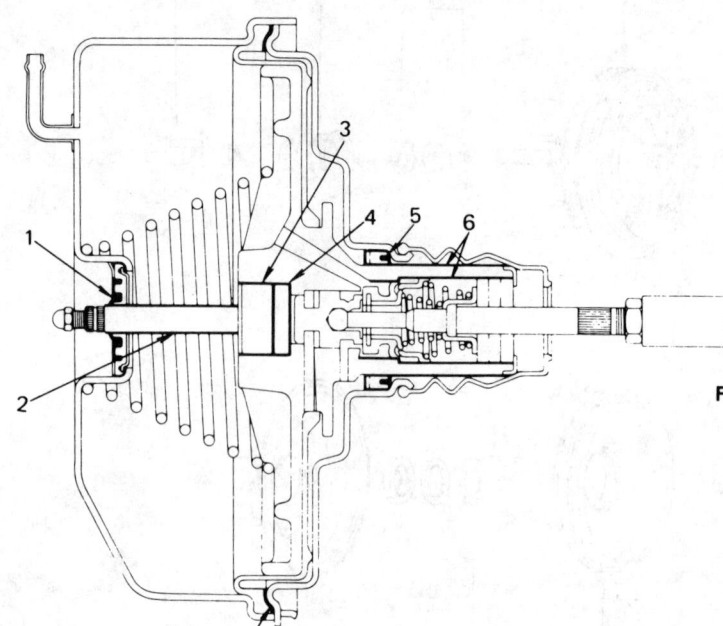

Fig. 9.21 Pre-assembly silicone greasing (JKK) – application areas

1 Rod seal lip
2 Push rod surface
3 Rod body surface
4 Reaction disc all surfaces
5 Bearing, valve body seal lip
6 Diaphragm plate inside and outside
7 Diaphragm-to-shell contacting surfaces

When the brake pedal is applied, the valve rod assembly moves forward until the control valve closes the vacuum port. Atmospheric pressure then enters the chamber to the rear of the diaphragm and forces the diaphragm plate forward to actuate the master cylinder pistons through the medium of the vacuum servo unit pushrod.

When pressure on the brake pedal is released, the vacuum port is opened and the atmospheric pressure in the rear chamber is extracted through the non-return valve. The atmospheric pressure inlet port remains closed as the operating rod assembly returns to its original position by action of the coil return spring. The diaphragm then remains in its position with vacuum conditions on both sides until the next depression of the brake pedal when the cycle is repeated.

The brake servo used may be either of ASCO or JKK manufacture and although there are slight differences in physical construction (Figs. 9.15 and 9.16) their principles of operation are the same. Either a 7.5 in or 9.0 in diameter booster may be fitted.

16 Booster unit – removal and installation

1 Exhaust the vacuum in the booster unit by repeated applications of the foot pedal.
2 Disconnect the vacuum hose from the booster.
3 Disconnect the hydraulic pipes from the master cylinder and plug or cap the ends of the pipes to prevent entry of dirt.
4 Disconnect the leads from the master cylinder pressure switches.
5 Inside the car, disconnect the pushrod from the pedal arm.
6 Unscrew the four booster securing nuts from their studs and remove the nuts.
7 Withdraw the booster unit complete with master cylinder from the engine compartment rear firewall.
8 Unbolt the master cylinder from the front face of the booster unit.
9 Installation is a reversal of removal but check the pedal height (Section 13) and bleed the system (Section 9).

17 Booster unit – dismantling and reassembly

1 If as a result of checking the fault diagnosis Section, the booster unit is found to be faulty, it is recommended that a factory exchange unit is obtained rather than servicing the unit yourself. However, where the necessary tools and skill are available, the servicing operations may be carried out as described in the following paragraphs.

2 Scratch alignment marks on the front and rear shells.
3 Hold the mounting studs of the front shell in a plate drilled to receive them and secured in the jaws of a vice.
4 Make up a suitable tool or lever which is drilled to fit over the four studs of the rear wheel and then turn the rear wheel anti-clockwise to separate it from the front shell.
5 Remove the diaphragm return spring.
6 Withdraw the diaphragm plate from the rear shell and remove the retainer, bearing and seal. The diaphragm plate is made of brittle plastic and should be handled carefully.
7 Remove the diaphragm from the diaphragm plate.
8 Remove the filter/silencer retainer, followed by the key. This is carried out by pointing the keyhole downwards and pressing the valve operating rod.
9 Withdraw the valve plunger assembly.
10 Remove the retainer, plate and seal and push the rod from the front shell.
11 Clean all the components in methylated spirit and check for cracks, corrosion and distortion. Discard the diaphragm and seals and obtain new ones.
12 Apply silicone grease to the bearing and seal and install them into the rear wheel. Press in the retainer so that it is carefully positioned between 0.26 and 0.28 in (6.7 to 7.0 mm) for 7.5 in boosters and 0.40 to 0.43 in (10.2 to 10.8 mm) for 9.0 in boosters, from the interior of the rear shell.
13 Apply silicone grease to the outer edge of the plunger disc and assemble to the diaphragm plate. Install the valve plunger key.
14 Install a new filter/silencer and press the retainer into the diaphragm plate.
15 Fit the diaphragm to the diaphragm plate and reaction disc first having smeared it with silicone grease.
16 Smear the outer edge of the diaphragm with silicone grease and then install the diaphragm plate assembly and the valve body to the rear shell.
17 Install the plate and seal assembly to the front wheel and then install the pushrod.
18 Secure the front shell in the drilled support plate and insert the diaphragm return spring.
19 Align the shell mating marks and engage the two shells using a clockwise twisting motion.
20 Using the tool made up for dismantling tighten the rear shell until it comes up against the stop on the front shell.
21 Before assembling the master cylinder to the booster, check the pushrod operating clearance as described in Section 8.

18 Fault diagnosis – braking system

Symptom	Reason/s
Brake grab	Pads or linings not bedded in Out of round drums Distorted discs Scored drums or discs Pads or linings contaminated with oil Booster unit faulty
Brake drag	Master cylinder faulty Pedal return impeded Blocked reservoir cap vent Reservoir overfilled Seized caliper or wheel cylinder Handbrake over adjusted Weak or broken shoe return spring Crushed or blocked pipe lines
Brake pedal feels hard	Friction surfaces contaminated with oil or grease Glazed surfaces to friction linings or pads Rusty discs Seized caliper or wheel cylinder Faulty booster unit
Excessive pedal travel	Low fluid level in reservoir Rear shoe automatic adjusters faulty Excessive disc run-out Worn front wheel bearings Hydraulic system requires bleeding Worn pads or linings
Pedal creep during sustained application	Fluid leak Faulty master cylinder Faulty booster unit
Pedal "spongy" or "springy"	System requires bleeding Perished flexible hose Loose master cylinder mounting Cracked brake drum Linings not bedded in Faulty master cylinder
Fall in master cylinder reservoir fluid level	Normal disc pad wear Leak in hydraulic system Internal leak into booster unit

Booster unit fault diagnosis

Hard pedal, lack of assistance with engine running	Lack of vacuum due to: Loose connections Restricted hose Blocked air filter/silencer Major fault in unit
Slow action of booster	Faulty vacuum hose Blocked air filter/silencer
Lack of assistance during heavy braking	Air leaks in: Check valve in vacuum line Vacuum hose or connections

Chapter 10 Electrical system

For modifications and information applicable to later models, refer to Supplements at end of manual

Contents

Alternator – dismantling, servicing and reassembly	8
Alternator – general description, maintenance and precautions	5
Alternator – removal and refitting	7
Alternator – testing in the car	6
Alternator regulator – testing and adjustment	9
Ammeter – testing	23
Battery charging	4
Battery maintenance	3
Battery – removal and installation	2
Brake warning light	40
Bulb replacement	31
Clock - adjustment	41
ESP system	44
Fault diagnosis – electrical system	45
Fuel level transmitter and gauge – testing	20
Fuses and fusible link	18
General description	1
Hazard warning and turn signal lamps	19
Headlamps – removal and refitting	29
Headlamps – beam alignment	30
Headlamp washer	39
Heated rear window – precautions and testing	37
Heated rear window – switch, relay and choke	38
Ignition switch – removal and refitting	27
Instrument cluster – removal and replacement	25
Neutral safety switch (automatic transmission) – removal and installation	28
Oil pressure switch and gauge – testing	22
Radio – removal and refitting	43
Rear window wiper – dismantling and reassembly	36
Rear window wiper – removal and installation	35
Relays	42
Starter motor (reduction gear type) – dismantling	15
Starter motor (standard type) – dismantling	12
Starter motor (reduction gear type) reassembly	17
Starter motor (standard type) – reassembly	14
Starter motor – removal and refitting	11
Starter motor (reduction gear type) – servicing and testing	16
Starter motor (standard type) – servicing and testing	13
Starter motor – testing in the car	10
Steering column and fascia-mounted switches – removal and installation	26
Tachometer – testing	24
Water temperature transmitter and gauge – testing	21
Windscreen washer and rear washer	34
Windscreen wiper – removal and refitting	32
Windscreen wiper motor – dismantling and reassembly	33

Specifications

System type .... 12 volt, negative earth

Battery .... 60 Amp-hour

Alternator
Output .... 45 amp
Brush exposed length (standard) .... 0.492 in (12.5 mm)
Brush exposed length (minimum) .... 0.217 in (5.5 mm)
Regulating voltage .... 13.8 to 14.8 volts
Relay operating voltage .... 4.0 to 5.8 volts

Starter motor

2T engine
Commutator diameter:
    Standard .... 1.18 in (30 mm)
    Limit .... 1.14 in (29 mm)
Brush length:
    Standard .... 0.55 in (14 mm)
    Limit .... 0.39 in (10 mm)

Chapter 10 Electrical system

18R and 20R engine

	Conventional type	Reduction type
Commutator diameter:		
Standard	1.287 in (32.7 mm)	1.18 in (30 mm)
Limit	1.22 in (31 mm)	1.14 in (29 mm)
Brush length:		
Standard	0.75 in (19 mm)	0.571 in (14.5 mm)
Limit	0.47 in (12 mm)	0.39 in (10 mm)
Pinion end-to-stop collar clearance	0.008 to 0.157 in (0.2 to 4.0 mm)	

Windscreen wiper motor
Commutator diameter:
- Standard .... 0.91 in (23.2 mm)
- Limit .... 0.86 in (21.8 mm)

Brush length:
- Standard .... 0.49 in (12.5 mm)
- LImit .... 0.24 in (6.0 mm)

Bulbs | Wattage

Bulb	Wattage
Headlight outer (sealed beam)	37.5/50W
Headlight inner (sealed beam)	37.5W
Headlight outer (semi-sealed beam)	45/40W
Headlight inner (semi-sealed beam)	45/40W
Front parking lights (Europe)	5W
Front parking lights (North America)	8W
Front turn signal (Europe)	21W
Front turn signal (North America)	27W
Front side turn signal (Europe)	5W
Front side turn signal (North America)	8W
Rear side turn signal (North America)	8W
Rear turn signal (Europe)	21W
Rear turn signal (North America)	27W
Stop and tail (Europe)	21/5W
Stop and tail (North America)	27/8W
Number plate light (Europe)	10W
Number plate light (North America)	7.5W
Reversing lights (Europe)	21W
Reversing lights (North America)	27W
Interior light	10W
Courtesy light	5W
Luggage compartment light	5W
Inspection light (GT)	7.5W
Heater panel light	2W
Transmission indicator light	3.4W

1 General description

The electrical system is of the negative earth type and is powered by an engine driven alternator (photo). A lead acid battery is charged by the alternator and provides the power during periods when alternator output is insufficient as well as providing current for engine starting.

The starter motor is of the pre-engaged type (photo) and on some cold climate models has a reduction gear between the motor armature and the drive pinion.

A fusible link in the battery lead (photo) protects the battery against overloading and each individual electrical circuit is protected by a fuse.

2 Battery – removal and installation

1 The battery is located at the front, on the left-hand side of the engine compartment.
2 Release the clamping bolts on the battery terminals and remove

1.1 The alternator

1.2 Starter motor

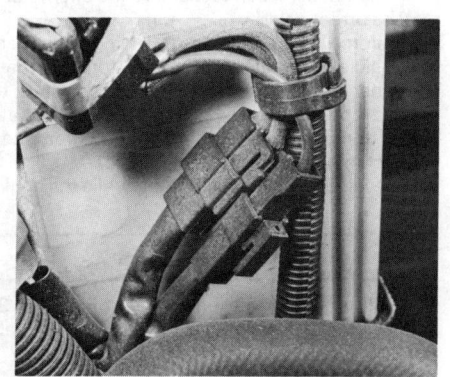

1.3 Battery fusible link

Chapter 10 Electrical system

the terminals from the battery clamp. Disconnect the negative lead first.
3   Remove the nuts and washers from the battery clamp and lift the clamp plate off.
4   Carefully lift the battery from its tray, keeping the battery level to avoid spilling any electrolyte.
5   Installation is the reverse of removal, but before connecting the battery, clean off any corrosion which may be present, clean the terminals and lugs with a wire brush and smear them with petroleum jelly, or a proprietary corrosion inhibitor. Refit the negative lead last.
6   Tighten the clamp bolts so that the terminals are tight and cannot be turned by hand, but do not distort them by excessive tightening.

3 Battery maintenance

1   The essentials of battery maintenance are to keep the electrolyte at the correct level, the terminals free from corrosion and tight enough to ensure good contact, the case of the battery clean and dry and the battery charged.
2   Inspect the battery weekly and top-up the electrolyte if necessary using distilled water.
3   At the first sign of any corrosion, clean it off and treat the surface with ammonia, or a proprietary acid corrosion neutraliser.

4 Battery charging

1   Under normal operating conditions during most of the year, the battery should remain fully charged, but during the winter when a cold engine calls for a higher starting current and a greater part of the alternator output is needed to supply lights and other accessories, there may be insufficient current to maintain a fully charged battery.
2   It is more difficult to damage a battery by excessive charging than by leaving it partially discharged and if possible the battery should be given a weekly overnight charge during the winter months.
3   A charging current of about three amps is ideal for this purpose, but a lower current for a longer time may also be used. The use of rapid boost chargers which are claimed to restore the battery in a very short time are not recommended. High rates of charge tend to reduce battery life.

5 Alternator – general description, maintenance and precautions

1   The alternator generates three-phase alternating current, which is rectified into direct current by six bridge-connected silicon diodes within the end frame of the alternator. Voltage control is by means of an externally mounted voltage regulator which controls the excitation of the alternator field.
2   Check the drivebelt tension every 6000 miles (9600 km) and adjust, as described in Chapter 2, by loosening the mounting bolts. Pull the alternator body away from the engine block; do not use a lever as it will distort the alternator casing.
3   No lubrication is required as the bearings are grease sealed for life.
4   Take extreme care when making circuit connections to a vehicle fitted with an alternator and observe the following.
5   When making connections to the alternator from a battery, always match correct polarity.
6   Before using electric-arc welding equipment to repair any part of the vehicle, disconnect the connector from the alternator and disconnect the earth battery terminal. Never start the car with a battery charger connected. Always disconnect the battery earth lead before using a mains charger. If boosting from another battery, always connect in parallel using heavy cable.
7   Never disconnect the battery while the engine is running.

6 Alternator – testing in the car

1   If an ammeter having a range of at least 30 amps DC and a DC voltmeter with a range of 0 to 15 volts is available, the performance of the alternator may be checked as follows:
2   First ensure that the alternator drivebelt is tensioned correctly and check that the alternator does not make any abnormal noise while it is running.
3   If the battery is suspect, check the specific gravity of the

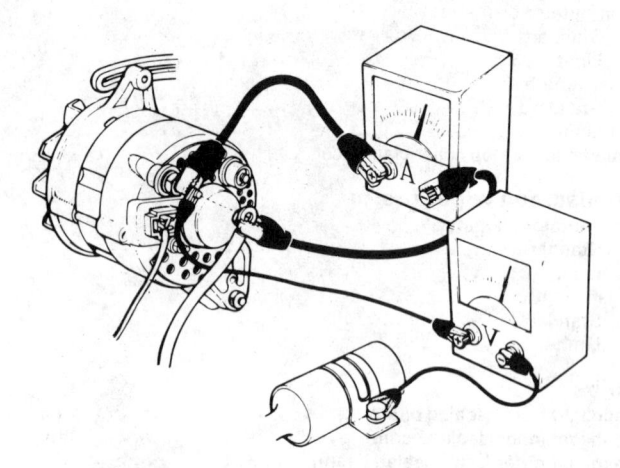

Fig. 10.1 Meter connections for alternator test

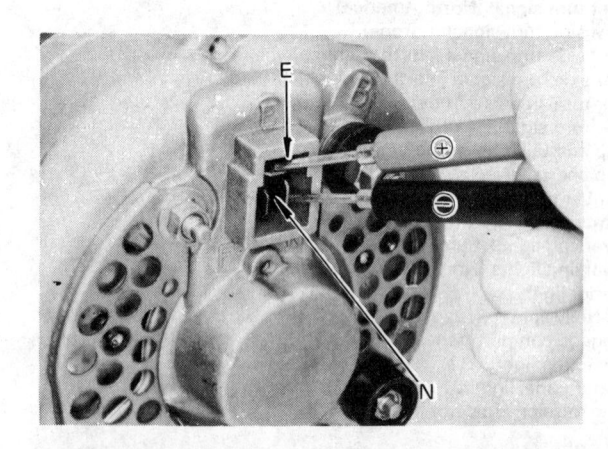

Fig. 10.2 Checking the negative side of the diodes

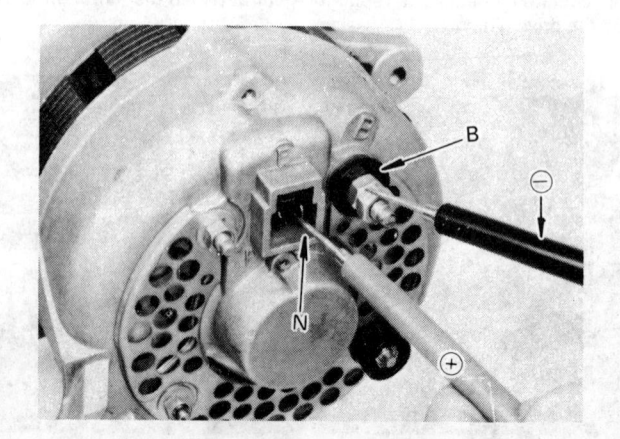

Fig. 10.3 Checking the positive side of the diodes

Chapter 10 Electrical system

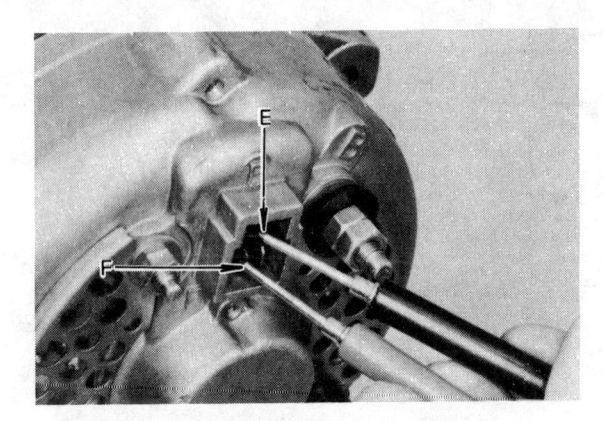

Fig. 10.4 Measuring alternator field resistance

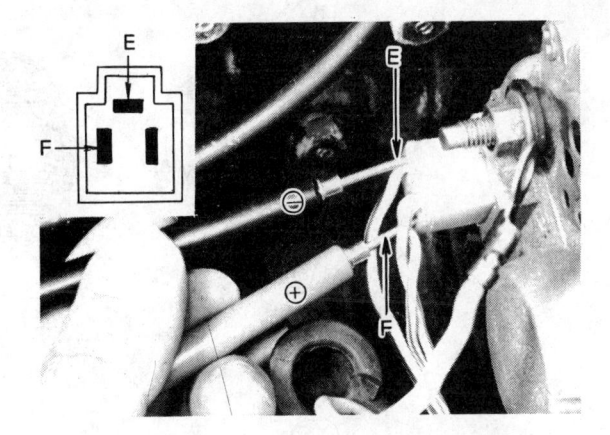

Fig. 10.5 Checking alternator field voltage

electrolyte in each cell, which should be in the range 1.25 to 1.27. Alternatively get the battery checked by a battery testing station.

4   Remove the cable from the large terminal of the alternator and connect the cable to the (−) terminal of the ammeter. Connect the large alternator terminal to the (+) terminal of the ammeter (Fig. 10.1).
5   Connect the (+) terminal of the voltmeter to the large terminal of the alternator and connect the (−) terminal of the voltmeter to a good connection on the vehicle's earth return.
6   With all accessories switched off, start the engine and run it at 2000 rpm. The voltage reading should be between 13.8 and 14.8 volts and the current less than 10 amps.
7   Turn on the headlamps and accessories. The voltage should steady at 12 volts and the current increase to more than 30 amps.
8   If the meter indications are not as specified, check the alternator diodes and the rotor coil resistance as follows:
9   Check the negative side of the diode bridge by connecting the (−) lead of an ohmeter to the N terminal and the (+) lead to the E terminal (Fig. 10.2). The diodes are satisfactory if the meter indicates infinity.
10   Check the positive side of the diode bridge by connecting the (−) lead of the ohmeter to the B terminal and the (+) lead to the N terminal (Fig. 10.3). The diodes are satisfactory if the meter indicates infinity.
11   Connect the ohmeter between the E and F terminals of the alternator (Fig. 10.4). The meter indication should be within the range 5 to 9 ohms.
12   With the alternator connecting plug inserted in the alternator and with the ignition switch turned ON, connect the (−) terminal of a voltmeter to the E terminal and the (+) terminal of the voltmeter to the F terminal of the alternator (Fig. 10.5). If the meter does not indicate battery voltage, check the 15 amp voltage regulator (engine) fuse.

7 Alternator – removal and refitting

1   Loosen the battery terminal clamp bolts and remove the terminal leads.
2   Loosen the alternator hinge bolt and the clamp bolt on the adjustment quadrant and push the alternator towards the cylinder block as far as it will go, so that the drivebelt can be removed from the pulley.
3   Remove the connecting plug from the alternator, remove the nut from the alternator output terminal and remove the cable from it.
4   Remove the bolt from the clamping bracket and the hinge bolt and spacer then lift the alternator clear.
5   Refit the alternator by reversing the removal operations, tensioning the drivebelt as described in Chapter 2.

8 Alternator – dismantling, servicing and reassembly

1   Remove the three tie bolts which secure the two end frames together.

2   Insert screwdrivers in the notches in the drive end frame and separate it from the stator.
3   Hold the rotor still by gripping it in a soft-jawed vice and remove the securing nut, pulley, fan and spacer.
4   Press the rotor shaft from the drive end frame.
5   Remove the bearing retainer, bearing, cover and felt ring from the drive end frame.
6   Remove the rectifier holder and brush holder securing screws and detach the stator from the rectifier end frame.
7   Remove the brush lead and stator coil N terminals from the brush holder by prising with a small screwdriver.
8   Test the rotor coil for an open circuit by connecting a circuit tester between the two slip rings located at the rear of the rotor. The indicated resistance should be from 4.1 to 4.3 ohms but if there is no resistance, the coil is open circuited and the rotor must be renewed as an assembly.
9   Connect the tester between each slip ring in turn and the rotor shaft. If the tester needle moves the rotor must be renewed because the insulation has broken down.
10   Inspect the rotor bearing for wear and renew if necessary by removing it from the shaft with a two-legged puller.
11   Clean the slip rings and rotor surfaces with a solvent moistened cloth.
12   Test the insulation of the stator coil by connecting the tester between the stator coil and the stator core. If the tester needle moves, the coil is grounded through a breakdown in the insulation and must be renewed.
13   To test the stator coil for open circuit, the coil leads must be disconnected from the rectifiers. Because the rectifiers are easily damaged by heat, hold the rectifier lead with a pair of long nosed pliers to prevent heat being conducted down the lead to the rectifier and apply the soldering iron to the joint for as little time as possible (Fig. 10.7).
14   Check the four stator coil leads for conductance. If the tester needle does not flicker, the coil has an open circuit and it must be renewed.
15   The testing of the rectifiers should be limited to measuring the resistance between their leads and holders in a similar manner to that described in paragraph 13. These tests will indicate short or open-circuited diodes, but not rectifying or reverse flow characteristics which can only be checked with specialised equipment.
16   If more than one of the preceding tests proves negative, it will be economically sound to exchange the alternator complete for a factory reconditioned unit, rather than renew more than one individual component.
17   Finally, examine the brushes for wear. If their exposed length is less than 0.217 in (5.5 mm), renew them. Remove the old brushes and insert the new ones in their holders, check to see they slide freely. Ensure that the brush does not project more than 0.5 in (12.5 mm) from its holder and then solder the brush lead, cutting off any surplus wire.

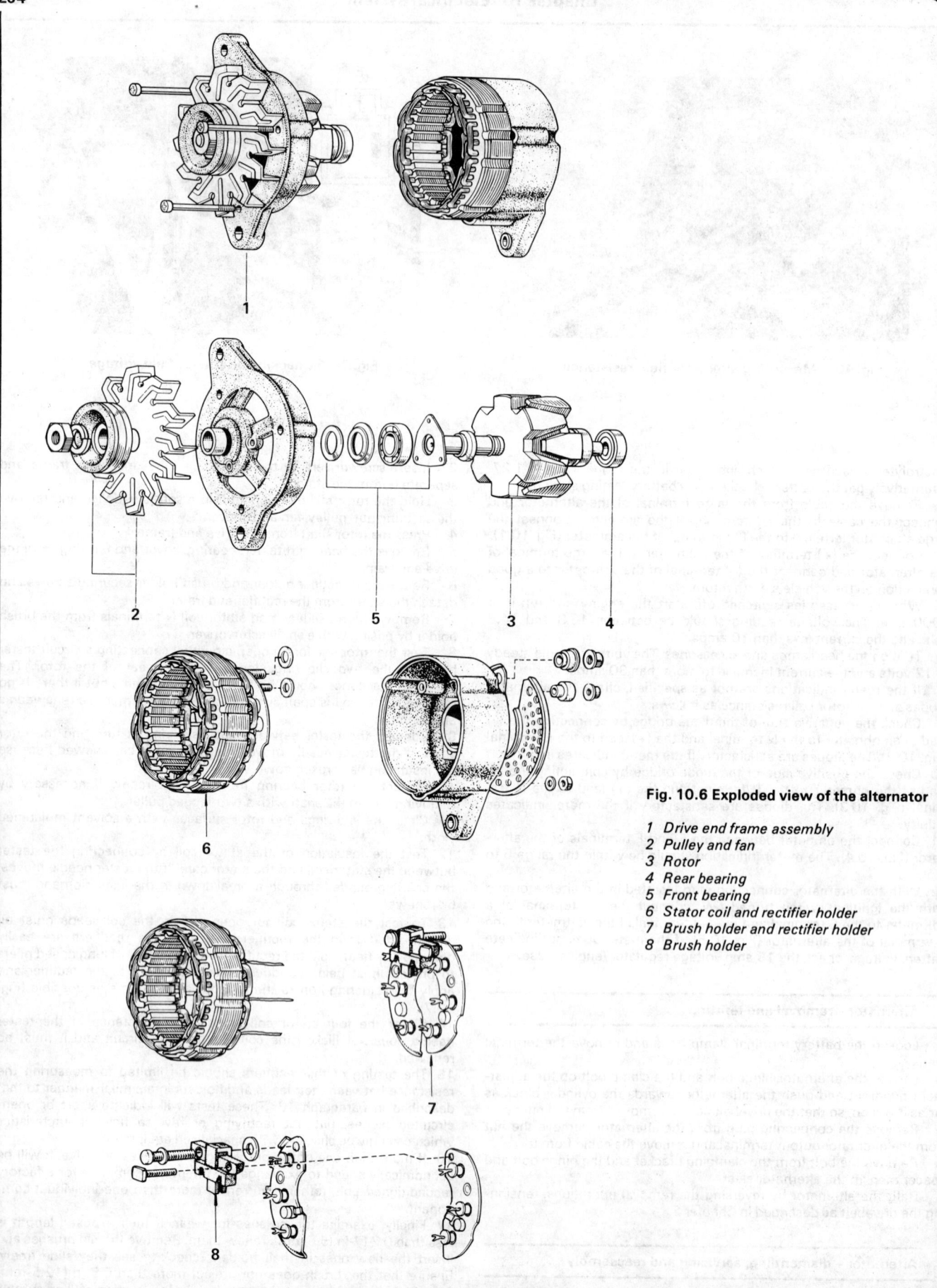

Fig. 10.6 Exploded view of the alternator

1. Drive end frame assembly
2. Pulley and fan
3. Rotor
4. Rear bearing
5. Front bearing
6. Stator coil and rectifier holder
7. Brush holder and rectifier holder
8. Brush holder

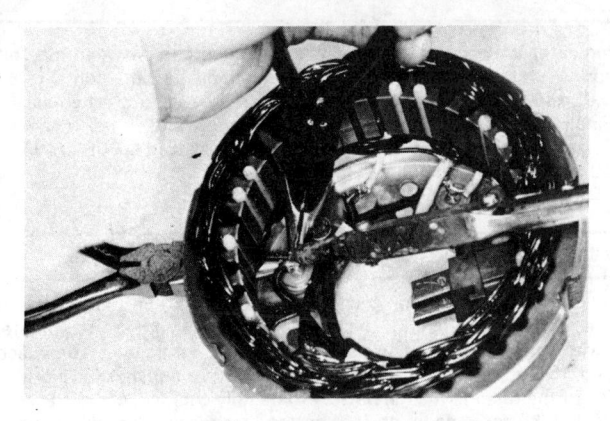

Fig. 10.7 Disconnecting the rectifiers

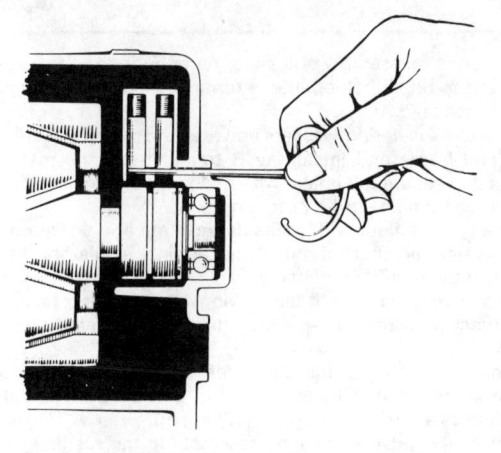

Fig. 10.8 Brushes supported by a piece of wire

"IG" terminal "N" terminal "F" terminal

"E" terminal "L" terminal "B" terminal

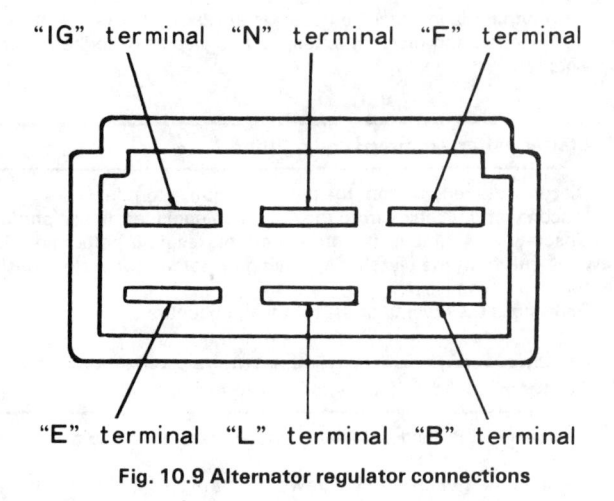

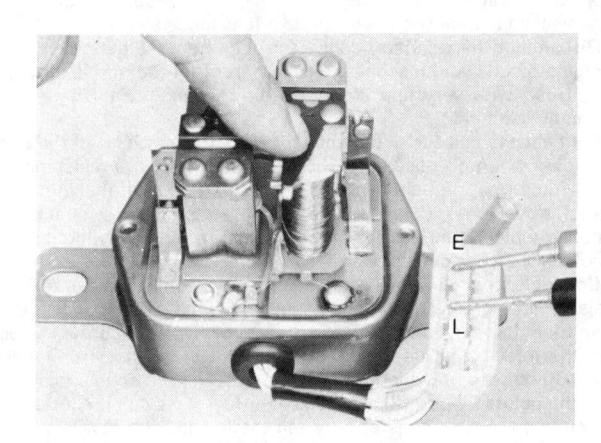

Fig. 10.9 Alternator regulator connections

Fig. 10.10 Voltage relay testing

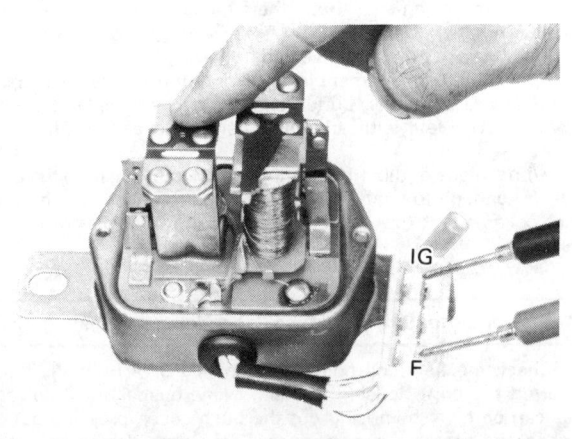

Fig. 10.11 Voltage regulator testing

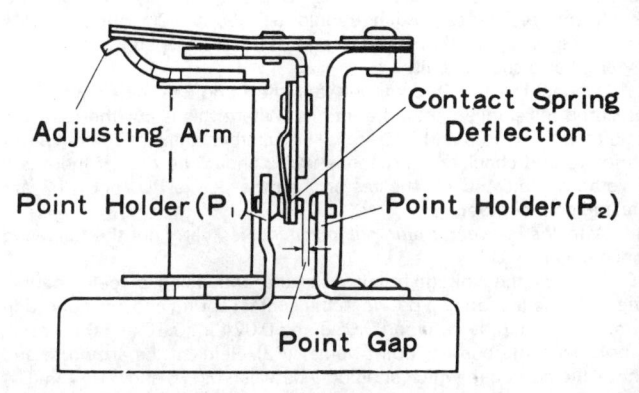

Adjusting Arm

Contact Spring Deflection

Point Holder (P_1)

Point Holder (P_2)

Point Gap

Fig. 10.12 Voltage relay contact identification

18 Commence reassembly of the alternator by fitting the stator coil N terminal to the brush holder, then a terminal insulator followed by the brush negative lead.
19 Fit the two insulating washers between the rectifier positive holder and the end frame and install the B terminal and the retaining bolt insulators and secure the holder with its four retaining nuts. Secure the negative rectifier holder with its four nuts.
20 Fit the brush holder with its insulating plate and tighten the securing screws passing them through the terminal insulators. Locate the stator coil in the rectifier end frame.
21 To the drive end frame fit the felt ring, cover (convex face to pulley) bearing (packed with multi-purpose grease) and bearing retainer (3 screws).
22 Fit the spacer ring to the rotor shaft and then press the drive end frame onto the shaft. Fit the collar, fan and pulley and tighten the securing nut to a torque of 35 lbf ft (4.9 kgf m).
23 Connect the drive end frame assembly to the rectifier end frame assembly and secure with the three tie bolts. Use a piece of wire to support the brushes in the raised position during this operation (Fig. 10.8).

9 Alternator regulator – testing and adjustment

1 The alternator regulator (photo) requires special equipment to check its voltage and current settings, but circuit testing and mechanical adjustments can be carried out as follows:
2 Disconnect the regulator connector plug. Remove the cover from the regulator unit and inspect the condition of the points, If they are pitted, clean with very fine emery cloth, otherwise clean them with methylated spirit.
3 Connect a circuit tester between the IG and F terminals of the connector plug when no resistance should be indicated. If a resistance is shown, then the regulator points PL1 and PL0 are making poor contact. Now press down the regulator armature and check the resistance which should be about 11 ohms. If it is much higher, the control resistance is defective and must be renewed.
4 Connect the circuit tester between the connector plug L and E terminals when no resistance should be indicated. If a resistance is shown then the relay points P1 and P0 are making poor contact. Press down the relay armature and check the resistance which should be about 100 ohms. If it is higher, the voltage coil has an open circuit or if lower, the points P1 and P0 are fused together or the coil is shorted.
5 Connect the circuit tester between the N and E terminals when a resistance of 25 ohms should be indicated. If the resistance is much higher, the pressure coil has an open circuit, if lower then it is short circuited.
6 Connect the circuit tester between the B and L terminals and depress the voltage relay armature. There should be no indicated resistance but if there is, this will show that the contact of the points P0 and P2 is poor.
7 Connect the circuit tester between the B and E terminals when the indicated resistance should be infinity. Where this is not so, the relay points P0 and P2 are fused together. Depress the relay armature and check the resistance which should be about 100 ohms. If the resistance is higher then the voltage coil has an open circuit and if lower it has a short circuit.
8 Connect the circuit tester between the F and E terminals when the indicated resistance should be zero. Where this is not the case the regulator points PL0 and PL2 are fused together. Depress the regulator armature and check the resistance which should be zero. If there is a resistance indicated on the tester then the points PL0 and PL2 are making poor contact.
9 *With the connector plug still disconnected* carry out the following mechanical checks.
10 Depress the voltage relay armature and using a feeler gauge, check the deflection gap between the contact spring and its supporting arm. This should be between 0.008 and 0.024 in (0.20 and 0.60 mm), if not, bend the contact point holder (P2). Release the armature and check the point gap which should be between 0.016 and 0.047 in (0.4 to 1.2 mm) if not, bend the contact point holder (P1).
11 Check the armature gap on the voltage regulator which should be in excess of 0.012 in (0.30 mm) otherwise bend the contact point holder PL2 to adjust it. Check the voltage regulator point gap which should be between 0.012 and 0.018 in (0.30 and 0.45 mm) otherwise bend the contact point holder PL2 to adjust. Depress the voltage regulator armature and check the deflection gap between the contact spring and its supporting arm. This should be between 0.008 and 0.024 in (0.2 and 0.6 mm). If not renew the regulator as an assembly. Finally depress the voltage regulator armature and check the angle gap at its narrowest point. This gap should not exceed 0.008 in (0.2 mm), otherwise renew the unit as an assembly.

10 Starter motor – testing in the car

1 If the starter motor fails to operate, check the state of charge of the battery by testing the specific gravity with a hydrometer or switching on the headlamps. If they glow brightly for several seconds and then gradually dim, then the battery is in an uncharged state.
2 If the test proves the battery to be fully charged, check the security of the battery leads at the battery terminals, scraping away any deposits which are preventing a good contact between the cable clamps and the terminal posts.
3 Check the battery negative lead at its body frame terminal, scraping the mating faces clean if necessary.
4 Check the security of the cables at the starter motor and solenoid switch terminals.
5 Check the wiring with a voltmeter for breaks or short circuits.
6 Check the wiring connections at the ignition/starter switch terminals.
7 If everything is in order, remove the starter motor as described in the next Section and dismantle, test and service as described later in this Chapter.

11 Starter motor – removal and refitting

1 Disconnect the lead from the battery negative terminal.
2 Disconnect the cables from the starter solenoid terminals (photo).
3 Unscrew and remove the starter motor securing bolts and withdraw the unit from the clutch bellhousing (or torque converter housing – automatic transmission).
4 Refitting is the reverse of the removal procedure.

12 Starter motor (standard type) – dismantling

1 Disconnect the field coil lead from the starter solenoid main terminal.
2 Remove the two securing screws from the solenoid and withdraw the solenoid far enough to enable it to be unhooked from the drive engagement lever fork.
3 Remove the end frame cover, the lockplate, washer, spring and seal.
4 Unscrew and remove the two tie bolts and withdraw the commutator end frame.
5 Pull out the brushes from their holders and remove the brush holder assembly.
6 Pull the yoke from the drive end frame.
7 Remove the engagement lever pivot bolt from the drive end frame and detach the rubber buffer and its backing plate. Remove the armature, complete with drive engagement lever from the drive end frame.
8 With a piece of tubing, drive the pinion stop collar up the armature shaft far enough to enable the circlip to be removed and then pull the stop collar from the shaft together with the pinion and clutch assembly.

13 Starter motor (standard type) – servicing and testing

1 Check the field coil pole shoes for signs of rubbing; if this has occurred the drive housing and end frame bushes must be renewed. This can be done by prising out the bush cover, pressing out the old bushes and pressing in new ones. These should then be reamed to provide the minimum specified armature shaft/bush clearance. Stake the bush in position.
2 Check the armature runout, if necessary, this can be skimmed on a lathe to bring it within the specified limits of .002 to .012 in (0.05 to 0.3 mm).

9.1 Alternator regulator

11.2 Starter motor and solenoid, with cables disconnected

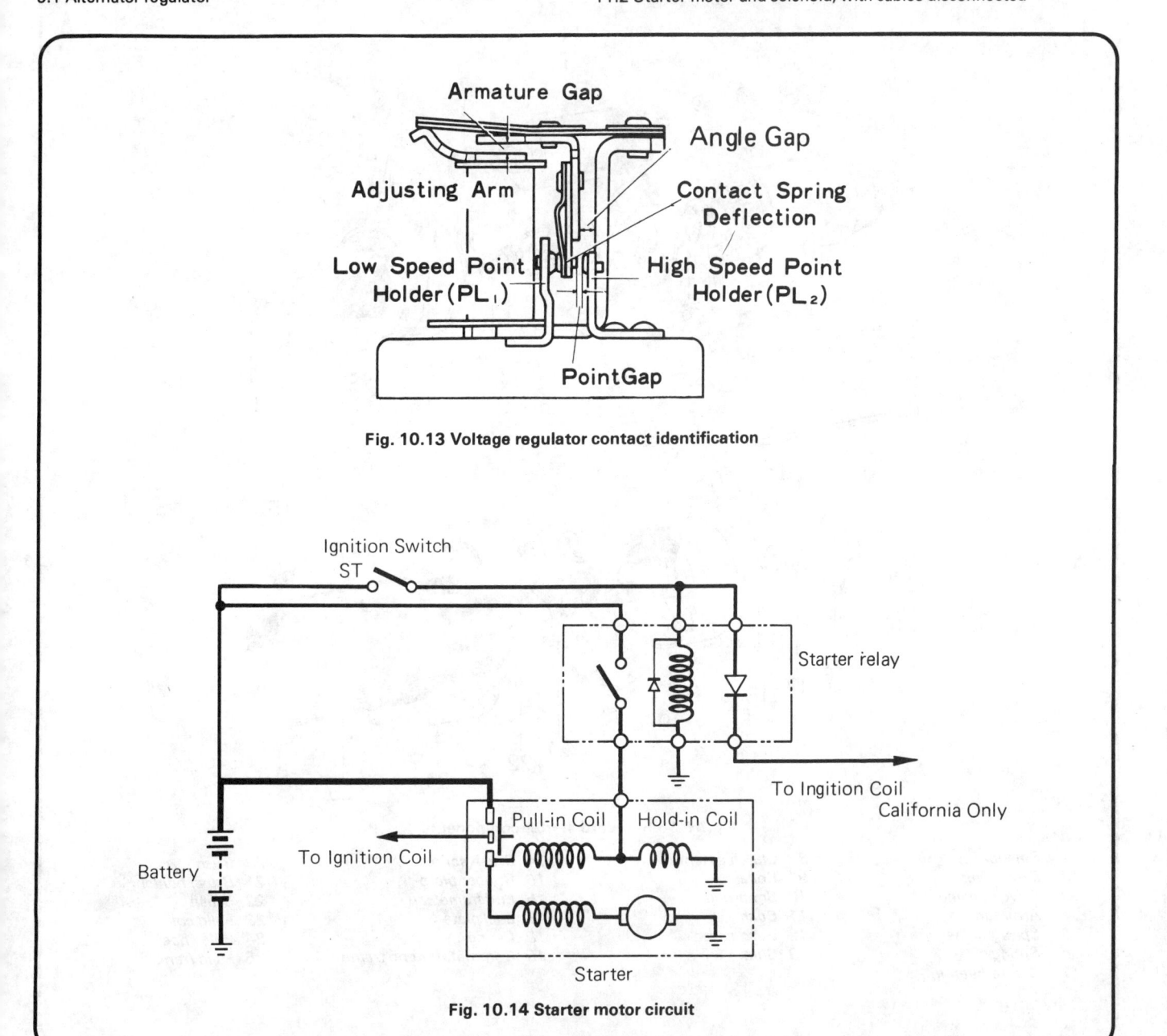

Fig. 10.13 Voltage regulator contact identification

Fig. 10.14 Starter motor circuit

Fig. 10.15 Standard starter (sectional view)

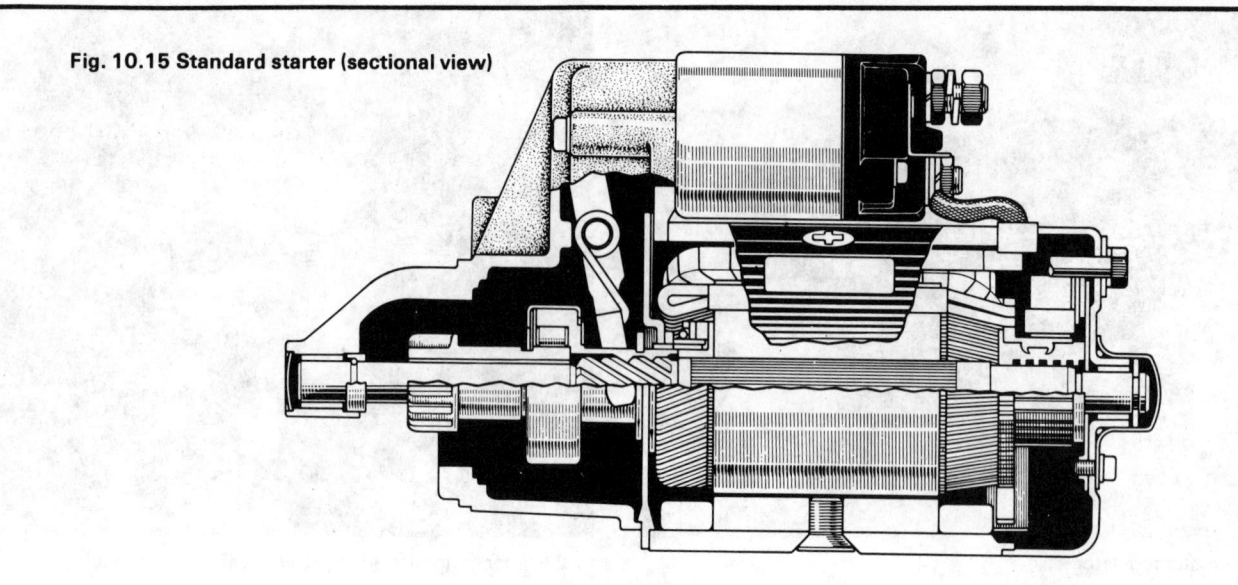

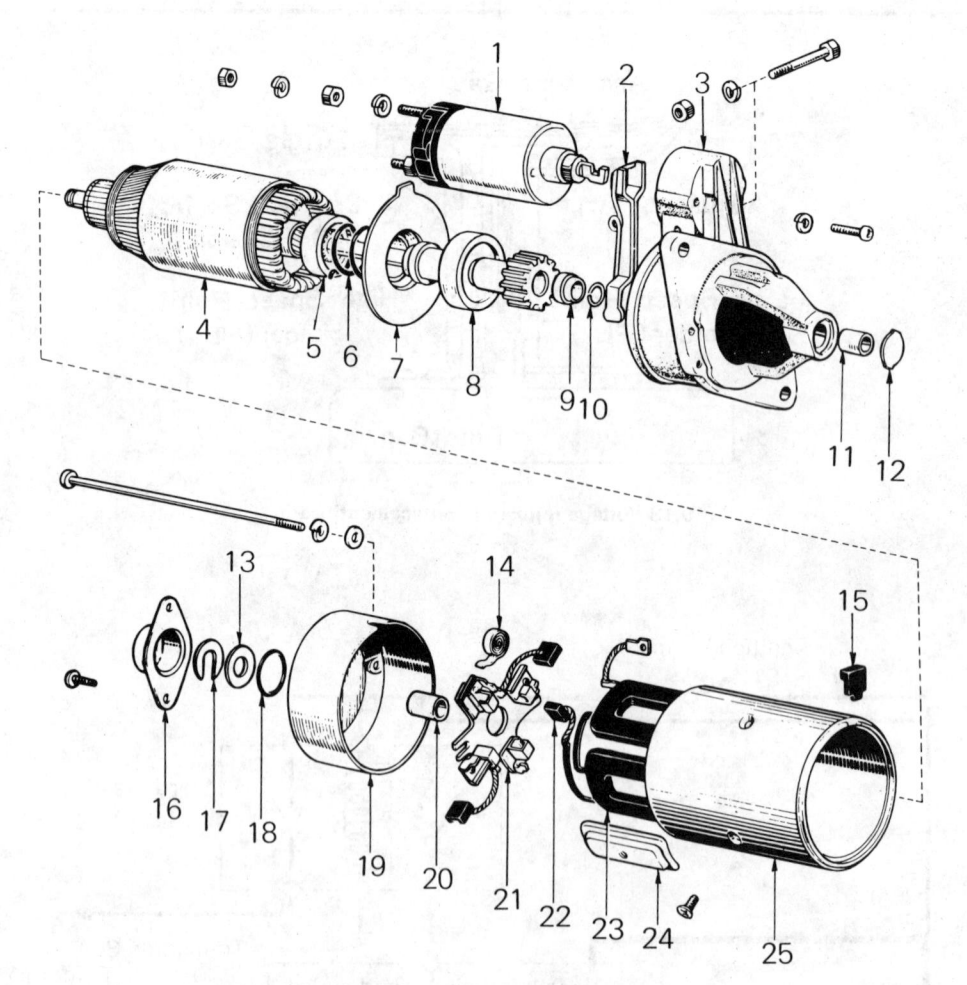

Fig. 10.16 Standard starter components

1 Solenoid	8 Clutch assembly	14 Brush spring	20 Bush
2 Drive lever	9 Collar	15 Rubber plate	21 Brush holder
3 Drive housing	10 Snap-ring	16 End frame cap	22 Brush
4 Armature	11 Bush	17 Lockplate	23 Field coil
5 Spring holder	12 Bush cover	18 O-ring	24 Pole shoe
6 Spring	13 Shim	19 Commutator end frame	25 Field frame
7 Centre bearing			

Chapter 10 Electrical system

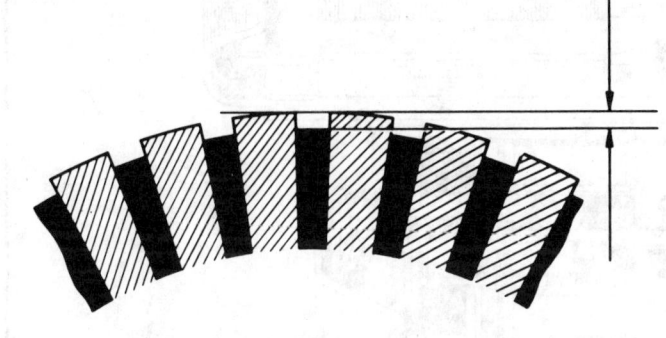

Fig. 10.17 Starter commutator undercut

Maximum .031 in (0.8 mm)
Minimum .008 in (0.2 mm)

Fig. 10.18 Testing the starter armature coils

Fig. 10.19 Testing starter armature insulation

3 Check the commutator segments and undercut the insulators if necessary, using a hacksaw blade ground to the correct thickness. If the commutator is burnt or discoloured, clean it with a piece of fine glass paper (not emery or carborundum) and finally wipe it with a solvent moistened cloth.

4 To test the armature is not difficult, but a voltmeter, or bulb and 12 volt battery are required. The two tests determine whether there may be a break in any circuit winding or if any wiring insulation is broken down. Fig. 10.18 and 10.19 shows how the battery, voltmeter and probe connectors are used to test whether, (a) any wire in the windings is broken or, (b) whether there is an insulation breakdown. In the first test the probes are placed on adjacent segments of a clean commutator. All voltmeter readings should be similar. If a bulb is used instead, it will glow very dimly or not at all if there is a fault. For the second test, any reading or bulb lighting indicates a fault. Test each segment in turn with one probe and keep the other on the shaft. Should either test indicate a faulty armature, the wisest action in the long run is to obtain a new starter. The field coils may be tested if an ohmmeter or ammeter can be obtained. With an ohmmeter the resistance (measured between the terminal and the yoke) should be 6 ohms. With an ammeter, connect it in series with a 12 volt battery, again from the field terminal to the yoke. A reading of 2 amps is normal. Zero (amps) or infinity (ohms) indicate an open circuit. More than 2 amps or less than 6 ohms indicates a breakdown of the insulation. If a fault in the field coils is diagnosed then a reconditioned starter should be obtained as the coils can only be removed and refitted with special equipment.

5 Check the insulation of the brush holders and the length of the brushes. If these have worn to below the minimum specified, renew them. Before fitting them to their holders, dress them to the correct contour by wrapping a piece of emery cloth round the commutator and rotating the commutator back and forth.

6 Check the starter clutch assembly for wear or sticky action, or chipped pinion teeth; renew the assembly, if necessary.

14 Starter motor (standard type) – reassembly

1 Fit the clutch assembly to the armature shaft followed by a new pinion stop collar and circlip. Pull the stop collar forward and stake the collar rim over the circlip. Grease all sliding surfaces.

2 Locate the drive engagement lever to the armature shaft with the spring towards the armature and the steel washer up against the clutch.

3 Apply grease to all sliding surfaces and locate the armature assembly in the drive end frame. Insert the drive engagement lever pivot pin, well greased.

4 Fit the rubber buffer together with its backing plate, then align and offer into position the yoke to the drive end frame.

5 Fit the brush holder to the armature and then insert the brushes.

6 Grease the commutator end frame bearing then fit the end frame into position. Insert and tighten the tie bolts.

7 Fit the seal, washer lockplate and end cover (half packed with multipurpose grease). Check the armature shaft endfloat, if this exceeds that given in the Specifications, remove the end cover and add an additional thrust washer.

8 Install the solenoid switch, making sure that its hook enages under the spring of the engagement lever fork.

9 Set up a test circuit, and check that the motor rotates smoothly at a current loading of less than 50 amps. With the solenoid switch energised, insert a feeler gauge between the end face of the clutch pinion and the pinion stop collar. This should be as given in the Specifications. If the clearance is incorrect, remove the solenoid switch and adjust the length of the adjustable hooked stub by loosening its locknut.

Note: *During this test, the starter motor must be adequately supported due to the very high starting torque.*

15 Starter motor (reduction gear type) – dismantling

1 Disconnect the leads from the solenoid.
2 Unscrew the two securing bolts and withdraw the field frame (yoke) complete with armature.
3 Extract the O-ring and the felt seal.
4 Remove the two securing screws and withdraw the reduction gear housing.

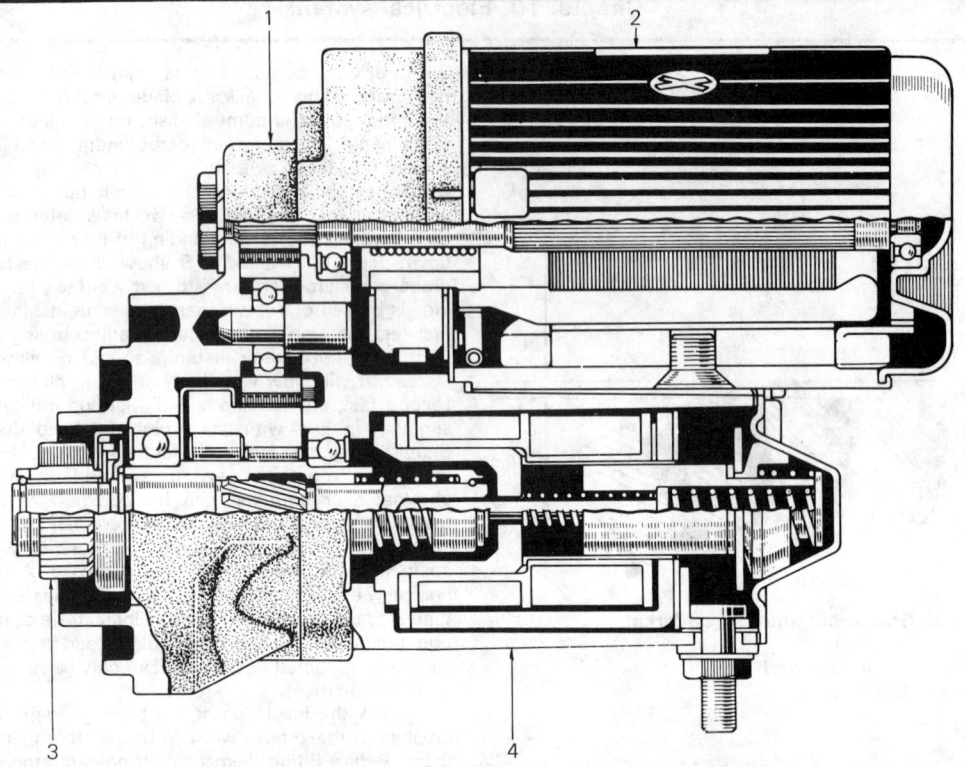

Fig. 10.20 Reduction gear type starter motor (sectional view)

1 Starter housing
2 Field frame assembly
3 Clutch assembly
4 Magnetic switch assembly

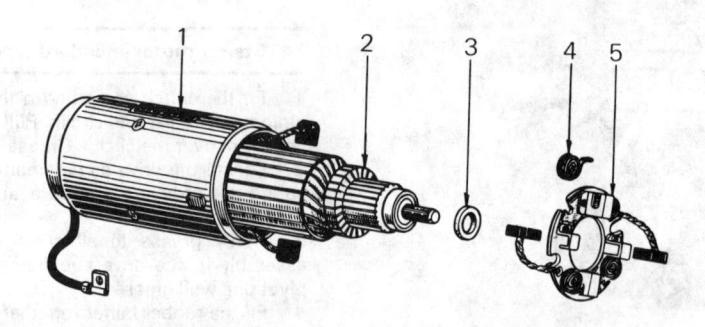

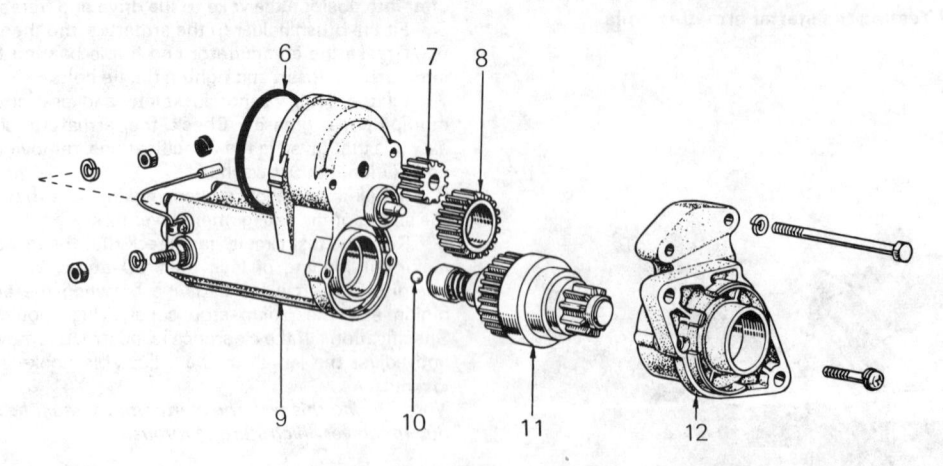

Fig. 10.21 Reduction gear type starter motor components

1 Field frame assembly
2 Armature
3 Felt seal
4 Brush spring
5 Brush holder
6 O-ring
7 Pinion gear
8 Idler gear
9 Solenoid assembly
10 Steel ball
11 Clutch assembly
12 Starter housing

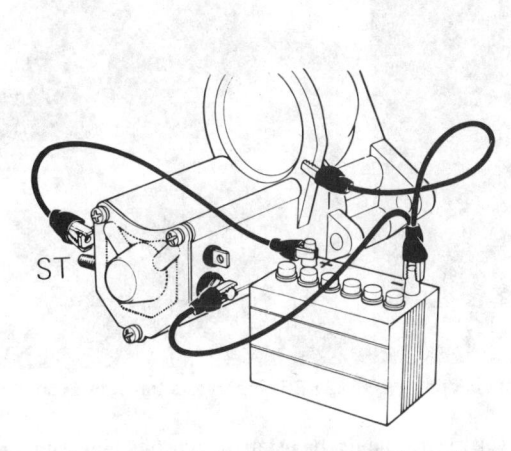

Fig. 10.22 Solenoid check for reduction gear type starter motor

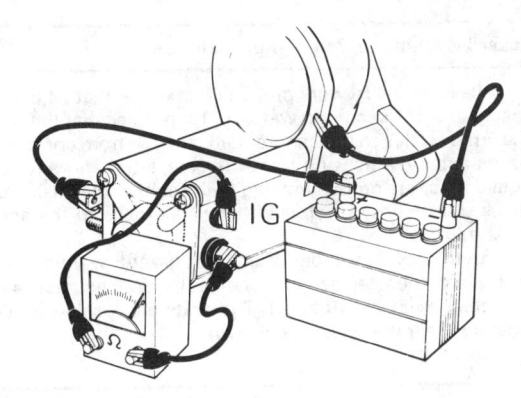

Fig. 10.23 Continuity check for reduction gear type starter motor

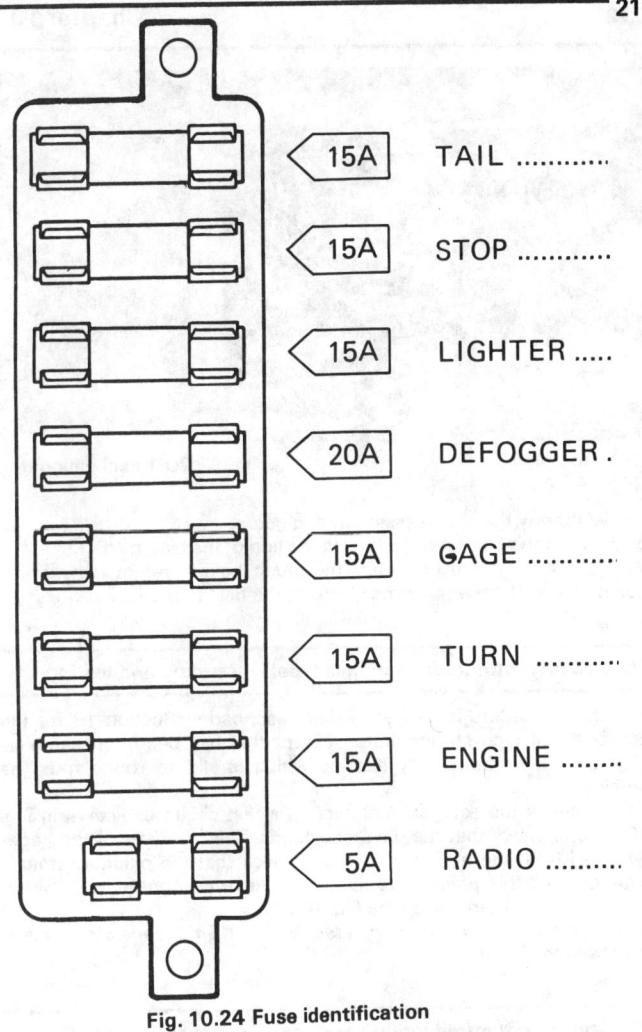

Fig. 10.24 Fuse identification

Fuse	Amp	Circuit
TAIL	15A	
STOP	15A	
LIGHTER	15A	
DEFOGGER	20A	
GAGE	15A	
TURN	15A	
ENGINE	15A	
RADIO	5A	

Tail .... Tail light, licence plate light, Clearance light, side marker light, Glove box light, Cigarette lighter light, Transmission indicator light, and Panel illumination (combination meter, clock and heater control panel)

Stop .... Stop light and Hazard warning light

Lighter .... Cigarette lighter, Interior light Door courtesy light and clock

Defogger .... Heated rear window

Gauge .... Reversing light, Fuel gauge, Fuel warning light, Water temperature gauge, Charge warning light, Oil pressure gauge, Low oil pressure warning light, Brake warning light, Heated rear window switch and heater relay

Turn .... Windscreen wiper, Screen-washer, Turn signal lights, Headlight cleaner and Rear wiper

Engine .... Voltage regulator (IG terminal), Fuel pump relay (for 20R engine) and Emission control computer (for 20R engine)

Radio .... Radio and Tape player

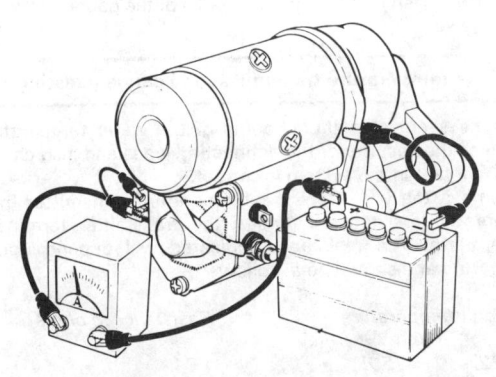

Fig. 10.25 Starter motor testing after assembly

Chapter 10 Electrical system

18.1 Fuse block

20.1 Fuel gauge transmitter connections

21.2 Temperature transmitter connection

5 Withdraw the clutch assembly and gears.
6 Extract the ball from the hole in the end of the clutch shaft.
7 Pull out the brushes from the brush holder, remove the brush holder and withdraw the armature from the field frame (yoke).

16 Starter motor (reduction gear type) – servicing and testing

1 This procedure is similar to that described in Section 13 for the conventional type starter motor, except that ball bearing assemblies are used throughout. Any bearing which is stiff to rotate must be renewed.
2 To check the solenoid, connect up a test circuit as shown in Fig. 10.22 and check that the plunger extends. Then disconnect the negative lead from the main terminal and check that the plunger remains extended. At this point there should be electrical continuity between the main and IG terminals (see Fig. 10.23).
Note: This test must be completed in less than 5 seconds to avoid overheating the coil.

17 Starter motor (reduction gear type) – reassembly

1 Apply high melting point grease to the armature rear bearing and then insert the armature into the field frame.
2 Install the brush holder, aligning its tab with the notch in the field frame.
3 Install the brushes into their holders and make sure that their leads are not earthed.
4 Fit the felt seal to the armature shaft and the O-ring to the field frame. Position the field coil leads towards the solenoid and install the field frame with armature to the solenoid housing, making sure that the raised bolt anchors are in alignment with the marks on the solenoid housing.
5 Apply grease and fit the starter pinion and idler gears.
6 Place a dab of grease into the clutch shaft hole and insert the ball.
7 Fit the clutch assembly.
8 Apply grease liberally to the gears, then install the reduction gear housing.
9 Set up a test circuit similar to that shown in Fig. 10.25 and check that the motor rotates smoothly at a current of less than 85 amps.
Note: During this test, the starter motor must be adequately supported due to the very high starting torque.

18 Fuses and fusible link

1 The fuse block is located on the fascia panel (photo). The fuse ratings and circuits protected vary according to model and date of manufacture but the fuse block is clearly marked and the cover incorporates two spare fuses.
2 In the event of a fuse blowing, always find the reason and rectify the trouble before fitting the new one. Always replace a fuse with one of the same amperage rating as the original.
3 Protection is provided for the electrical harness by a fusible link installed in the lead running from the battery positive terminal. The fusible link must never be bypassed and should it melt, the cause of the circuit overload must be established before renewing the link with one of similar type and rating. The electrical equipment described in the remainder of this Chapter may only be fitted in part to vehicles according to date of manufacture and operating territory. Only the very latest vehicles marketed in North America are equipped with the more sophisticated warning devices. Readers should check the Specifications of their particular vehicle to ascertain which Sections of this Chapter are relevant.

19 Hazard warning and turn signal lamps

1 If the flashers fail to work properly first check that all the bulbs are serviceable and of the correct wattage. Then check that the nuts which hold the lamp bodies to the car are tight and free from corrosion. These are the means by which the circuit is completed and any resistance here could affect the proper working of the coils in the flasher unit.
2 Check the security of all leads after reference to the appropriate wiring diagram.
3 If everything is in order then the hazard warning or flasher indicator units themselves must be faulty and as they cannot be repaired, they must be renewed. The units are located between the fascia panels, near the brake pedal pivot.

20 Fuel level transmitter and gauge – testing

1 In the event of malfunction of the fuel gauge, first check that there is in fact fuel in the tank and then that the connecting leads are secure and unbroken (photo).
2 Using a circuit tester, measure the resistance which should correspond approximately with the amount of fuel known to be in the tank. If the tank is empty, the resistance should be 110 ohms, if half full 33 ohms and if empty 3 ohms.
3 If the transmitter unit is satisfactory, test the fuel gauge. Pull the connector out of the transmitter and connect the wire from the transmitter to the gauge to earth through a 3.4W bulb (Fig. 10.28).
4 Turn the ignition ON and if the gauge and circuit are satisfactory the bulb will flash rapidly and the needle of the gauge will vibrate.

21 Water temperature transmitter and gauge – testing

1 In the event of malfunctioning of the water temperature gauge, first check the security of the connecting leads and also check that the level of coolant in the system is correct.
2 Remove the wire from the temperature transmitter (photo), and measure the resistance between the transmitter terminal and the vehicle earth. The resistance indicated will depend upon coolant temperature and should be as follows:

Water temperature		Resistance in ohms
°F	(°C)	
122	(50)	154
176	(80)	52
212	(100)	27.5
248	(120)	16

213

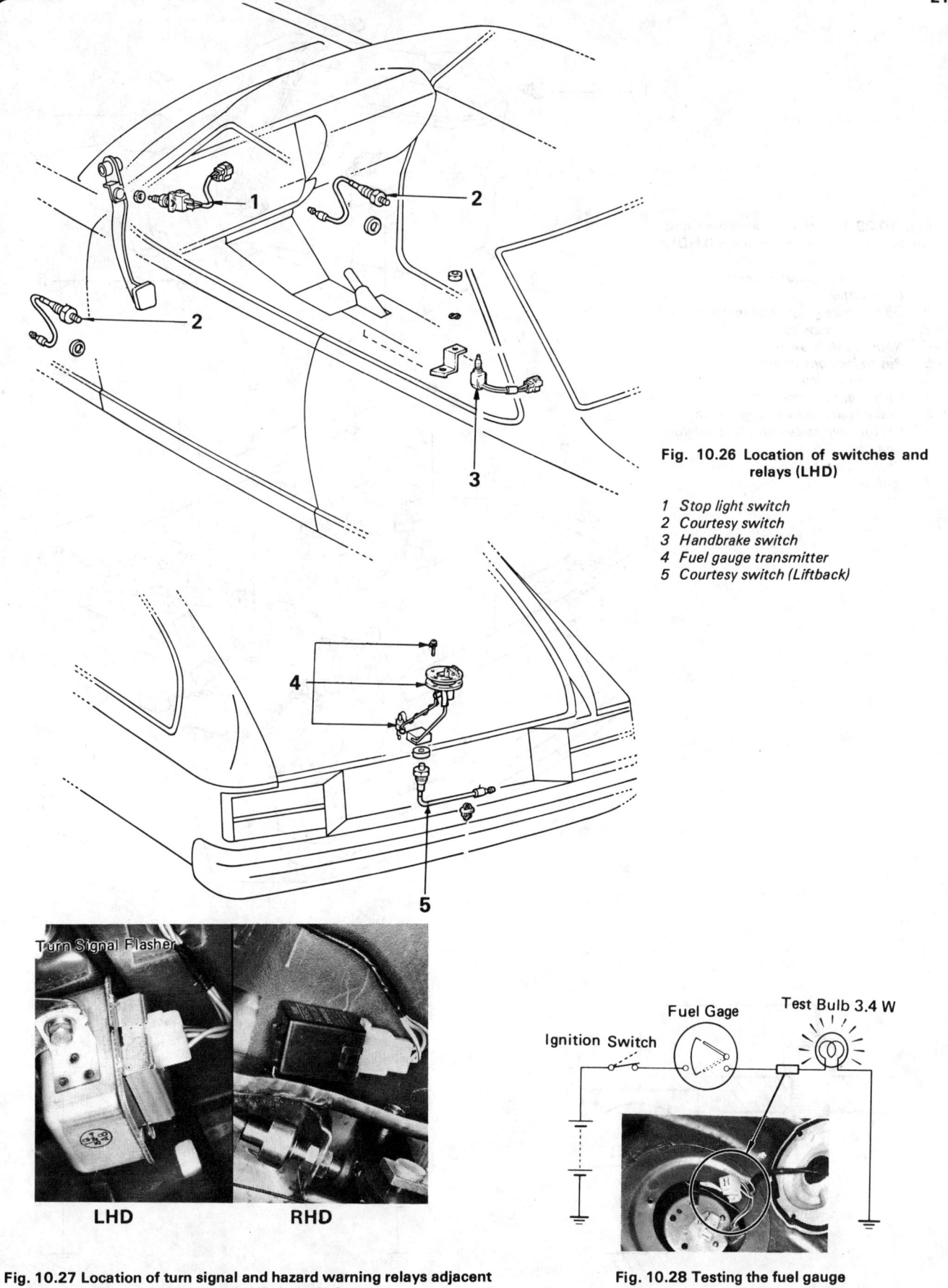

Fig. 10.26 Location of switches and relays (LHD)

1 Stop light switch
2 Courtesy switch
3 Handbrake switch
4 Fuel gauge transmitter
5 Courtesy switch (Liftback)

LHD   RHD

Fig. 10.27 Location of turn signal and hazard warning relays adjacent to brake foot pedal pivot shaft

Fig. 10.28 Testing the fuel gauge

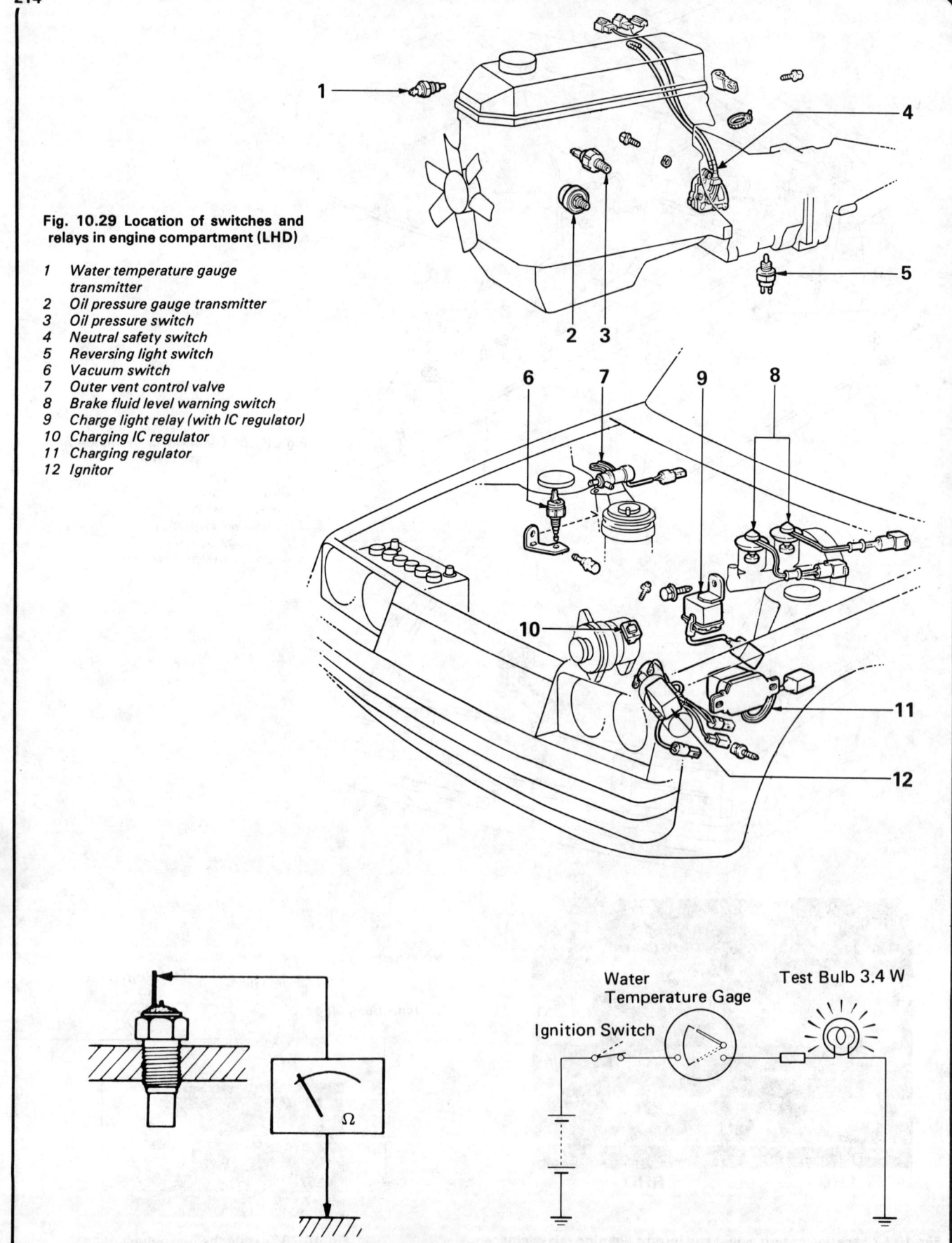

Fig. 10.29 Location of switches and relays in engine compartment (LHD)

1. Water temperature gauge transmitter
2. Oil pressure gauge transmitter
3. Oil pressure switch
4. Neutral safety switch
5. Reversing light switch
6. Vacuum switch
7. Outer vent control valve
8. Brake fluid level warning switch
9. Charge light relay (with IC regulator)
10. Charging IC regulator
11. Charging regulator
12. Ignitor

Fig. 10.30 Testing the water temperature transmitter

Fig. 10.31 Testing the water temperature gauge

Chapter 10 Electrical system

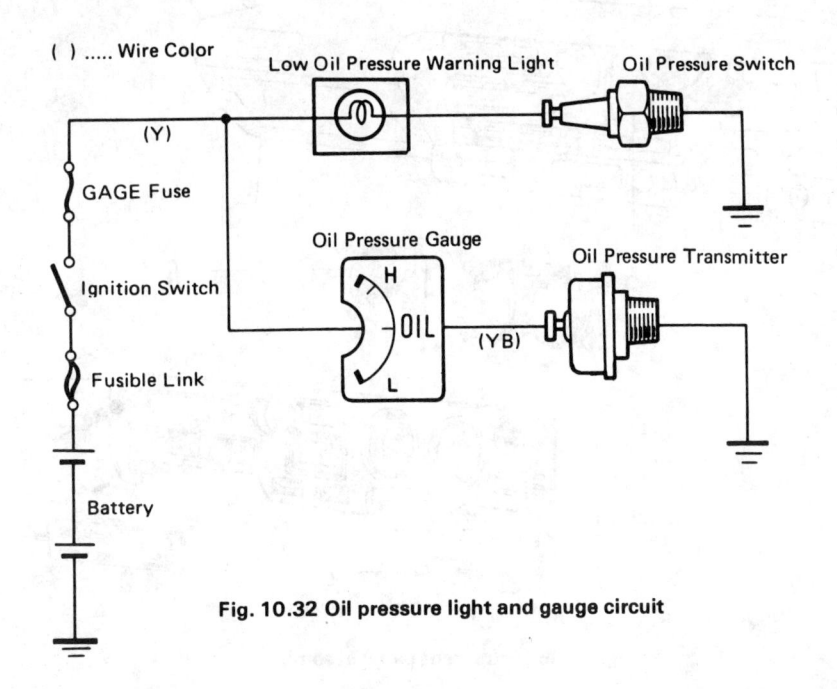

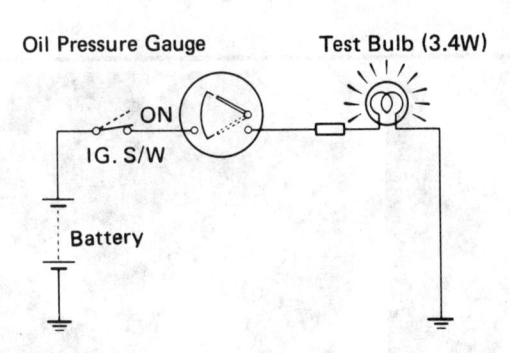

Fig. 10.32 Oil pressure light and gauge circuit

Fig. 10.33 Testing the oil pressure gauge

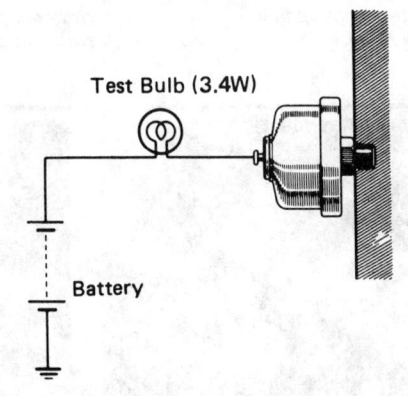

Fig. 10.34 Testing the oil pressure transmitter

3 If the transmitter is satisfactory, check the operation of the gauge by connecting the transmitter wire to earth through a 3.4W bulb and proceeding as in paragraph 4 of the preceding Section.

22 Oil pressure switch and gauge – testing

1 If the oil pressure gauge shows no reading, or the oil warning light does not operate, check that the engine oil level is correct and that there is no broken or disconnected electrical lead.
2 To test the pressure switch, connect a circuit tester between the switch terminal and a good electrical earth. The switch should make contact when the ignition is switched ON and the engine is stationary. When the engine is running, the switch should break the circuit. The operating oil pressure for the switch is a pressure of over 4.3 lbf in^2 (0.3 kg/cm^2).
3 To test the gauge, disconnect the gauge from the transmitter and earth the gauge wire through a 3.4W bulb (Fig. 10.33). With the ignition turned ON, the bulb should light and the gauge pointer should deflect.
4 To test the oil pressure transmitter (photo), disconnect the cable from it and apply battery voltage to the transmitter terminal through a 3.4W bulb (Fig. 10.34). With the engine switched off the bulb may light for an instant only and then go out. If the engine is started the bulb should flash on and off with the speed of flashing varying with the speed of the engine.

23 Ammeter – testing

1 In the event of malfunction of the ammeter, first check that the alternator drivebelt is not broken or slipping, then that the charging system is operating correctly.
2 Check the ammeter by connecting battery voltage (with a 12V, 8W bulb incorporated) between the gauge terminals. If the ammeter indicates 30 amps then the ammeter is serviceable. Do not perform this test with a bulb of more than 10W in series with the ammeter and do not apply battery voltage to the ammeter without a bulb in series or the ammeter will be damaged and may burn out.

24 Tachometer – testing

1 The tachometer can only be checked by connecting a tachometer of known calibration to the engine and comparing readings at a range of speeds. The expected error is about ± 10% at the bottom end of the

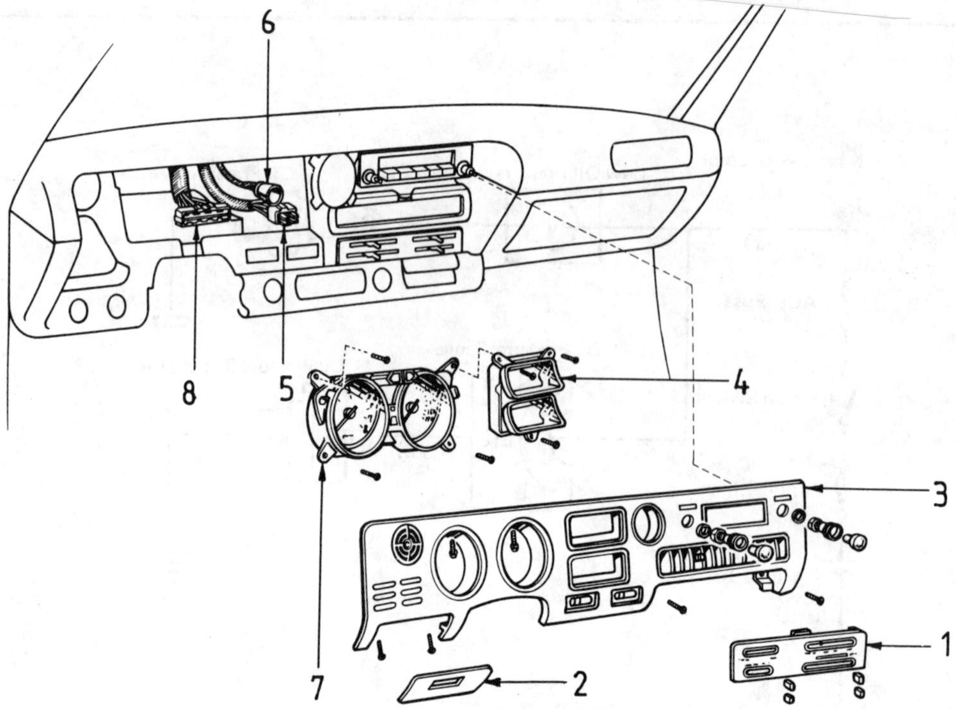

216

Fig. 10.35 Instrument panel assembly

1 Heater control panel
2 Fuse cover
3 Instrument panel
4 Instrument cluster
5 Gauge wiring connector
6 Speedometer cable
7 Instrument cluster
8 Instrument wiring connector

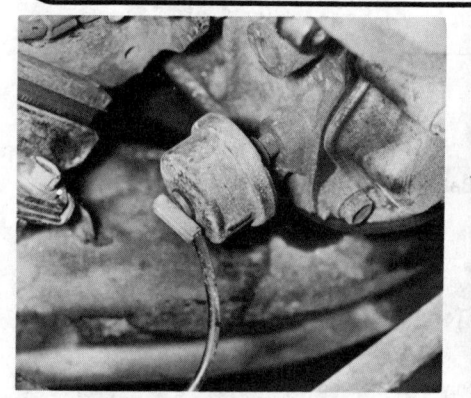

22.4 Oil pressure transmitter

26.1 Removing the horn switch assembly

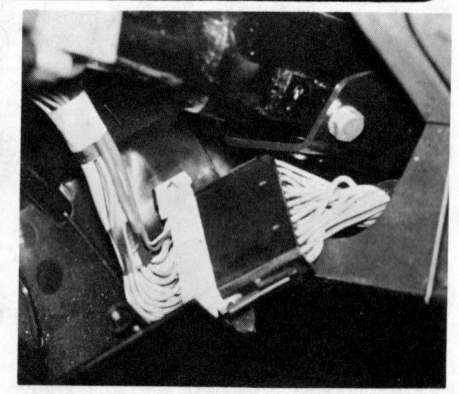

26.4 Combination switch connector

26.5 Combination switch assembly fixing screws (arrowed)

27.3 Ignition switch fixing screw (arrowed)

Chapter 10 Electrical system

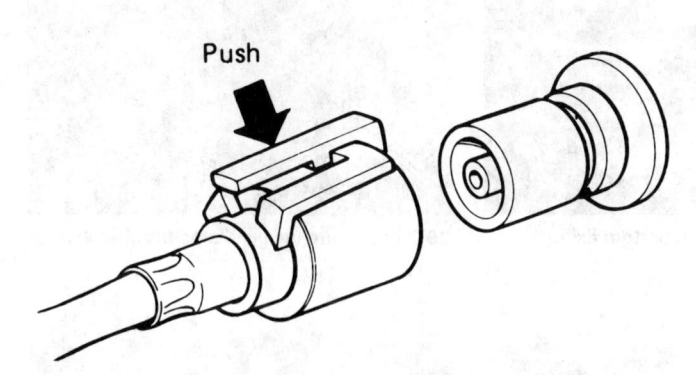

Fig. 10.36 Disconnecting the speedometer cable

scale and about ± 5% for the mid and upper range. A tachometer cannot be adjusted and if the error is not acceptable, a new instrument must be fitted.
2 Take care to ensure that the polarity of the tachometer connections is never reversed, otherwise the transistors and diodes inside it will be damaged.
3 When removing or installing a tachometer it must be handled with care so as not to drop it, or subject it to severe shocks.

25 Instrument cluster – removal and replacement

1 Disconnect the battery by removing the lead from the negative terminal.
2 Pull off the heater control knobs. Remove the heater panel fixing screws and lift the heater panel out.
3 Remove the cover from the fuse panel, take out the four fixing screws from the bottom of the instrument panel and the two in the tachometer and speedometer shrouds. Remove the knobs and spindle nuts from the radio and pull the instrument panel forward.
4 Remove the three fixing screws, detach the wiring connector and take out the instrument cluster.
5 Disconnect the speedometer cable by pushing in the cable lock and pulling off the connector while holding the lock depressed (Fig. 10.36).
6 Remove the three screws from the speedometer and tachometer cluster, disconnect the meter wiring connector and remove the assembly.

26 Steering column and fascia-mounted switches – removal and installation

1 Disconnect the battery. From the back of the steering wheel remove the three screws securing the horn button assembly (photo). Remove the assembly and disconnect it from the wiring harness.
2 Remove the steering wheel, as described in Chapter 11.
3 Remove the steering column trim and the upper and lower parts of the switch cover.
4 Disconnect the combination switch leads at the connector (photo) by pressing down on the connector latches and then separating the plug and socket parts.
5 Unscrew and remove the switch assembly fixing screws (photo). Note the position of the switch assembly in relation to the cancelling pawl and then remove the switch assembly.
6 When refitting the switch assembly (by reversing the removal operation), check that the switch and the automatic cancelling pawl are in their correct relative positions before fitting the steering wheel. After reconnecting the wiring harness plug check that all the switch functions operate correctly before refitting the switch cover and column trim.
7 Fascia mounted switches are held in position either by a bezel nut or by sprung fixing tags. Inspection will determine the method of fixing.

27 Ignition switch – removal and refitting

1 The ignition switch may be removed without disturbing the steering column lock assembly.
2 Disconnect the battery and remove the steering column trim and the upper and lower parts of the switch cover.
3 Remove the switch assembly fixing screw (photo) and pull the switch out of the lock assembly.
4 Before refitting the switch, make sure that the slot in the switch is in line with the tongue of the steering column lock and that the recess in the switch casing engages with the lug of the lock housing. For removal of the steering column lock, see Chapter 11.

28 Neutral safety switch (automatic transmission) – removal and installation

1 Remove the nut and washer and pull off the control shaft lever.
2 Bend down the tabs of the lockwasher, unscrew and remove the nut and take off the lockwasher and grommet.

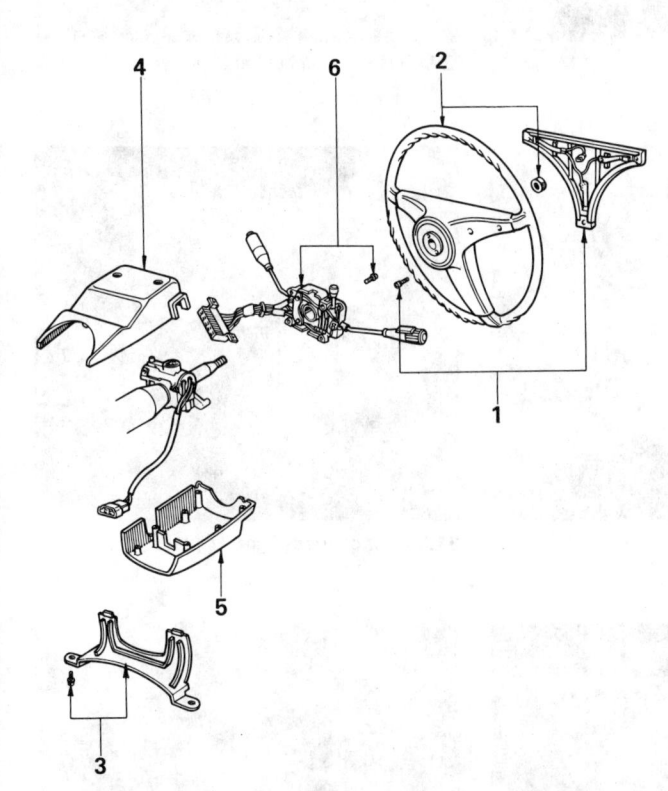

Fig. 10.37 Combination switch assembly

1 Horn button
2 Steering wheel
3 Steering column trim
4 Upper cover
5 Lower cover
6 Switch assembly

29.2a Headlamp housing top fixings

29.2b Headlamp housing bottom fixings

29.3 Loosening the headlamp bezel screws

29.4 Headlamp plug connection

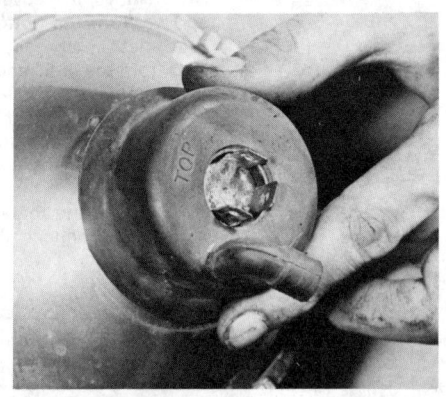

29.5a Headlamp boot

29.5b Headlamp bulb and carrier

30.1 Headlamp adjusting screws

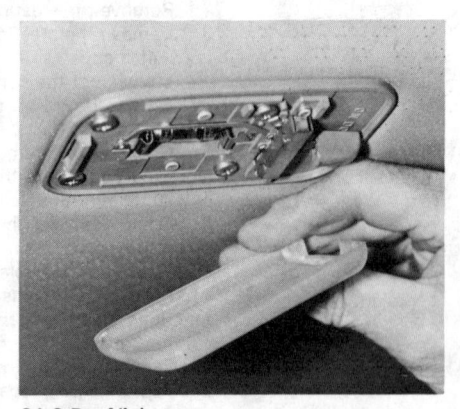

31.2 Roof light

31.3 Door courtesy light

31.4 Luggage compartment light

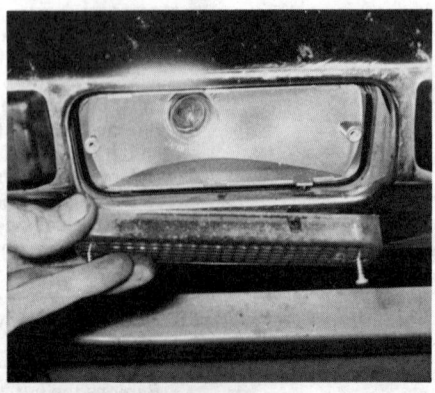

31.5a Front flasher light

31.5b Front side marker light

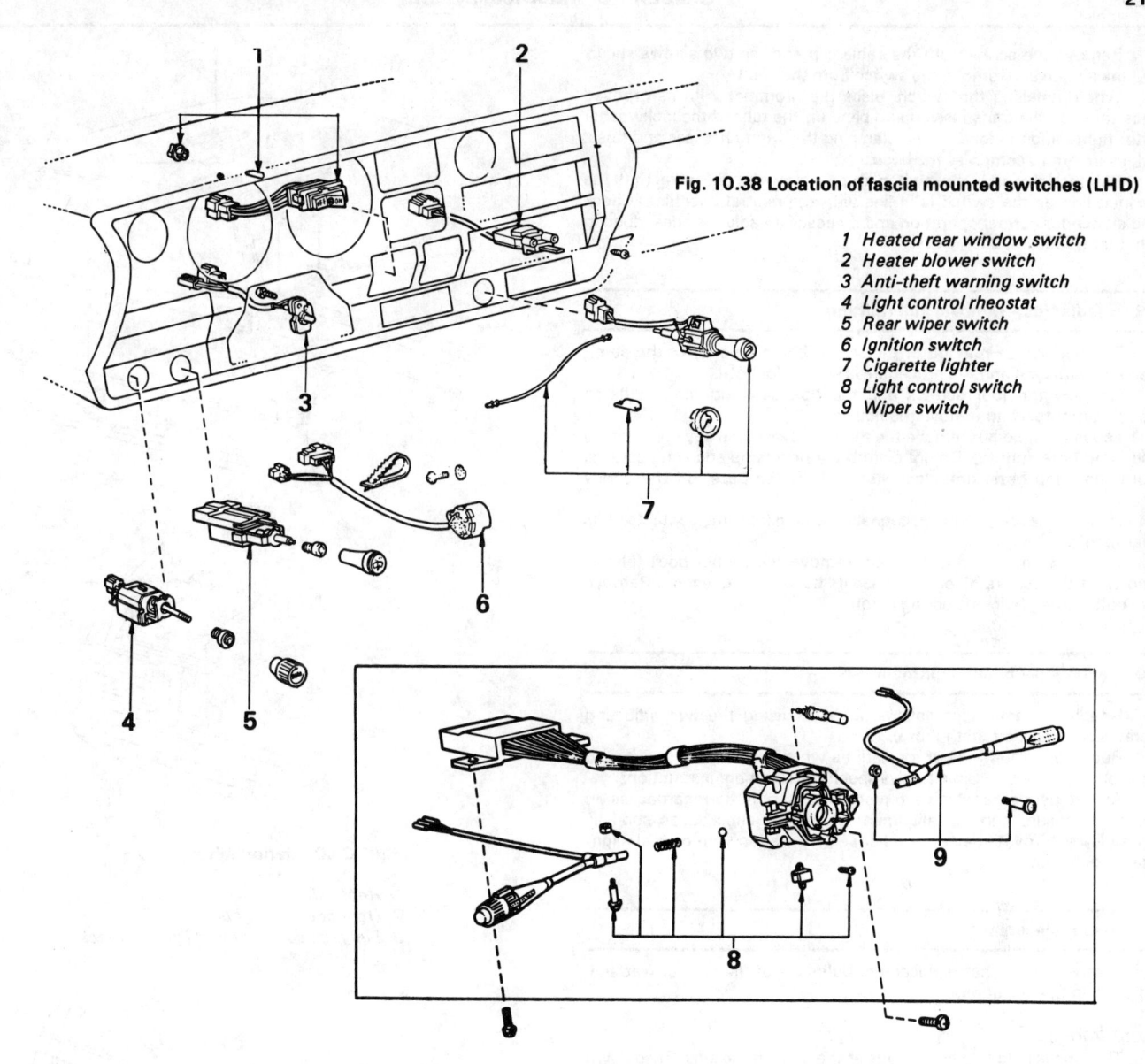

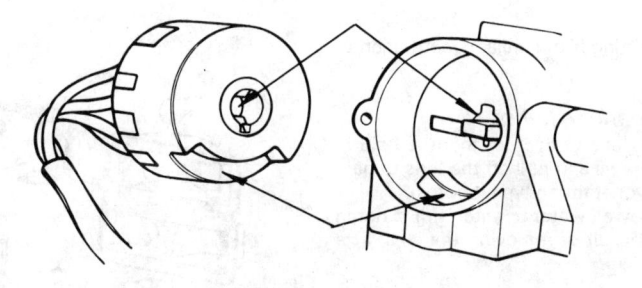

Fig. 10.38 Location of fascia mounted switches (LHD)

1 Heated rear window switch
2 Heater blower switch
3 Anti-theft warning switch
4 Light control rheostat
5 Rear wiper switch
6 Ignition switch
7 Cigarette lighter
8 Light control switch
9 Wiper switch

Fig. 10.39 Correct relative positions of steering column lock and switch

3 Remove the screw from the cable clip and the two screws securing the switch and remove the switch from the shaft
4 When installing the switch, place the grommet with its grooved face towards the transmission and bend up the tabs of the lockwasher after tightening the lock. When clamping the wiring harness and hose, take care not to compress the hose.
5 Before tightening the switch fixing screws, make sure that the scribed line on the switch is in line with the neutral base line. Check the switch for correct operation and if necessary adjust as described in Chapter 6, Section 21.

29 Headlamps – removal and refitting

1 The headlamps may be of the sealed beam type, or of the semi-sealed beam type and the procedure is similar for both.
2 Remove the four screws and the bolt securing the headlamp housing (photos) and remove the housing.
3 Loosen, but do not remove the three screws in the key hole slots in the lamp bezel (photo). Do not disturb the headlamp adjusting screws. Turn the lamp bezel anti-clockwise and lift the bezel off the screw heads.
4 Hold the headlight and disconnect the plug from the contacts at its rear (photo).
5 On the semi-sealed beam unit, remove the rubber boot (photo) and twist the bulb retainer to release its bayonet connection. Remove the bulb carrier, bulb and spring (photo).

30 Headlamp – beam alignment

1 Headlight beam alignment is achieved using the two adjusting screws on each light unit (photo).
2 Because of the range of regulations which are in force in different territories, it is not possible to give specific adjustment instructions.
3 Any adjustments which are made should only be regarded as an interim measure and the alignment should be checked as soon as possible by a Toyota dealer or other service station with optical alignment equipment.

31 Bulb replacement

1 Always ensure that replacement bulbs are of the correct wattage, as given in Specifications.

Roof light

2 The lens is held in place by lugs at the ends of the lens. Prise down one end of the lens and then unhook the other end to expose the festoon bulb and the fitting fixing screws (photo).

Door courtesy lights

3 Remove the two screws from the lens and prise out the lens to expose the festoon bulb. The lens screws also secure the fitting to the door and with the lens removed, the light fitting can be pulled out of the door (photo).

Luggage compartment light

4 The luggage compartment light fitting is of similar construction to the door courtesy lights (photo).

Front flasher, side marker and parking lights

5 The lights are similar in that the lens of each of them is held in place by two screws. Remove the screws and pull off the lens to gain access to the bulb, which is of the bayonet fitting type (photos).
6 Check that the gasket has not allowed water to enter. If it is not in good condition, or is damaged as a result of removing the lens, fit a new gasket.

Rear combination light

7 There are differences between the combination light assembly on the Liftback and Coupe (Fig. 10.41 and 10.42).
8 Remove the screws securing the boot trim over the light assembly and remove the trim. On the Liftback the individual bulb carriers can be removed by twisting them to release the bayonet fitting (photo).

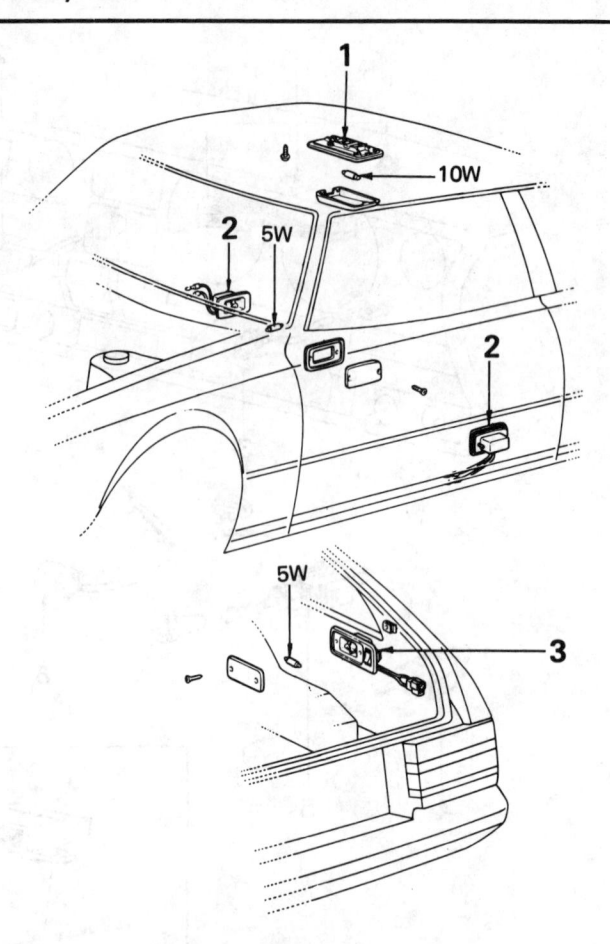

Fig. 10.40 Interior lights

1 Roof light
2 Door courtesy lights
3 Luggage compartment light (Liftback)

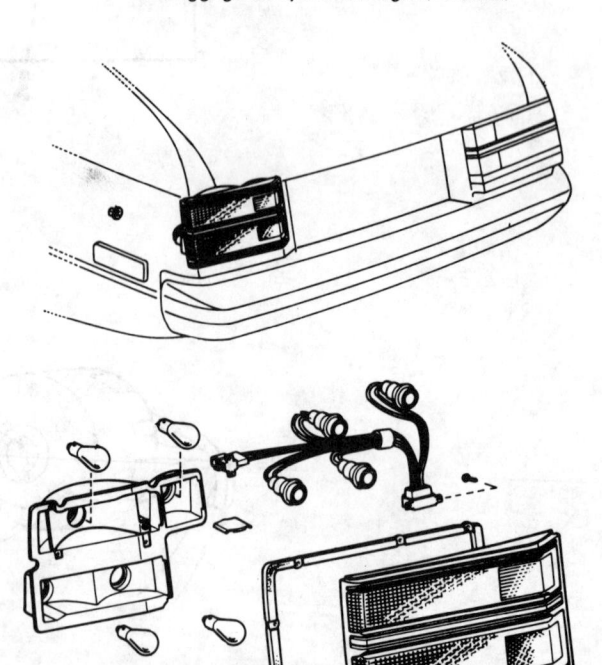

Fig. 10.41 Rear combination light (Liftback)

Chapter 10 Electrical system

9 On the Coupe there is a common bulb carrier. Remove its fixing screws and pull the carrier off.

Rear side marker lamps
10 Rear side marker lamps, when fitted, are shown at Fig. 10.43.
11 Lift the rear door and lift up the luggage compartment lining to expose the bulb carrier. Pull out the bulb carrier to gain access to the bulb.

Number plate light
12 Remove the screw from each end of the light cover and lift off the cover to expose the bulbs (photo).
13 When refitting the cover, take care that it engages in the groove of the sealing gasket.

Panel warning lamps
14 Remove the instrument panel as described in the first three paragraphs of Section 25.

Instrument panel lamps
15 Remove the instrument panel as above and remove the appropriate bulb carrier from the printed circuit board.

Heater control indicator lamp
16 Pull off the heater control knob and remove the four fixing screws from the heater control panel. Lift off the panel to gain access to the festoon bulb.

Transmission indicator lamp
17 Remove the four screws from the control lever escutcheon plate and lift the plate to expose the indicator lamp.

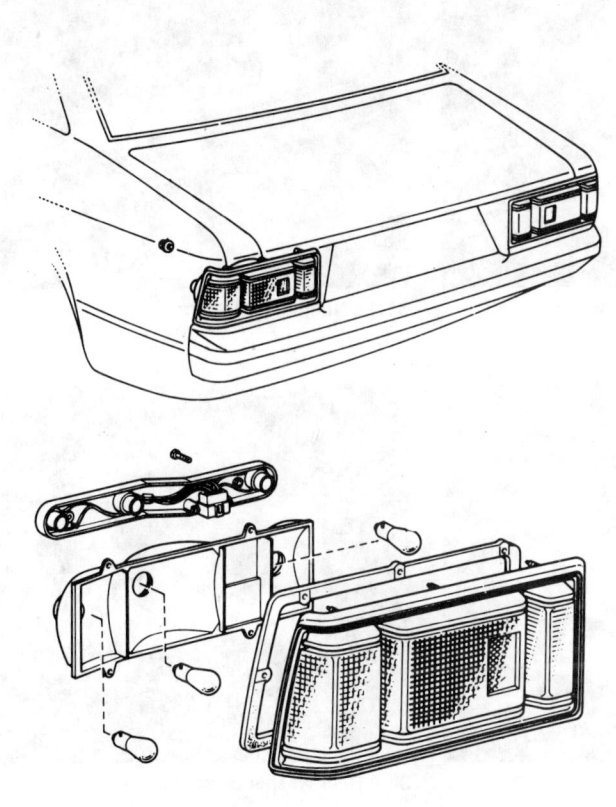

Fig. 10.42 Rear combination light (Coupe)

32 Windscreen wiper – removal and refitting

1 Lift up the covers on the wiper arms, unscrew and remove the nuts then exposed and pull off the wiper arms (photo).
2 Remove the six screws underneath the sealing strip of the louvred panel and remove the panel.
3 Remove the screws from the two cover panels and remove the panels.
4 Using a large screwdriver, prise the wiper linkage off the wiper crankarm.
5 Disconnect the electrical leads of the motor by separating the plug and socket connection, then remove the three motor fixing screws (photo) and lift the motor out.
6 To remove the linkage, remove the three fixing screws from each of the wiper drive spindle mountings and push the drive spindles downwards. Remove the linkage through the access hole.
7 Refit the motor and linkage by reversing the removal operations, but before fitting the wiper arms, ensure that the wiper motor is in the auto-stop positions.
8 Fit the wiper arms so that the wiper blades are parallel to the bottom edge of the screen and note that the straight arm is on the passenger side and the cranked arm is on the driver's side.

33 Windscreen wiper motor – dismantling and reassembly

1 Remove the three screws from the gear housing cover plate and remove the plate, but do not detach the soldered wires unless the brush holder, or gear housing is being renewed.
2 Put an alignment mark on the crankarm and driveshaft. Detach the crank from the wiper motor by removing the retaining nut and pulling the crankarm from the splined tapered shaft of the drive gear. Retain the thrust and wave washers. Detach the rubber seat.
3 Remove the gear housing from the stator by withdrawing the two securing screws. Retain the ball from the end of the armature shaft. Remove the thrust adjuster screw from the gear housing.
4 Pull the armature from the stator using enough force to overcome the magnetic attraction. Retain the ball from the other end of the armature shaft.
5 Clean the grease from all components and inspect for wear or damage. Check the driveshaft endfloat and if it exceeds 0·008 in (0·2 mm) renew the thrust washer.

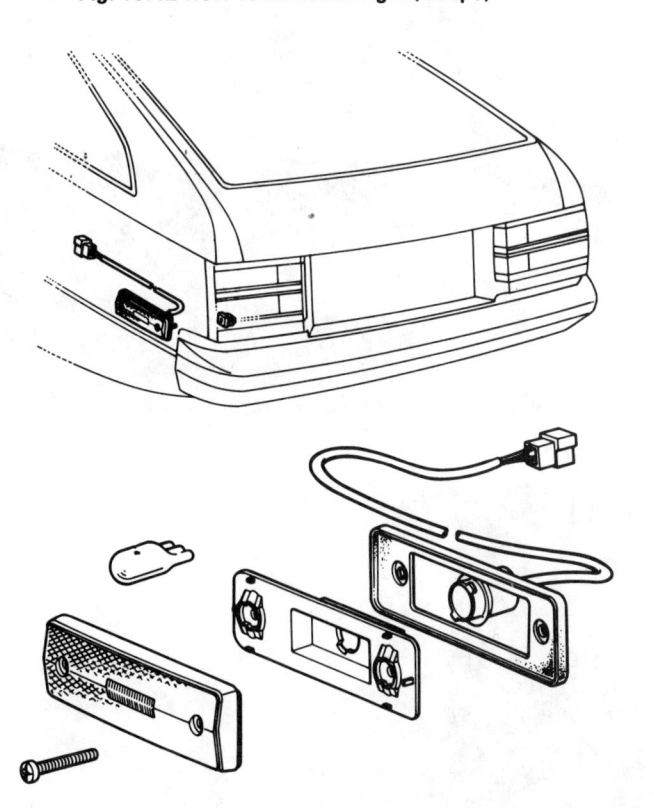

Fig. 10.43 Rear side marker lamp

31.5c Front parking light

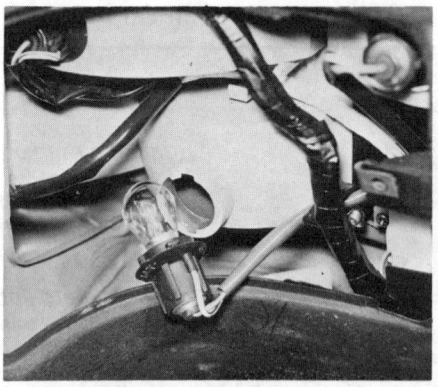

31.8 Replacing a rear bulb (Liftback)

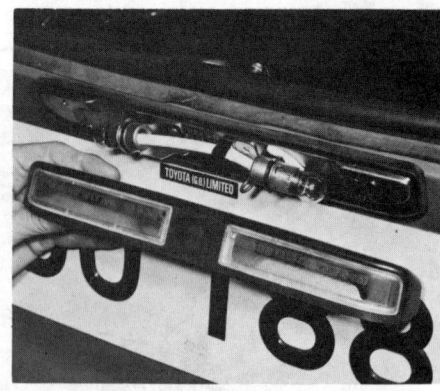

31.12 Number plate light

32.1 Windscreen wiper arm securing nut

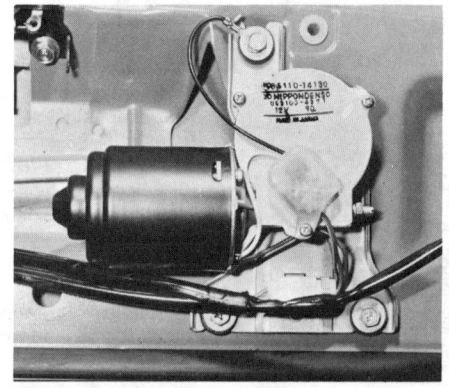

32.5 Windscreen wiper motor fixings

34.4 Rear washer reservoir

34.5a Washer feed hole in wiper arm spindle

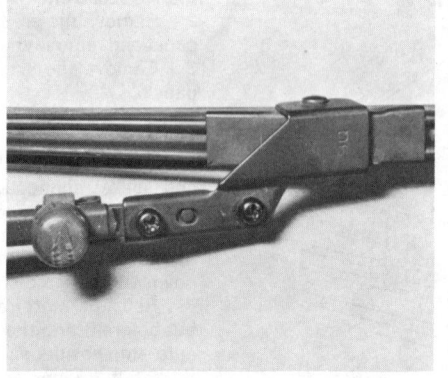

34.5b Washer nozzle and blade attachment

34.5c Washer nozzle attachment

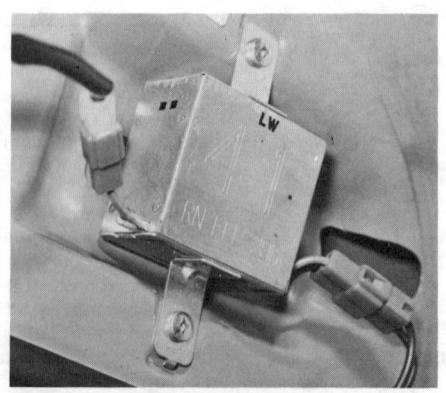

38.4 Demister panel choke

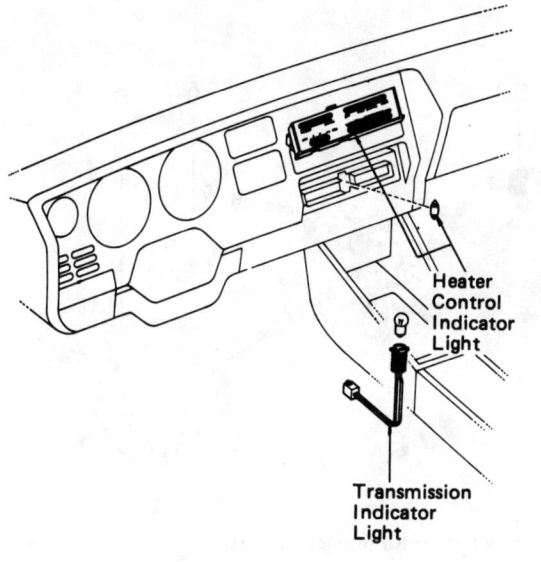

Fig. 10.44 Heater and transmission indicators

Heater Control Indicator Light

Transmission Indicator Light

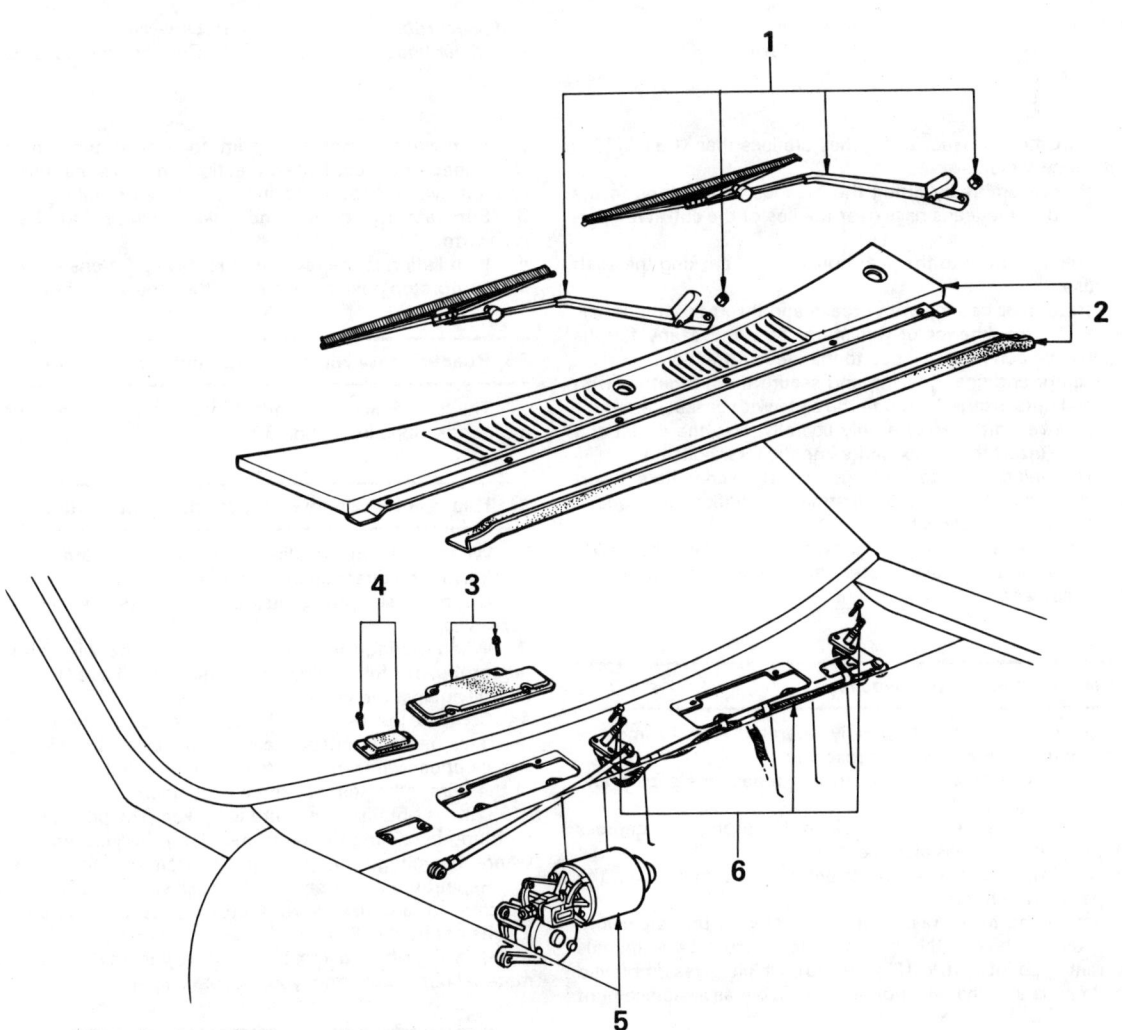

Fig. 10.45 Windscreen wiper motor removal sequence

1 Wiper arm and blade
2 Louvred panel
3 Cover panel
4 Cover panel
5 Wiper motor
6 Wiper linkage

223

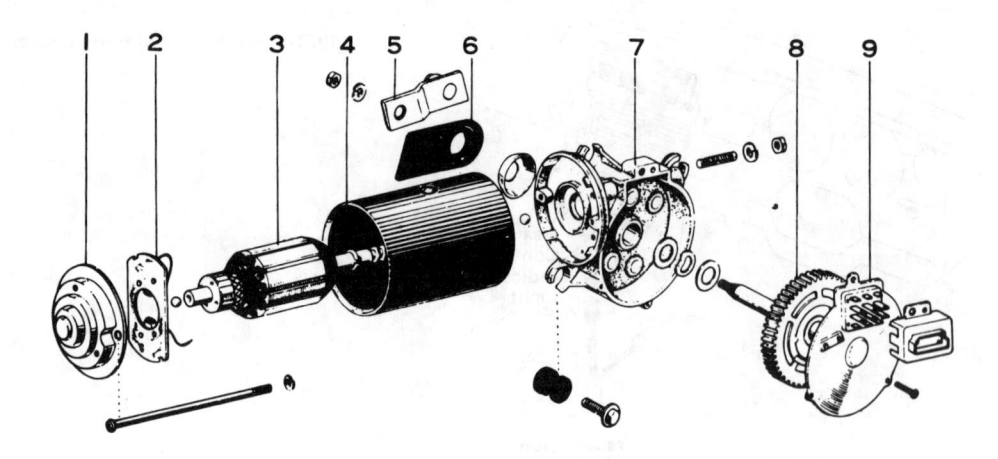

Fig. 10.46 Windscreen wiper motor (exploded view)

1 Endframe
2 Flange holder
3 Armature
4 Stator
5 Crankarm
6 Seat rubber
7 Gear housing
8 Driveshaft
9 Gear housing cover plate

6 Check the brushes for wear and if they are less than 0·31 in (8·0 mm) in length, renew them.
7 Commence reassembly by fitting the brushes and springs to the brush holders so that the leads pass over the lips of the cut-away portions.
8 Assemble the armature to the gear housing first packing the shaft bush with grease.
9 Pack the stator rear bearing with grease and insert the thrust ball. Carefully wipe out the interior of the stator to remove any ferrous filings which may be adhering to it due to magnetic attraction.
10 Align the stator and gear housing and secure them together with the two bolts and nuts inserted through the slots with gasket cement.
11 Grease the driveshaft/gear assembly liberally, also the thrust and wave washers and insert the components into the gear housing.
12 Drop the remaining ball into the thrust adjuster screw hole, insert the adjuster screw and tighten it until all armature endfloat disappears. Secure the screw with the locknut.
13 Check the height of the automatic switch lever from the inside surface of the gear housing cover plate. This should be 0·39 in (9·9 mm) adjust if necessary.

34 Windscreen washer and rear washer

1 Both washers are of the electrically operated type having the pump mounted within the washer fluid reservoir.
2 Keep the reservoirs clean, the electrical connections secure and the washer connecting pipes tight.
3 Never operate the washer without fluid in the reservoir, or depress the switch for periods in excess of 20 seconds.
4 The rear window washer reservoir is behind the trim panel in the luggage compartment (photo).
5 The nozzles of the front washer are integral with the wiper arm and are fed through a hole in the drive spindle (photos). Later models have a different type of nozzle (Fig. 10.48) which gives improved washer performance and the new nozzle is available as a replacement part.
6 The rear window washer nozzle is not incorporated in the wiper arm and may be removed by pushing up one lockpin with a screwdriver and then pulling the nozzle out of the hose (Fg. 10.49).

35 Rear window wiper – removal and installation

1 Remove the wiper arm assembly as described in Section 32, and remove the two bolts securing the wiper drive spindle mounting.

2 Remove the rear door trim to expose the motor and linkage, Disconnect the electrical connections, remove the motor fixing screws and remove the motor and linkage as an assembly.
3 Separate the motor and linkage by prising the link from the crankarm.
4 Installation is the reverse of removal, but ensure that the motor is in its auto-stop position before refitting the wiper arm.

36 Rear window wiper – dismantling and reassembly

The procedure is identical with that for the windscreen wiper motor described in Section 33.

37 Heated rear window – precautions and testing

1 When cleaning the glass, use a soft dry cloth and wipe the glass carefully in the direction of the wires.
2 Do not use detergents, or any glass cleaner which contains abrasives.
3 When testing the wire; do not touch the wire with the tip of the test probe, but fold a piece of aluminium foil over the probe and press the foil against the wire.
4 Connect the (+) probe of a voltmeter to the (+) terminal of the demister panel. Connect the (–) terminal to the mid-point of each of the six upper wires in turn. A voltage of about 2 volts at the mid-point of the wire indicates that the wire is unbroken. If a reading of zero, or 4·5 volts is obtained, the wire is broken. The position of the break can be found by sliding the (–) probe along the wire until a point is found where the voltage changes suddenly from zero to several volts.
5 Repeat the check on the lower six wires. A voltage of about 7 volts indicates an unbroken wire. A broken wire will give a reading of about 2 volts or about 4·5 volts.
6 It is possible to repair a broken wire by using a special repair material obtainable from your Toyota dealer.

38 Heated rear window – switch, relay and choke

1 If it is found when testing the rear window panel that there is no electrical supply to it, test the switch and relay.
2 To remove the switch, remove the instrument fascia panel and prise out the switch. Test the switch for continuity between terminals B and R when in the ON position. If this is so, the switch is satisfactory.
3 Test the relay by switching the ignition ON and testing for voltage

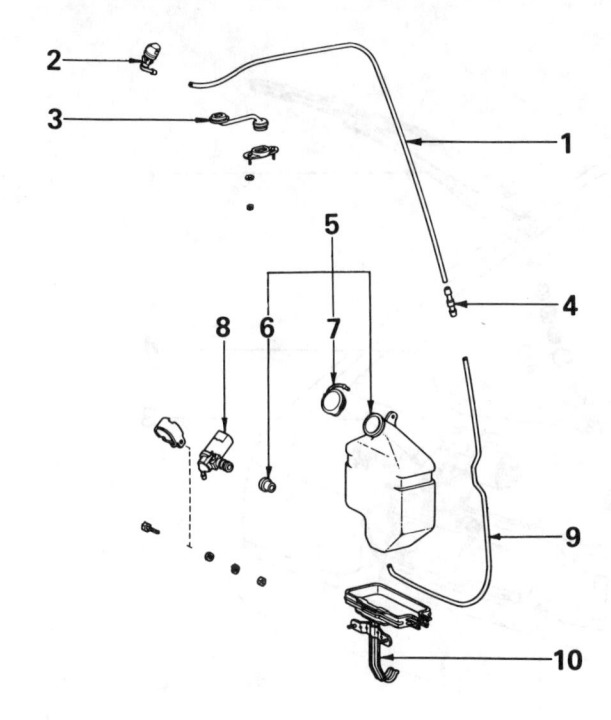

Fig. 10.47 Rear washer components

1 Hose
2 Nozzle
3 Grommet
4 Joint
5 Fluid reservoir
6 Bush
7 Cap
8 Motor & pump
9 Hose
10 Bracket

WASHER NOZZLE

PREVIOUS

WASHER NOZZLE

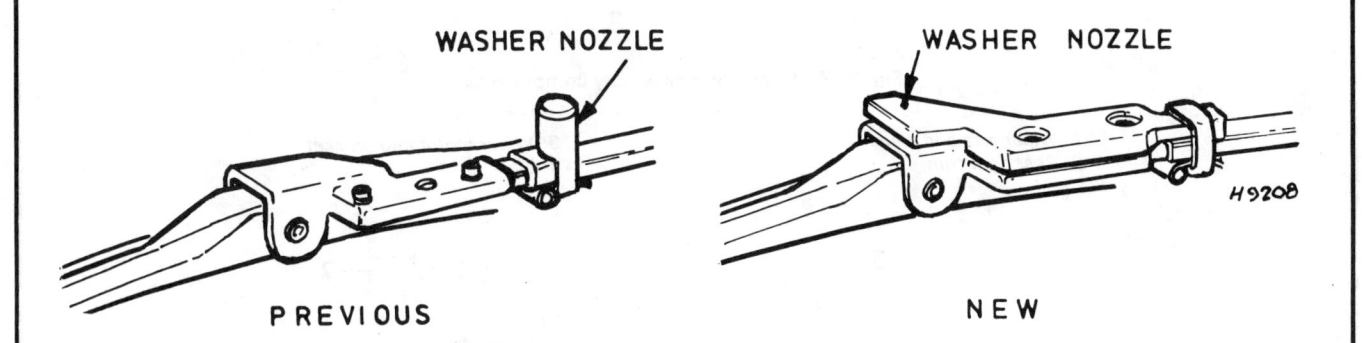

NEW

Fig. 10.48 Old and new style washer nozzles

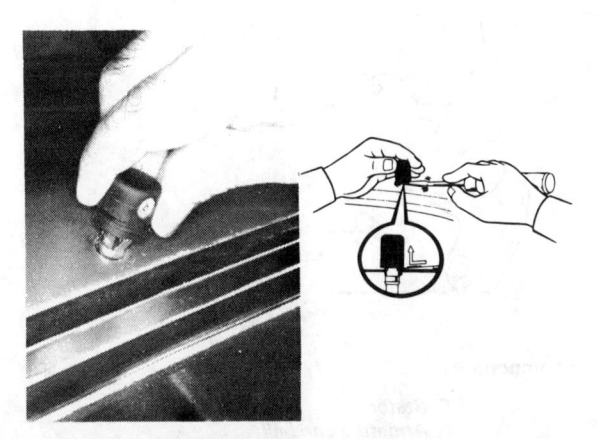

Fig. 10.49 Removing the rear washer nozzle

Check Valve

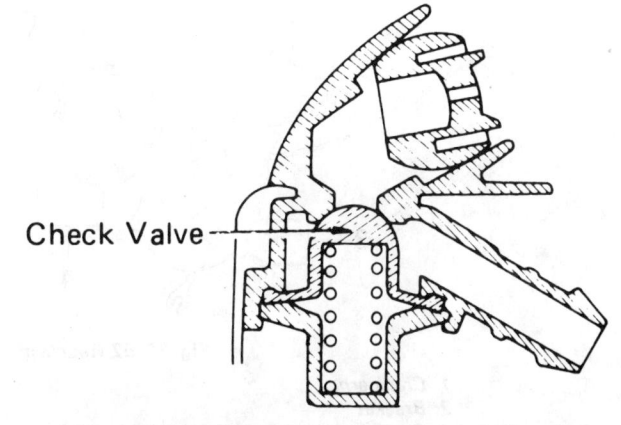

Fig. 10.50 Headlamp washer nozzle and check valve

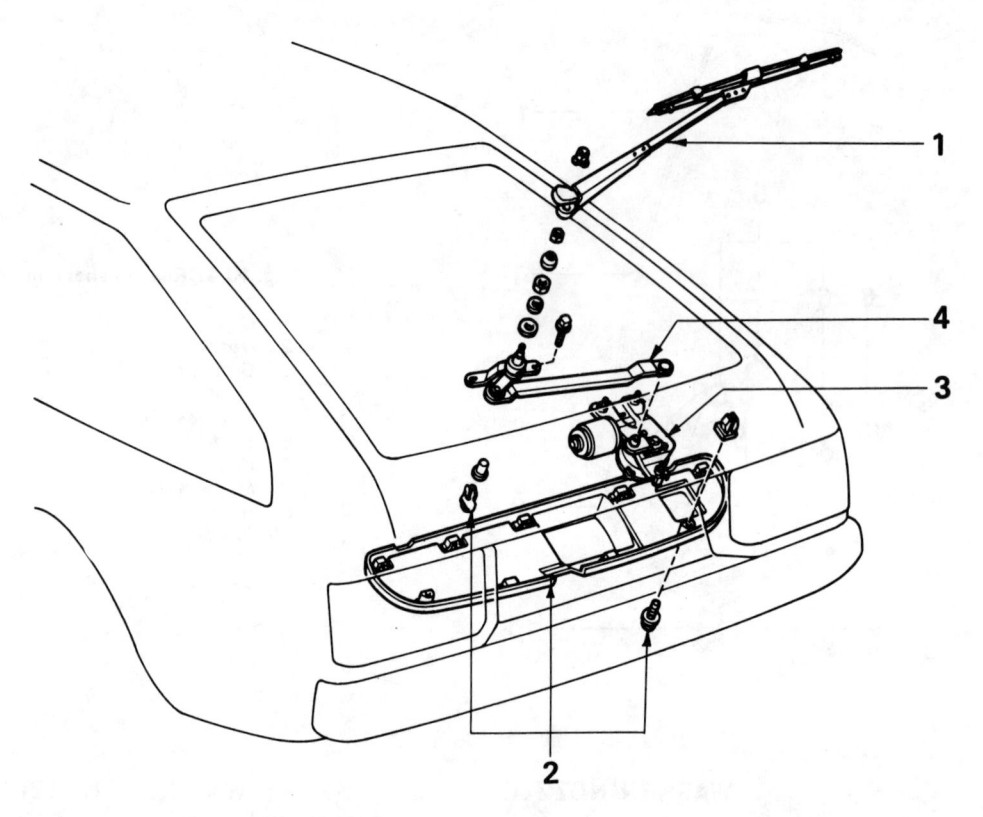

Fig. 10.51 Rear wiper assembly components

1 Wiper arm and blade
2 Rear door trim
3 Wiper motor and bracket
4 Link

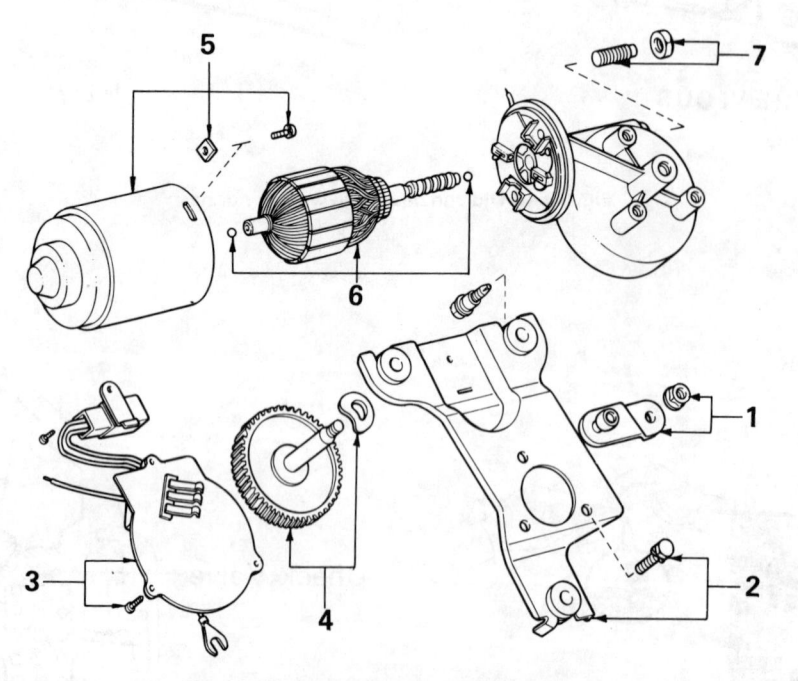

Fig. 10.52 Rear wiper motor components

1 Crank arm
2 Bracket
3 Crank housing cover plate
4 Driveshaft and wavy washer
5 Stator
6 Armature and ball
7 Screw and nut

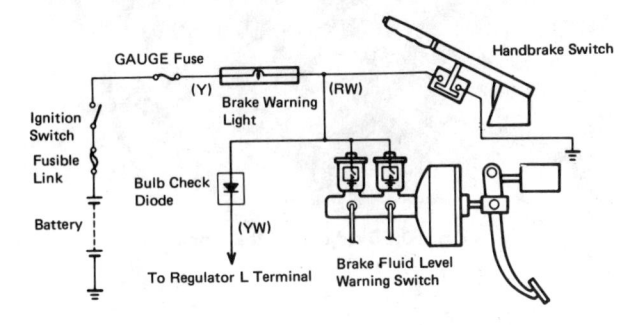

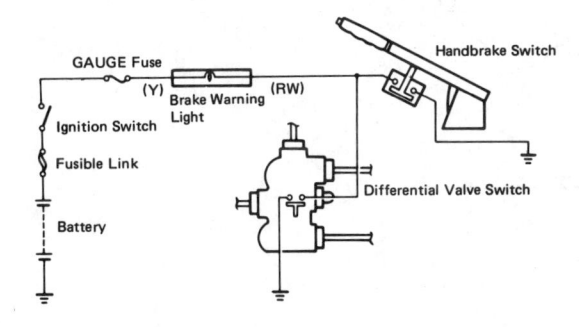

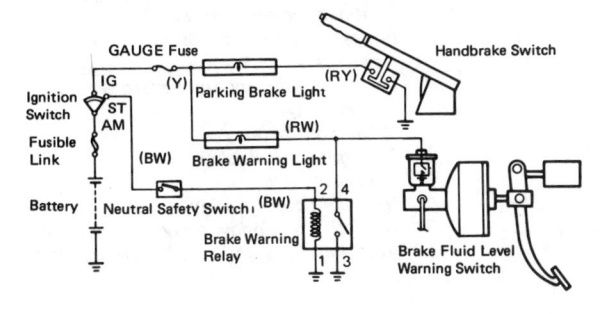

Chapter 10 Electrical system 227

on terminal 4. Join terminals 4 and 2 with a piece of wire and the relay should operate. There should then be voltage on terminal 3.

4 The demister panel has a choke in its circuit which is mounted on the rear door (photo). If both the switch and relay are satisfactory, check that there is continuity between one of the panel leads and the incoming lead and between the other panel lead and the case of the choke.

39 Headlamp washer

1 The headlamp washer, when fitted, is a separate system, independent of the windscreen and rear window washers.
2 Each nozzle incorporates a check valve, which should open at a pressure in the range 24·2 to 29·9 lbf in^2 (1·7 to 2·1 kg/cm^2) and the difference between left and right sides must not exceed 2·8 lbf in^2 (0·2 kg/cm^2).
3 Adjust the nozzles by bending the brackets and aim them so that the jet strikes the headlight 0·2 to 0·4 in (5 to 10 mm) above the centre line.

40 Brake warning light

1 The type of brake warning circuit which is employed is dependent upon the territory for which the car was manufactured (Fig. 10.53).
2 All systems incorporate a switch on the handbrake and may in addition have a float switch on the reservoir, or a switch in the differential valve (see Chapter 9, Section 14).

41 Clock – adjustment

1 The type of clock fitted may be either a quartz one, or an electronically driven mechanical movement. The quartz type should be accurate to within ± 2 seconds a day and no adjustment is provided. A suspect quartz clock should be checked by your Toyota dealer.
2 Mecahnical clocks have an allowable error of ± 2 minutes a day and can be adjusted.
3 Remove the fascia (see Section 25) and remove the clock from the instrument panel. Turning the adjustment screw (Fig. 10.55) 10° anti-clockwise will cause the clock to lose 1 minute a day and a 10° clockwise turn will cause it to gain 1 minute a day.

42 Relays

1 In addition to the relays mentioned earlier which control the direction indicator and hazard warning circuits, some models have additional relays to control the the following devices:

Heater
Seat belt warning
Wiper control
Tail light control
Fuel pump
Headlight control
Heated rear window

2 Relays are sealed units and when a failure is suspected have the unit tested by your Toyota dealer and then renew if necessary with one of a similar type.

43 Radio – removal and refitting

1 To remove the radio fitted as standard to these models, first disconnect the battery negative lead.
2 Pull off the heater control lever knobs and then remove the control panel.
3 Pull off the radio control knobs and unscrew the nuts from the

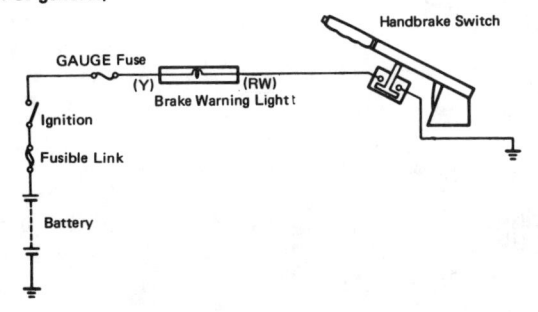

Fig. 10.53 Brake warning system circuits

Nozzle Adjusting

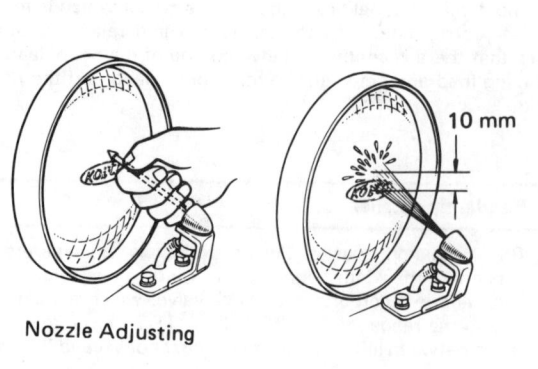

Fig. 10.54 Headlamp washer nozzle aiming diagram

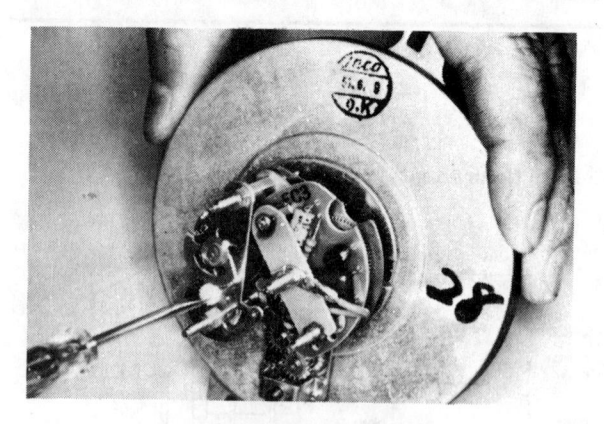

Fig. 10.55 Adjusting the clock

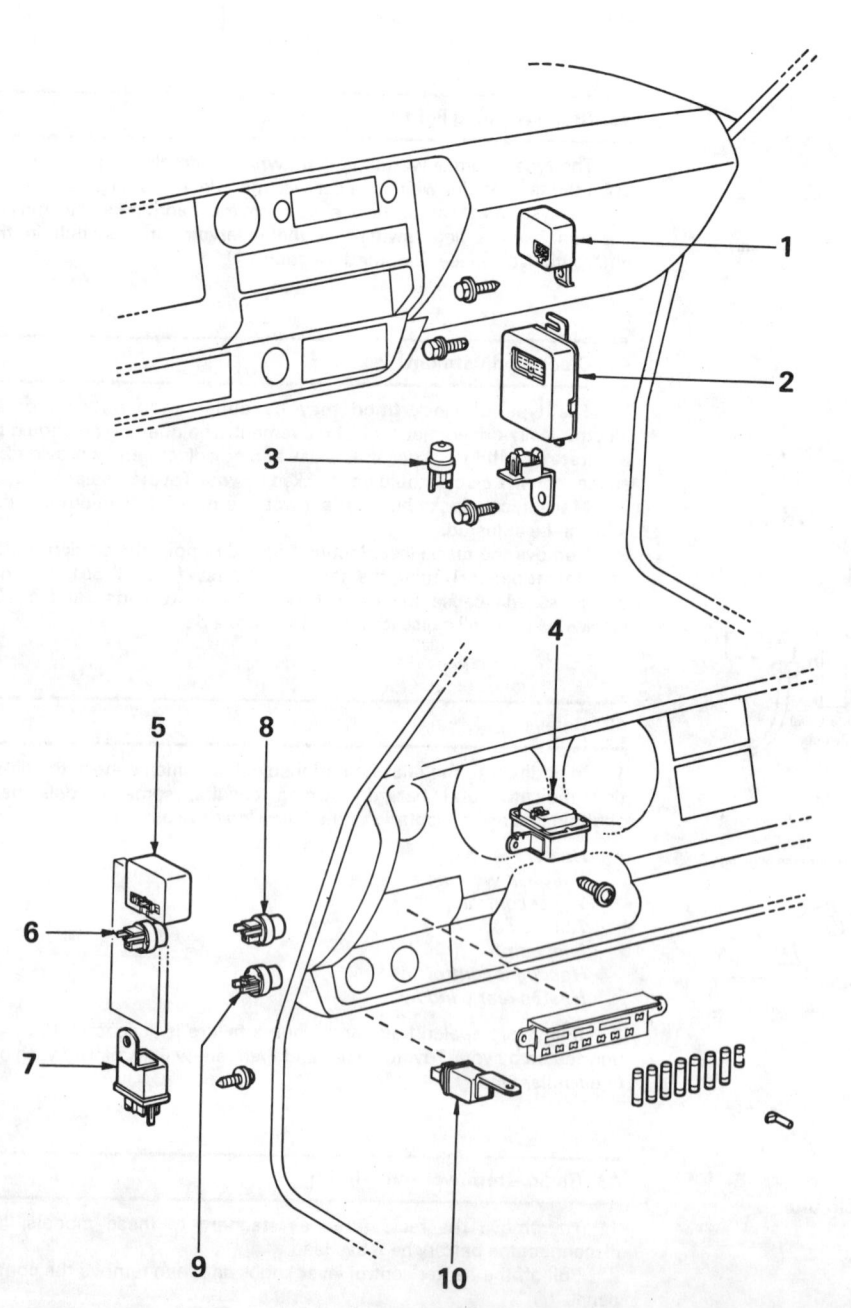

Fig. 10.56 Location of relays and other components (LHD, North America)

1   Seat belt warning relay
2   Emission control computer
3   Heater relay
4   Direction indicator and hazard warning flasher relay
5   Wiper control relay
6   Tail light control relay
7   Fuel pump relay
8   Headlight control relay
9   Heated rear window relay
10 Anti-theft (door unlocked) warning buzzer

Chapter 10 Electrical system

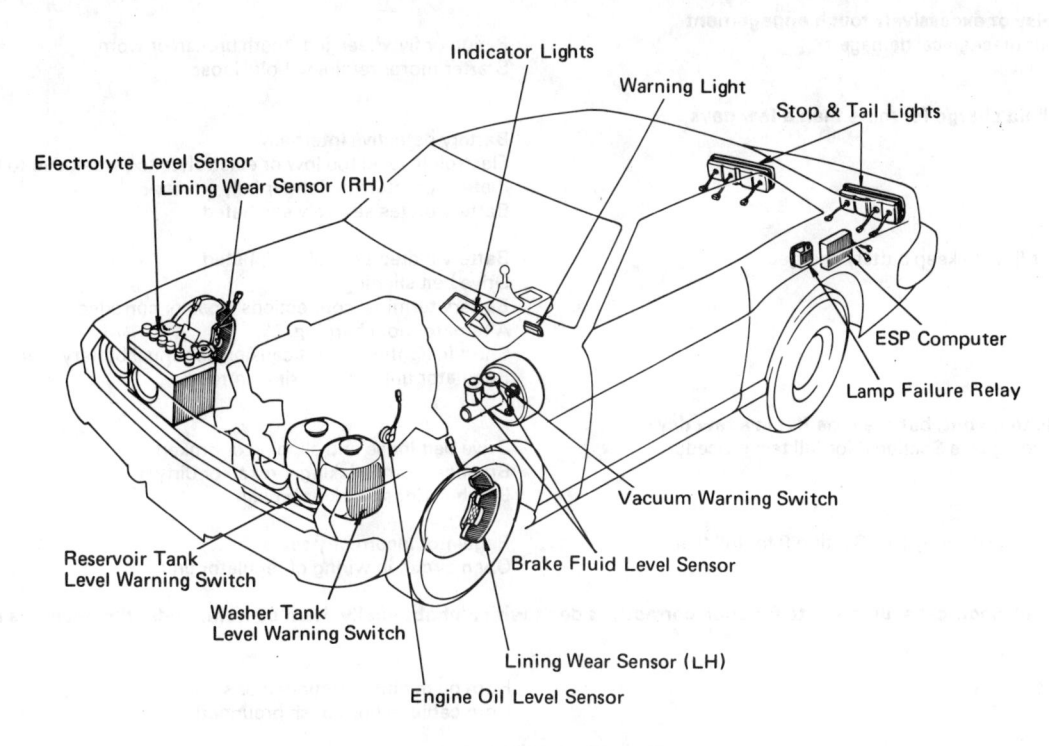

Fig. 10.57 ESP warning system

control shaft sleeves.
4 Remove the cover from the fuse block.
5 Extract the fixing screws and withdraw the cluster finish panel.
6 Remove the radio mounting screws and withdraw the radio far enough to be able to disconnect the power lead (at the in-line fuse holder), the aerial lead, and the speaker wires. Remove the radio completely.
7 Install by reversing the removal operations. If a new aerial or reciever has been fitted, the aerial must be trimmed once the receiver is installed. To do this, fully extend the aerial and tune in to a weak station near 1400 kHz. Then turn the trim spindle on the receiver until maximum volume and clarity is obtained.
8 Fit the remaining control knobs and panels.

44 ESP system

1 The Electrical Sensor Panel System is fitted to some later models, depending upon operating territory.
2 The system is basically a means of constantly monitoring certain vehicle components in order to warn the driver of wear, bulb failure or or low fluid level conditions. A relay device (designated the ESP computer) operates indicator lights showing the prevailing conditions of various components which could have an effect on safety. The system also incorporates a warning light which alerts the driver when an unsafe indication occurs.
3 An integral bulb checking device is incorporated to enable the driver to periodically check that a circuit bulb has not blown.

45 Fault diagnosis – electrical system

Symptom	Cause
Starter motor fails to turn engine	
No electricity at starter motor	Battery discharged
	Battery defective internally
	Battery terminal leads loose or earth lead not securely attached to body
	Loose or broken connections in starter motor circuit
	Starter motor switch or solenoid faulty
Electricity at starter motor: faulty motor	Starter brushes badly worn, sticking or brush wires loose
	Commutator dirty, worn or burnt
	Starter motor armature faulty
	Field coils earthed
Starter motor turns engine very slowly	
Electrical defects	Battery in discharged condition
	Starter brushes badly worn, sticking, or brush wires loose
	Loose wires in starter motor circuit
Starter motor operates without turning engine	
Mechanical damage	Pinion or flywheel gear teeth broken or worn

Chapter 10 Electrical system

Starter motor noisy or excessively rough engagement
Lack of attention or mechanical damage

    Pinion or flywheel gear teeth broken or worn
    Starter motor retaining bolts loose

Battery will not hold charge for more than a few days
Wear or damage

    Battery defective internally
    Electrolyte level too low or electrolyte too weak due to leakage
    Plate separators no longer fully effective
    Battery plates severely sulphated

Insufficient current flow to keep battery charged

    Battery plates severely sulphated
    Drive belt slipping
    Battery terminal connections loose or corroded
    Alternator not charging
    Short in lighting circuit causing continual battery drain
    Regulator unit not working correctly

Ignition light fails to go out, battery runs flat in a few days
Alternator not charging (see Section 6 for full test procedure)

    Drive belt loose and slipping or broken
    Brushes worn, sticking, broken or dirty
    Brush springs weak or broken

Regulator fails to work correctly (see Section 9 for full test procedure)

    Regulator incorrectly set
    Open circuit in wiring of regulator unit

Failure of individual electrical equipment to function correctly is dealt with alphabetically, item by item, under the headings listed below:

Horn
Horn operates all the time

    Horn push either grounded or stuck down
    Horn cable to horn push grounded

Horn fails to operate

    Blown fuse
    Cable or cable connection loose, broken or disconnected
    Horn has an internal fault

Horn emits intermittent or unsatisfactory noise

    Cable connections loose
    Horn incorrectly adjusted

Lights
Lights do not come on

    If engine not running, battery discharged
    Light bulb filament burnt out or bulbs broken
    Wire connections loose, disconnected or broken
    Light switch shorting or otherwise faulty

Lights come on but fade out

    If engine not running, battery discharged

Lights give very poor illumination

    Lamp glasses dirty
    Lamp badly out of adjustment
    Incorrect bulb with too low wattage fitted
    Existing bulbs old and badly discoloured

Lights work erratically – flashing on and off, especially over bumps

    Battery terminals or earth connection loose
    Light not grounding properly
    Contacts in light switch faulty

Wipers
Wiper motor fails to work

    Blown fuse
    Wire connection loose, disconnected, or broken
    Brushes badly worn
    Armature worn or faulty
    Field coils faulty

Wiper motor works very slowly and takes excessive current

    Commutator dirty, greasy or burnt
    Armature bearings dirty or unaligned
    Armature badly worn or faulty
    Armature thrust adjuster screw overtightened

Wiper motor works slowly and takes little current

    Brushes badly worn
    Commutator dirty, grease or burnt
    Armature badly worn or faulty

Grid location	Components	Grid location	Components
D-2	Alternator	G-7	Side light
D-1	Ammeter	G-7	Rear light
G-2	Battery	G-7	Indicator light
E-9	Choke coil	F-13	Magnet clutch (air conditioner)
C-8	Cigarette lighter	D-14	Front washer motor
G-8	Clock	C-13	Front wiper motor
C-10	Instrument cluster	G-14	Head light cleaner motor
C-10	Charge warning light	D-12	Heater blower motor
G-10	Fuel gauge	E-14	Rear washer motor
D-10	Oil pressure warning light	C-15	Rear wiper motor
E-10	Oil pressure gauge	G-1	Starter motor
D-10	Tachometer	E-10	Oil pressure sender unit
F-10	Water temperature gauge	D-16	Radio
G-9	Heated rear window	G-2	Regulator
E-3	Distributor	E-12	Resister
C-7	Flasher (RHD)	G-6	Rheostat (for Australia)
D-7	Flasher (LHD)	D-12	Air conditioner relay
G-10	Fuel sender	D-9	Heated rear window relay
B-4	Fuse box	C-14	Headlight cleaner relay
C-2, D-2	Fusible link	C-13	Heater main relay
D-4	Horn	C-4	Light control relay (for head light)
G-13	Idle up solenoid	C-5	Light control relay (for tail light)
D-3	Ignition coil	C-11	Valve check relay (for Australia)
E-8	Interior light and switch	D-13	Wiper control relay
F-12	Back reversing lights, LH.RH	E-16	Stereo tape player
C-11	Brake warning light	F-16	Speaker, LH
D-6	Cigarette lighter light	G-16	Speaker, RH
D-8	Door courtesy light	F-9	Back door switch
D-5, E-5	Front clearance lights	E-12	Reversing lamp switch
D-6	Glove box light	D-11	Brake fluid level warning switch (for Australia)
E-4	Head lights	C-9	Heated rear window switch
G-4	High beam indicator lights	E-11	Differential switch (for EEC)
F-6	Illumination lights	G-4	Dimmer switch
G-5	Number plate lights	E-9	Door switch
F-11	Parking brake warning light (for Australia)	F-14	Front wiper control switch
F-8	Interior light	C-6	Hazard warning light switch
D-7	Red hazard indicator light (for W. Germany and France)	G-12	Heater blower switch
		F-3	Horn switch
G-8	Stop lights, LH.RH	A-1	Ignition switch
E-5	Tail lights, LH.RH	F-5	Light control switch
E-6	Transmission indicator light	E-1	Neutral safety switch
	Turn signal (LH):	D-10	Oil pressure switch
G-6	Front light	E-11	Handbrake switch (except Australia)
G-6	Side light	F-11	Handbrake switch (for Australia)
G-7	Rear light	F-15	Rear wiper control switch
G-7	Indicator light	D-8	Stop light switch
	Turn signal (RH):	E-7	Turn signal switch
G-7	Front light	F-10	Water temperature sender unit

Colour code

B	Black
G	Green
L	Light blue
O	Orange
R	Red
LG	Light green
W	White
Y	Yellow

The first letter indicates the basic colour and the second letter the spiral colour.

Broken lines in the wiring diagram are for varied models or optional equipment.

Fig. 10.58 Key to wiring diagram – all early models except North America

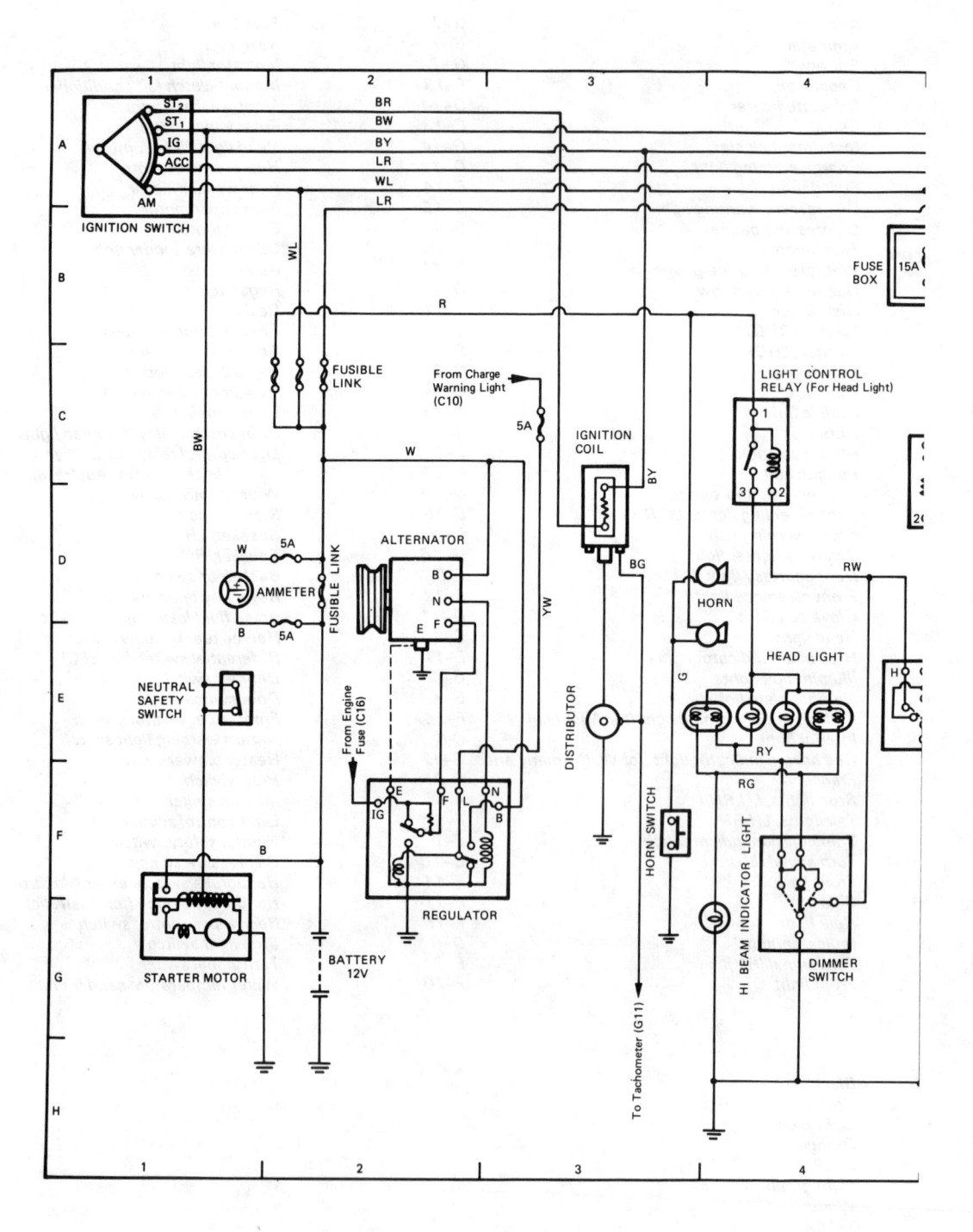

Fig. 10.58 Wiring diagram – all early models except North America

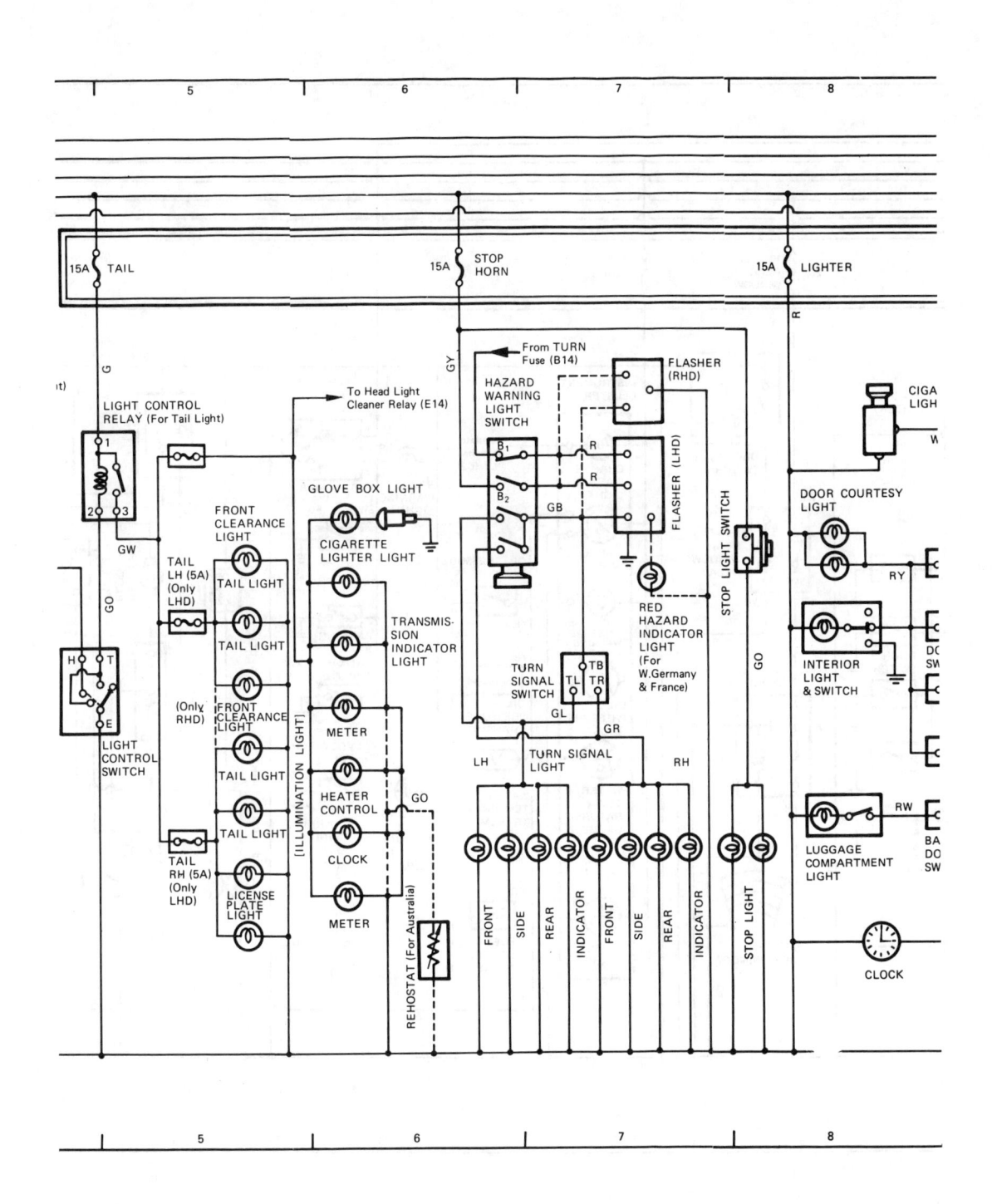

Fig. 10.58 Wiring diagram – all early models except North America – continued

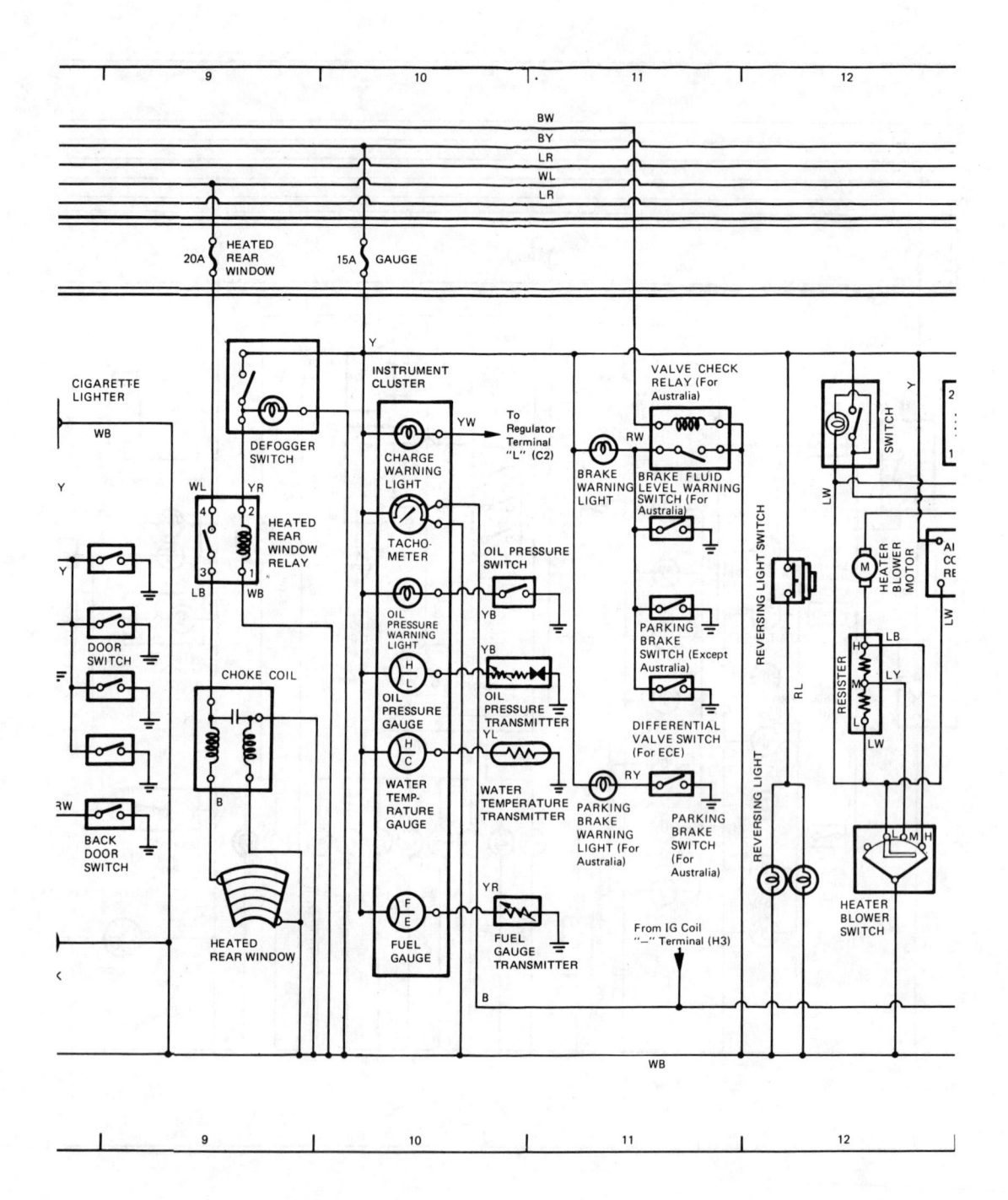

Fig. 10.58 Wiring diagram – all early models except North America – continued

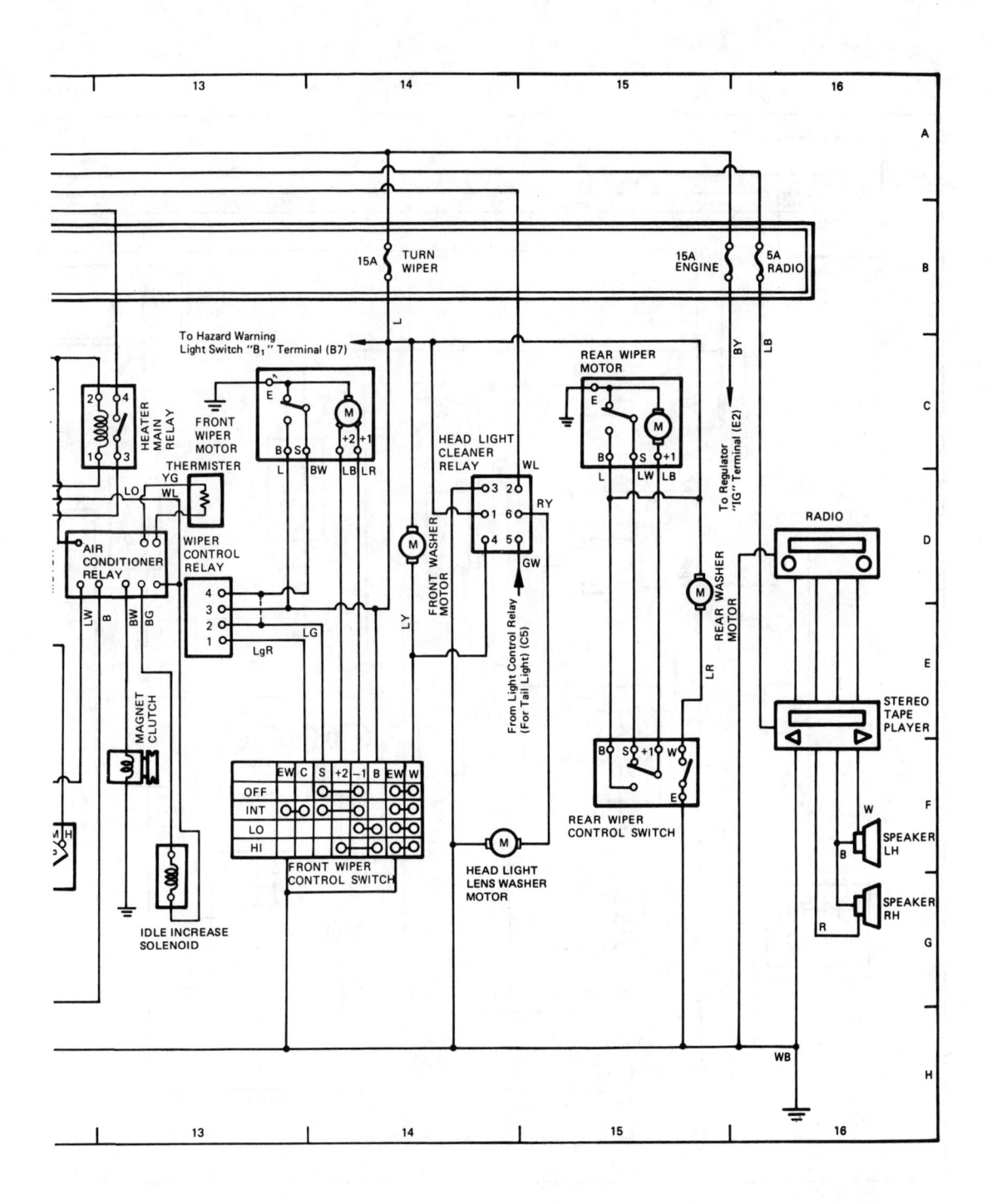

Fig. 10.58 Wiring diagram – all early models except North America – continued

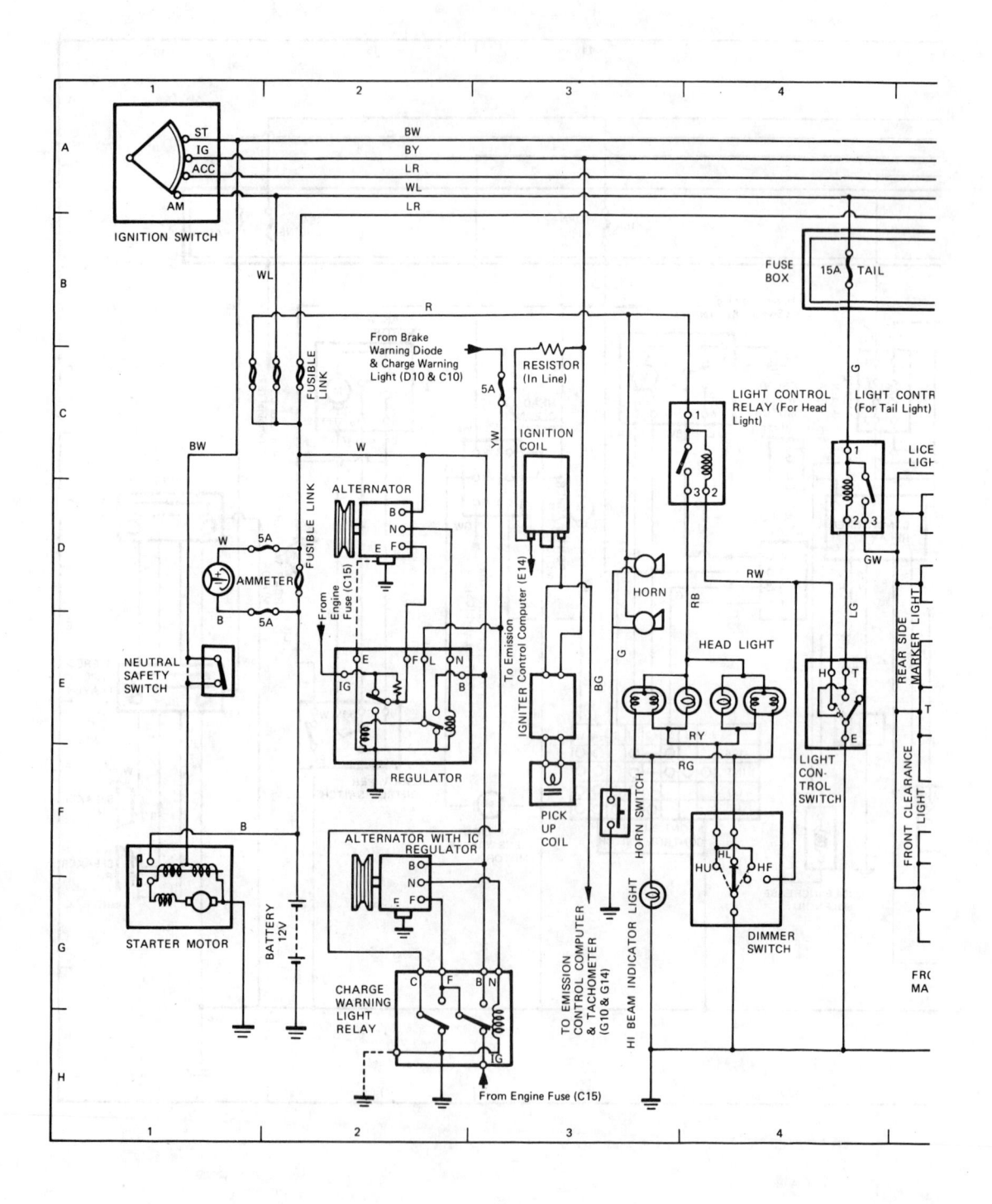

Fig. 10.59 Wiring diagram – North American models

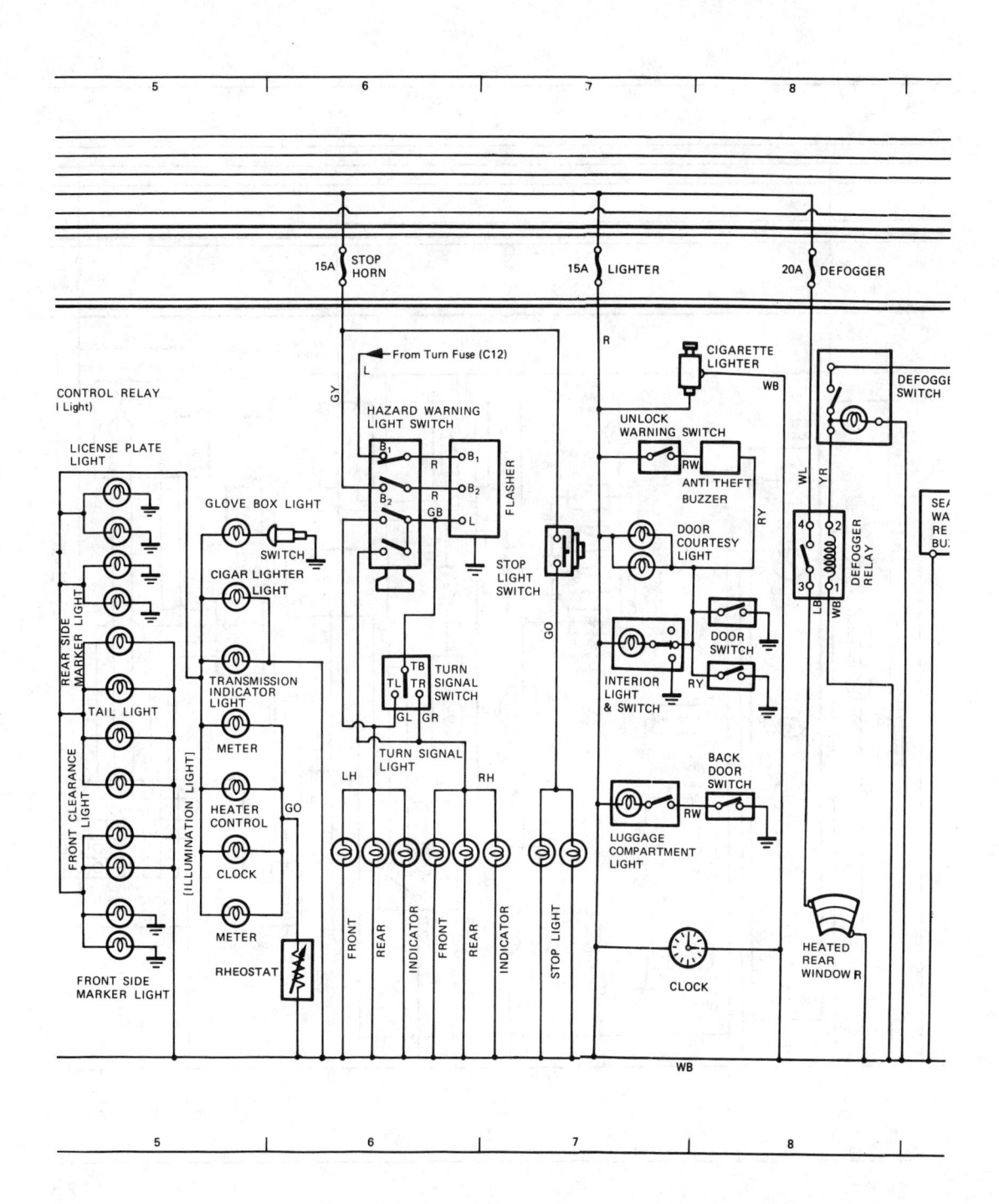

Fig. 10.59 Wiring diagram – North American models – continued

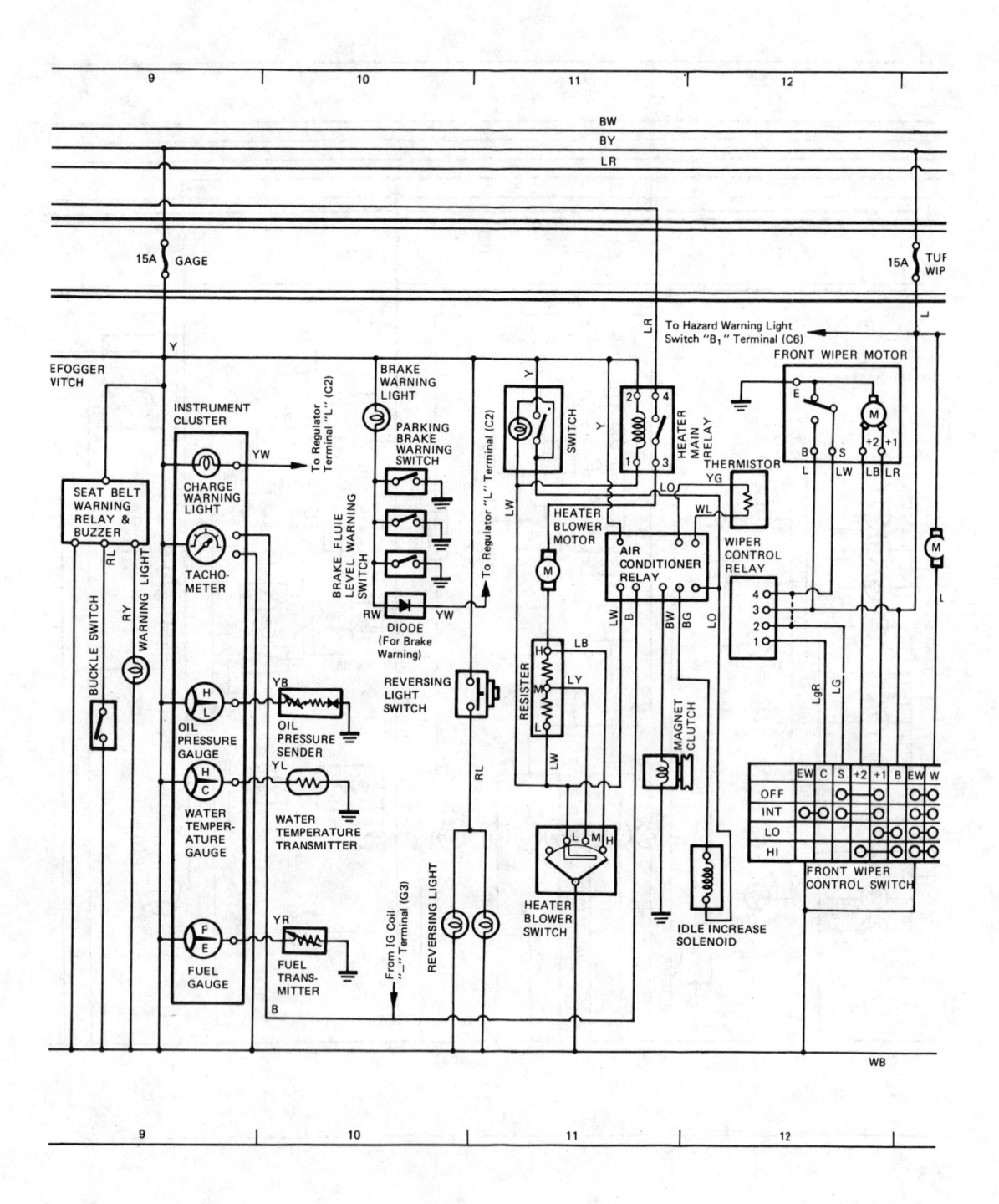

Fig. 10.59 Wiring diagram – North American models – continued

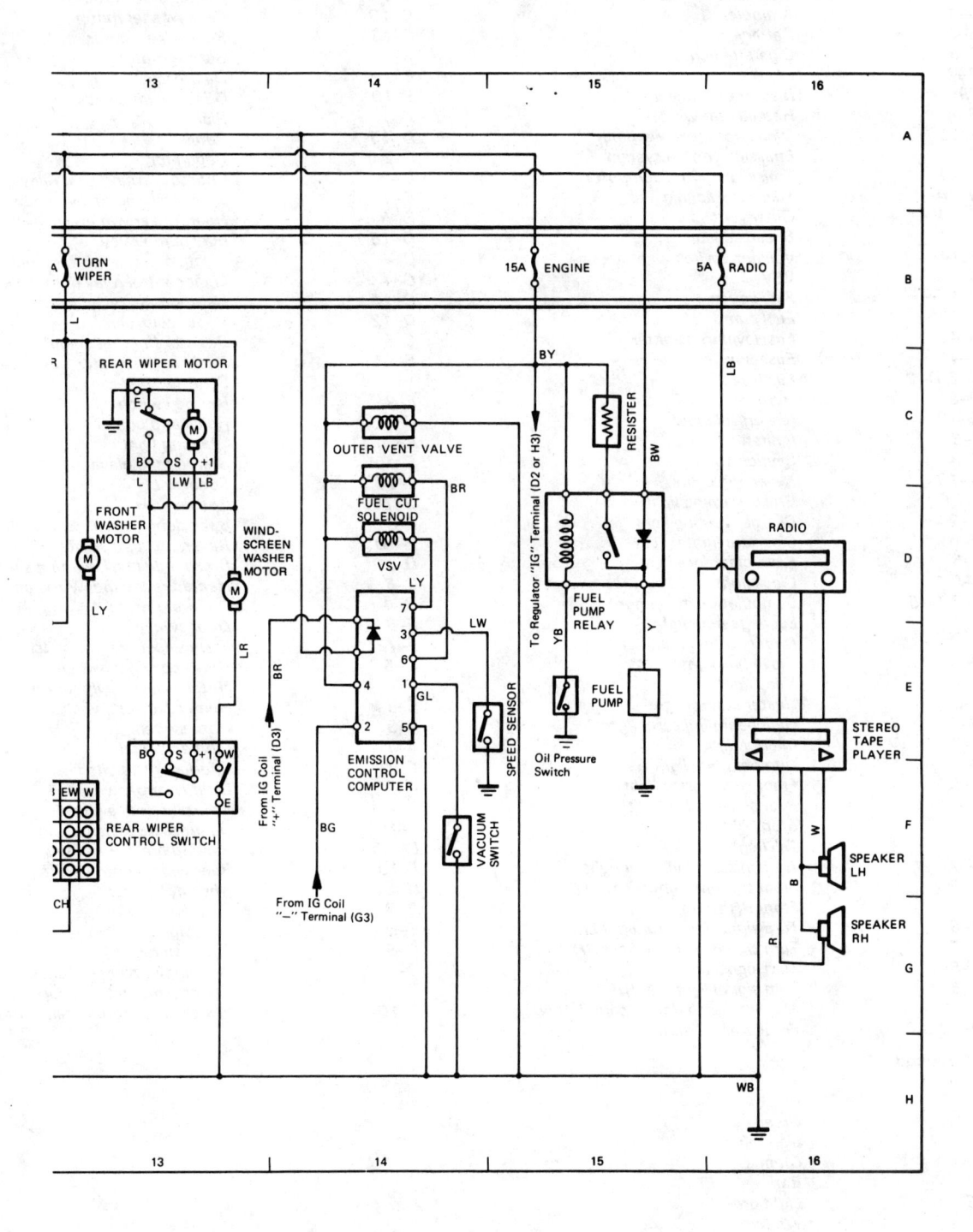

Fig. 10.59 Wiring diagram – North American models – continued

240

Grid location	Components		
D–2	Alternator	C–12	Front wiper motor
F–2	Alternator with IC regulator	D–11	Heater blower motor
D–1	Ammeter	D–13	Rear washer motor
G–2	Battery	C–13	Rear wiper motor
C–8	Cigarette lighter	G–1	Starter motor
G–7	Clock	E–9	Oil pressure gauge
C9	Instrument cluster	E–10	Oil pressure switch
G–8	Heated rear window	F–3	Pick up coil
E–10	Diode (for brake warning)	D–16	Radio
	Emission control system:	F–2	Regulator
E–14	Emission control computer	G–2	Charge warning light relay
D–14	Fuel cut solenoid	D–11	Air conditioner relay
C–14	Outer vent valve	D–8	Heated rear window relay
E–14	Speed sensor	D–15	Fuel pump relay
F–14	Vacuum switch	C–4	Light control relay (for head lights)
D–14	VSV	C–4	Light control relay (for tail lights)
G–9	Fuel gauge	C–11	Heater main relay
E–15	Fuel pump	D–12	Wiper control relay
G–10	Fuel level transmitter	C–5	Resister (for fuel pump)
B–4	Fuse box	E–11	Resister (for heater blower motor)
C–2, D–2	Fusible link		Seat belt system:
D–3	Horn	E–9	Buckle switch
G–12	Idle up solenoid	D–9	Relay and buzzer
E–3	Igniter	E–9	Warning light
C–3	Ignition coil	E–16	Stereo tape player
G–10	Reversing lights	F–16	Speaker LH
C–10	Brake warning light	G–16	Speaker RH
D–9	Charge warning light	F–8	Back door switch
D–5	Cigarette lighter light	E–10	Reversing light switch
F–5	Front clearance light	D–10	Brake fluid level warning switch
F–5	Clock light	C–8	Heated rear window switch
E–5, G–5	Combination meter light	G–4	Dimmer switch
F–7	Luggage boot light	E–8	Door switch
D–7	Door courtesy light	F–12	Front wiper control switch
D–5	Glove box light	D–5	Glove box light switch
E–4	Head lights	C–6	Hazard warning light switch
F–5	Heater control light	G–11	Heater blower switch
G–3	High beam indicator light	F–3	Horn switch
E–7	Interior light	A–1	Ignition switch
C–5	Number plate light	E–7	Interior light switch
G–5	Front side marker light	F–4	Light control switch
E–5	Rear side marker light	E–1	Neutral safety switch
G–7	Stop light	F–15	Oil pressure switch
E–5	Tail light	C–10	Parking brake warning switch
E–5	Transmission indicator light	F–13	Rear wiper control switch
G–6	Front LH turn signal light	G–5	Rheostat switch
G–6	Front RH turn signal light	D–7	Stop light switch
G–6	Turn signal indicator light LH	E–6	Turn signal switch
G–7	Turn signal indicator light RH	D–9	Tachometer
G–6	Turn signal light rear LH	C–8	Warning buzzer (anti-theft)
G–6	Turn signal light rear RH	F–9	Water temperature gauge
F–11	Magnet clutch (for air conditioner)	F–10	Water temperature transmitter
D–13	Front washer motor		

Colour code

B	Black
G	Green
L	Light blue
O	Orange
R	Red
LG	Light green
W	White

The first letter indicates the basic colour and the second the spiral colour.

Broken lines in the wiring diagram are for varied models or optional equipment.

Fig. 10.59 Key to wiring diagram – North American models

Chapter 11 Suspension and steering

For modifications and information applicable to later USA models, refer to Supplement at end of manual

Contents

Fault diagnosis – suspension and steering	28
Front crossmember – removal and installation	9
Front hubs – servicing and adjustment	4
Front anti-roll bar – removal and installation	5
Front suspension strut – removal and installation	8
Front suspension radius rod – removal and installation	6
Front wheel alignment	26
General description	1
Maintenance and inspection	2
Power steering – bleeding	20
Power steering – fluid replacement	21
Power steering – maintenance	19
Power steering pump – removal and refitting	22
Rear anti-roll bar – removal and refitting	14
Rear lateral control rod – removal and installation	13
Rear shock absorbers – removal, testing and installation	3
Rear suspension coil spring – removal and installation	10
Rear suspension lower control arm – removal and installation	11
Rear suspension upper control arm – removal and installation	12
Steering column lock – removal and installation	25
Steering column and shaft – removal, servicing and installation	17
Steering linkage – removal and installation	15
Steering gear (manual) – dismantling and reassembly	23
Steering gear (power-assisted) – dismantling and reassembly	24
Steering gear housing – removal and installation	18
Steering wheel – removal and installation	16
Suspension lower arm – removal and installation	7
Wheels and tyres	27

Specifications

Front suspension

Type .... Independent, MacPherson strut with coil springs and anti-roll bar

Rear suspension

Type .... Live rear axle with coil springs, telescopic shock absorbers, upper and lower suspension arms, and transverse control rod. GT has anti-roll bar

Steering

Type .... Worm and nut, recirculating ball with collapsible column. Power steering optional

Steering angles
Camber .... 1° ± 30' positive
Castor .... 1° 45' ± 30' positive
Steering axis inclination .... 7° 40' ± 30'
Toe-in .... 0.04 in ± 0.04 in (1.0 mm ± 1.0 mm)

Steering gear oil capacity
Manual steering .... 380 to 400 cc
Power steering:
    Pump .... 300 cc
    Pump and gear .... 800 cc

Wheels

Type and size
Coupe .... 5.0 J x 13
Liftback .... 5.5 J x 14

Chapter 11 Suspension and steering

Tyres

Type and size
Coupe	165 SR 13 steel braced radial
Liftback ST and XT	165 SR 14 steel braced radial
Liftback GT	185 x 70 HR 14 steel braced radial

Pressures

	Front	Rear
Coupe	24 lbf in² (1.7 kgf cm²)	27 lbf/in² (1.9 kgf cm²)
Liftback ST and XT	24 lbf in² (1.7 kgf cm²)	27 lbf in² (1.9 kgf cm²)
Liftback GT	24 lbf in² (1.7 kgf cm²)	24 lbf in² (1.7 kgf cm²)

Torque wrench settings

	lbf ft	Nm
Lower balljoint-to-stub axle carrier	65	90
Suspension lower arm-to-crossmember	65	90
Anti-roll bar drop link nuts	16	22
Radius rod-to-suspension lower arm	40	55
Radius rod-to-body bracket	75	104
Crossmember fixing bolts	75	104
Suspension strut piston rod-to-upper support	40	55
Suspension strut upper mounting nuts	23	32
Suspension strut gland nut	95	132
Suspension strut-to-stub axle carrier	80	111

Rear suspension

	lbf ft	Nm
Shock absorber upper mounting	22	30
Shock absorber lower mounting	32	44
Control arm pivot bolts-to-body	100	138
Control arm pivot bolts-to-rear axle	75	104
Lateral control rod-to-rear axle	40	55
Lateral control rod-to-body bracket	50	69

Steering (manual)

	lbf ft	Nm
Steering wheel nut	25	35
Steering column upper bracket	22	30
Steering shaft flexible coupling pinch-bolt	20	28
Worm bearing locknut	70	97
Sector shaft end cover bolts	15	21
Sector shaft adjuster screw locknut	20	28
Steering gear housing-to-body	50	69
Steering drop arm to sector shaft	90	125
Idler arm nut	50	69
Idler arm-to-relay rod	50	69
Track-rod ends-to-body rod	45	62
Steering drop arm-to-relay rod	50	69
Track-rod ends-to-stub axle carrier	50	69
Idler arm-to-body	40	55
Worm bearing adjuster plug locknut	100	138

Steering (power assisted)
As for manual steering where applicable plus the following:

	lbf ft	Nm
Pump pulley nut	38	53
Worm bearing adjuster plug locknut	38	53
Sector shaft screw locknut	38	53
End cover bolts	38	53
Piston/valve assembly cover bolts	38	53
Fluid hose unions	30	41

1 General description

The front suspension is by means of MacPherson struts and an anti-roll bar. The rear suspension is by coil springs with upper and lower control arms and a lateral control rod. Telescopic double-acting shock absorbers are incorporated and the GT model also has an anti-roll bar.

The steering system is of the recirculating-ball, worm and nut type, with a collapsible steering column. Power steering is available as an option.

2 Maintenance and inspection

1 Refer to 'Routine Maintenance' at the beginning of this manual and carry out the maintenance tasks and checks at the specified intervals.
2 Regularly inspect the condition of the balljoint rubber covers for splits or deterioration and renew if necessary.
3 Check the condition of the rear suspension control arm bushes for wear and renew the bushes if there is any evidence of movement.
4 Check the security of all suspension nuts and bolts.

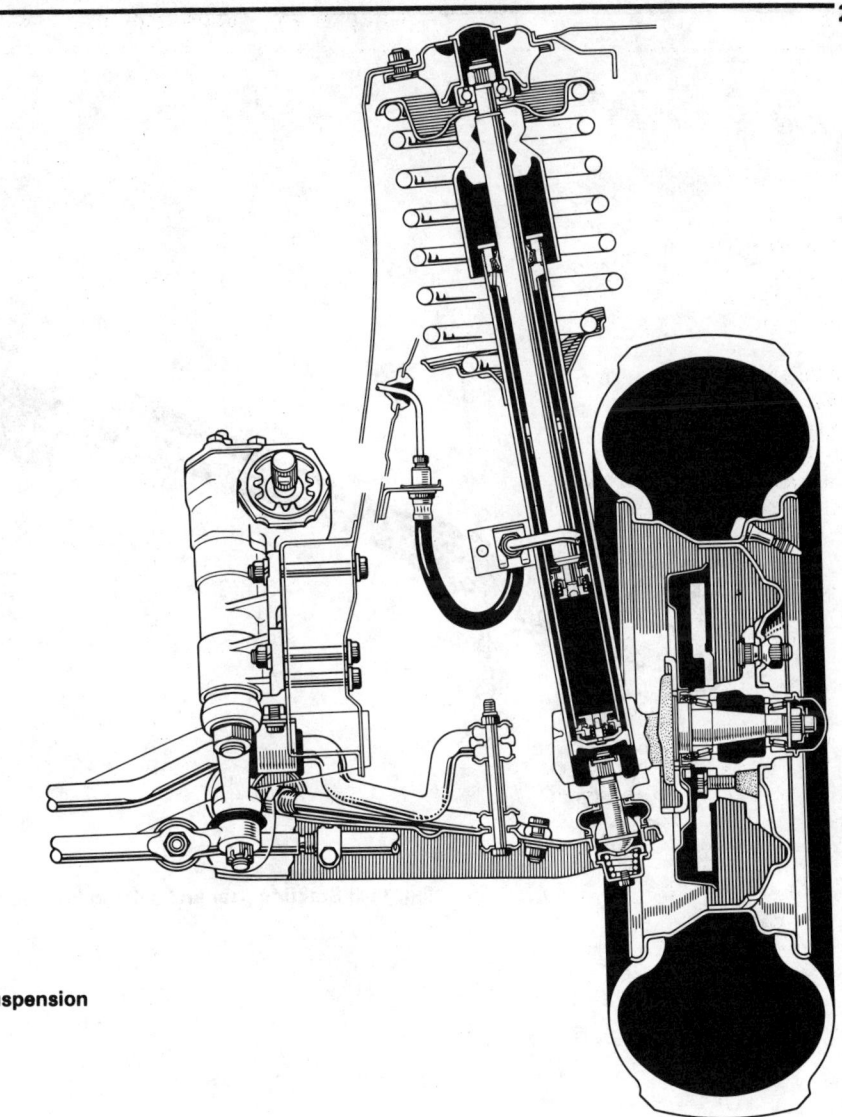

Fig. 11.1 Cross-section view of the front suspension

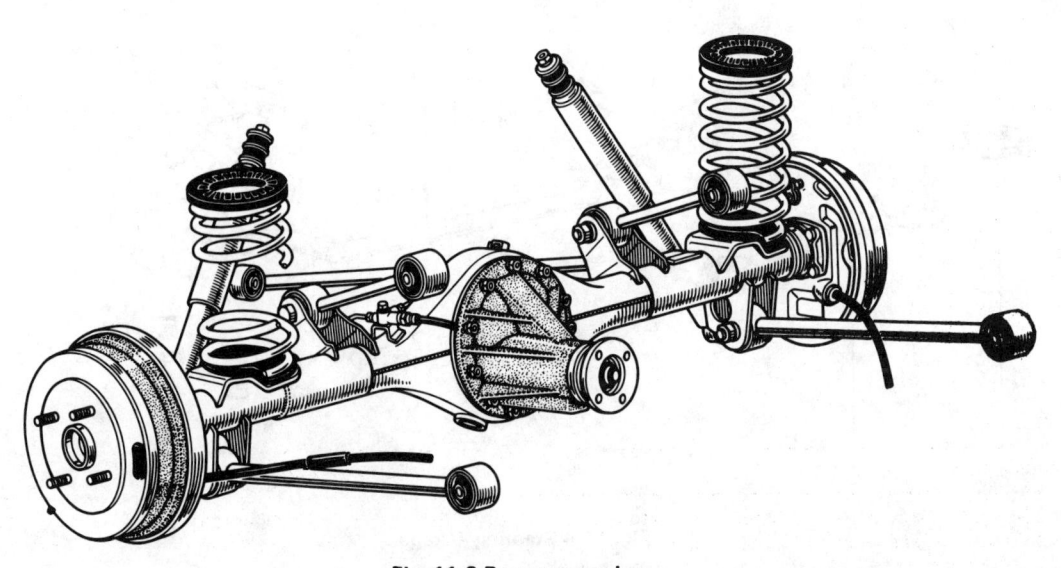

Fig. 11.2 Rear suspension

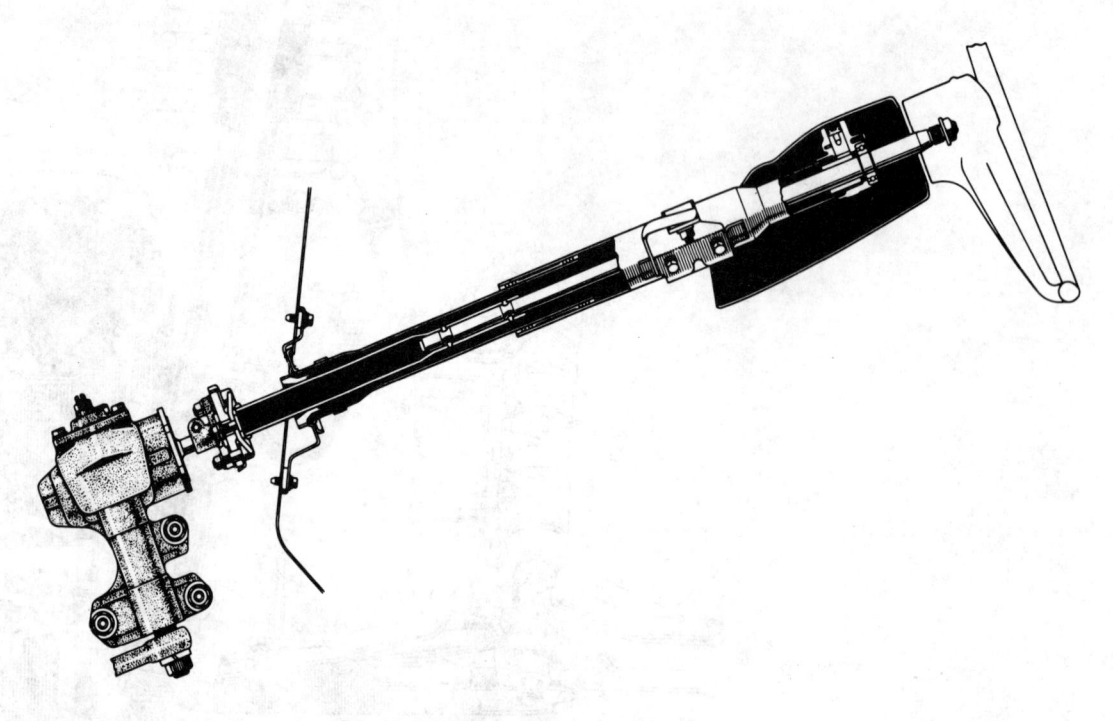

Fig. 11.3 Steering gear and column (manual)

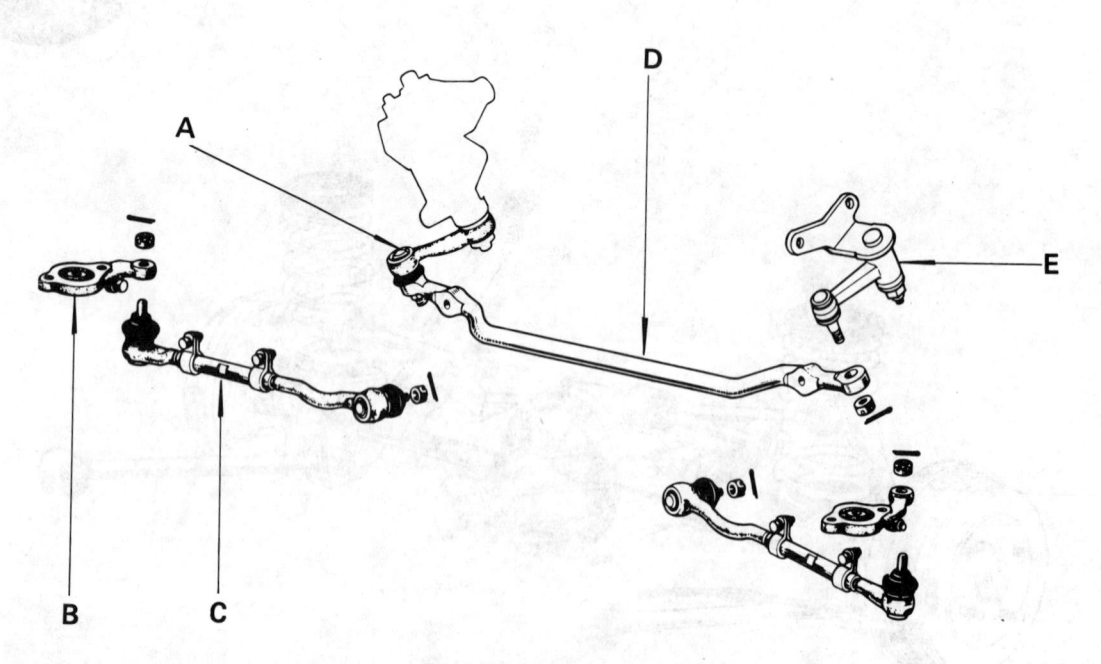

Fig. 11.4 Steering linkage

A Drop arm   B Steering arm   C Track-rod   D Relay rod   E Idler arm

Chapter 11 Suspension and steering

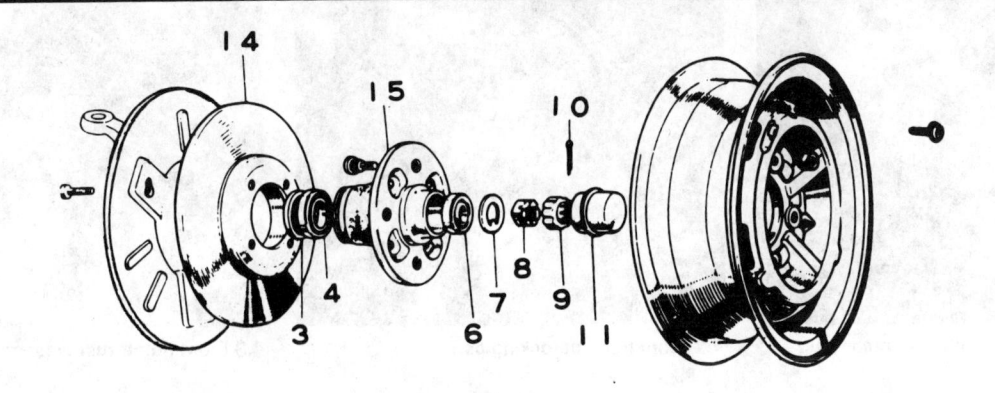

Fig. 11.5 Front hub components

3 Oil seal
4 Inner taper-roller bearing
6 Outer taper-roller bearing
7 Thrust washer
8 Nut
9 Nut retainer
10 Cotter pin
11 Grease cap
14 Disc
15 Hub

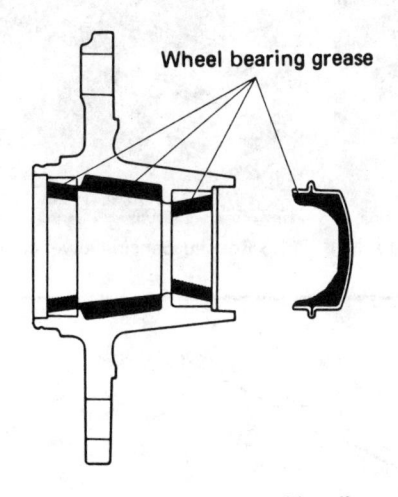

Fig. 11.6 Front hub grease packing diagram

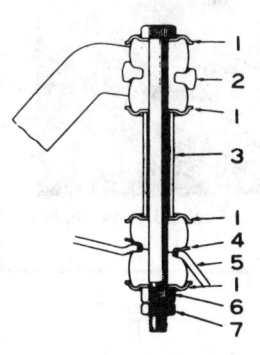

Fig. 11.7 Anti-roll bar drop link attachment

1 Retainer
2 Anti-roll bar
3 Collar
4 Retainer
5 Lower arm
6 Nut
7 Locknut

5 At the intervals specified in 'Routine Maintenance' remove the threaded plug from the front suspension swivel balljoints, screw in a nipple and apply several strokes of a grease gun (photo).

3 Rear shock absorbers – removal, testing and installation

1 The shock absorbers are attached at their lower ends by pivot bolts which pass through rubber bushes inside the shock absorber eye.
2 The upper mountings comprise the threaded shock absorber rod, various rubber cushions and washers and a securing nut and locknut.
3 Removal is simply a matter of releasing the mountings and withdrawing the unit from the car. Do not attempt to remove a shock absorber while the car is jacked-up and the roadwheel is hanging free. If working space is required, jack-up the car under the differential to keep the coil springs under compression.
4 To test the efficiency of a shock absorber, grip the lower mounting in a vice so that the unit is held vertically. Extend and contract the shock absorber (to the full extent of its travel) about ten times. There should be a definite resistance in both directions, otherwise the unit must be renewed. Any sign of oil leakage around the operating rod seal will also indicate the need for a new unit, as no repair is possible.
5 Refitting is a reversal of removal, but tighten the mountings to the specified torque.

4 Front hubs – servicing and adjustment

1 Jack-up the front of the vehicle and remove the roadwheel. Remove the disc pads and then detach the caliper and tie it up out of the way without straining the flexible hose.
2 Tap off the grease cap, remove the split cotter and nut retainer (photo).
3 Unscrew the retaining nut and remove the thrust washer (photo). Pull the hub forward an inch or two and then push back. This will expose the hub outer bearing which may then be removed.
4 Pull the hub assembly straight off the stub axle.
5 Wipe out all old grease from the bearings and hub interior, taking care not to damage the oil seal. Check the bearings and tracks for wear, damage or scoring.
6 If they are in good condition, repack the inside of the hub as shown.
7 If there is evidence of grease seepage onto the discs, drift out the old seal and tap in a new one using a tubular drift.
8 If either the inner or outer bearings require renewal, drift out the tracks with a brass drift and press in the new ones. Where both front hubs are being serviced at the same time, do not mix the bearing components as the race and the track are matched in production.
9 The disc should not be removed from the hub assembly unless it is to be renewed, or refaced as described in Chapter 9.
10 Reassembly is a reversal of dismantling, but the bearings must be adjusted *before* the disc pads are fitted.
11 Tighten the hub nut to 22 lbf ft (30 Nm) rotating the hub at the same time. Unscrew the nut and then tighten it again using finger pressure only.
12 All endfloat should have now been eliminated and the nut retainer and a new cotter pin can be installed.
13 Tap the grease cap into position, install the caliper and roadwheel.

2.5 Front suspension balljoint grease plug

4.2 Front hub nut locking cap

4.3 Front hub thrust washer and outer bearing

6.2 Front suspension radius rod body anchor bracket

7.5 Front suspension track control arm (pivot bolt arrowed)

11.3 Rear suspension lower control arm

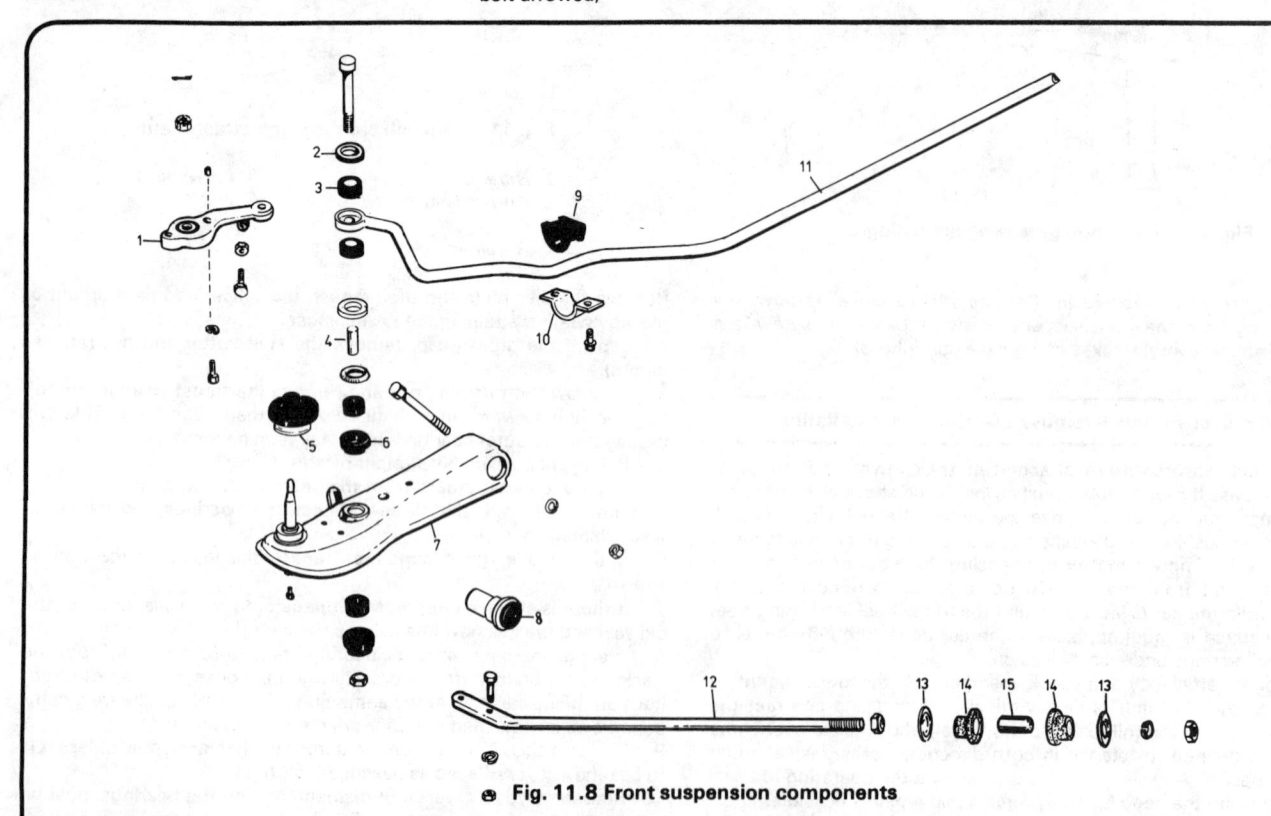

Fig. 11.8 Front suspension components

1	Steering knuckle arm	5	Dust excluding boot	9	Anti-roll bar bracket bush
2	Retainer	6	Retainer	10	Bracket
3	Rubber bush	7	Suspension lower arm	11	Anti-roll bar
4	Collar	8	Suspension arm bush	12	Radius rod

13	Retainer
14	Rubber bush
15	Collar

Fig. 11.9 Radius rod attachment to body bracket

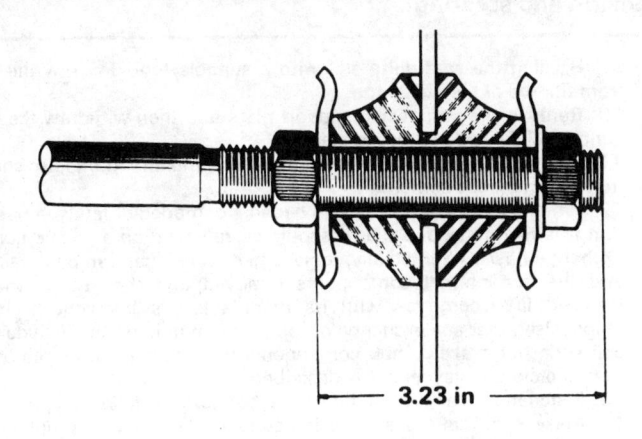

Fig. 11.10 Disconnecting the suspension strut from the stub axle carrier

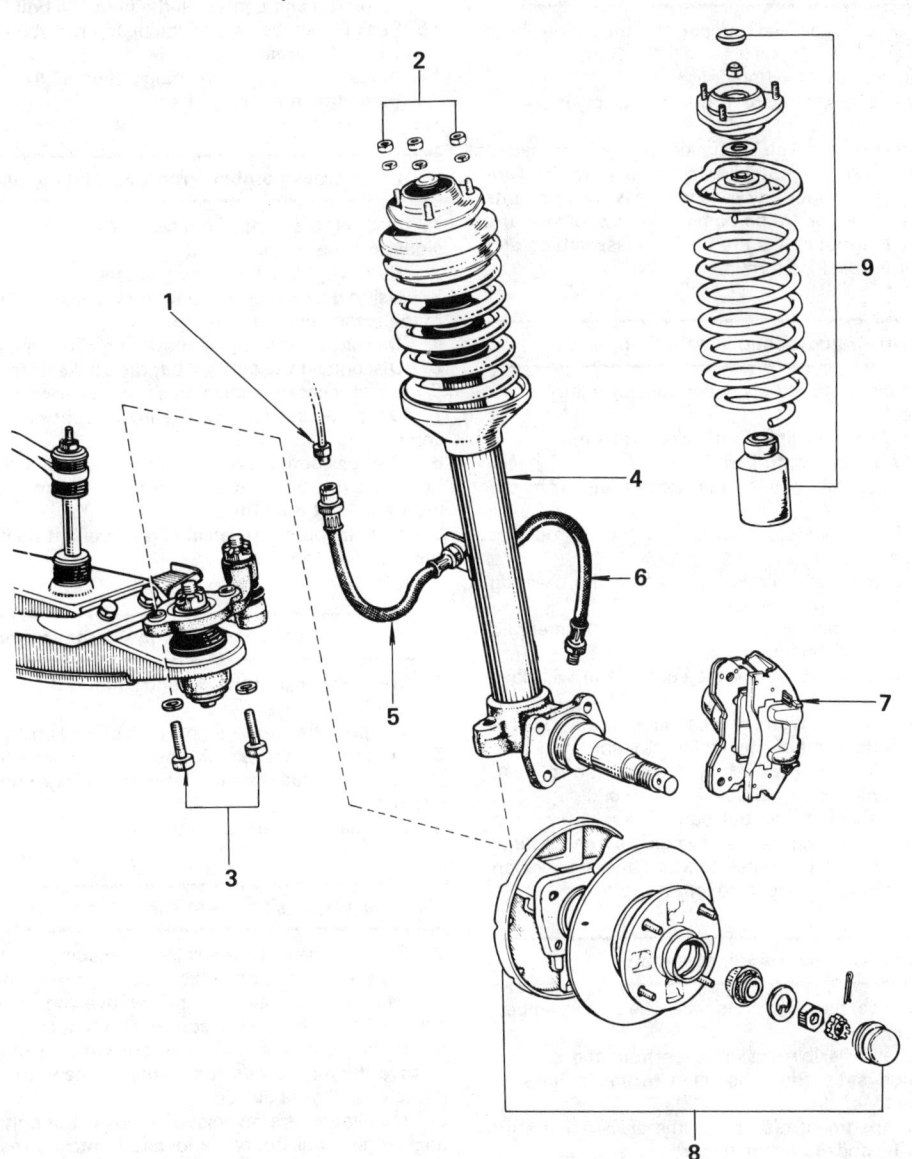

Fig. 11.11 Front suspension strut attachment

1 Brake pipe	4 Shock absorber	7 Brake caliper
2 Nuts	5 Brake hose	8 Front axle hub & backing plate
3 Bolts	6 Brake hose	9 Suspension support, spring seat, spring & dust cover

14 Repeat the operations on the opposite front hub and then lower the car to the ground.

5 Front anti-roll bar – removal and installation

1 Jack-up the front of the car and support it securely under the crossmember.
2 Remove the splash shield from under the engine.
3 At each end of the anti-roll bar, disconnect the drop link from the lower control arms.
4 Remove the anti-roll bar brackets and withdraw the anti-roll bar.
5 Installation is a reversal of removal but tighten the nuts and bolts to the specified torque and ensure that the drop link mountings are correctly assembled.

6 Front suspension radius rod – removal and installation

1 Jack-up the front of the car and support it under the front crossmember.
2 Disconnect the radius rod from the bodyframe (photo).
3 Unbolt the opposite end of the radius rod from the lower control arm and remove the rod.
4 Installation is a reversal of removal. The inner nut on the threaded end of the radius rod is normally staked in position to faciliate reassembly. If it has been moved for any reason or new components fitted, set the connection of the rod to the bodyframe as shown in the diagram. Make sure that the mountings are correctly assembled and tighten nuts and bolts to the specified torque.

7 Suspension lower arm – removal and installation

1 Jack-up the car and support it securely under the crossmember.
2 Remove the roadwheel.
3 Disconnect the anti-roll bar drop link from the control arm.
4 Unbolt the radius rod from the control arm.
5 Remove the pivot (photo) and release the control arm from the suspension crossmember.
6 Using a suitable balljoint separator, disconnect the tie-rod end from the steering knuckle arm.
7 Unscrew and remove the two bolts which secure the steering knuckle arm to the base of the suspension leg.
8 If required, the control arm can be disconnected from the steering knuckle arm again using a suitable extractor.
9 If the control arm balljoint is worn then the control arm will have to be renewed complete.
10 If the control arm inner bush is worn, this can be renewed by pressing out the bush towards the front of the car (control arm in its normally installed position). Install the new bush from the front of the car. A vice and suitable distance pieces are suitable for this work.
11 Installation is a reversal of removal but tighten the control arm pivot bolt finger-tight only until the car is lowered to the ground. Bounce the car up and down several times to settle the suspension and then tighten the pivot bolt to the specified torque.

8 Front suspension strut – removal and installation

1 Jack-up the front of the car and support it under the crossmember.
2 Remove the roadwheel.
3 Disconnect the rigid and flexible brake hoses from the support bracket on the suspension strut. Plug the open hydraulic lines to prevent loss of fluid and entry of dirt.
4 Within the engine compartment disconnect the suspension strut top mounting by unscrewing and removing the three nuts.
5 Unscrew and remove the two bolts which retain the steering knuckle arm to the base of the suspension strut.
6 Withdraw the suspension strut complete with hub assembly from under the fender. The strut will have to be lifted slightly to separate it from the knuckle arm due to the use of positioning collars installed between the two components.
7 Using spring compressors, compress the spring until it is loose within the spring pans.
8 Prise out the bearing dust cover.
9 Holding the seat quite still with a suitable tool, unscrew the nut from the top of the piston rod.
10 Remove the suspension support plate and then withdraw the coil spring (still in its compressed state).
11 Remove the hub and brake components and unbolt the disc shield from the stub axle carrier.
12 When a suspension strut has become damaged or faulty in operation it is recommended that it should be renewed on a reconditioned exchange basis. Alternatively, replacement cartridges can be obtained and the old internal components removed and the new cartridge installed in accordance with its manufacturer's instructions. It is emphasised that any exchange or replacement unit will not include the coil spring or brake or hub components and these must be removed from the old unit as previously described.
13 Installation is a reversal of removal but note the following points.
14 Make sure that the suspension support plate locates properly on the end of the piston rod.
15 Always use a new self-locking nut for securing the support plate to the piston rod and tighten all the nuts and bolts to the specified torque.
16 Pack the space around the piston rod above the support plate with multi-purpose grease.
17 Adjust the front hub bearings (Section 4).
18 Bleed the front brake circuit.

9 Front crossmember – removal and installation

1 Jack-up the front of the car and support it securely under the body sideframe members.
2 Remove the two front roadwheels.
3 Using a jack and wooden block as an insulator, support the weight of the engine under the sump.
4 Disconnect the engine mountings from the crossmember.
5 Disconnect the anti-roll bar drop links from the lower control arms.
6 Disconnect the radius rods from the lower control arms.
7 Unscrew and remove the lower control arm pivot bolts from the crossmember.
8 Depress both suspension lower control arms to release them from the crossmember and then remove the four crossmember securing bolts and lift it from the car.
9 Installation is a reversal of removal but tighten all bolts and nuts to the specified torque.

10 Rear suspension coil spring – removal and installation

1 Raise the rear of the car and support the bodyframe side-members on stands.
2 Support the rear axle on a jack placed under the differential.
3 Disconnect the shock absorber lower mountings and lower the jack under the rear axle until the coil springs and their insulators can be withdrawn.
4 Installation is a reversal of removal.

11 Rear suspension lower control arm – removal and installation

1 Remove the coil springs (see preceding Section).
2 Unscrew and remove the control arm-to-rear axle pivot bolts.
3 Jack-up the rear axle and remove the control arm-to-bodyframe pivot bolts. Withdraw the control arm (photo).
4 If the control arm bushes are worn, a press will be required to remove the old bushes and install the new; this is probably a job best left to your Toyota dealer.
5 Installation is a reversal of removal, but only tighten the pivot bolts finger-tight until the car is lowered onto its roadwheels, then tighten them to specified torque – body end first then rear axle housing end.

12 Rear suspension upper control arm – removal and installation

1 Remove the rear coil springs, as described in Section 10.
2 Remove the control arm pivot bolt at the rear axle casing end.
3 Place a jack under the differential and raise the rear axle a few inches.

249

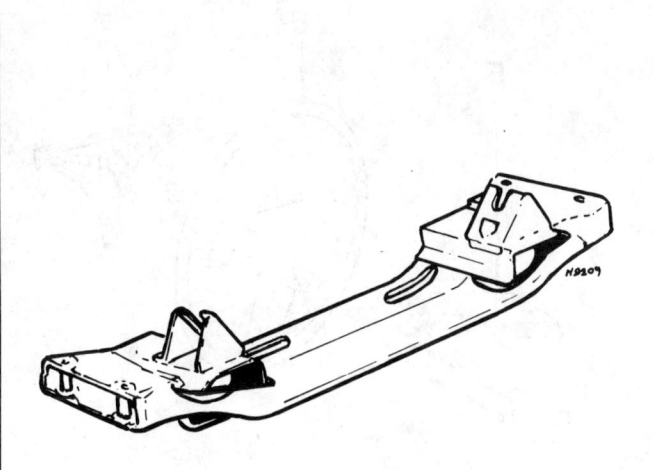

Fig. 11.12 Front crossmember

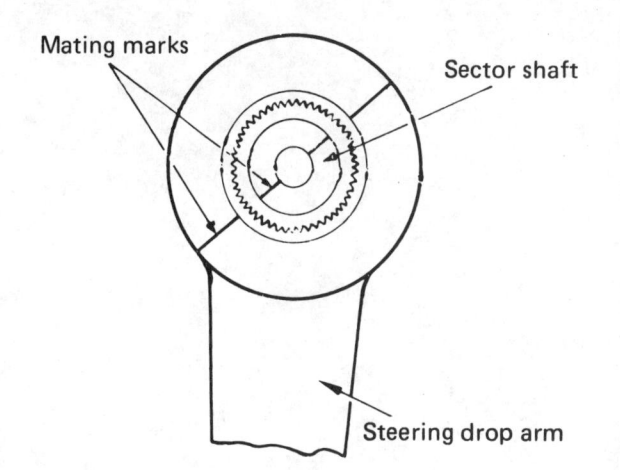

Fig. 11.13 Mating marks on steering drop arm and sector shaft

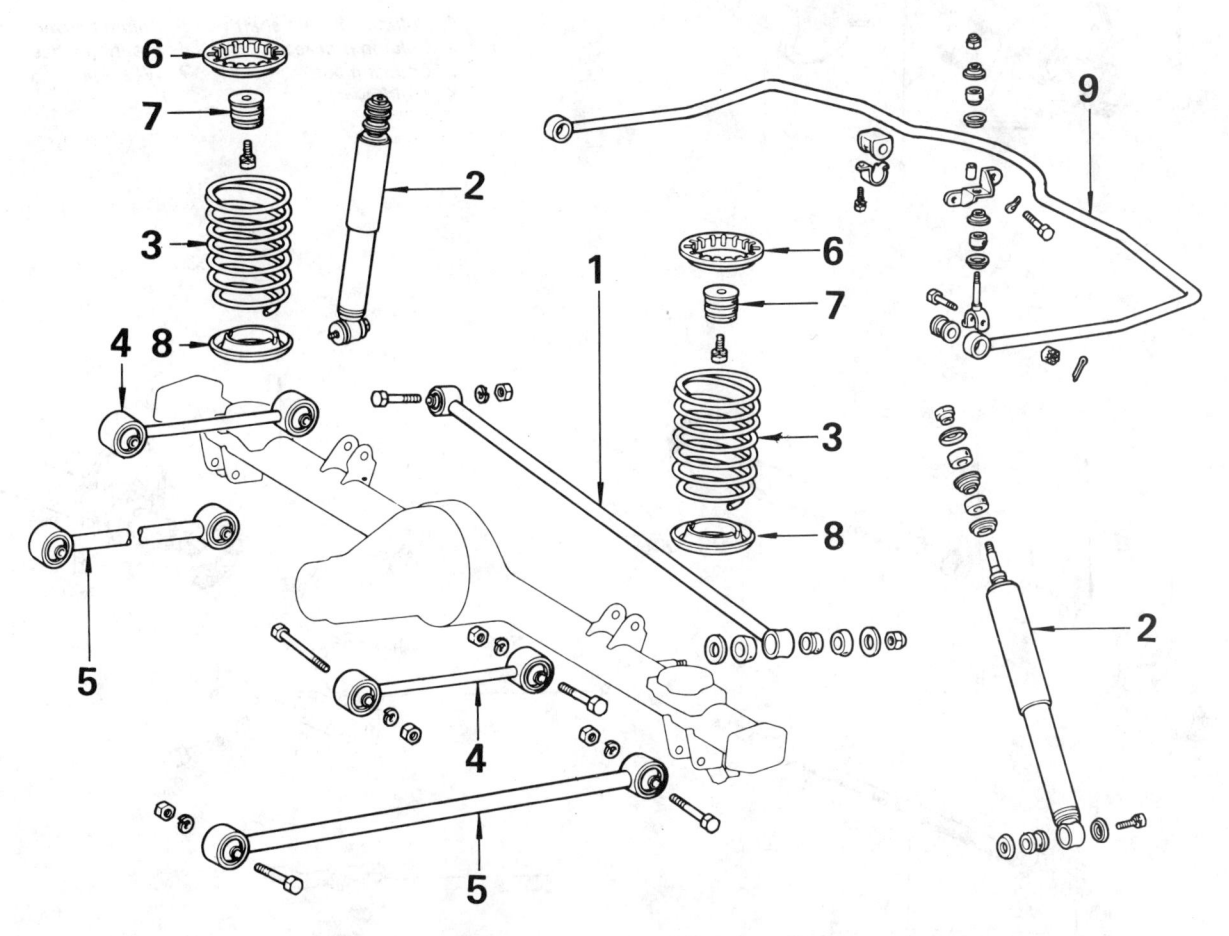

Fig. 11.14 Rear suspension components

1 Lateral control rod
2 Shock absorber
3 Coil spring
4 Upper control arm
5 Lower control arm
6 Upper insulator
7 Spring bumper
8 Lower insulator
9 Rear anti-roll bar (GT models)

Fig. 11.15 Steering column

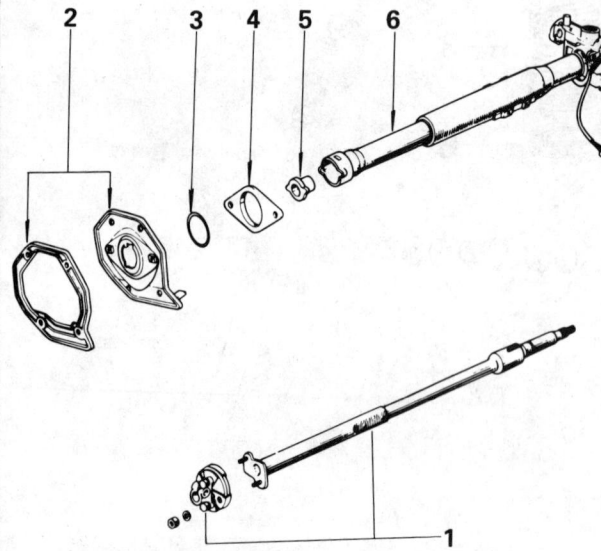

1 Column & main shaft
2 Column bracket
3 Coupling bolt
4 Multi-switch
5 Column cover
6 Steering wheel
7 Wheel pad

Fig. 11.16 Steering column dismantled

1 Shaft and flexible coupling
2 Cover plate and gasket
3 O-ring
4 Plate
5 Seal
6 Column with upper bracket

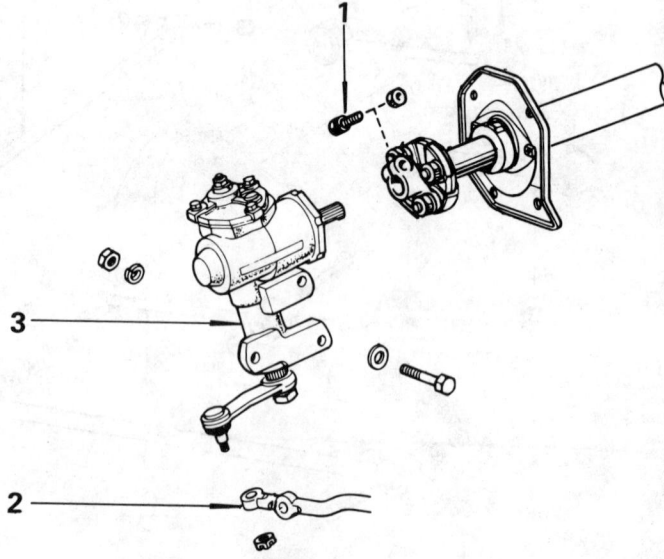

Fig. 11.17 Manual steering gear with flexible coupling

1 Pinch-bolt
2 Relay rod
3 Gearbox

Chapter 11 Suspension and steering

4 Remove the pivot bolt from the other end of the control arm and withdraw the control arm.
5 Repeat the operations described for the lower control arms in paragraphs 4 and 5 of the preceding Section.

13 Rear lateral control rod – removal and installation

1 Remove the rear coil springs, as described in Section 10.
2 At the rear axle attachment end of the control rod remove the nut, washer and bush (photo).
3 Jack-up the rear axle a few inches and then disconnect the lateral control rod from the bodyframe mounting.
4 Renewal of the lateral control rod flexible bushes is carried out in a similar manner to that described for the upper and lower suspension control arms.
5 Installation is a reversal of removal, but tighten the mounting bolts to the specified torque as described in Section 11, paragraph 5.

14 Rear anti-roll bar – removal and refitting

1 It is recommended that the rear anti-roll bar fitted to GT versions is removed with the car standing on its roadwheels. If it is necessary to raise the vehicle to provide working clearance, the rear should be jacked-up evenly and axle-stands placed under the rear axle.
2 Disconnect the end fittings and then the bar clips and rubber insulators. Lift the bar away.
3 Refitting is a reversal of removal, renew the rubber insulators or bushes if they have deteriorated.

15 Steering linkage – removal and installation

1 Make alignment marks on the end of the steering sector shaft and the steering drop arm and after unscrewing the nut (photo) use an extractor to pull the steering drop arm from the sector shaft (photo).
2 Unbolt the idler arm bracket from the bodyframe.
3 Using a balljoint separator, disconnect the tie-rod ends from the knuckle arms.
4 The steering linkage can now be withdrawn and separated into individual components, again using the balljoint separator.
5 Renew the tie-rod end balljoints (photo) if they are worn. Wear in the relay rod balljoints will necessitate renewal of the complete relay rod.
6 The steering idler can be dismantled by removing the self-locking nut and pulling the idler arm from the mounting pivot bolt. Renew the flexible bushes if worn and the idler arm complete, if the balljoint is worn.
7 Reassembly is a reversal of dismantling but tighten the idler nut to the specified torque.
8 Installation is a reversal of removal, but the front wheel alignment (toe-in) must be checked and adjusted, as described in Section 24.

16 Steering wheel – removal and refitting

1 Set the steering wheel in the 'straight-ahead' position.
2 Working from the rear of the steering wheel spokes, unscrew and remove the three screws which secure the horn button assembly.
3 Unscrew and remove the steering wheel securing nut.
4 Make alignment marks on the end of the steering shaft and on the splined centre collar of the wheel so that it can be installed in the same relative position.
5 A puller will almost certainly be needed to remove the steering wheel. Take care that the legs of the tool do not damage the plastic surface of the steering wheel. *Never attempt to jar the wheel from the splined shaft, as the collapsible type steering column may be damaged.*
6 Installation is a reversal of removal but remember to align the marks (made before removal) with the roadwheels in the 'straight-ahead' position.

17 Steering column and shaft – removal, servicing and installation

1 Set the roadwheels in the 'straight-ahead' position.
2 Make alignment marks on the coupling yoke and worm shaft so that they may be refitted in the same relative position.
3 Unscrew and remove the pinch-bolt from the coupling.
4 Remove the steering wheel, as described in the preceding Section.
5 Remove the column upper shrouds and withdraw the multi-switch from the column. Disconnect the ignition switch leads at the connector plug.
6 From the lower end of the steering column release the cover plate.
7 From the upper end of the steering column unbolt the clamp bracket.
8 Withdraw the steering column/shaft assembly into the car interior.
9 To dismantle the assembly, remove bracket from the upper end of the steering column and then extract the circlip and withdraw the bearing.
10 Withdraw the shaft from the outer column (ignition key must be in ON position).
11 Disconnect the column lower flange plate from the cover plate (two bolts) and remove the cover plate.
12 Remove the flexible coupling and tap the coupling yoke from the shaft.
13 Renew any worn or defective components and check particularly the plastic shear pins on the collapsible steering shaft for security and rigidity.
14 Reassembly and installation are reversals of removal and dismantling but make sure that the following operations are carried out.
15 Install the flexible coupling making sure that the ground connection is correctly made.
16 To the lower end of the steering column fit the O-ring and flange plate and install them to the cover.
17 Smear the inside of the dust seal with multi-purpose grease and install it onto the mainshaft.
18 Insert the shaft into the steering column tube.

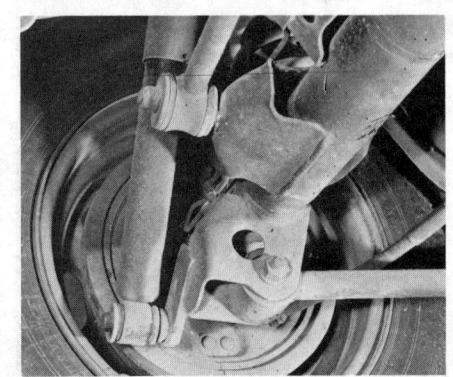

13.2 Rear shock absorber lower mounting

15.1 Steering box

15.5 Relay rod and track-rod inner balljoints

252　Chapter 11 Suspension and steering

19 To the upper end of the steering column fit the bearing (well coated with multi-purpose grease) and the circlip.
20 Fit the column bracket to the column tube (bolts finger-tight only).
21 Install the steering column shaft assembly making sure that the shaft, coupling and worm alignment marks (made before removal) are correctly mated. *Do not tighten the coupling pinch bolt at this stage.*
22 Fasten the steering column upper bracket to the instrument panel (bolts finger-tight only) and install the column cover plate.
23 Tighten the steering column bracket (8.0 mm bolts), and then tighten the bracket to the instrument panel (10.0 mm bolts).
24 Tighten the coupling yoke pinch-bolt.
25 Fit the multi-switch, shrouds and steering wheel.

18 Steering gear housing – removal and installation

1 Mark the relative position of the steering drop arm to the sector shaft, unscrew the nut and then pull off the steering drop arm using a suitable extractor.
2 Mark the relative position of the coupling yoke to the worm shaft and then slacken the pinch-bolt.

3 Unscrew and remove the steering gear housing-to-bodyframe bolts and withdraw the housing.
4 Installation is a reversal of removal but make sure that the marks made before removal are in alignment.
5 On vehicles fitted with power steering it is necessary to disconnect the plug the two oil pipes before starting to remove the steering gearbox. After refitting, the steering system must be bled as described in Section 19.

19 Power steering – maintenance

1 Check that the drivebelt to the pump is in good condition and that when the mid point of the top run is pressed with the thumb, it will deflect by 0.3 to 0.5 in (8 to 12 mm).
2 With the engine warm and idling at 1000 rpm or less, turn the steering from lock-to-lock several times to raise the oil temperature to at least 100°F (40°C). Check the fluid level in the reservoir with the built-in dipstick and also look for emulsification or foaming of the fluid. Any tendency to emulsification or foaming is an indication of air in the system, or of insufficient fluid.

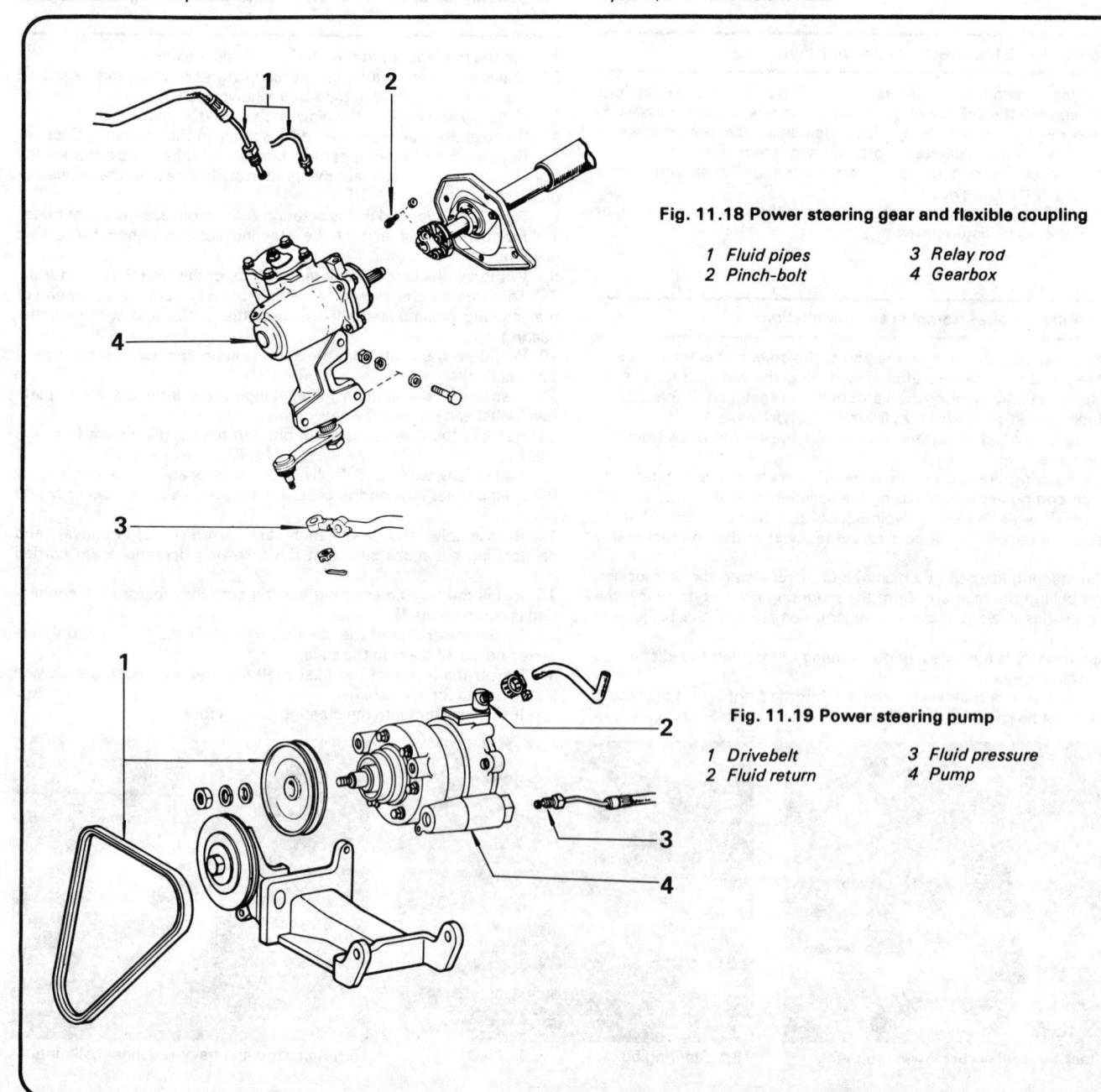

Fig. 11.18 Power steering gear and flexible coupling

1 Fluid pipes
2 Pinch-bolt
3 Relay rod
4 Gearbox

Fig. 11.19 Power steering pump

1 Drivebelt
2 Fluid return
3 Fluid pressure
4 Pump

253

Fig. 11.20 Sectional view of the power steering pump

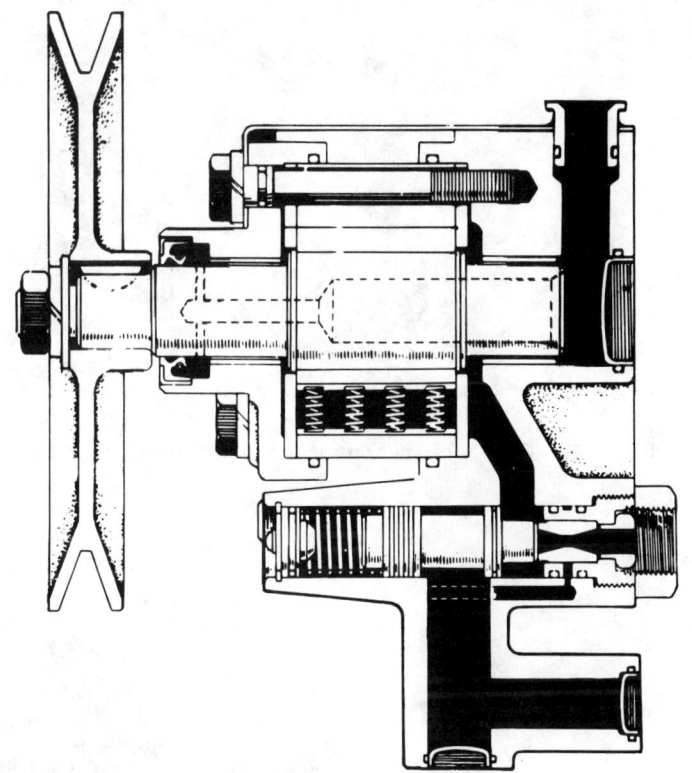

Fig. 11.22 Exploded view of the manual steering gear

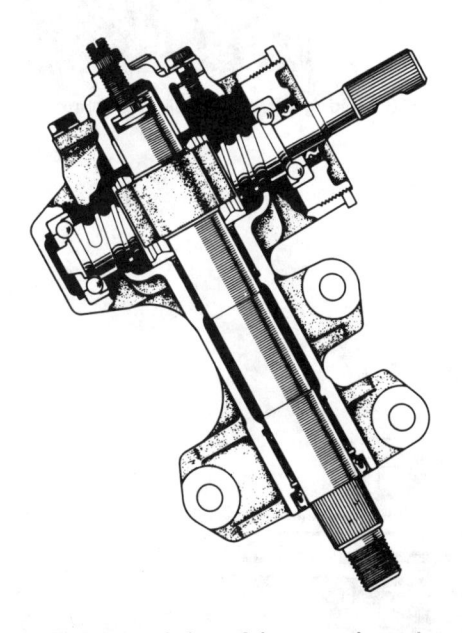

1 Drop arm
2 Locknut
3 Sector shaft and end cover
4 Worm bearing adjuster
5 Wormshaft

Fig. 11.21 Sectional view of the manual steering gear

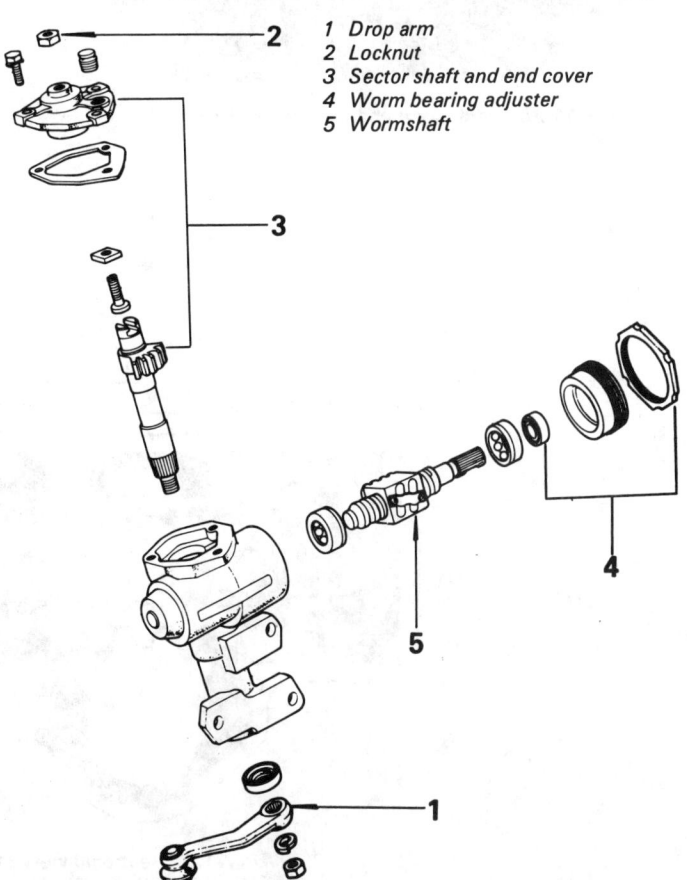

254

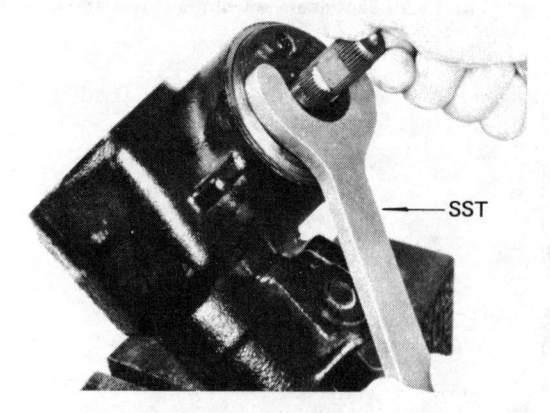

Fig. 11.23 Unscrewing worm bearing adjuster (manual)

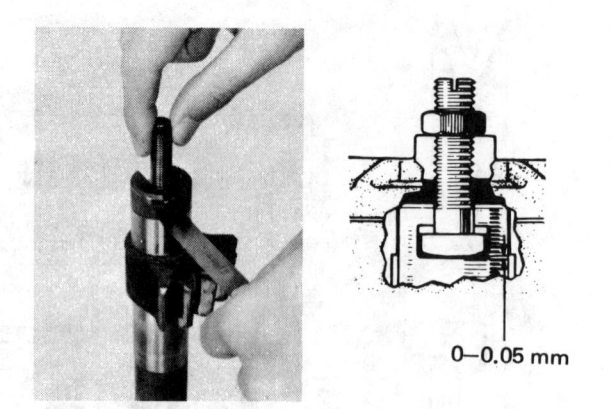

Fig. 11.24 Checking sector shaft thrust clearance (manual)

0–0.05 mm

Fig. 11.25 Sector shaft and nut centred (manual)

Fig. 11.26 Locking end cover adjuster screw

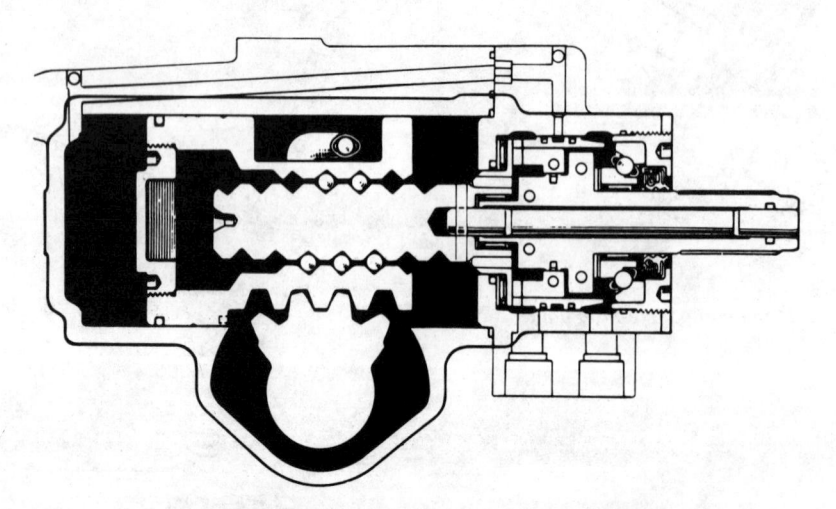

Fig. 11.27 Sectional view of power steering gear

Chapter 11 Suspension and steering

3 Top-up with approved fluid after checking the system for leaks.

20 Power steering – bleeding

1 Check that the fluid reservoir is at the correct level and add approved fluid if necessary.
2 Jack-up the front of the vehicle so that the wheels are clear of the ground and support the body with axle-stands.
3 Turn the steering from lock-to-lock two or three times, then check the fluid level.
4 Start the engine and with it idling, again turn the steering from lock-to-lock two or three times.
5 Stop the engine and lower the front wheels to the ground.
6 Start the engine again and with it idling turn the steering wheel from lock-to-lock several times, centre the steering wheel and stop the engine.
7 Bleeding is complete if the level of fluid in the reservoir has not risen morer than 0.2 in (5 mm) and there is no sign of foaming when the engine has stopped.
8 If there is foaming, or an excessive rise in oil level, repeat the procedure in paragraph 6.

21 Power steering – fluid replacement

1 Raise the front wheels of the vehicle clear of the ground and support the body on axle-stands.
2 Remove the fluid return hose from the oil reservoir, drain the fluid from the reservoir into a suitable container and put the end of the hose into a container to collect the fluid expelled.
3 Turn the steering wheel from lock-to-lock until no more fluid is expelled.
4 Reconnect the return hose to the fluid reservoir, add fresh fluid and bleed the system.

22 Power steering pump – removal and refitting

1 Push down hard on the drivebelt to prevent the pump from rotating and loosen the nut securing the pulley to the pump, remove the pulley and drivebelt.
2 Loosen the clamp on the return hose, prepare to catch the fluid running out and then pull off the return hose.
3 Prepare to catch the fluid running out, then unscrew the pump pressure line connection.
4 Remove the pump fixing bolts and lift the pump clear.
5 Discard the fluid drained from the system.
6 Refitting is the reverse of removal. After fitting the drivebelt and tightening the pulley, tension the belt. Add fluid to the system as necessary and bleed the system.
7 It is not recommended that a faulty pump is overhauled. Renew it with a new or factory reconditioned assembly.

23 Steering gear (manual) – dismantling and reassembly

1 Release the sector shaft adjuster screw locknut.
2 Remove the sector shaft end cover bolts and withdraw the end cover/sector shaft assembly.
3 Pour the oil from the housing into a suitable container.
4 Remove the worm bearing adjuster nut locking ring and then unscrew and remove the adjuster nut from the housing. A special type of wrench will be required for this, having two pins to engage in the holes in the nut.
5 Withdraw the worm/ball nut assembly complete with bearings from the steering housing. *Do not attempt to dismantle the ball nut from the worm.*
6 Inspect all components for wear or damage. If the worm or ball nut is damaged or worn, the complete assembly must be renewed.
7 Renew the two oil seals as a matter of routine.
8 If the bushes or bearing tracks in the housing are worn or damaged, the bearing tracks can be removed using suitable extractors, but renewal of the bushes in best left to your Toyota dealer as they require honing after installation.
9 Commence reassembly by greasing the lips of the oil seals and then dipping each component in clean gear oil.

10 Insert the worm/ball nut assembly into the steering box, followed by the adjuster nut and locking ring.
11 Wind a thin cord around the splined section of the worm shaft and attach a spring balance to the cord. Tighten the adjuster nut until the spring balance registers between 9 and 13 lbs when given an even pull, this is the required worm bearing preload. Without moving the adjuster nut, tighten the locking ring to the specified torque
12 Fit the sector shaft to the end cover. Using a feeler gauge, measure the clearance between the convex face of the adjuster screw and the sector shaft contact face. If the clearance exceeds 0.002 in (0.05 mm) change the thrust washer (sizes available – 0.079, 0.080, 0.082, 0.083, 0.085, 0.087 in – 2.0, 2.04, 2.08, 2.12, 2.16 mm). Insert the sector shaft into the housing, not forgetting to fit a new cover gasket.
13 Set the ball nut to its center position and engage the center tooth of the sector shaft in the center groove of the ball nut.
14 Smear gasket cement into the end cover bolts and install them.
15 Unscrew the sector half adjuster screw fully.
16 Again using the cord and spring balance method, described earlier in paragraph 11, tighten the adjuster screw until the preload registered on the spring balance is between 18 and 24 lbs. Tighten the adjuster screw locknut.
17 With the sector shaft in mesh with the ball nut at its center position, there should be no backlash between the two components. This can be checked by temporarily installing the steering drop arm to the shaft.
18 With reassembly complete, the steering gear housing should be filled with oil of the correct grade, but this operation is best left until the housing is installed in the car.

24 Steering gear (power-assisted) – dismantling and reassembly

1 With the steering gear removed as described in Section 17, clean away external dirt and grease.
2 Extract the end cover bolts and then screw in the adjuster screw which will push off the cover.
3 Using a plastic faced hammer, tap out the sector shaft.
4 Unscrew the power piston/valve housing bolts and remove them. Now hold the piston nut from turning and rotate the wormshaft in a clockwise direction. Withdraw the valve body and power piston assembly. Do not allow the piston nut to run off the wormshaft.
5 *Do not dismantle the valve body or remove the power piston from the wormshaft.*
6 If the valve body assembly exhibits more than an almost imperceptible shake on the wormshaft then the assembly must be renewed.
7 Renew the O-ring in the end cover and check the shaft and tooth surfaces for damage or wear.
8 Check the adjusting screw for thrust (axial) clearance. If it exceeds 0.002 in (0.05 mm) then the staking of the locknut must be relieved and the screw turned while the locknut is held stationary. When the thrust clearance is between 0.001 and 0.002 in (0.03 and 0.05 mm) stake the locknut.
9 Remove the worm bearing plug with a pin wrench, renew the O-ring seal and bearings if necessary.
10 If the needle roller bearings in the main housing are worn, extract the oil and dust seals and withdraw the bearings. When installing the new bearings, make sure that the longer rims of their outer races are facing outward.
11 The top bearing must be flush with the end of the housing bearing recess, while the lower one should be 0.076 in (19.4 mm) from the lower end of the housing. Fit a new O-ring, Teflon ring and spacer.
12 Commence reassembly by fitting the power piston/valve body with a new O-ring. Tighten the cover bolts evenly in diagonal sequence to the specified torque.
13 Now adjust the worm bearing pre-load by turning the threaded plug. The turning torque of the splined wormshaft should be between 3.5 and 5.6 lbf in (40 and 65 Ncm) in both directions. Ideally a torque meter should be used but a cord and spring balance will give a reasonably accurate alternative provided the cord is wound round the shaft a sufficient number of times so that it leaves the shaft 1 in (25.4 mm) from the centre point of the shaft. Tighten the locking ring with a C-spanner or other suitable wrench.
14 Tape over the splines on the sector shaft to prevent damage to the oil seal.

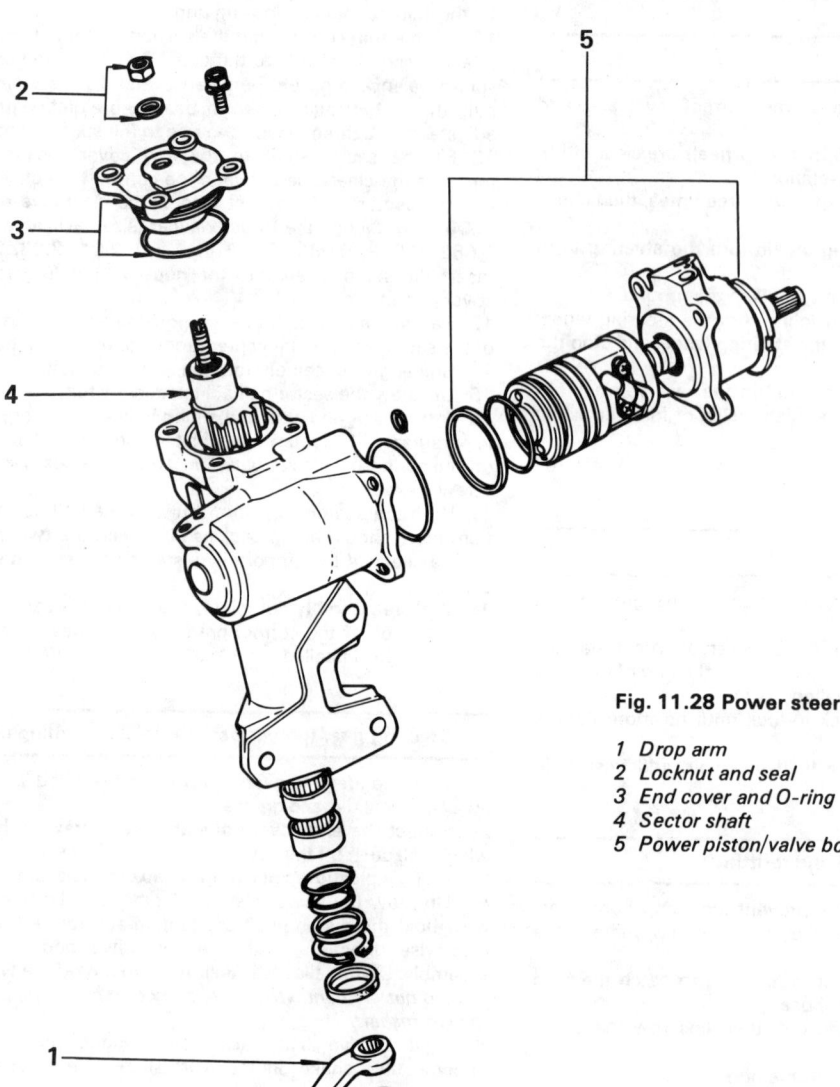

Fig. 11.28 Power steering gear components

1 Drop arm
2 Locknut and seal
3 End cover and O-ring
4 Sector shaft
5 Power piston/valve body

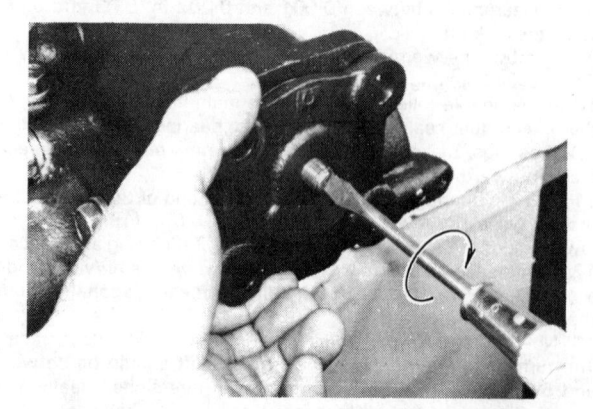

Fig. 11.29 Tightening adjuster screw to remove end cover (power steering)

Fig. 11.30 Tapping out sector shaft (power steering)

Fig. 11.31 Removing power piston/valve body (power steering)

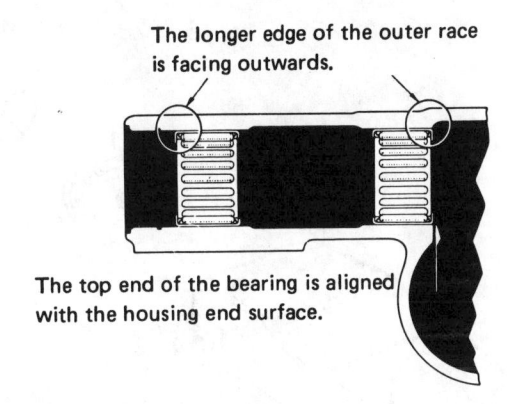

The longer edge of the outer race is facing outwards.

The top end of the bearing is aligned with the housing end surface.

Fig. 11.32 Needle bearing installation diagram (power steering)

Fig. 11.33 Sector shaft and nut centred (power steering)

Fig. 11.34 Steering gear centralising marks (power steering)

Fig. 11.35 Tightening end cover adjuster screw locknut (power steering)

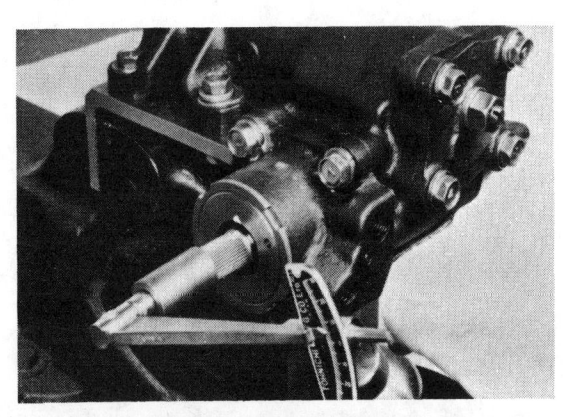

Fig. 11.36 Checking wormshaft bearing pre-load with a torque gauge

258

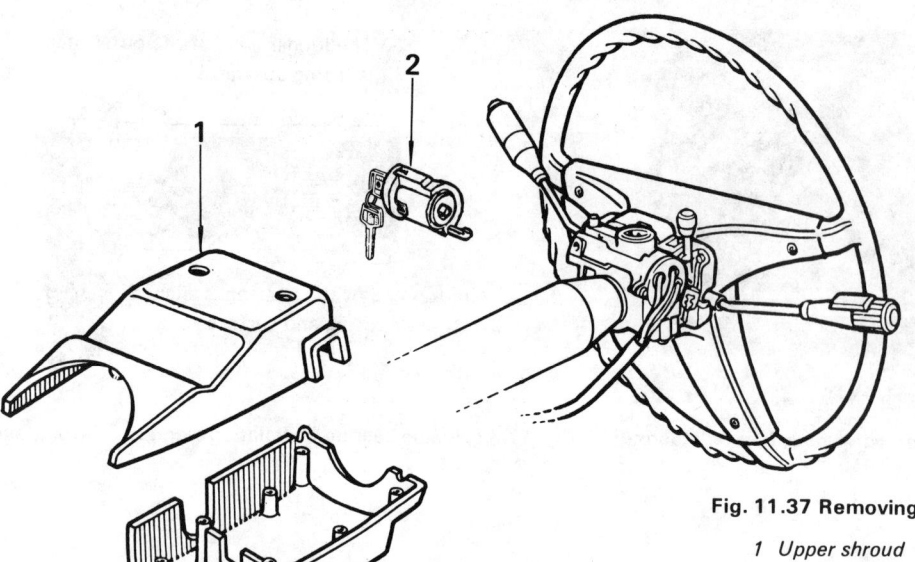

Fig. 11.37 Removing steering column lock cylinder

1 Upper shroud   2 Cylinder

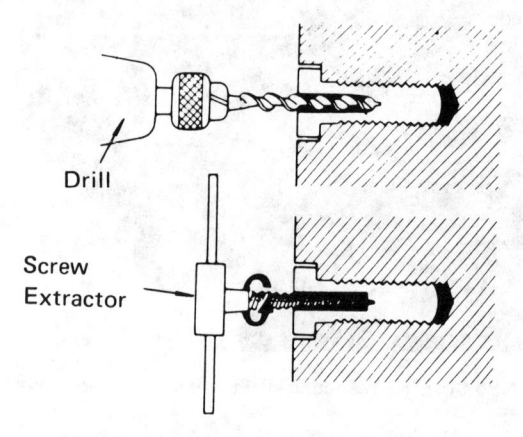

Drill

Screw Extractor

Fig. 11.38 Removing the column lock shear bolts

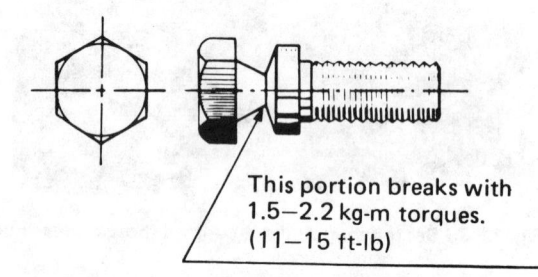

This portion breaks with 1.5—2.2 kg-m torques. (11—15 ft-lb)

Fig. 11.39 Column lock shear bolts before tightening

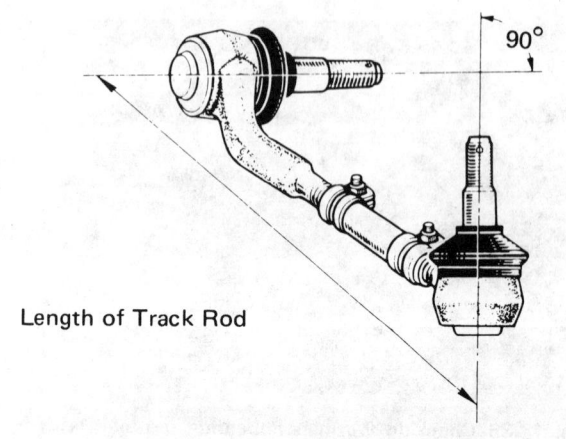

90°

Length of Track Rod

Fig. 11.40 Track-rod end balljoint setting

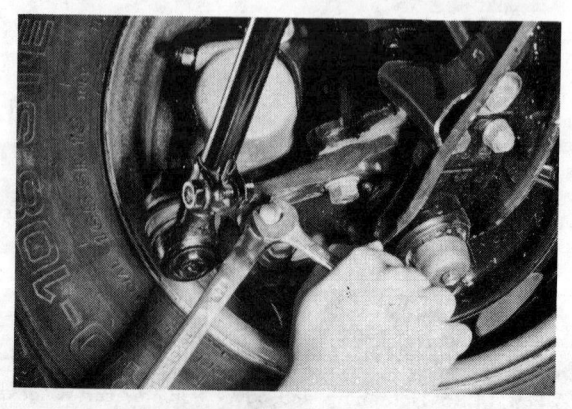

Fig. 11.41 Steering stop bolts

15 Align the centre spline and groove of the sector shaft and power piston nut gears. Push the shaft into the housing without rotating it. Tighten the cover bolts evenly in diagonal sequence to the specified torque.
16 Centre the steering gear. To do this turn the sector shaft from lock-to-lock and count the number of turns. Now turn the shaft back by half this number of turns. Mark the relative position of the wormshaft to the adjuster plug.
17 With the gear centralised, turn the slotted adjusting screw in the end cover until the starting torque at the wormshaft, measured with a torque meter or using a cord and spring balance, is between 2.6 and 3.5 lbf in (30 and 40 Ncm).
18 Use a new sealing washer and fit the locknut without disturbing the setting of the adjuster screw. Re-check the pre-load (starting torque) in both directions of travel.
19 Stake the worm plug locking ring nut in three places.

25 Steering column lock – removal and installation

1 The steering column lock is part of a composite mechanism which is incorporates the ignition starter switch.
2 The lock is mechanically operated by the ignition key which inserts a tongue into a mortice attached to the upper part of the steering shaft.
3 A safety feature is that the ignition key can only be turned to the *lock* position after the button has been depressed.
4 The steering column lock will normally only require removal if the lock has become faulty. There is no need to remove the lock when dismantling the steering column.
5 Access to the lock is attained after removing the steering column shrouds.
6 Set the ignition key to the *ACC* position and then depress the lock pin and extract the lock cylinder.
7 Disconnect the leads to the ignition switch at the connector plug and remove the switch.
8 The two shear-head type bolts which retain both halves of the steering column lock housing must be center punched and drilled, and a screw extractor used to remove the bolts.
9 When installing the new steering column lock, do not shear the heads from the securing bolts until the lock has been operated two or three times to check for smooth operation and engagement of the lock tongue.

26 Front wheel alignment

1 Accurate front wheel alignment is essential to ensure good steering and uniform wear. Before considering the steering angles, check that the tyres are correctly inflated, that the front wheels are not buckled, the hub bearings are not worn or incorrectly adjusted and that the steering linkage is in good order.
2 The steering angles are:
Camber, which is the angle at which the front wheels are set from the vertical when viewed from the front of the car. Positive camber is the number of degrees that the roadwheels are tilted outwards at their tops from the vertical. The camber angle is set in production and cannot be adjusted.
Camber, which is the angle at which the front wheels are set from the vertical when viewed from the front of the car. Positive camber is the number of degrees that the roadwheels are tilted outwards at their tops from the vertical. The camber angle is set in production and cannot be adjusted.
Steering axis inclination is the angle, when viewed from the front of the car, between the vertical and an imaginary line drawn between the suspension strut upper and lower swivel points.
Castor is the angle between the steering axis and a vertical line when viewed from each side of the car. Positive castor is when the steering axis is inclined rearward at the top. The castor angle is set in production and cannot be adjusted.
Toe-in is the amount by which the inside rim to inside rim measurement at the front of the roadwheels is less than the similar measurement taken at the rear of the roadwheels. The measurements should be taken at hub height.
3 Due to the need for precision gauges to measure the front wheel alignment accurately, it is best to leave this work to a service station. It is possible for the home mechanic to check and adjust the toe-in, for those who wish to carry out this work, first place the car on level ground with the front wheels in the straight-ahead position.
4 Obtain or make a toe-in gauge. One can easily be mde from tubing cranked to clear the engine oil pan and clutch bellhousing. Install an adjustable nut and setscrew at one end of the tube.
5 With the gauge, measure the distance between the two inner rims (at hub height) at the rear of the roadwheel.
6 Push or pull the car so that the roadwheel turns through 180° (half-a-turn) and measure the distance between the two inner rims, again at hub height at the front the the roadwheel. This measurement should be less than the measurement taken first by the amount quoted in Specifications. This represents the correct toe-in of the front wheels.
7 Where the toe-in is found to be incorrect, slacken all four tie-rod end clamp bolts and rotate each tubular tie-rod equally to increase or decrease the effective lengths of the tie-rods as required. Before rotating the tie-rods it is best to make a reference mark on them and only turn them so that the reference marks move through about 45° ($\frac{1}{8}$ turn) before rechecking the toe-in. Where new tie-rods or ends have been installed, set the tie-rods initially by measuring between the centres of the balljoints studs on one tie-rod and then adjust the opposite one to the same dimension (about 12.6 in – 320 mm).
8 When the toe-in is correct, make sure that the tie-rod end balljoints are in their correct attitudes before tightening the clamps.
9 When the steering wheel is turned to full lock the angle of the roadwheel on the inside of the turning circle should be 38° 10' and the outside one 31° 15'. Obviously, these angles can only be measured on specialized equipment but as a temporary expedient to prevent a tyre rubbing the steering lock stop bolts can be adjusted after releasing their locknuts.

27 Wheels and tyres

1 The roadwheels are of pressed steel type. Periodically, remove the wheels, clean dirt and mud from the inside and outside surfaces and examine for signs of rusting or rim damage and rectify as necessary.
2 Apply a smear of light grease to the wheel studs before screwing on the nuts.
3 The tyres fitted as original equipment are of steel-braced, radial-ply construction. It is advisable not to substitute crossplies. Don't mix tyres of different construction on the same axle.
4 Check the tyre pressures (including the spare) weekly.
5 If the wheels have been balanced on the car then it is important that the wheels are not moved round the car in an effort to equalize tyre wear. If a wheel is removed then the relationship of the wheel studs to the holes in the wheel should be marked to ensure exact replacement, otherwise the balance of the wheel, hub and tyre will be upset.
6 Where the wheels have been balanced off the car, then they may be moved round to equalize wear. Include the spare in any rotational pattern. With radial tyres it is preferable not to move the tyres round the car but only to interchange them on the same side.
7 Balancing of the wheels is an essential factor in good steering and roadholding. When the tyres have been in use for about half their useful life, the wheels should be rebalanced to compensate for the lost tread rubber.
8 Inspect the tyre walls and treads regularly for cuts and damage and where evident, have them professionally repaired.

28 Fault diagnosis – suspension and steering

Symptom	Cause
Steering wheel can be moved without corresponding movement in roadwheels	Wear in steering linkage Wear in steering gear Wear in column flexible coupling
Wander – car difficult to hold in straight line	Wear in linkage, gear or flexible coupling Incorrect wheel alignment Worn or seized balljoints Worn suspension strut lower swivel Front hub bearings worn or require adjustment
Steering stiff and heavy	Incorrect wheel alignment Worn or seized balljoints or swivels Lack of oil in steering gear Excessive wear in steering gear
Wheel wobble and vibration	Wheels require balancing Roadwheels buckled Incorrect wheel alignment Broken front spring or faulty suspension strut Wear in steering linkage balljoints and swivels
Excessive pitching and rolling on corners and during braking	Defective suspension struts Defective rear dampers Broken or weak road springs Under inflated tyres

Special to vehicles equipped with power steering

Symptom	Cause
Noisy operation	Loose pump drivebelt Worn pump Low fluid level Air in system Restricted pipes or hoses
Steering wheel kick-back	Air in system Incorrect steering gear adjustment Loose steering shaft flexible coupling Worn steering linkage
Steering wheel jerks during parking	Low fluid level Slack pump drivebelt Worn pump
Lack of assistance when turning steering wheel	Worn pump Slack pump drivebelt Low fluid level Worn steering gear

Chapter 12 Bodywork and fittings

Contents

Air conditioning system – description and maintenance	38
Bonnet lid – adjustment	10
Bonnet lid – removal and installation	9
Bonnet lock – adjustment	11
Boot lid (Coupe) – removal and installation	23
Boot lid lock – removal, installation and adjustment	24
Door glass – adjustment	18
Door glass and regulator – removal and installation	17
Door lock – removal, refitting and adjustment	15
Door striker – adjustment	16
Doors – removal, installation and adjustment	14
Doors – tracing rattles and their rectification	7
Front bumper – removal and installation	8
Front wing – removal and refitting	12
General description	1
Heater and ventilation system – description	33
Heater blower motor – removal and installation	34
Heater control – cable adjustment	36
Heater control – removal and installation	35
Heater unit – removal and installation	37
Instrument panel – removal and installation	32
Instrument panel crash pad – removal and installation	31
Liftback rear door damper stays – precautions	28
Maintenance – bodywork and underframe	2
Maintenance – hinges and locks	6
Maintenance – upholstery and carpets	3
Minor body damage – repair	4
Major body damage – repair	5
Quarter window (Coupe) – removal and installation	19
Quarter window (Liftback) – removal and installation	20
Rear bumper (Coupe) – removal and installation	29
Rear bumper (Liftback) – removal and installation	30
Rear door (Liftback) – removal and installation	25
Rear door glass (Liftback) – removal and installation	27
Rear door lock (Liftback) – removal, installation and adjustment	26
Rear window (Coupe) – removal and installation	22
Sunroof – removal and refitting	21
Windscreen – removal and refitting	13

1 General description

The body and underframe is a unitary all-welded steel construction. There are two basic body styles, the Liftback and Coupe, each being an essentially two-door style. The ST, XT and GT versions of the Liftback have the same body, but there are differences in interior fittings and trim.

Heating and ventilating is standard equipment on all models and the XT models is also fitted with air conditioning.

2 Maintenance – bodywork and underframe

1 The general condition of a car's bodywork is the one thing that significantly affects its value. Maintenance is easy but needs to be regular. Neglect, particularly after minor damage, can lead quickly to further deterioration and costly repair bills. It is important also to keep watch on those parts of the car not immediately visible, for instance the underside, inside all the wheel arches and the lower part of the engine compartment.
2 The basic maintenance routine for the bodywork is washing – preferably with a lot of water, from a hose. This will remove all the loose solids which may have stuck to the car. It is important to flush these off in such a way as to prevent grit from scratching the finish. The wheel arches and underbody need washing in the same way to remove any accumulated mud which will retain moisture and tend to encourage rust. Paradoxically enough, the best time to clean the underbody and wheel arches is in wet weather when the mud is thoroughly wet and soft. In very wet weather the underbody is usually cleaned of large accumulations automatically and this is a good time for inspection.
3 Periodically it is a good idea to have the whole of the underside of the car steam cleaned, engine compartment included, so that a thorough inspection can be carried out to see what minor repairs and renovations are necessary. Steam cleaning is available at many garages and is necessary for removal and accumulation of oily grime which sometimes is allowed to cake thick in certain areas near the engine, gearbox and back axle. If steam facilities are not available, there are one or two excellent grease solvents available which can be brush applied. The dirt can then be simply hosed off.
4 After washing paintwork, wipe off with a chamois leather to give an unspotted clear finish. A coat of clear protective wax polish will give added protection against chemical pollutants in the air. If the paintwork sheen has dulled or oxidised, use a cleaner/polisher combination to restore the brilliance of the shine. This requires a little effort, but is usually caused because regular washing has been neglected. Always check that the door and ventilator opening drain holes and pipes are completely clear so that water can drain out. Bright work should be treated the same way as paintwork. Windscreens and windows can be kept clear of the smeary film which often appears if a little ammonia is added to the water. If they are

This sequence of photographs deals with the repair of the dent and paintwork damage shown in this photo. The procedure will be similar for the repair of a hole. It should be noted that the procedures given here are simplified — more explicit instructions will be found in the text

In the case of a dent the first job — after removing surrounding trim — is to hammer out the dent where access is possible. This will minimise filling. Here, the large dent having been hammered out, the damaged area is being made slightly concave

Now all paint must be removed from the damaged area, by rubbing with coarse abrasive paper. Alternatively, a wire brush or abrasive pad can be used in a power drill. Where the repair area meets good paintwork, the edge of the paintwork should be 'feathered', using a finer grade of abrasive paper

In the case of a hole caused by rusting, all damaged sheet-metal should be cut away before proceeding to this stage. Here, the damaged area is being treated with rust remover and inhibitor before being filled

Mix the body filler according to its manufacturer's instructions. In the case of corrosion damage, it will be necessary to block off any large holes before filling — this can be done with aluminium or plastic mesh, or aluminium tape. Make sure the area is absolutely clean before ...

... applying the filler. Filler should be applied with a flexible applicator, as shown, for best results; the wooden spatula being used for confined areas. Apply thin layers of filler at 20-minute intervals, until the surface of the filler is slightly proud of the surrounding bodywork

Initial shaping can be done with a Surform plane or Dreadnought file. Then, using progressively finer grades of wet-and-dry paper, wrapped around a sanding block, and copious amounts of clean water, rub down the filler until really smooth and flat. Again, feather the edges of adjoining paintwork

The whole repair area can now be sprayed or brush-painted with primer. If spraying, ensure adjoining areas are protected from over-spray. Note that at least one inch of the surrounding sound paintwork should be coated with primer. Primer has a 'thick' consistency, so will find small imperfections

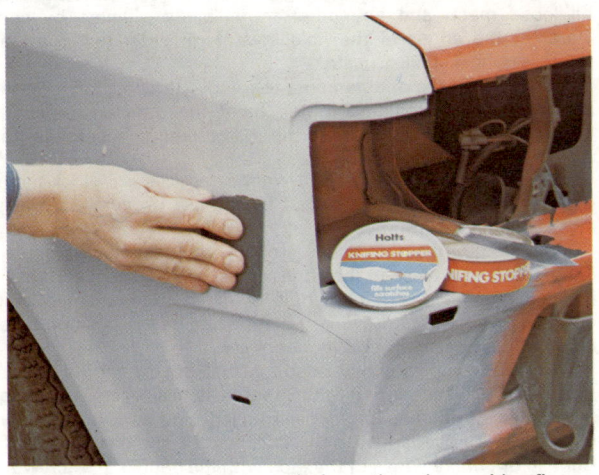

Again, using plenty of water, rub down the primer with a fine grade wet-and-dry paper (400 grade is probably best) until it is really smooth and well blended into the surrounding paintwork. Any remaining imperfections can now be filled by carefully applied knifing stopper paste

When the stopper has hardened, rub down the repair area again before applying the final coat of primer. Before rubbing down this last coat of primer, ensure the repair area is blemish-free – use more stopper if necessary. To ensure that the surface of the primer is really smooth use some finishing compound

The top coat can now be applied. When working out of doors, pick a dry, warm and wind-free day. Ensure surrounding areas are protected from over-spray. Agitate the aerosol thoroughly, then spray the centre of the repair area, working outwards with a circular motion. Apply the paint as several thin coats

After a period of about two weeks, which the paint needs to harden fully, the surface of the repaired area can be 'cut' with a mild cutting compound prior to wax polishing. When carrying out bodywork repairs, remember that the quality of the finished job is proportional to the time and effort expended

scratched, a good rub with a proprietary metal polish will often clear them. Never use any form of wax or other body or chromium polish on glass.

3 Maintenance – upholstery and carpets

1 Mats and carpets should be brushed or vacuum cleaned regularly to keep them free of grit. If they are badly stained remove them from the car for scrubbing or sponging and make quite sure they are dry before replacement. Seats and interior trim panels can be kept clean by a wipe over with a damp cloth. If they do become stained (which can be more apparent on light coloured upholstery) use a little liquid detergent and a soft nail brush to scour the grime out of the grain of the material. Do not forget to keep the head lining clean in the same way as the upholstery. When using liquid cleaners inside the car do not over-wet the surfaces being cleaned. Excessive damp could get into the seams and padded interior causing stains, offensive odours or even rot. If the inside of the car gets wet accidentally it is worthwhile taking some trouble to dry it out properly, particularly where carpets are involved. **Do not** leave oil or electric heaters inside the car for this purpose.

4 Minor body damage – repair

The photographic sequences on pages 262 and 263 illustrate the operations detailed in the following sub-sections.

Repair of minor scratches in the car's bodywork

If the scratch is very superficial, and does not penetrate to the metal of the bodywork, repair is very simple. Lightly rub the area of the scratch with a paintwork renovator, or a very fine cutting paste, to remove loose paint from the scratch and to clear the surrounding bodywork of wax polish. Rinse the area with clean water.

Apply touch-up paint to the scratch using a thin paint brush; continue to apply thin layers of paint until the surface of the paint in the scratch is level with the surrounding paintwork. Allow the new paint at least two weeks to harden; then blend it into the surrounding paintwork by rubbing the paintwork, in the scratch area with a paintwork renovator or a very fine cutting paste. Finally, apply wax polish.

Where the scratch has penetrated right through to the metal of the bodywork, causing the metal to rust, a different repair technique is required. Remove any loose rust from the bottom of the scratch with a penknife, then apply rust inhibiting paint to prevent the formation of rust in the future. Using a rubber or nylon applicator fill the scratch with bodystopper paste. If required, this paste can be mixed with cellulose thinners to provide a very thin paste which is ideal for filling narrow scratches. Before the stopper-paste in the scratch hardens, wrap a piece of smooth cotton rag around the top of a finger. Dip the finger in cellulose thinners and then quickly sweep it across the surface of the stopper-paste in the scratch; this will ensure that the surface of the stopper-paste is slightly hollowed. The scratch can now be painted over as described earlier in this Section.

Repair of dents in the car's bodywork

When deep denting of the car's bodywork has taken place, the first task is to pull the dent out, until the affected bodywork almost attains its original shape. There is little point in trying to restore the original shape completely, as the metal in the damaged area will have stretched on impact and cannot be reshaped fully to its original contour. It is better to bring the level of the dent up to a point which is about $\frac{1}{8}$ in (3 mm) below the level of the surrounding bodywork. In cases where the dent is very shallow anyway, it is not worth trying to pull it out at all. If the underside of the dent is accessible, it can be hammered out gently from behind, using a mallet with a wooden or plastic head. Whilst doing this, hold a suitable block of wood firmly against the impact from the hammer blows and thus prevent a large area of the bodywork from being 'belled-out'.

Should the dent be in a section of the bodywork which has double skin or some other factor making it inaccessible from behind, a different technique is called for. Drill several small holes through the metal inside the area – particularly in the deeper section. Then screw long self-tapping screws into the holes just sufficiently for them to gain a good purchase in the metal. Now the dent can be pulled out by pulling on the protruding heads of the screws with a pair of pliers.

The next stage of the repair is the removal of the paint from the damaged area, and from an inch or so of the surrounding 'sound' bodywork. This is accomplished most easily by using a wire brush or abrasive pad on a power drill, although it can be done just as effectively by hand using sheets of abrasive paper. To complete the preparation for filling, score the surface of the bare metal with a screwdriver or the tang of a file, or alternatively, drill small holes in the affected area. This will provide a really good 'key' for the filler paste.

To complete the repair see the Section on filling and respraying.

Repair of rust holes or gashes in the car's bodywork

Remove all paint from the affected area and from an inch or so of the surrounding 'sound' bodywork, using an abrasive pad or a wire brush on a power drill. If these are not available a few sheets of abrasive paper will do the job just as effectively. With the paint removed you will be able to gauge the severity of the corrosion and therefore decide whether to renew the whole panel (if this is possible) or to repair the affected area. New body panels are not as expensive as most people think and it is often quicker and more satisfactory to fit a new panel than to attempt to repair large areas of corrosion.

Remove all fittings from the affected area except those which will act as a guide to the original shape of the damaged bodywork (eg headlamp shells etc). Then, using tin snips or a hacksaw blade, remove all loose metal and any other metal badly affected by corrosion. Hammer the edges of the hole inwards in order to create a slight depression for the filler paste.

Wire brush the affected area to remove the powdery rust from the surface of the remaining metal. Paint the affected area with rust inhibiting paint; if the back of the rusted area is accessible treat this also.

Before filling can take place it will be necessary to block the hole in some way. This can be achieved by the use of aluminium or plastic mesh, or aluminium tape.

Aluminium or plastic mesh is probably the best material to use for a large hole. Cut a piece to the approximate size and shape of the hole to be filled, then position it in the hole so that its edges are below the level of the surrounding bodywork. It can be retained in position by several blobs of filler paste around its periphery.

Aluminium tape should be used for small or very narrow holes. Pull a piece off the roll and trim it to the approximate size and shape required, then pull it off the backing paper (if used) and stick the tape over the hole; it can be overlapped if the thickness of one piece is insufficient. Burnish down the edges of the tape with the handle of a screwdriver or similar, to ensure that the tape is securely attached to the metal underneath.

Bodywork repairs – filling and respraying

Before using this Section, see the Sections on dent, deep scratch, rust holes and gash repairs.

Many types of bodyfiller are available, but generally speaking those proprietary kits which contain a tin of filler paste and a tube of resin hardener are best for this type of repair. A wide, flexible plastic or nylon applicator will be found invaluable for imparting a smooth and well contoured finish to the surface of the filler.

Mix up a little filler on a clean piece of card or board – use the hardener sparingly (follow the maker's instructions on the packet) otherwise the filler will set very rapidly.

Using the applicator apply the filler paste to the prepared area: draw the applicator across the surface of the filler to achieve the correct contour and to level the filler surface. As soon as a contour that approximates the correct one is achieved, stop working the paste – if you carry on too long the paste will become sticky and begin to 'pick up' on the applicator. Continue to add thin layers of filler paste at twenty-minute intervals until the level of the filler is just proud of the surrounding bodywork.

Once the filler has hardened, excess can be removed using a Surform plane or Dreadnought file. From then on, progressively finer grades of abrasive paper should be used, starting with a 40 grade production paper and finishing with 400 grade wet-and-dry paper. Always wrap the abrasive paper around a flat rubber, cork, or wooden block – otherwise the surface of the filler will not be completely flat. During the smoothing of the filler surface the wet-and-dry paper should be periodically rinsed in water. This will ensure that a very smooth finish is imparted to the filler at the final stage.

At this stage the 'dent' should be surrounded by a ring of bare

Chapter 12 Bodywork and fittings

metal, which in turn should be encircled by the finely 'feathered' edge of the good paintwork. Rinse the repair area with clean water, until all of the dust produced by the rubbing-down operation has gone.

Spray the whole repair area with a light coat of primer – this will show up any imperfections in the surface of the filler. Repair these imperfections with fresh filler paste or bodystopper, and once more smooth the surface with abrasive paper. If bodystopper is used, it can be mixed with cellulose thinners to form a really thin paste which is ideal for filling small holes. Repeat this spray and repair procedure until you are satisfied that the surface of the filler, and the feathered edge of the paintwork are perfect. Clean the repair area with clean water and allow to dry fully.

The repair area is now ready for final spraying. Paint spraying must be carried out in a warm, dry, windless and dust free atmosphere. This condition can be created artificially if you have access to a large indoor working area, but if you are forced to work in the open, you will have to pick your day very carefully. If you are working indoors, dousing the floor in the work area with water will help settle the dust which would otherwise be in the atmosphere. If the repair area is confined to one body panel, mask off the surrounding panels; this will help to minimise the effects of a slight mis-match in paint colours. Bodywork fittings (eg chrome strips, door handles etc) will also need to be masked off. Use genuine masking tape and several thicknesses of newspaper for the masking operations.

Before commencing to spray, agitate the aerosol can thoroughly, then spray a test area (an old tin, or similar) until the technique is mastered. Cover the repair area with a thick coat of primer; the thickness should be built up using several thin layers of paint rather than one thick one. Using 400 grade wet-and-dry paper, rub down the surface of the primer until it is really smooth. While doing this, the work area should be thoroughly doused with water, and the wet-and-dry paper periodically rinsed in water. Allow to dry before spraying on more paint.

Spray on the top coat, again building up the thickness by using several thin layers of paint. Start spraying in the centre of the repair area and then using a circular motion, work outwards until the whole repair area and about 2 inches of the surrounding original paintwork is covered. Remove all masking material 10 to 15 minutes after spraying on the final coat of paint.

Allow the new paint at least two weeks to harden, then, using a paintwork renovator or a very fine cutting paste, blend the edges of the paint into the existing paintwork. Finally, apply wax polish.

5 Major body damage – repair

Where serious damage has occurred or large areas need renewal due to neglect, it means certainly that completely new sections or panels will need welding in and this is best left to professionals. If the damage is due to impact it will also be necessary to completely check the alignment of the bodyshell structure. Due to the principle of construction the strength and shape of the whole can be affected by damage to a part. In such instances the services of a Toyota agent with specialist checking jigs are essential. If a body is left misaligned it is first of all dangerous as the car will not handle properly and secondly uneven stresses will be imposed on the steering, engine and transmission, causing abnormal wear or complete failure. Tyre wear may also be excessive.

6 Maintenance – hinges and locks

1   Oil the hinges of the bonnet, boot and doors with a drop or two of light oil periodically. A good time is after the car has been washed.
2   Oil the bonnet release catch pivot pin and the safety catch pivot pin periodically.
3   Do not over lubricate door latches and strikers. Normally a little oil on the rotary cam spindle alone is sufficient.

7 Doors – tracing rattles and their rectification

1   Check first that the door is not loose at the hinges and that the latch is holding the door firmly in position. Check also that the door lines up with the aperture in the body.
2   If the hinges are loose or the door is out of alignment it will be necessary to reset the hinge positions, as described in Section 14.
3   If the latch is holding the door properly it should hold the door tightly when fully latched and the door should line up with the body. If it is out of alignment it needs adjustment as described in Section 14. If loose, some part of the lock mechanism must be worn out and requiring renewal.
4   Other rattles from the door would be caused by wear or looseness in the window winder, the glass channels and sill strips or the door buttons and interior latch release mechanism.

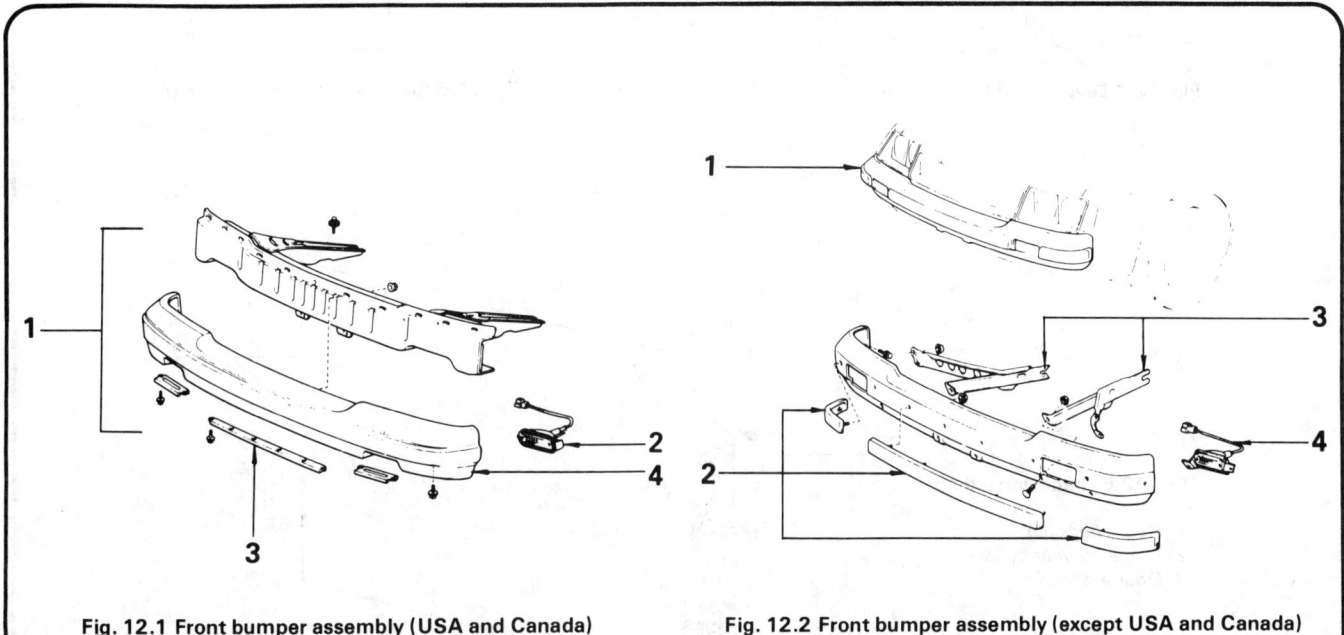

Fig. 12.1 Front bumper assembly (USA and Canada)

1 Front bumper assembly
2 Front direction indicator lamp
3 Retainer
4 Bumper bar

Fig. 12.2 Front bumper assembly (except USA and Canada)

1 Front bumper assembly
2 Bumper insert
3 Bumper stay
4 Front direction indicator lamp

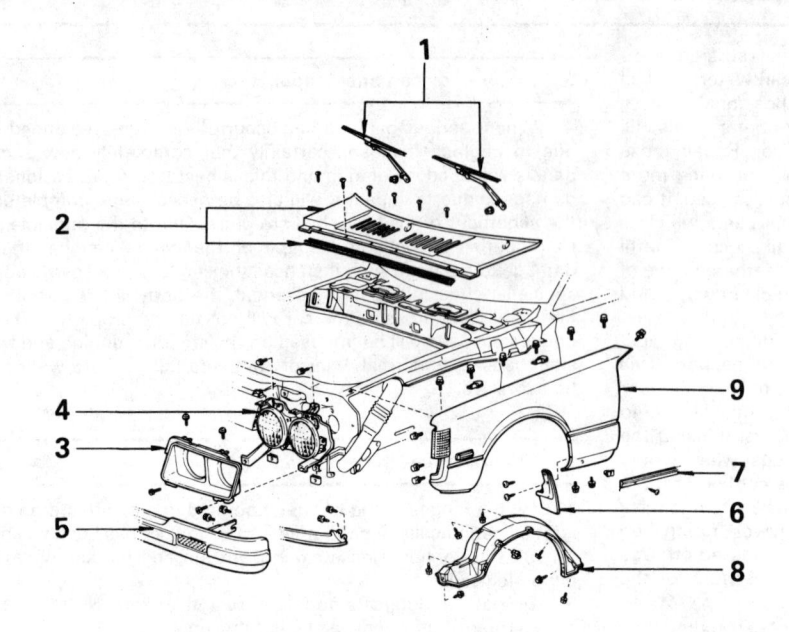

Fig. 12.3 Front wing removal

1 Wiper arms
2 Front louvred panel
3 Headlamp housing
4 Headlamp mounting assembly
5 Front bumper
6 Mud flap
7 Sill trim
8 Wheel arch
9 Front wing

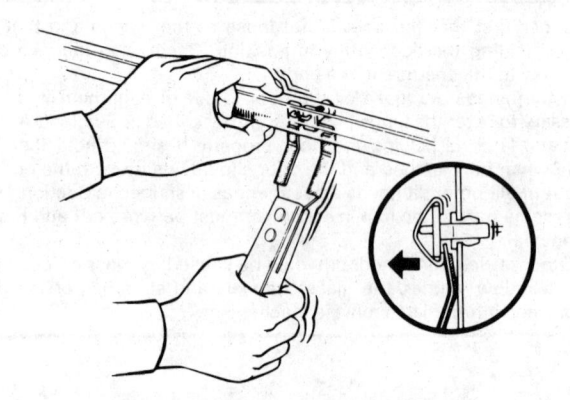

Fig. 12.4 Levering off the sill trim

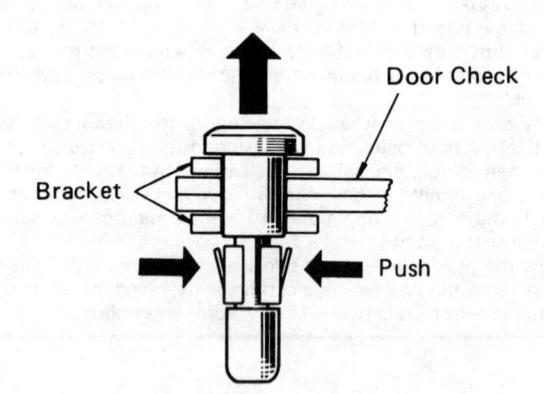

Fig. 12.5 Removing the door check pin

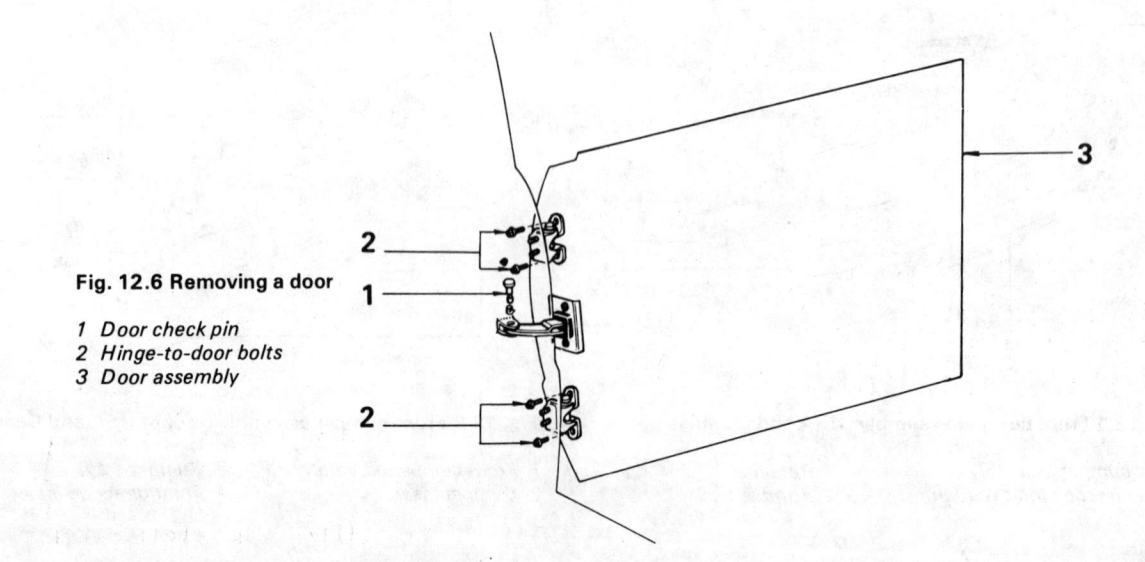

Fig. 12.6 Removing a door

1 Door check pin
2 Hinge-to-door bolts
3 Door assembly

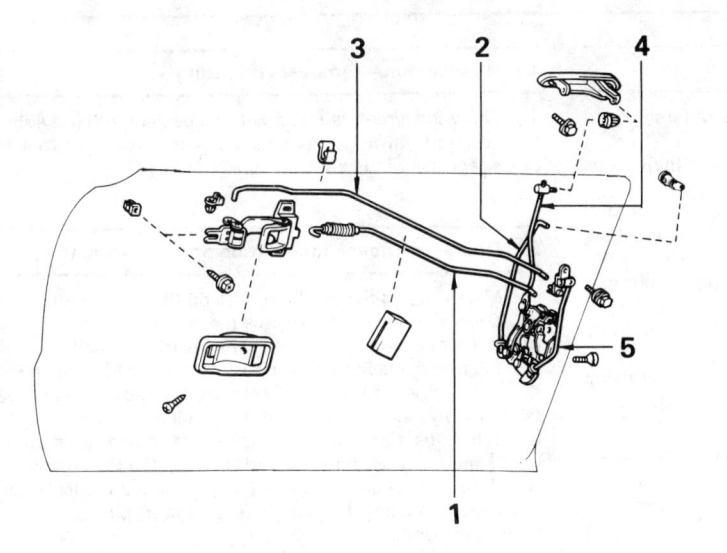

Fig. 12.7 Door lock and linkage

1 Inside handle opening link
2 Door cylinder lock link
3 Inside lock link
4 Outside handle opening link
5 Door lock assembly

Fig. 12.8 Door inside lock adjustment

Fig. 12.9 Door outside handle adjustment

1 Door handle operating rod
2 Screwed adjuster

Fig. 12.10 Door striker adjustment

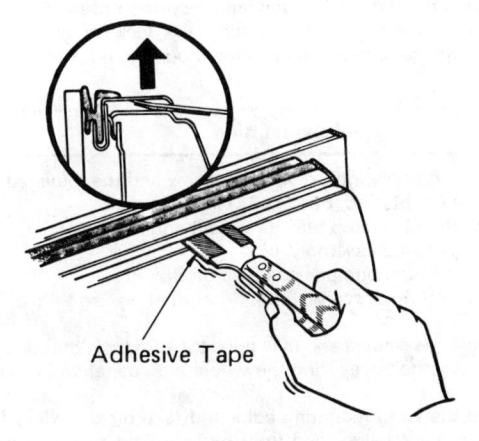

Fig. 12.11 Removing the belt moulding

8 Front bumper – removal and installation

1 The front bumper of models for North America is different from that for other markets (Fig. 12.1 and 12.2).
2 Disconnect each of the front turn indicators by pulling their plugs out of the wiring harness sockets.
3 With the exception of the North American models, remove the bolt from each end of the bumper, remove the four nuts and washers securing the bumper to the bumper stay and lift the bonnet off.
4 On North American models, remove the nuts securing the bumper to the bumper reinforcement.
5 To remove the bumper insert, when fitted, remove the fixing nuts attaching each part of the bumper insert to the bumper.
6 When refitting a bumper, insert the bolts and screw on all fixing nuts finger tight to align the bumper and then tighten the fixings.

9 Bonnet lid – removal and installation

1 To avoid the risk of damaging the paintwork, this operation requires two people.
2 Open the bonnet and prop it with the bonnet stay.
3 Mark round the hinges where they are attached to the bonnet lid, so that the lid can be refitted in the same position.
4 With a person at each side supporting the bonnet lid, remove the two fixing bolts from each side and lift the bonnet lid clear.
5 Install the bonnet lid by reversing the removal operations but initially only tighten the bolts finger tight. Adjust the lid so that the hinges are in the same place as before removal and then tighten the fixing bolts fully.

10 Bonnet lid – adjustment

1 Open the bonnet lid and loosen the four hinge fixing bolts until they just grip.
2 Lower the bonnet lid and adjust it to give an even gap at the two sides and the rear edge.
3 Lift the lid carefully so that the adjustment is not disturbed and tighten the bolts fully.
4 Close the bonnet lid and re-check, re-adjusting if necessary.
5 To adjust the height of the rear edge of the bonnet lid, loosen the three bolts connecting the lower end of the hinge to the front bulkhead.
6 Check that the bonnet locks satisfactorily and if not, adjust the lock as described in the following Section.

11 Bonnet lock – adjustment

1 Check that the bonnet lid closes easily and that when the catch engages the lid is secured firmly without rattling.
2 To adjust both the vertical and lateral position of the lock, slacken the two lock mounting bolts, position the lock to give the most satisfactory engagement and tighten the mounting bolts firmly.

12 Front wing – removal and refitting

1 Remove the two windscreen wiper arms and the louvered bonnet top panel (see Chapter 10, Section 32).
2 Remove the headlamp stay as described in Chapter 10, and also remove the headlamp assembly.
3 Remove the front bumper (see Section 8).
4 Remove the two screws from the mudflap and take off the mudflap.
5 Remove three screws and then prise the trim from the sill.
6 Remove the bolts securing the wheel arch panel and remove the panel.
7 Remove the wing mounting bolts and take off the wing. Disconnect the side marker light and front parking light wiring and lift the wing clear.
8 When refitting the wing, note that a retainer is fitted to the second bolt from the front and a damper to the fourth bolt.

13 Windscreen – removal and fitting

1 The windscreen is fitted with a special urethane adhesive and the removal and refitting of windscreens should be left to a Toyota dealer or a specialist windscreen service station.

14 Doors – removal, installation and adjustment

1 Mark the position of the hinges on the leading edge of the door.
2 Open the door fully, place a piece of wood under the bottom edge of the door and support the door by inserting a jack beneath the wood.
3 Push in the claws of the door check pin and remove the pin.
4 While holding the door to steady it, remove the four bolts securing the hinge to the door and lift the door clear.
5 When installing the door, take care to line up the hinges with the marks made before removal and then tighten the bolts.
6 If the door requires moving in, out, or in a vertical direction, loosen the bolts attaching the hinge to the door, position the door and re-tighten the bolts.
7 To position the door forwards or backwards, loosen the bolts attaching the hinges to the door pillar and fit packing pieces between the hinge and the door pillar.

15 Door lock – removal, refitting and adjustment

1 Remove the armrest from the door interior panel (photos).
2 Using a hooked piece of wire, extract the spring clip from behind the window regulator handle (photo) and remove the handle and escutcheon plate (photo).
3 Unscrew and remove the door interior handle bezel (photo).
4 Insert a thin blade or steel rule between the lower edge of the door interior trim panel and the door frame and prise away one or two of the panel securing clips. Now insert the fingers and release the remaining clips and remove the trim panel far enough to disconnect the courtesy lamp leads.
5 Carefully peel the waterproof sheet from the door frame.
6 Temporarily install the window regulator handle and wind the window glass fully up.
7 Disconnect the two lock control link rods and the rods from the cylinder lock and exterior door handle (photo).
8 Unscrew the bellcrank and frame mounting bolts and then unscrew the lock mounting screws from the edge of the door frame.
9 Withdraw the door lock assembly through the aperture in the door interior panel.
10 Before refitting the lock, check that all its operating parts perform satisfactorily and smear grease over all the sliding surfaces.
11 Fit the lock by reversing the removal operation and then make the following adjustments.
12 Adjust the inside door handle by slackening the three fixing screws (photo). Move the handle assembly forward until strong resistance is felt, then move it back 0·04 in (1 mm) before tightening the screws.
13 Adjust the door inside lock by loosening the lock adjusting bolt (Fig. 12.8), moving the bolt upwards until strong resistance is felt and then lowering it 0·04 in (1 mm) before tightening.
14 Adjust the door outside handle by unscrewing and removing the nut which secures the operating rod to the outside door handle. With the link hanging under its own weight, lift the outside handle about 0·04 in (1 mm) and with the handle held in this position and the link still hanging naturally, turn the adjuster on the top of the rod until it can be entered into the hole in the door handle lever. After fitting the adjuster into its hole, fit the nut and tighten it.

16 Door striker – adjustment

1 After making sure that the door hinges and the door lock have been adjusted properly, check that the door closes easily and firmly.
2 If it is not satisfactory, draw round the door striker so that its original position can be seen. Loosen the two door striker fixing bolts and move the striker vertically, or laterally as required. When the setting is satisfactory, tighten the two screws securely.

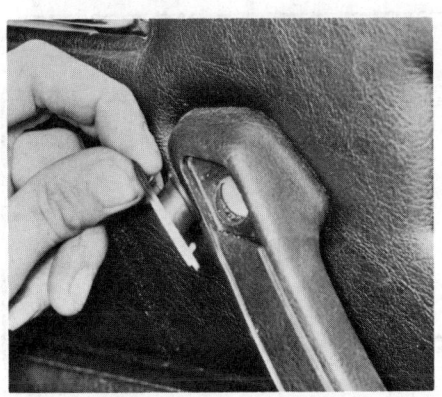

15.1a Removing an arm rest screw cover

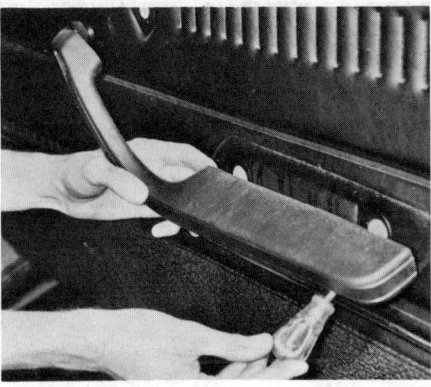

15.1b Removing the arm rest lower screws

15.2a The window regulator handle securing clip

15.2b Removing the window regulator and escutcheon plate

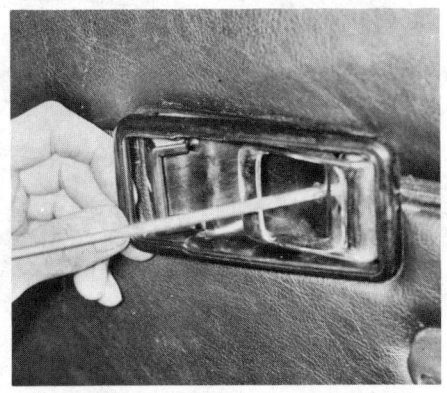

15.3 Removing the interior handle bezel

15.7 Lock mechanism and control rods

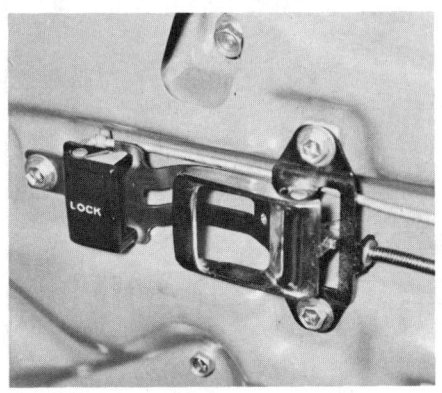

15.12 Inside door handle fixing screws

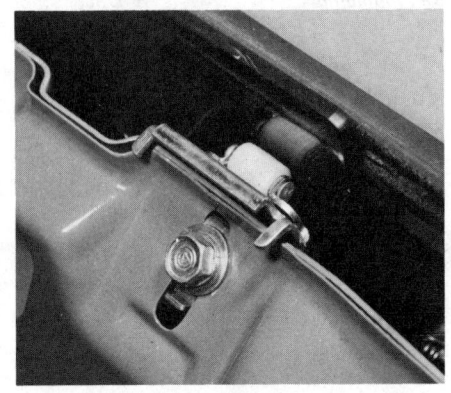

17.4 Guide roller for door glass

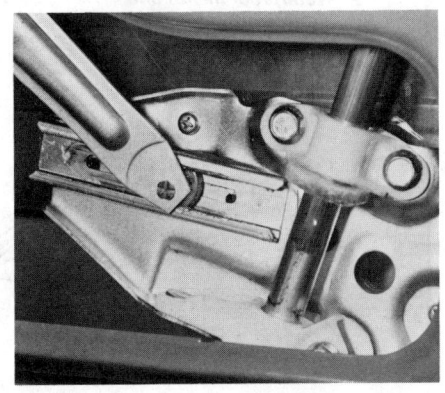

17.7 Door glass regulator roller in channel

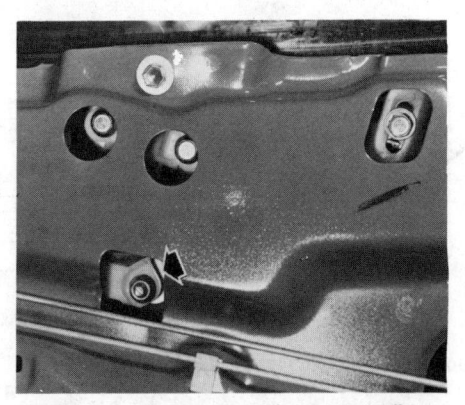

18.7 Door glass weatherstrip contact adjuster (arrowed)

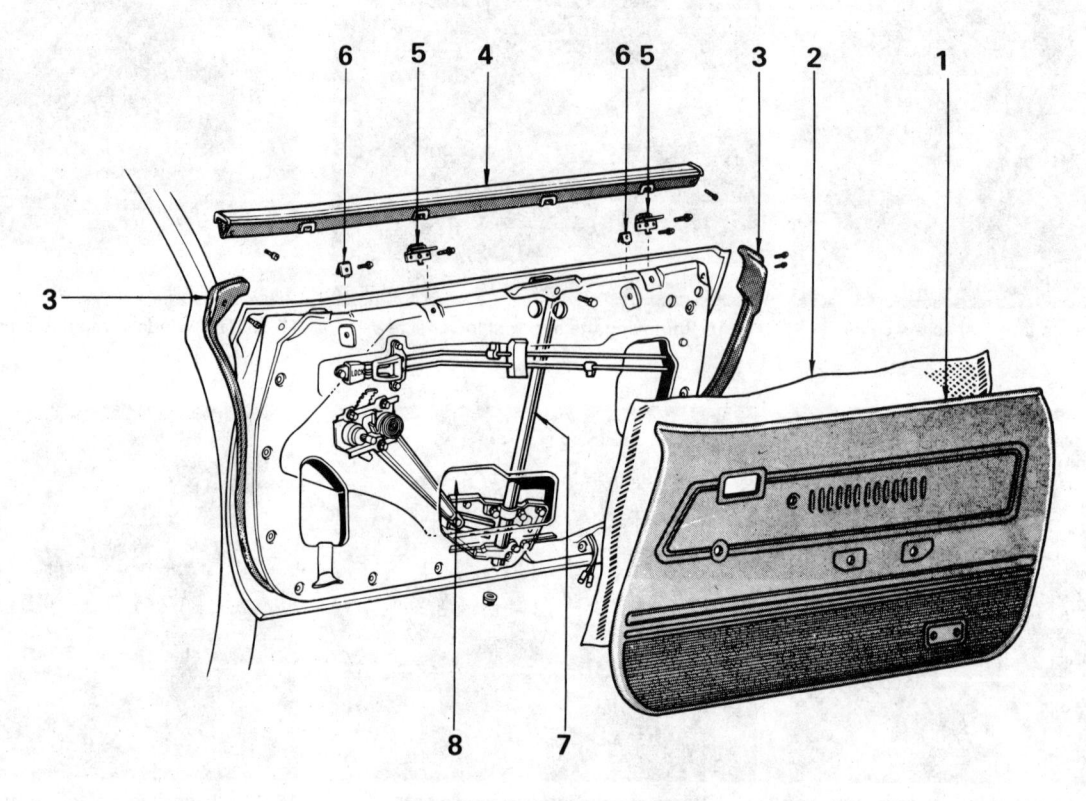

Fig. 12.12 Door glass removal

1 Door trim panel	3 Door weatherstrip	5 Guide rollers	7 Window guide assembly
2 Waterproof membrane	4 Belt moulding	6 Upper stops	8 Glass

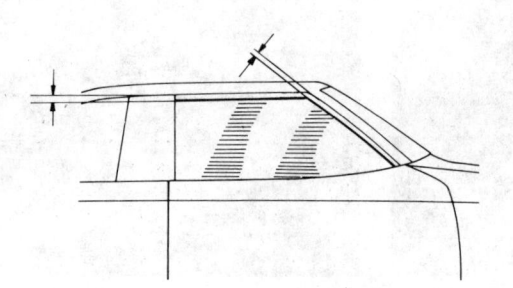

Fig. 12.13 Door glass too low

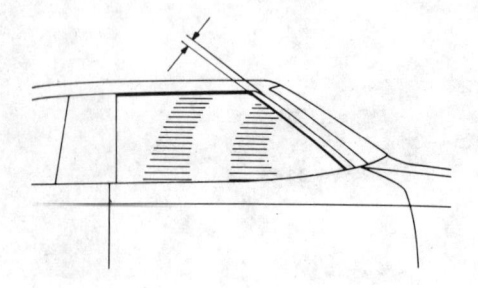

Fig. 12.14 Gap along front edge of door glass

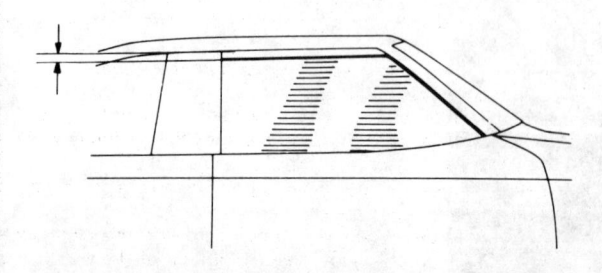

Fig. 12.15 Gap at top edge of door glass

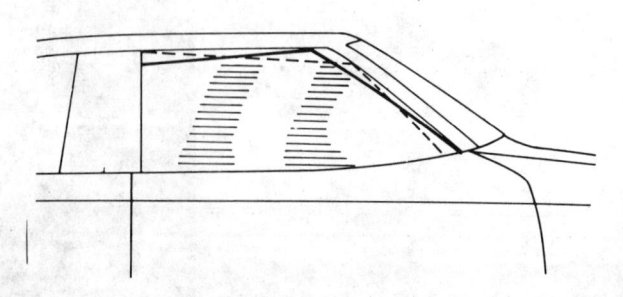

Fig. 12.16 Door glass tilted

17 Door glass and regulator – removal and installation

1 Lower the window to its fullest extent, then remove the door trim panel as detailed in Section 15.
2 Peel off the waterproof membrane and pull off the upper ends of the door weatherstrip.
3 Using a flat chisel or wide screwdriver, prise off the belt moulding. Wrap the chisel or screwdriver blade with adhesive tape, or place a piece of cardboard underneath it to avoid marking the paintwork.
4 Unscrew and remove the two guide rollers (photo).
5 Unscrew and remove the two upper stops.
6 Raise the window about 1 in (25 mm), remove the nut from the bottom of the window guide rod and the bolt from the lower clamp and pull the guide rod out upwards.
7 Remove the door regulator fixing screws, remove the regulator roller from the channel in the door glass bracket (photo) and remove the regulator assembly through the bottom hole in the door.
8 Remove the door glass by pulling it upwards.
9 Refit the glass and regulator by reversing the removal operations, but do not fit the waterproof membrane and door trim until the adjustment of the glass is satisfactory.

Fig. 12.17 Inadequate contact with door glass weatherstrip

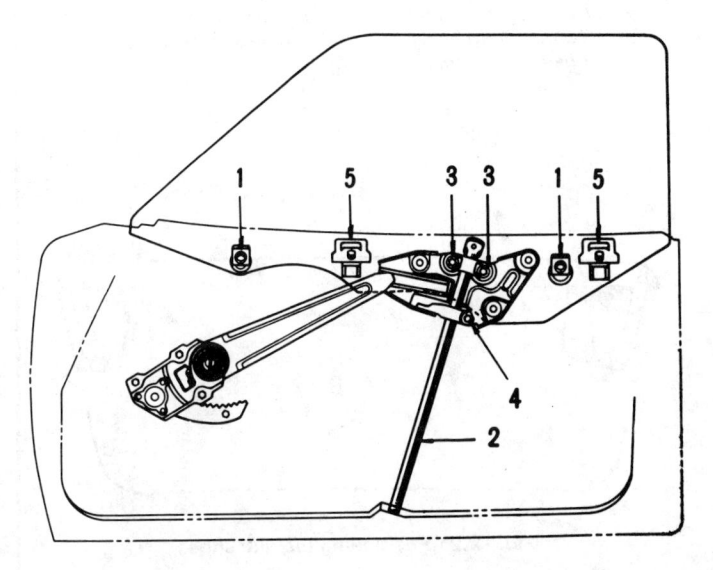

Fig. 12.18 Door glass adjustment points

1 Top stops
2 Guide rod
3 Guide adjusting bolt (upper)
4 Guide adjusting bolt (lower)
5 Rollers

18 Door glass – adjustment

1 The various adjustment points are shown in Fig. 12.18 and various combinations of them can be used to adjust the glass as follows:

Glass too low
2 If the closed position of the window is not high enough (Fig. 12.13), loosen the securing bolts of the top stops (1) and raise the stops until the window closes fully.

Gap along the front edge
3 If the window closes to the correct height, but leaves a gap along the front edge (Fig. 12.14) it requires lateral adjustment by moving the guide rod (2) forward.

Gap at top edge of glass
4 If the front edge of the glass is hard against the frame, leaving a gap at the top edge (Fig. 12.15), the upper guide adjusting bolts need to be loosened and the guide assembly moved sideways to eliminate the gap. It may also be necessary to adjust the top stops (1).

Glass is tilted
5 When the glass is not symmetrical in its frame (Fig. 12.16), the basic correction is obtained by loosening the upper guide adjusting bolts and rotating the guide assembly until the edges of the glass are parallel with the corresponding edges of the frame. It may also be necessary to adjust the top stops (1).

Excessive lateral play
6 If there is excessive lateral movement of the glass, resulting in window rattles, raise the glass to its highest extent, loosen the bolts of the rollers and push the rollers into firm contact with the glass.

Inadequate contact with weatherstrip
7 If there is a tendency for draughts or leaks because the top of the glass is not contacting the wide flange of the weatherstrip (Fig. 12.17), release the locknut which is visible through the inner door panel hole above the lock rods (photo). Turn the adjuster screw to obtain maximum contact between the glass and weatherstrip and while holding the screw in this position, tighten the locknut.

19 Quarter window (Coupe) – removal and installation

1 Prise up the front of the rear cushion with a screwdriver to disengage the securing clips, then remove the rear seat.
2 Remove the bolts from the rear seat belts, then lift off and remove the rear seat back.
3 Remove the two screws from the seat belt guide after prising out the base with a screwdriver.
4 Insert a screwdriver behind the quarter trim panel and prise it off.
5 Remove the bolt from the shoulder anchor of the front seat belt. Remove the two screws and remove the centre pillar carrier trim. Remove three nuts and take off the centre pillar outer trim.
6 Remove the shoulder anchor bolt of the rear seat belt.
7 Remove the two screws and take off the side trim of the rear window.
8 Pull off the roof trim, drip moulding and belt moulding.
9 Remove the trim moulding.
10 Remove the screws from the window assembly, push the weatherstrip lip from outside to inside the body flange and remove the glass and its frame from inside the car.
11 If a new glass is required, this is a job which is best left to a car glass specialist.
12 Before refitting the window, fit a pull cord into the weatherstrip groove (Fig. 12.20). Apply soapy water to the lip of the weatherstrip and to the body flange.
13 When fitting the window, start at the center pillar. While holding the window pressed against the body frame from inside the car, pull the cord from outside to fit the weatherstrip lip over the body flange.
14 Before fitting the quarter window retaining screws, tap the window into place using the palm of your hand.

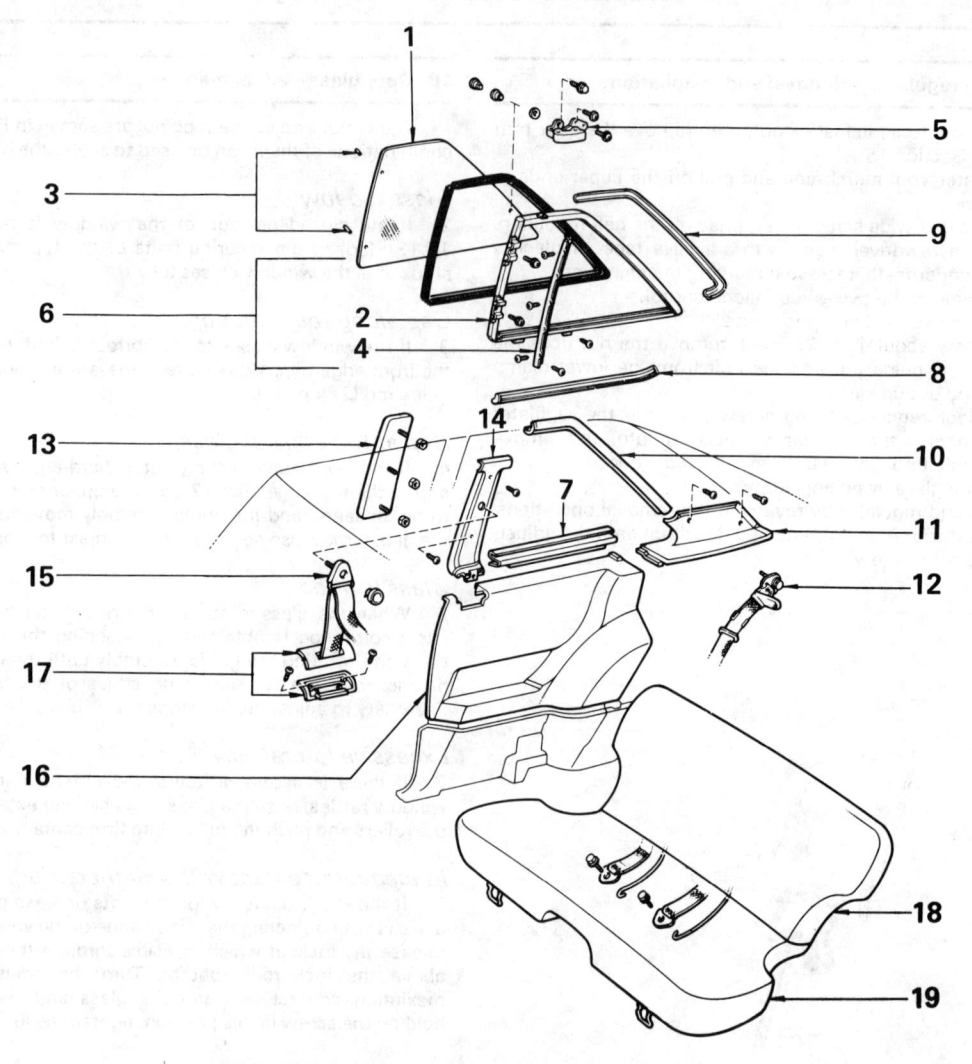

Fig. 12.19 Rear quarter window components (Coupe)

1	Glass	6	Quarter window assembly	11	Rear window trim plate
2	Retainer	7	Trim	12	Rear seat belt anchor bolt
3	Hinge pin	8	Moulding	13	Centre pillar outer trim panel
4	Retainer	9	Drip moulding	14	Centre pillar inner trim panel
5	Lock	10	Side rail trim	15	Front seat belt anchor bolt

16 Quarter trim panel
17 Seat belt housing
18 Seat back
19 Rear seat cushion

Fig. 12.20 Fitting a pull cord to the quarter window

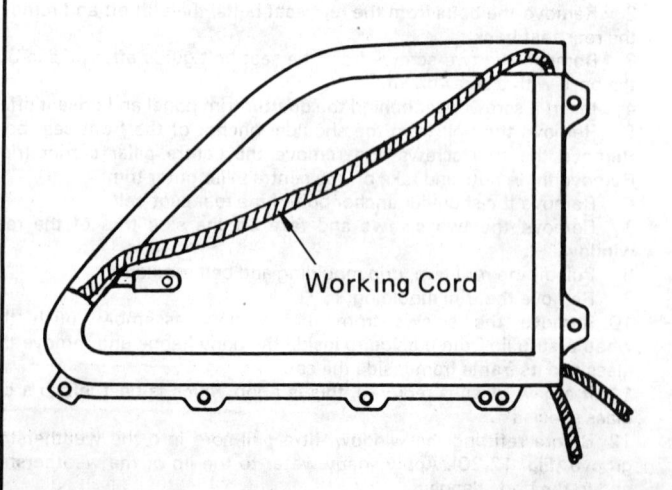

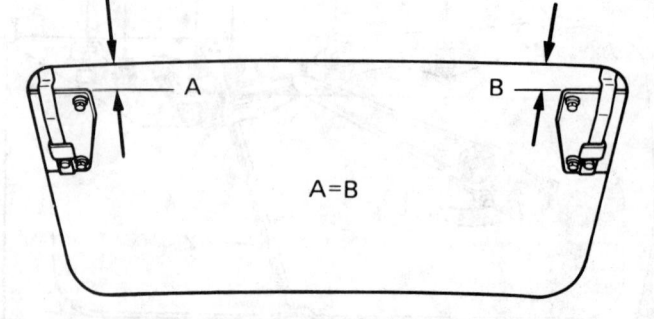

Fig. 12.21 Positioning the rear shoes

Chapter 12 Bodywork and fittings

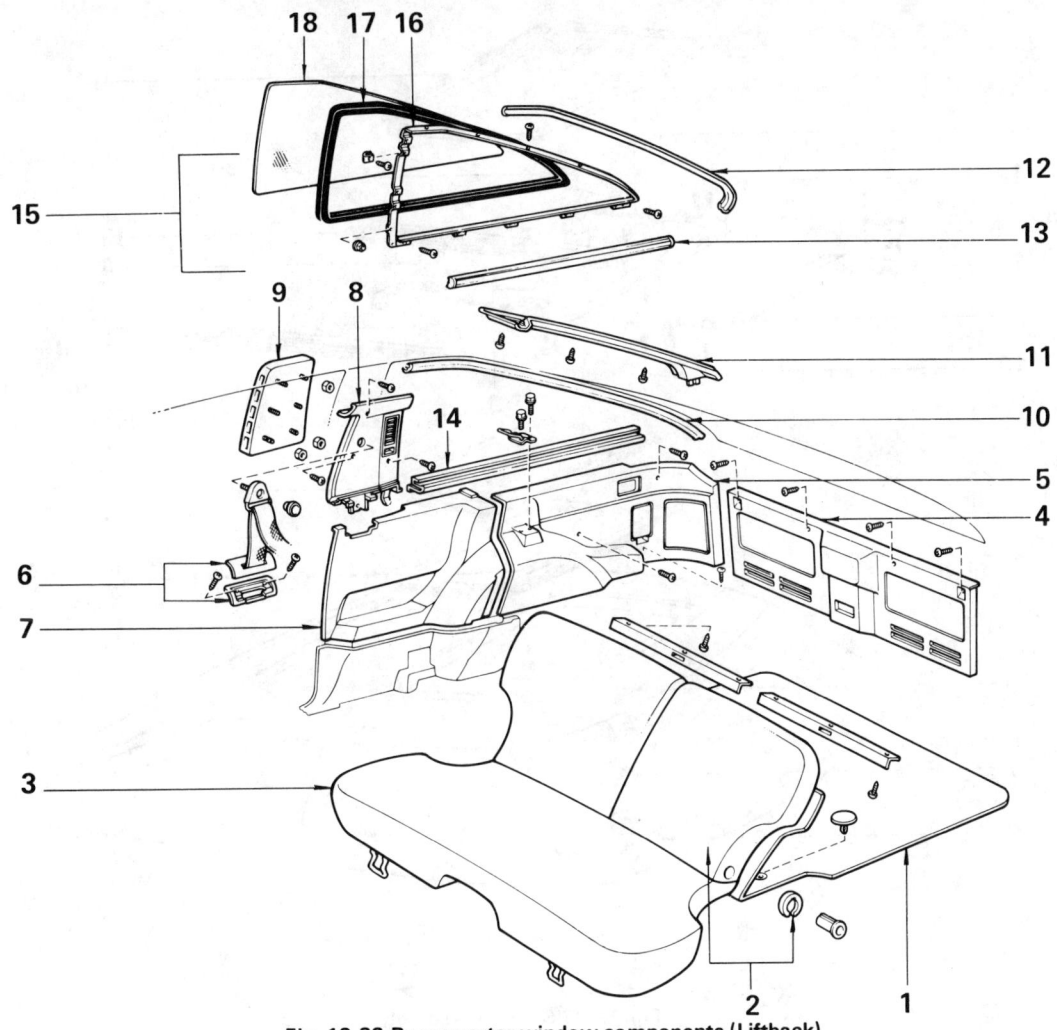

Fig. 12.22 Rear quarter window components (Liftback)

1. Luggage space carpet
2. Rear seat back
3. Rear seat cushion
4. Luggage space rear trim
5. Luggage space side trim
6. Seat belt guide and base
7. Quarter trim panel
8. Center pillar inner trim
9. Center pillar outer trim
10. Roof trim
11. Rear window side trim
12. Drip moulding
13. Belt moulding
14. Trim moulding
15. Quarter window assembly
16. Glass retainer
17. Weatherstrip
18. Quarter window glass

20 Quarter window (Liftback) – removal and installation

1 The procedure for the Liftback is very similar to that for the Coupe described in the previous Section.
2 The items which need to be removed and the order in which the work should be done is shown in Fig. 12.22.

21 Sunroof – removal and refitting

1 Remove the two screws and take off the deflector strip.
2 Insert a screwdriver between the sunroof and its head lining, prise off the lining and remove it through the roof opening.
3 Remove the two bolts from each front shoe and remove the shoe and its shim.
4 Remove the bolt from each of the rear shoes and take out the shim.
5 Remove the sliding panel assembly, then remove the screw from the centre of the operating handle and remove the handle and bezel.
6 Remove part of the trim round the opening to gain access to the roof operating gear assembly. Remove the two bolts and take off the gear assembly.
7 Remove the remaining two screws in each of the slide rail covers and remove the covers.
8 Remove the slide rails and then take out the cable assembly.
9 Before refitting the operating rails, cables and gear, make sure that they are clean and then smear them with grease.
10 When fitting the rear shoes, make sure that their ends are equidistant from the end of the roof panel (Fig. 12.21).
11 Adjust the height of the roof panel so that it is flush with the roof line. Adjustment is achieved by using shims under the front and rear shoes.
12 If adjustment is required to make the sliding panel square with the roof opening, loosen the bolts securing the front and rear shoes, move the roof panel to the required position and tighten the shoe bolts.
13 After making the adjustments, close the panel and fit the handle so that when the panel is closed and the handle is in its parked position, pointing forward.

22 Rear window (Coupe) – removal and installation

1 Remove the rear seat cushion by prising up the two fixing clips with a screwdriver.
2 Remove the back of the rear seat by lifting it up off its fixing hooks.
3 Remove the rear seat belt lower and shoulder anchor fixings.
4 Prise up and remove the rear parcel shelf.
5 Disconnect the rear window demister panel at its plug and socket connection.
6 Use a screwdriver to prise out the rear window trim and then use

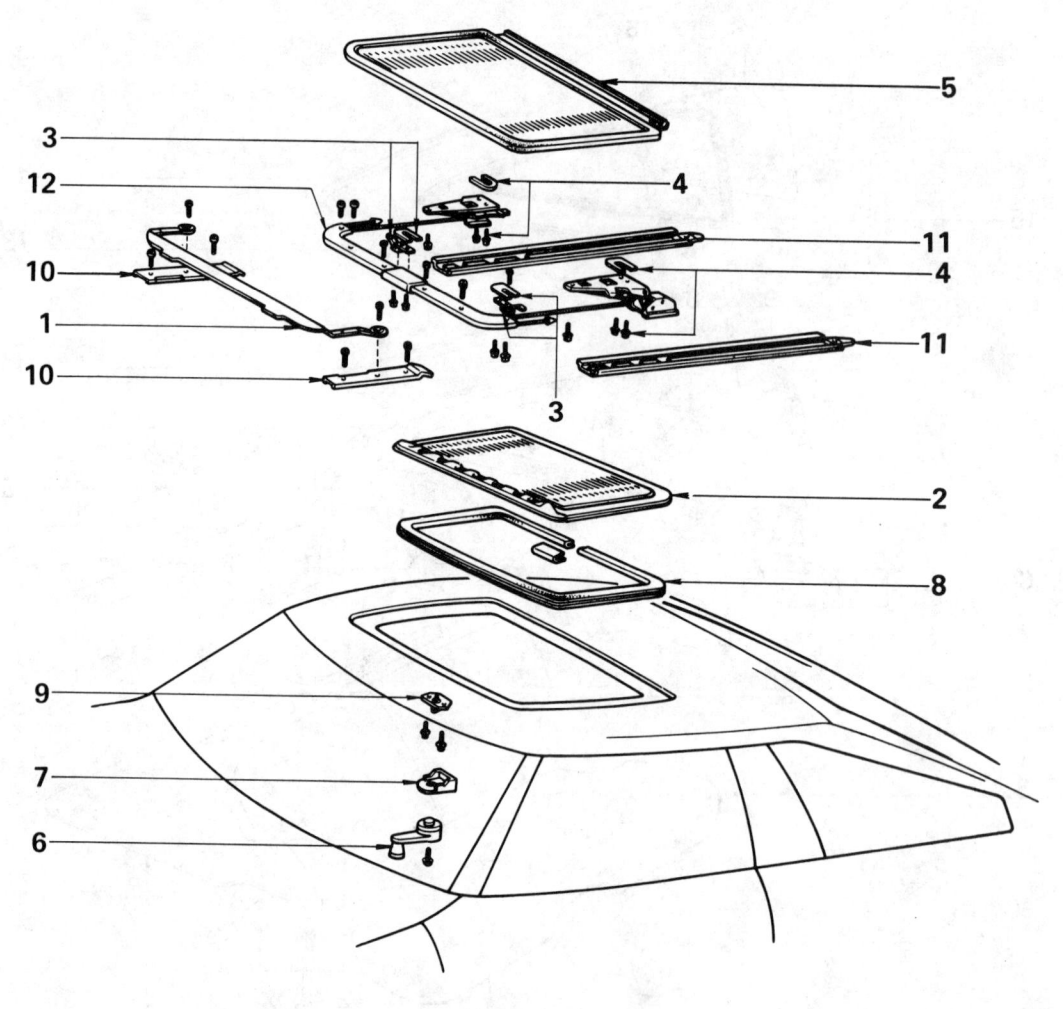

Fig. 12.23 Sunroof assembly

1 Deflector	4 Rear shoe, bolt and shim	7 Bezel
2 Head lining	5 Sliding panel assembly	8 Trim
3 Front shoe and shim	6 Handle	9 Gear

10 Rail cover
11 Rail
12 Cable assembly

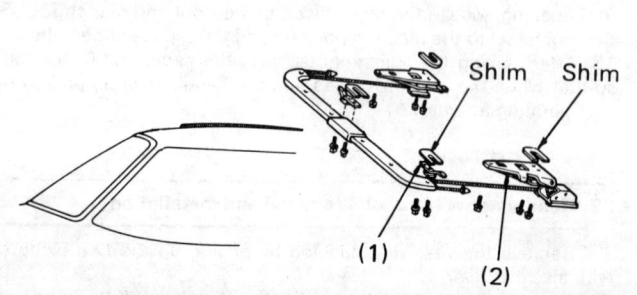

Fig. 12.24 Adjusting the roof panel height

1 Front shoe
2 Rear shoe

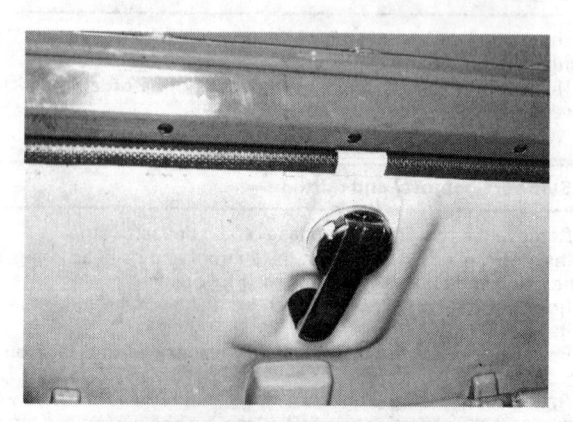

Fig. 12.25 Parked position of operating handle

Chapter 12 Bodywork and fittings

the screwdriver to separate the rear glass weatherstrip from the body flange.
7 Lever the lip of the weatherstrip over the body flange from inside to outside, pull the glass outwards and remove it with the weatherstrip.
8 If the weatherstrip has hardened, it is advisable to fit a new one before refitting the glass, because a non-pliable weatherstrip will be difficult to fit and may leak.
9 Fit a strong cord into the weatherstrip groove (Fig. 12.29).
10 Apply soapy water to the contact faces of the weatherstrip and the body flange.
11 With an assistant holding the glass in its properly aligned position and pushing it against the body, pull the cord from inside the car to pull the weatherstrip flange over the body flange. Start the operation in the middle of the bottom edge of the glass and gradually work round until the entire flange is in place.
12 Settle the glass by tapping the outside with the palm of the hand.
13 Fit masking tape along the edge of the weatherstrip and body on the outside surface of the glass and squeeze sealant under the weatherstrip. On completion of the sealing, peel off the masking tape.

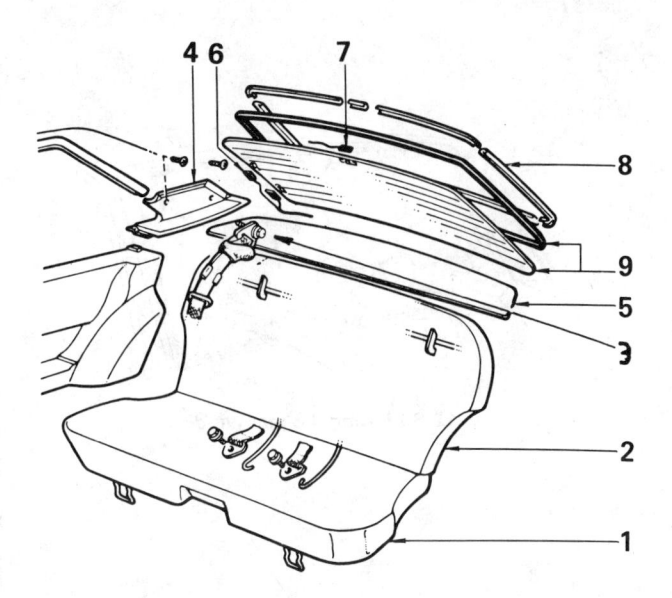

Fig. 12.26 Rear window removal

1 Rear seat cushion
2 Rear seat back
3 Rear seat belt shoulder anchor
4 Rear window side trim
5 Rear parcel shelf
6 Rear window demister wire
7 Radio aerial wire
8 Weatherstrip trim
9 Glass and weatherstrip

23 Boot lid (Coupe) – removal and installation

1 Open the boot lid and mark round the hinges so that they can be refitted in the same position.
2 Release the torsion bars, by prising their cranked ends clear of the hinge. This operation requires a large screwdriver, or a bar to use as a lever.
3 With an assistant supporting the lid, remove the four screws attaching the hinge to the lid and remove the lid.
4 After fitting the boot lid as a reversal of removal, check that the lid closes easily and securely without rattling. If necessary adjust the lock striker to achieve this.
5 If the position of the hinges is marked before removing the lid, the lid should be positioned correctly when refitted. If the lid does not have an even gap between its edge and the adjacent body, loosen the four hinge to body screws, push the lid into the correct position and re-tighten the screws.

24 Boot lid lock – removal, installation and adjustment

1 Remove the two screws securing the latch to the boot lid (photo) and remove the latch assembly.
2 Pull off the spring clip securing the lock barrel. Remove the packing piece from under the clip and push the lock cylinder assembly out of the boot lid panel.
3 The lock striker is attached to the rear panel by two screws. Loosening the screws allows the striker to be moved vertically and laterally to adjust the latching of the boot lid.

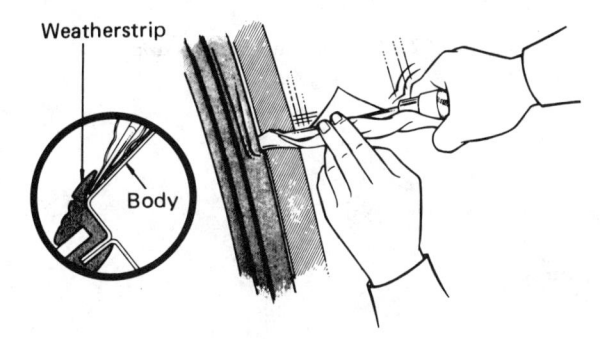

Fig. 12.27 Prising the weatherstrip away from the body

25 Rear door (Liftback) – removal and installation

1 Lift the rear door and support it in the open position.
2 Unscrew the damper stay from each side of the door panel.
3 Mark round the hinges and with an assistant supporting the door, remove the four bolts attaching the hinges to the door.
4 Installation is the reversal of removal and door adjustment is similar to that for the Coupe boot lid (Section 23).

26 Rear door lock (Liftback) – removal, installation and adjustment

1 The removal of the lock is similar to that of the Coupe boot lid (Section 24), but it is first necessary to remove the trim over the latch.
2 Remove the two screws from the cover trim, insert a screwdriver between the trim and the panel and prise the trim off.
3 Adjustment of the lock striker is similar to that for the Coupe, although the striker is of different appearance and incorporates a switch to operate the luggage compartment light (photo).

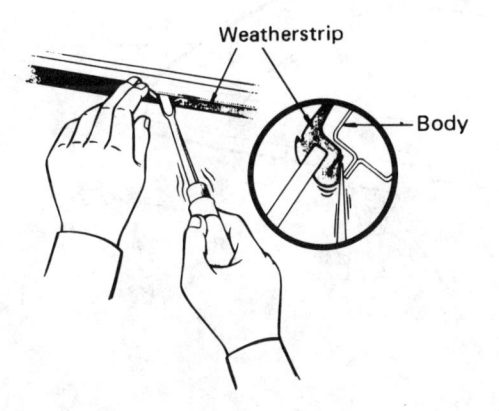

Fig. 12.28 Prising the weatherstrip flange over the body flange

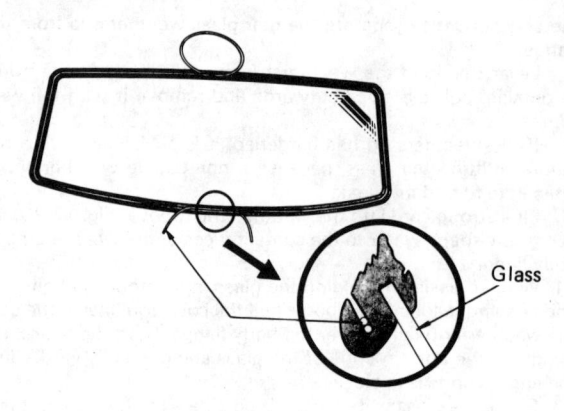

Fig. 12.29 Fitting a pull cord

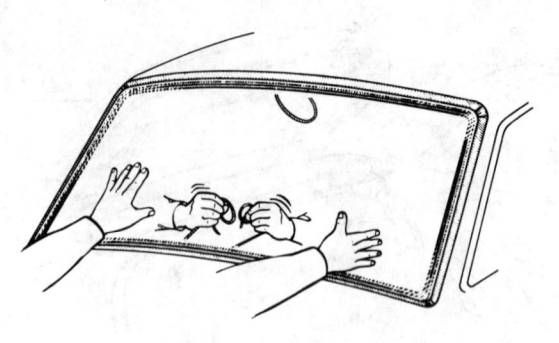

Fig. 12.30 Fitting the rear window

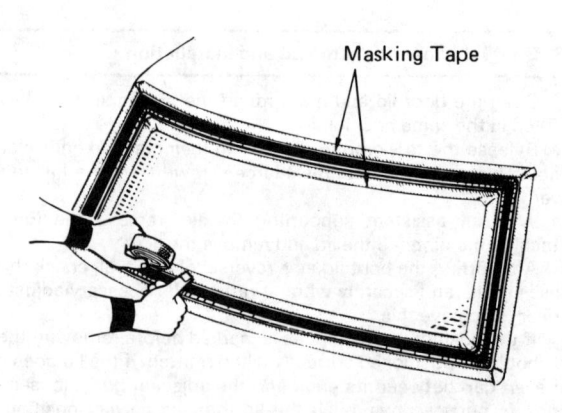

Fig. 12.31 Masking prior to sealing

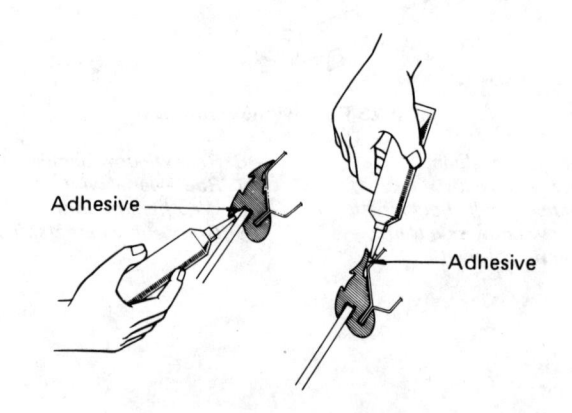

Fig. 12.32 Sealing the weatherstrip

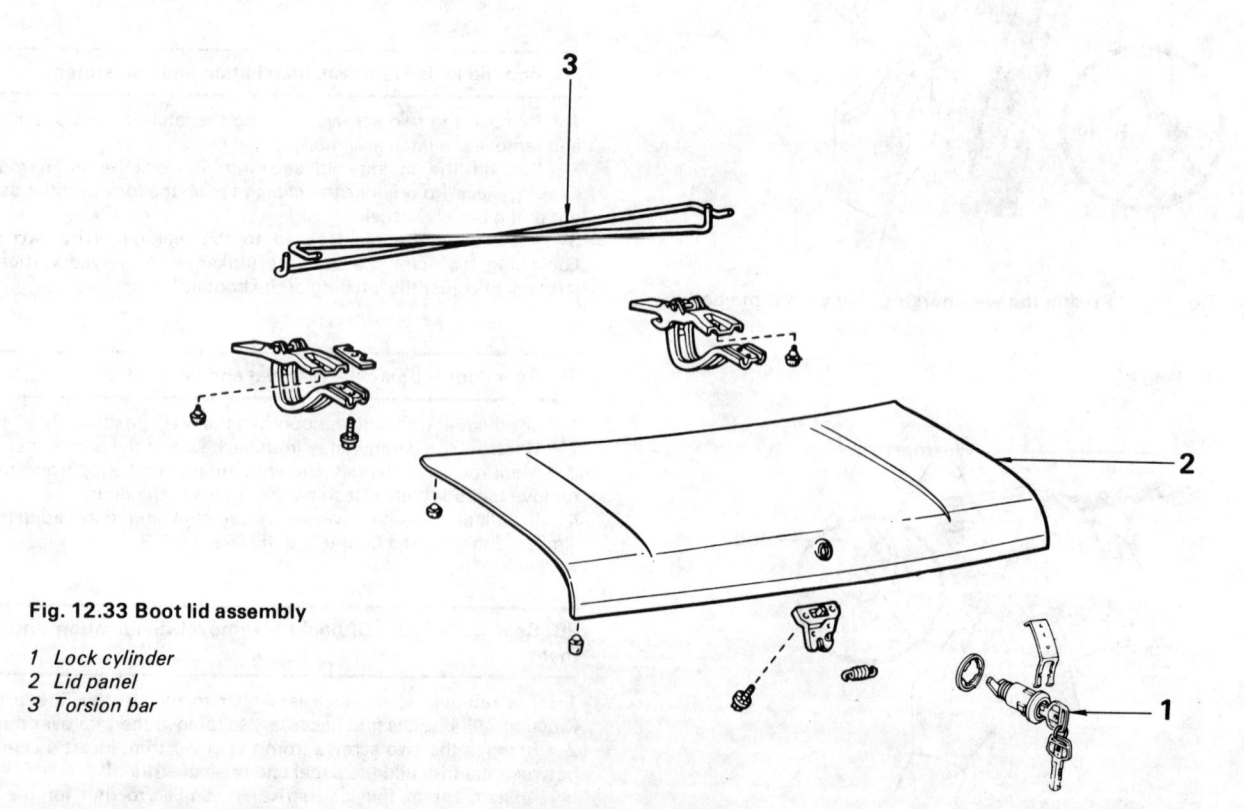

Fig. 12.33 Boot lid assembly

1 Lock cylinder
2 Lid panel
3 Torsion bar

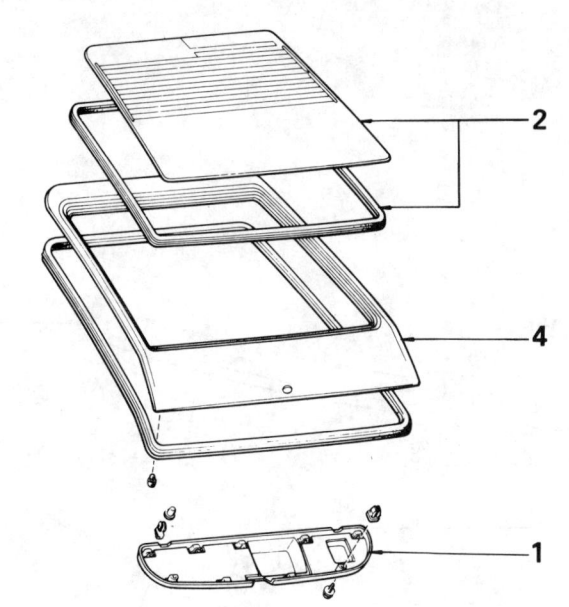

Fig. 12.34 Liftback rear door

1 Latch cover trim
2 Glass and weatherstrip
3 Damper stay
4 Door panel

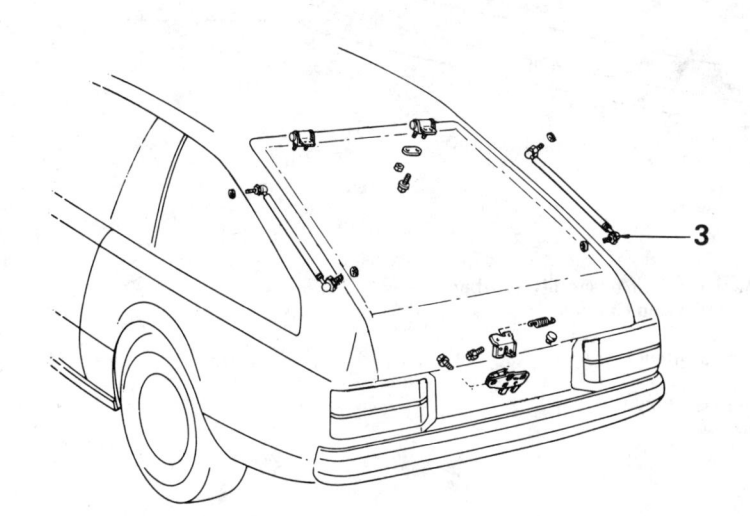

1 Grommet
2 Rear bumper assembly
3 Retainer
4 Bumper insert

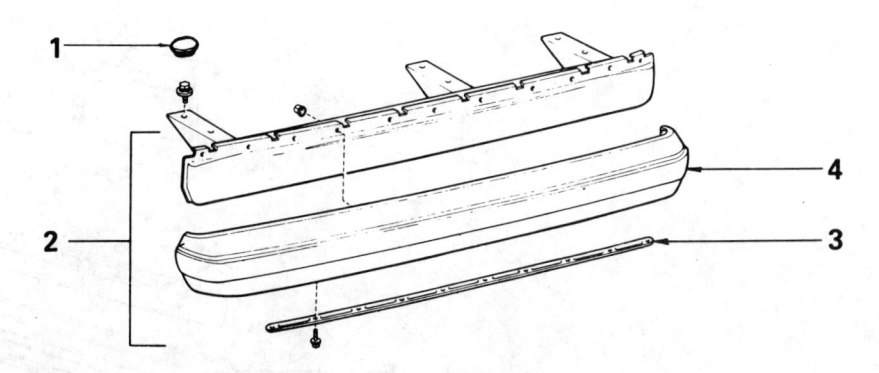

Fig. 12.35 Rear bumper assembly (Coupe) – USA and Canada

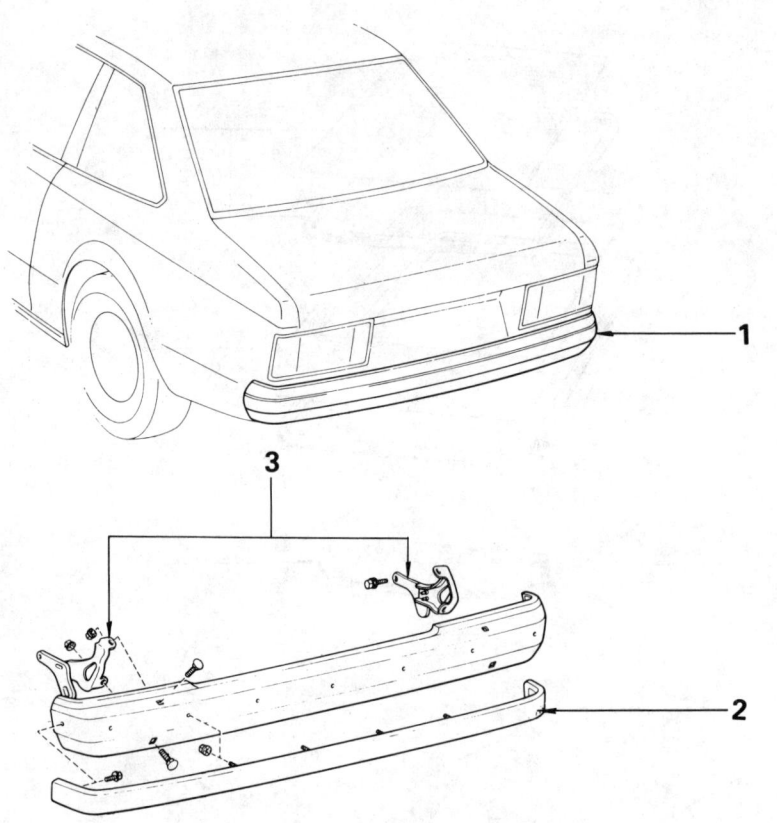

Fig. 12.36 Rear bumper assembly (Coupe) – except USA and Canada

1 Rear bumper assembly
2 Rear bumper insert
3 Bumper stays

Fig. 12.37 Rear bumper assembly (Liftback) – USA and Canada

1 Luggage compartment rear trim
2 Grommet
3 Rear bumper assembly
4 Bumper retainer
5 Bumper bar

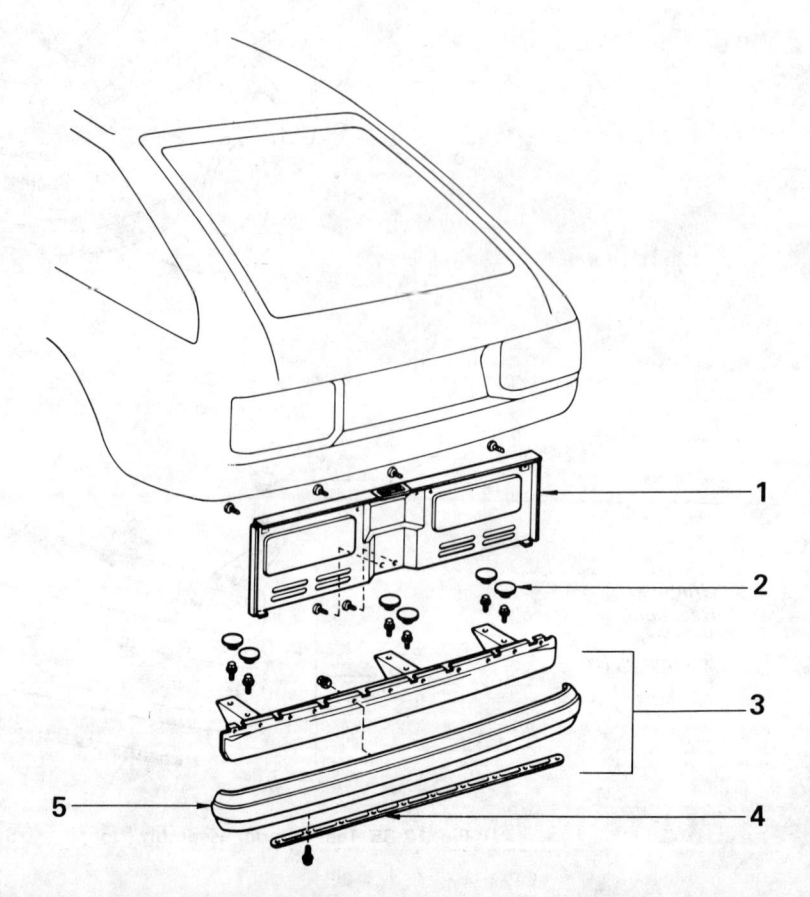

Chapter 12 Bodywork and fittings

24.1 Boot latch

26.3 Liftback lock striker and interior lamp switch

33.7 Heater temperature control valve

27 Rear door glass (Liftback) – removal and installation

The procedure is similar to that for the rear window of the Coupe, starting at paragraph 5 of Section 22.

28 Liftback rear door damper stays – precautions

1   The rear door dampers are filled with an inert gas under pressure and no attempt should be made to dismantle them.
2   Take care not to scratch or damage the piston rod. The rod must not be lubricated and care must be taken not to get paint on it.
3   Avoid turning the piston rod relative to the cylinder.

29 Rear bumper (Coupe) – removal and installation

1   The rear bumper of models for North America is different from that for other markets (Figs. 12.35 and 12.36).
2   With the exception of North American models, remove the bolt from each end of the bumper and remove the four nuts securing the bumper bar to the bumper brackets.
3   To remove the bumper insert, remove the four nuts securing the insert to the bumper, after removing the bumper.
4   When installing the bumper, insert the two end bolts and start their threads. Fit the four nuts and bolts attaching the bumper to the bumper stay, but do not tighten any nuts or bolts until all the fixings have been inserted.
5   On North American models, the bumper is removed as an assembly with its mounting brackets.
6   Lift the rear edge of the boot carpet to expose the grommets of the bumper attachment bolt access holes and remove the grommets.
7   Unscrew and remove the six bolts and remove the bumper assembly.
8   Separate the bumper and bumper insert by removing the bolts from the bumper retainer and removing the retainer.

30 Rear bumper (Liftback) – removal and installation

1   The rear bumper of the Liftback is similar to that of the Coupe and removal and installation procedures are similar to those detailed in the previous Section.
2   When removing the bumper assembly of models for North America, the rear panel of the luggage compartment trim has to be removed to reveal the access holes for the bumper bolts. After the rear panel has been removed, the quarter trim panels can be prised forward to get access to the end bolts, using a socket spanner on a long extension.

31 Instrument panel crash pad – removal and installation

1   Isolate the battery by removing the negative terminal.
2   Pull the knobs off the heater controls and then use a screwdriver to prise the panel out.
3   Remove the fascia panel (Chapter 10, Section 25).
4   Disconnect the speaker and aerial from the radio. Remove the four fixing screws and take out the radio.
5   Disconnect the clock, remove the two fixing screws and remove the clock, taking care not to jar it.
6   Remove the two fixing screws from each of the two side trim panels and take the trim panels off.
7   Remove the end of the glove box. Remove the two screws securing the glove box lining and remove it.
8   Remove the screws from the ends of the air duct and remove it.
9   Remove the screws and nuts securing the crash pad to the instrument panel and lift the crash pad off.
10   Installation is the reverse of removal.

32 Instrument panel – removal and installation

1   Remove the crash pad as detailed in the previous Section.
2   Unscrew and remove the gear lever knob.
3   Remove the three screws from the rear part of the console and the two screws from the front part and remove the console.
4   Remove the top and bottom screws from the console attachment bracket and remove the bracket.
5   Remove the two cowl side trim panels.
6   Remove the tray from under the glove box.
7   Disconnect and remove the instrument clusters (refer to Chapter 10, Section 25).
8   Remove the two screws from the bracket supporting the bonnet lock.
9   Remove the trim panel and the two air ducts.
10   Remove the two screws to free the fuse block assembly and disconnect and remove the tape player, if fitted.
11   Remove the air duct from the top of the heater and the two-piece duct from the front of the heater.
12   Disconnect the fout heater control wires, remove the four screws from the heater control panel and take out the panel and cable assembly.
13   Loosen the nut at each end of the instrument panel, remove eight screws and pull the panel out towards the rear of the car.
14   When refitting the panel, which is the reversal of removal, ensure that all electrical connections are firm and that the heater controls operate properly over the full range of control.

33 Heater and ventilation system – description

1   The system draws fresh air into the car interior through the grilles mounted just in front of the windscreen and stale air is expelled through the flap valves located to the rear of the quarter windows.
2   When the car is moving slowly, an electrically operated booster fan can be switched on to increase the airflow.
3   In heavy traffic conditions, the entry of air can be stopped and the air in the car interior recirculated to prevent any fumes being drawn in.
4   Fully adjustable face-level louvres are provided for driver and passenger in addition to the heater main louvre.

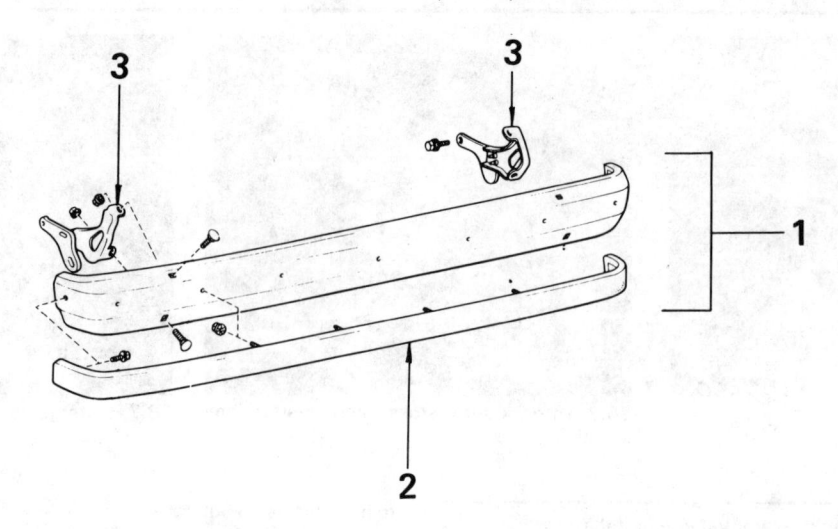

Fig. 12.38 Rear bumper assembly (Liftback) – except USA and Canada

1 Rear bumper assembly   2 Bumper insert   3 Bumper stays

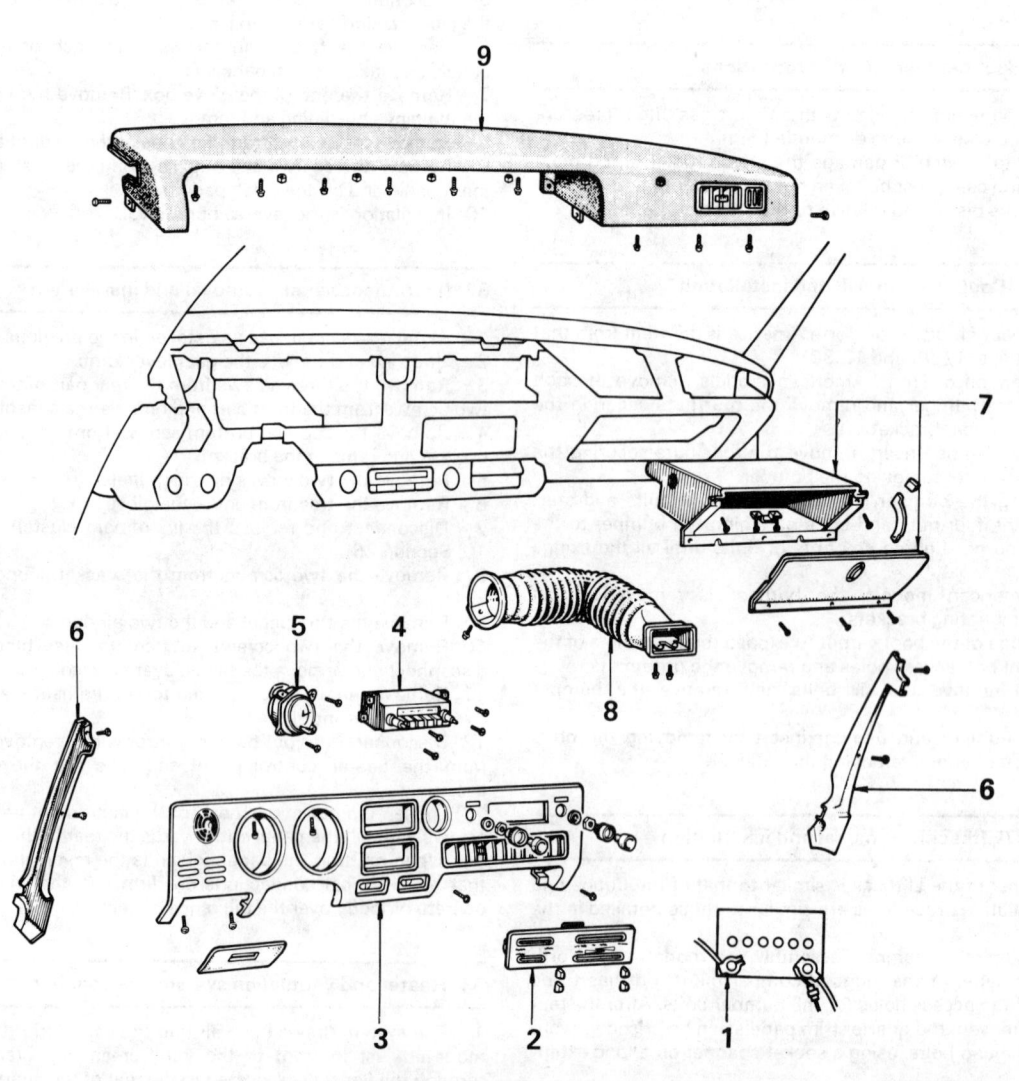

Fig. 12.39 Instrument panel components (1)

1 Negative battery terminal
2 Heater knobs and panel
3 Fascia panel
4 Radio
5 Clock
6 Front pillar trim
7 Glove box
8 Air duct
9 Crash pad

Chapter 12 Bodywork and fittings

281

5 The control levers enable a wide selection of air temperature, airflow direction and volume to be made also a choice of fresh or recirculated air.
6 The heater assembly comprises a matrix, heated by water from the engine cooling system and a booster fan controlled by a three position switch. During normal forward motion of the car, air is forced into the air intake and passes through the heater matrix absorbing heat and then carrying it to the car interior.
7 Temperature control is achieved by a valve (photo) which controls the flow of coolant through the heater matrix.

34 Heater blower motor – removal and installation

1 The blower unit may be removed without disturbing the main heater assembly.
2 Isolate the battery by disconnecting the negative terminal.
3 Remove the dashboard under tray.
4 Prise the air inlet control outer cable from its clip and unhook the inner wire from the damper arm.
5 Disconnect the motor electrical connection from the wiring harness and unclamp the blower air duct.
6 Remove the four fixings from the blower unit and remove it from beneath the dashboard.
7 Remove the three mounting screws and withdraw the motor and impeller assembly from the casing.

35 Heater control – removal and installation

1 Isolate the battery by disconnecting the negative terminal.
2 Remove the three screws from the dashboard under tray and remove the tray.
3 Remove the lid from the glove box, then unscrew the glove box and remove it.
4 Pull the knobs from the heater controls and prise off the control panel with a screwdriver.
5 Remove the cover from the fuse box.
6 Remove the knobs from the radio and the six screws from the instrument panel. Pull the panel out at the bottom and remove it.
7 Disconnect and remove the instrument cluster and the clock.
8 Disconnect the aerial and speaker wires from the radio. Remove the four radio fixing screws and remove the radio.
9 Remove the cranked air duct from the front of the heater assembly and the angled duct from the top of the heater.
10 Disconnect the four heater control cables and remove the connections from the heater blower switch.
11 Remove the four fixing screws and remove the heater control assembly with the control cables attached.
12 The control panel may be dismantled by removing the clips from

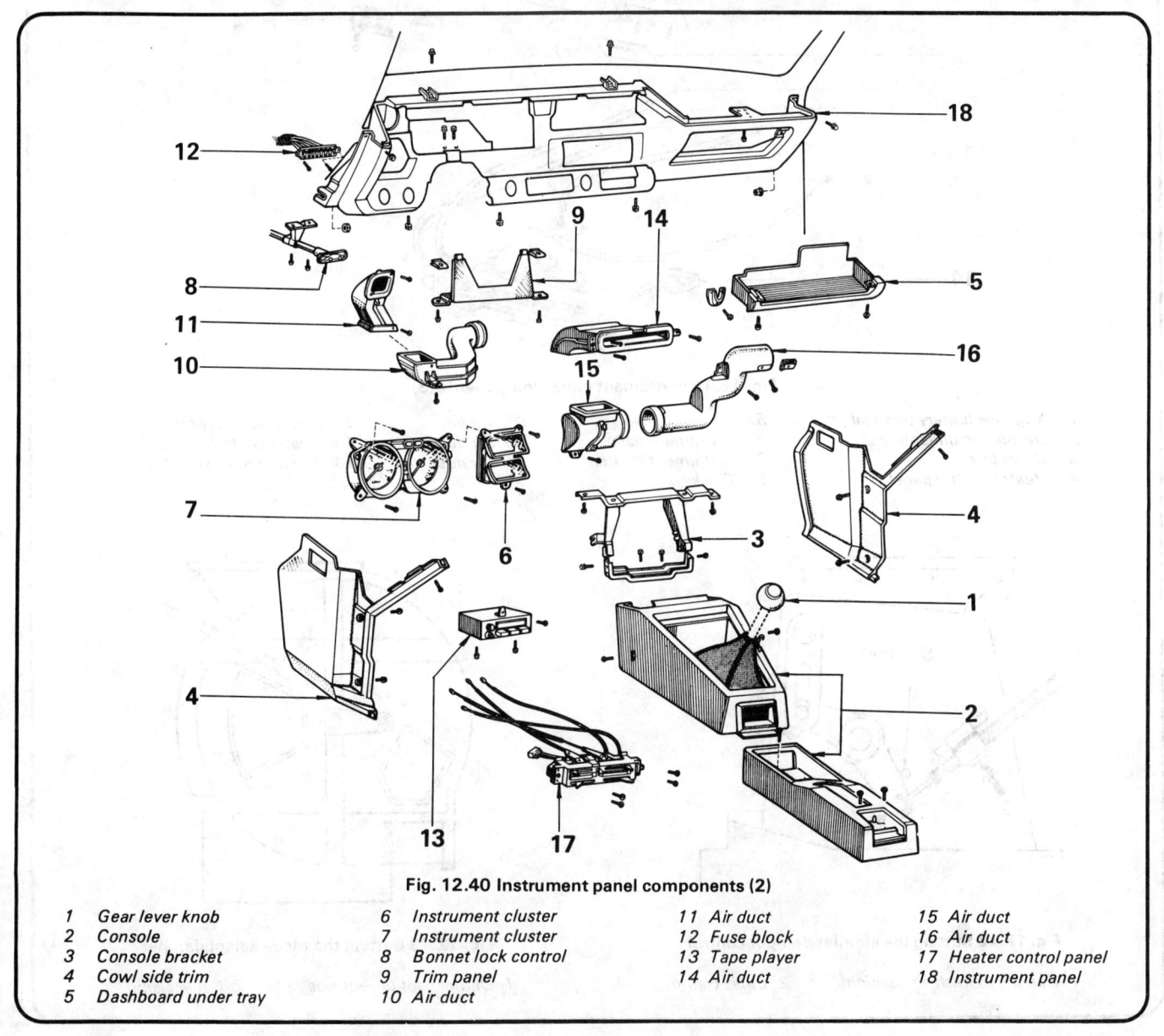

Fig. 12.40 Instrument panel components (2)

1	Gear lever knob	6	Instrument cluster	11	Air duct	15	Air duct
2	Console	7	Instrument cluster	12	Fuse block	16	Air duct
3	Console bracket	8	Bonnet lock control	13	Tape player	17	Heater control panel
4	Cowl side trim	9	Trim panel	14	Air duct	18	Instrument panel
5	Dashboard under tray	10	Air duct				

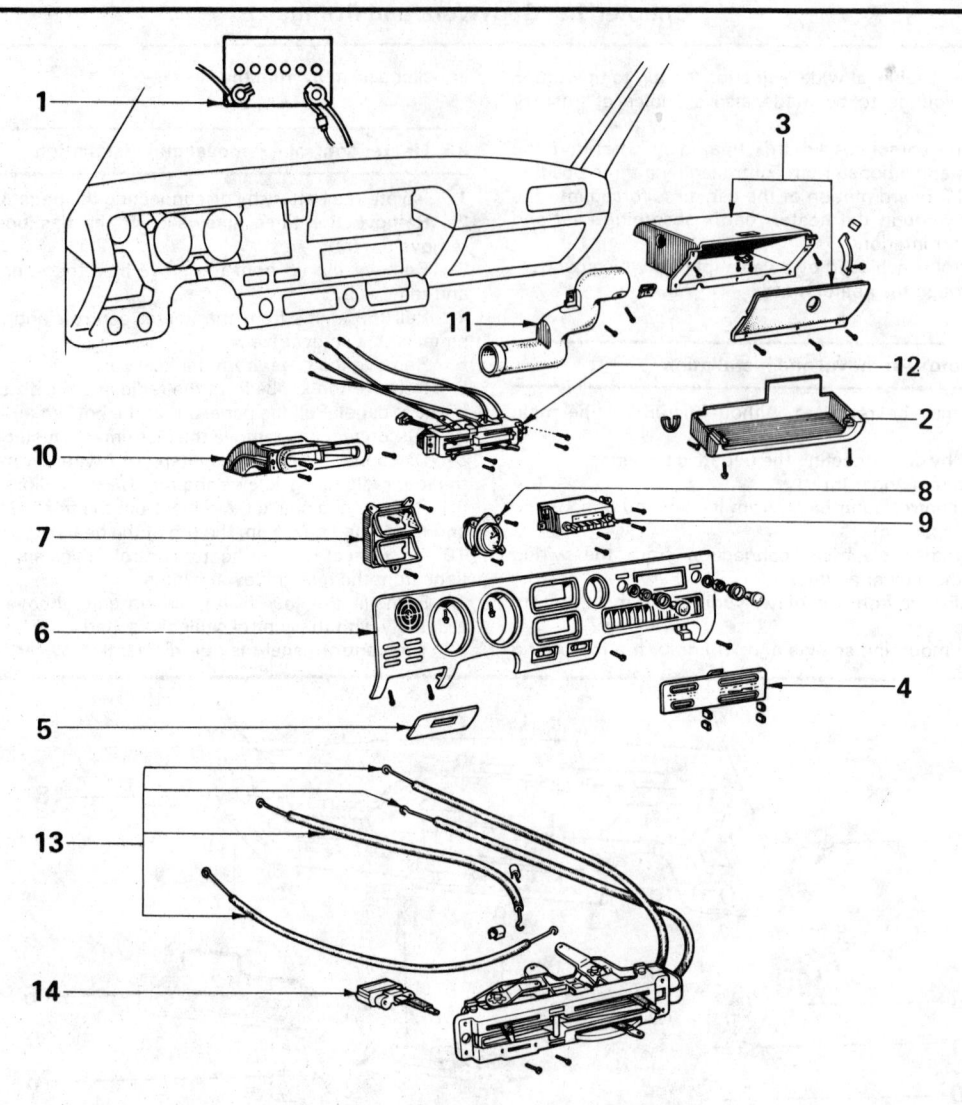

Fig. 12.41 Instrument panel components (3)

1	Negative battery terminal	5 Fuse box cover	9 Radio	12 Heater control assembly
2	Dashboard under tray	6 Instrument panel	10 Air duct	13 Heater control cables
3	Glove box	7 Instrument cluster	11 Air duct	14 Blower motor switch
4	Heater control panel	8 Clock		

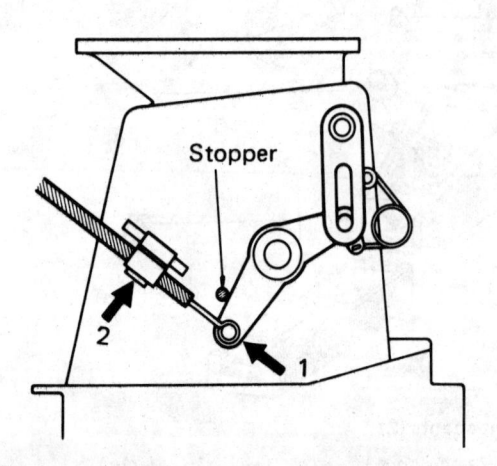

Fig. 12.42 Setting the air inlet damper control

1 Damper in fresh air position   2 Cable clamp

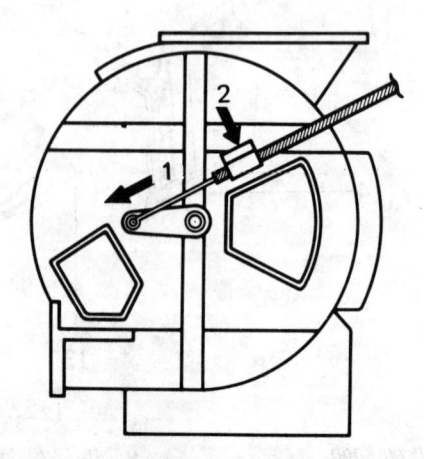

Fig. 12.43 Setting the mode select damper

1 Damper set to vent side   2 Cable clamp

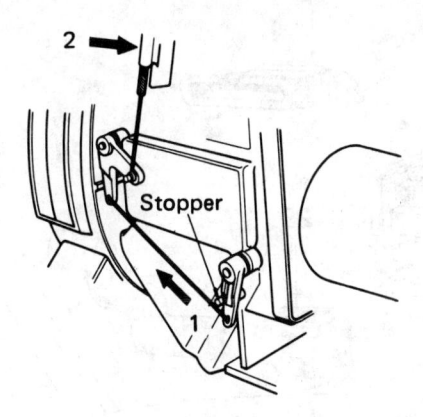

Fig. 12.44 Setting the air mix damper

1 Damper set to cool
2 Cable clamp

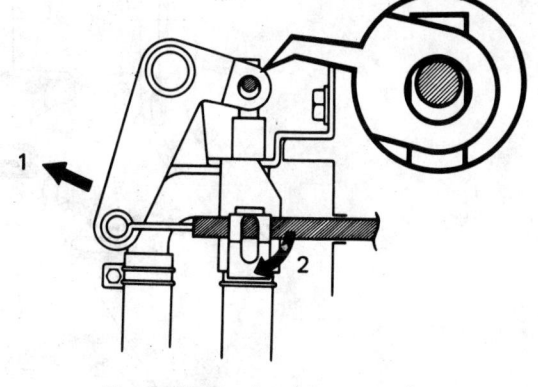

Fig. 12.45 Setting the water valve

1 Valve set to cold
2 Cable clamp

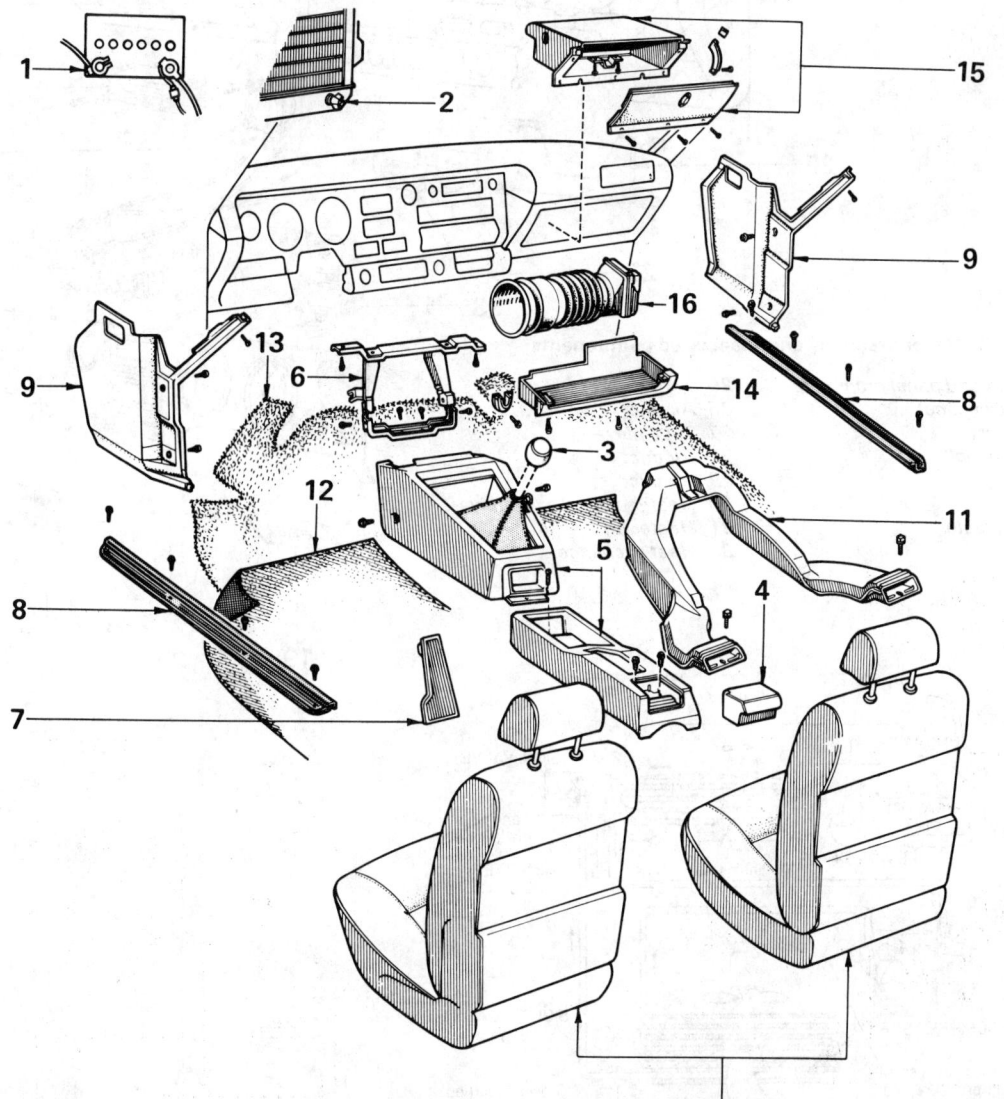

Fig. 12.46A Heater assembly and associated components

1 Battery negative terminal	5 Console	9 Cowl side trim	13 Carpet underfelt
2 Radiator drain tap	6 Console support bracket	10 Front seats	14 Dashboard under tray
3 Gear lever knob	7 Accelerator pedal	11 Heater rear duct	15 Glove box
4 Rear ash tray	8 Scarf plates	12 Front carpets	16 Blower duct

Fig. 12.46B Heater assembly and associated components

17 Heater hoses and panel grommets
18 Heater control panel
19 Fuse box cover
20 Instrument panel
21 Radio
22 Clock
23 Instrument cluster
24 Air duct
25 Trim panel
26 Air duct
27 Tape player
28 Air duct
29 Air duct
30 Heater control assembly
31 Heater unit
32 Heat exchanger

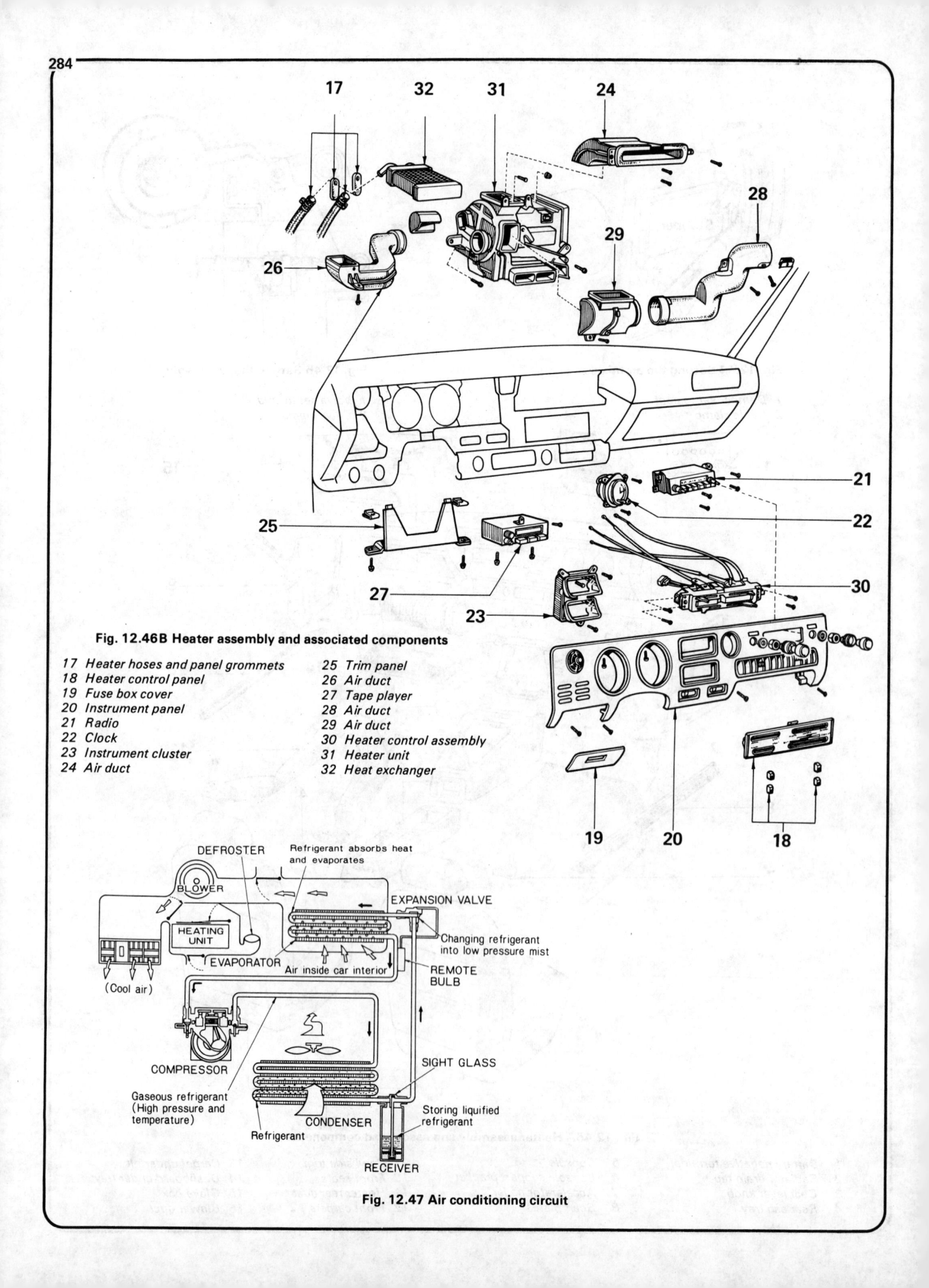

Fig. 12.47 Air conditioning circuit

the outer cables and unhooking the inner ones. The blower switch may be taken off after removing its two fixing screws.
13 Installing the panel is the reversal of the removal operations, but on fitting the panel, connect the control cables and adjust as described in the following Section.

36 Heater control – cable adjustment

1 The heater control panel operates four control cables, each of a different length. The shortest one operates the *Mode selection damper* and the next longest the *Water valve*. The longest cable operates the *Air inlet damper* and the remaining one the *Air mix damper*.
2 To set the *Air inlet damper* cable, push the damper against its stop on the fresh air side and connect the inner wire to the damper lever.
3 Hold the lever in this position, slide the outer cable as far away from the damper as possible and fix the outer cable in the cable clip (Fig. 12.42).
4 To set the *Mode select damper,* push the damper until its lever is horizontal in the vent position and connect the inner wire to the damper lever.
5 Hold the lever in this position, slide the outer cable as far away from the damper as possible and fix the outer cable in the cable clip (Fig. 12.43).
6 To set the *Air mix damper,* push the damper against its stop in the cool position and connect the inner wire to the damper lever.
7 Hold the lever in this position, slide the outer cable as far away from the damper as possible and fix the outer cable in the cable clip (Fig. 12.44).
8 To set the *Water valve,* lift the valve control lever up as far as possible and connect the inner wire to the lever.
9 Hold the lever in this position, slide the outer cable as far away from the lever as possible and fix the outer cable in the cable clip (Fig. 12.45).
10 After connecting the cables, operate each of the control levers and check that the lever to which it is connected operates smoothly over its full range of travel.

37 Heater unit – removal and installation

1 Isolate the battery by disconnecting the negative terminal.
2 Drain the cooling system, referring to Chapter 2 if necessary.
3 Remove the knob from the gear lever.
4 Remove the rear ash tray to expose the rear fixing screws of the centre console. Remove the five screws securing the two parts of the console and remove the console.
5 Remove the console support bracket.
6 Disconnect the accelerator linkage from the pedal and remove the pedal.
7 Remove the scarf plate from each side of the car.
8 Remove the three screws from each of the cowl side trims and remove the trims.
9 Remove the four bolts attaching the seat tracks to the floor and remove both the front seats.
10 Remove the rear heater duct.
11 Take out the front carpet and its underfelt.
12 Remove the dashboard under tray, the glove box lid and then the glove box.
13 Remove the two clamps from the blower air duct and remove the duct.
14 Slacken the clips on the two water hoses to the heater, place a tray underneath to catch any coolant which runs out and remove the hoses. Remove the two panel grommets from the heat exchanger pipes.
15 Remove the heater control, as described in Section 35.
16 Remove the two bolts, the nut and the screw retaining the heater unit and remove the unit from the passenger side.
17 Pull the heat exchanger out of the heater unit.
18 Refitting the heater is a reversal of removal.

38 Air conditioning system – description and maintenance

1 This optionally specified system (standard on XT models) comprises a heating and cooling unit, a belt driven compressor, a condenser and a receiver together with the necessary temperature controls and stabilizers.
2 The oil filled compressor is driven by belt from the crankshaft pulley and incorporates a magnetic type clutch.
3 The air conditioning control is mounted above the ashtray on the centre ventilation louvre panel.
4 A layout of the air conditioning system is shown from the point of view of interest, but servicing is beyond the scope of the home mechanic. Special equipment is needed to purge or recharge the system with refrigerant gas and dismantling of any part or component must not be undertaken without having first discharged the system pressure. This is a job for the professional refrigeration engineer.
5 To maintain optimum performance the owner should limit his operations to the following items.

   (a) Checking the drivebelt tension. There should be a total deflexion of $\frac{1}{2}$ in. at the midway point of the longest run of the belt. Where adjustment is required, release the locknut on the idler pulley adjuster bolt and screw the bolt in or out as required.
   (b) Checking the security of all hoses and unions.
   (c) Checking the security of the electrical connections.
   (d) Removing and washing (periodically) the air filter screen. This is located in the air inlet in front of the evaporator, which is mounted below the instrument panel.
   (e) Use a soft brush to remove accumulations of dust and dirt from the fins of the condenser which is mounted ahead of the radiator. During the cold season, operate the air conditioner for a few minutes each week to lubricate the interior of the compressor.

Chapter 13 Supplement:
Revisions and information on later USA models

Contents

Introduction... 1	Manual transmission................................. 6
Specification... 2	W55 - dismantling
Engine (22R)... 3	W55 - reassembly
Engine/transmission - removal	Electrical system..................................... 7
Sump and timing gear - removal	Alternator - disassembly
Timing components	Alternator - reassembly
Fuel, exhaust and emission control systems........ 4	IC Regulator - testing
Carburetor - dismantling	O_2 sensor (22R) - cancel switch
Carburetor - inspection	O_2 sensor (22R) - reset procedure
Carburetor - reassembly	Suspension and steering............................. 8
Emission control systems	Tilt steering column and shaft - installation
Ignition system.. 5	Tilt steering column and shaft - removal
Distributor - general	

1 Introduction

This supplement contains specifications and changes that apply to Celicas produced in 1980 and 1981. Also included is information related to previous models that was not available at the time of original publication of this manual.

Where no differences (or very minor differences) exist between 1980 and 1981 models and previous models, no information is given. In those instances, the original material included in Chapters 1 through 12 pertaining to the 20R engine and the W50 transmission should be used.

2 Specifications

Dimensions and weights
1980

	Coupe	Liftback
Length..	175.6 in (4460 mm)	176.2 in (4475 mm)
Width...	64.5 in (1640 mm)	64.5 in (1640 mm)
Height..	51.2 in (1300 mm)	50.8 in (1290 mm)
Wheelbase.......................................	98.4 in (2500 mm)	98.4 in (2500 mm)
Ground clearance...............................	6.7 in (170 mm)	6.7 in (170 mm)
Front track......................................	53.1 in (1350 mm)	53.1 in (1350 mm)
Rear track.......................................	53.7 in (1365 mm)	53.7 in (1365 mm)

1981
(Same as above except for the following:)

Height..	50.8 in (1290 mm)	50.6 in (1285 mm)

Engine (22R)
General

Type..	4 Cylinder, inline, single OH
Displacement.....................................	2366 cc (144.4 cu. in)
Compression pressure...........................	171 psi (12.0 Kg/cm^2)
Compression limit................................	128 psi (9.0 Kg/cm^2)
Permissible difference between cylinders.....	14 psi (1.0 Kg/cm^2)
Compression ratio................................	9.0 : 1
Bore..	3.62 in (92.0 mm)
Ignition timing....................................	8° BTC at 950 rpm max (vacuum adv. off)
Oil capacity	
With oil filter change........................	4.9 US qts (4.6 liters)
Without filter change........................	4.0 US qts (3.8 liters)

Chapter 13/Supplement: Revisions and information on later USA models

Valves
Overall length
    Inlet . . . . 4.468 in (113.5 mm)
    Exhaust . . . . 4.425 in (112.4 mm)
Valve stem diameter
    Inlet . . . . 0.3188 to 0.3145 in (7.970 to 7.985 mm)
    Exhaust . . . . 0.3136 to 0.3142 in (7.965 to 7.980 mm)
Valve head thickness - limit . . . . 0.024 in (0.6 mm)

Valve spring
Free length . . . . 1.803 in (45.8 mm)
Installed length . . . . 1.594 in (40.5 mm)

Manifold
Surface warpage - limit . . . . 0.0276 in (0.70 mm)

Cylinder block
Cylinder bore - standard . . . . 3.6220 to 3.6232 in (92.0 to 92.3 mm)
Wear limit . . . . 0.008 in (0.2 mm)
Taper and out-of-round-limit . . . . 0.0008 in (0.02 mm)

Pistons and rings
Outside diameter - standard . . . . 3.6196 to 3.6208 in (91.938 to 91.968 in)
Piston oversizes . . . . 0.50 mm and 1.00 mm
Cylinder-to-piston-clearance . . . . 0.0020 to 0.0028 in (0.052 to 0.072 mm)
Piston ring endgap (compression)
    No. 1 . . . . 0.0094 to 0.0142 in (0.24 to to 0.36 mm)
    No. 2 . . . . 0.0071 to 0.0154 in (0.18 to 0.39 mm)
Piston pin installing temperature . . . . 176°F (80°C)

Crankshaft
Journal clearance
    Standard . . . . 0.0006 to 0.0020 in (0.016 to 0.05 mm)
    Limit . . . . 0.0031 in (0.08 mm)
Journal taper and out-of-round-limit . . . . 0.0004 in (0.01 mm)

Torque wrench settings
Crankshaft pully-to-crankshaft . . . . 115 ft/lbs (16 kg-m)
Flywheel-to-crankshaft . . . . 80 ft/lbs (11 kg-m)
Thermo switch-to-inlet manifold . . . . 22 ft/lbs (3 kg-m)

Fuel, exhaust and emission control system
Carburetor (1980 20R)
Carburetor part numbers 21100-38311, 38331, 38351, 38361, 38440
Throttle valve full-open angle
    Primary . . . . 90° from horizontal plane
    Secondary . . . . 75° from horizontal plane
Secondary touch angle . . . . 59° from horizontal plane
Fast idle . . . . 24° from horizontal plane
Unloader angle . . . . 50° from horizontal plane
Choke breaker angle
    First . . . . 38° from horizontal plane
    Second (ex. Canada) . . . . 60° from horizontal plane
Float raised position . . . . 0.28 in (7 mm)
Float lowered position . . . . 0.04 in (1 mm)
Kick-up . . . . 0.008 in (0.2 mm)
Idle mixture adjusting screw presetting
    21100-38331 . . . . Screw out 1-1/3 turns
    All others . . . . Screw out 1-1/2 turns
Accelerator pump stroke . . . . 0.154 in (3.9 mm)
Throttle positioner angle . . . . 16.5° from horizontal plane

Carburetor (1981 22R)
Carburetor part numbers 21100-35010, 35020, 35030, 35040, 35050
Trottle valve closed angle
    Primary . . . . 9° from horizontal plane
    Secondary . . . . 20° from horizontal plane

Carburetor (1981 22R) continued

Throttle valve full-open angle
    Primary . . . . 90° from horizontal plane
    Secondary . . . . 90° from horizontal plane
Secondary touch angle . . . . 59° from horizontal plane
Fast idle angle . . . . 24° from horizontal plane
Unloader angle . . . . 45° from horizontal plane
Float level
    Raised (float top to air horn) . . . . 0.413 in (10.5 mm)
    Lowered (float bottom to air horn) . . . . 1.89 in (48 mm)
Float lip clearance (at float lowered) . . . . 0.04 in (1 mm)
Idle mixture presetting . . . . Screw out 2-1/2 turns
Idle mixture speed
    Manual transmission . . . . 740 rpm
    Automatic transmission (USA) . . . . 790 rpm
    Automatic transmission (Canada) . . . . 890 rpm
Accelerator pump stroke . . . . 0.161 in (4.1 mm)
Throttle positioner angle . . . . 16° from horizontal plane

Ignition system
General
System type . . . . Transistorized ignition
Ignition timing . . . . 8° BTDC at max 950 rpm
Distributor air gap . . . . 0.008 to 0.016 in (0.2 to 0.4 mm)

Ignition coil
1980
	Federal	California
Primary coil resistance	0.5 to 0.6 ohms	0.8 to 1.0 ohms
Secondary coil resistance	1.5 to 15.5 K ohms	11.5 to 15.5 K ohms
Resistor resistance	1.2 to 1.4 ohms	

1981
Primary coil resistance . . . . 0.8 to 1.0 ohms
Secondary coil resistance . . . . 11.5 to 15.5 K ohms

Clutch
Pedal height . . . . 6.28 to 6.67 in (159.5 to 169.5 mm)
Pedal freeplay . . . . 0.51 to 0.91 in (13 to 23 mm)

Transmission (W55)
Oil Capacity . . . . 2.5 qts (2.4 liters)

Brakes
Front discs
Disc thickness (minimum) . . . . 0.453 in (11.5 mm)
Disc runout (maximum) . . . . 0.006 in (0.15 mm)
Pad thickness (1980 minimum) . . . . 0.04 in (1.0 mm)
Pad thickness (1981 minimum) . . . . 0.118 in (3.0 mm)

Suspension and steering
Torque wrench settings - tilt steering
Steering pawl set bolt . . . . 11 to 15 ft/lbs (1.5 to 2.2 kg–m)
Tilt lever retainer . . . . 11 to 15 ft/lbs (1.5 to 2.2 kg–m)
Support-to-breakaway bracket nut . . . . 11 to 15 ft/lbs (1.5 to 2.2 kg–m)
Breakaway bracket-to-column tube . . . . 11 to 15 ft/lbs (1.5 to 2.2 kg–m)
Main shaft-to-intermediate shaft . . . . 15 to 21 ft/lbs (2.0 to 3.0 kg–m)
Support stopper bolt . . . . 70 to 104 in/lbs (0.8 to 1.2 kg–m)
Upper bracket-to-tilt steering support . . . . 52 to 78 in/lbs (0.6 to 0.9 kg–m)

Chapter 13/Supplement: Revisions and information on later USA models

3 Engine

Engine/transmission - removal (22R)
1 Use the same procedure given in Chapter 1, Section 5 with these additions before paragraph 23.
2 Remove the power steering pump and drive belt if so equipped. Attach the pump to the wheelwell, out of the way, using mechanic's wire. The hoses don't need to be disconnected.
3 Remove the engine shock absorber from the left motor mount.
4 Remove the motor mount bolts from each side of the engine.
5 From beneath the car, remove the two engine shock absorber bolts.
6 Remove the two bolts securing the engine shock absorber cover and remove the cover.
7 Disconnect the Reverse light switch wire from the transmission.
8 Disconnect the electrical connectors to the Neutral start switch and the overdrive solenoid, if the vehicle has an automatic transmission.
9 Continue on with paragragh 23 of Section 5.

Sump and timing gear
10 If the sump is to be removed with the engine still in the car, it will be necessary to remove the engine undercover, engine shock absorber cover, engine shock absorber, motor mounts and jack up the engine about an inch (25mm) before the sump can be removed.

Timing Components - examination and renovation
11 Refer to Section 27 of Chapter 1.
12 In addition, measure 17 links pulled tight. This measurement should be no more than 5.787 in (147.0mm). Measure at random, in at least three different places, and replace the chain if over the limit anywhere along its length.
13 Wrap the chain around the sprocket and measure the outer sides of the chain rollers using a vernier caliper. Measure the camshaft sprocket to be at least 4.480 in. (113.8mm) and the crankshaft sprocket to be at least 2.339 in. (59.4mm). Replace if necessary.
14 Inspect the chain tensioner for wear. Measure the tensioner with a vernier caliper. Minimum thickness is 0.43 in. (11mm).
15 Check the chain dampeners for wear and measure them with a micrometer. The No.1 dampener has a minimum of 0.197 in. (5.0mm) and the No.2 has a minimum of 0.177 in. (4.5mm). Replace if the part is worn or less than the minimum.

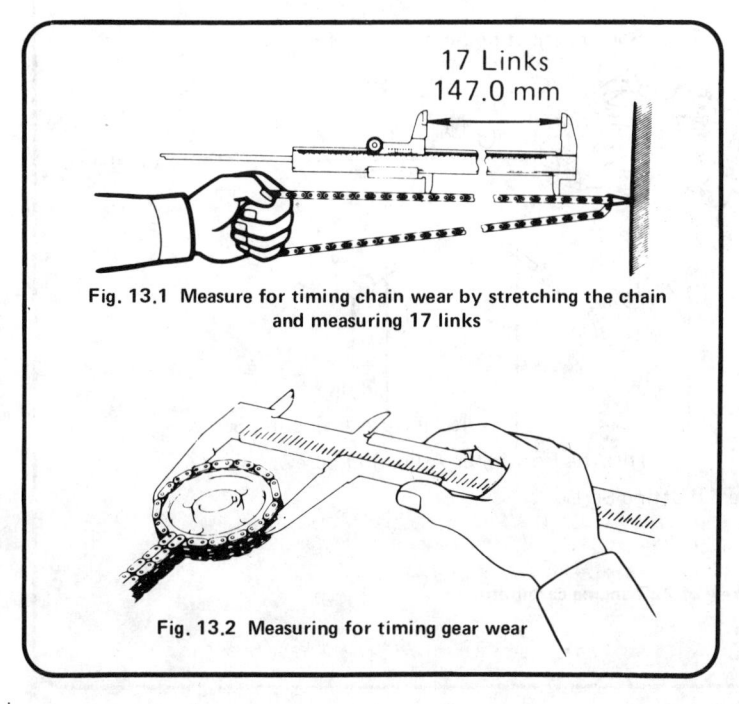

4 Fuel, exhaust and emission control systems

Carburetor (22R) - dismantling
1 Remove the carburetor from the vehicle as described in Chapter 3, Section 15.
2 The carburetor should not be dismantled unnecessarily. If a carburetor has had prolonged use and is likely to be badly worn, it is better to purchase a new, or exchange unit which has been tested and calibrated, rather than try and obtain a lot of small replacement parts.
3 Before starting to dismantle a carburetor, it is vital to have a clean work place, screwdrivers and spanners which are the correct fit on the parts to be removed and several clean containers in which to put the parts of each sub-assembly.
4 Clean the outside of the carburetor before starting to dismantle it.
5 The idle mixture screw is adjusted and sealed by the manufacturer to conform to U.S. emission standards.

Air horn
6 Loosen the screw and remove the metering needle.
7 Disconnect the fast idle link from the throttle shaft lever.
8 Remove the 5 screws that hold the air horn to the carburetor body and lift away the air horn and gasket.
9 Loosen the solenoid valve and remove it by rotating the body counterclockwise.
10 If the air horn is to be dismantled, remove the float pivot pin and float; remove the needle valve, spring, plunger and seat.
11 Pull out the pump plunger. Remove the power piston retainer, the power piston and spring.
12 Remove the outer vent control valve.

Carburetor body
13 On U.S. vehicles equipped with automatic transmissions, remove the dash pot.
14 Remove the slow jet, power valve with jet, the metering needle guide, secondary main jet, plug, and the primary main jet.
15 Remove the accelerator pump housing, diaphragm and spring.
16 Remove the auxiliary accelerator pump housing, diaphragm and spring.

Fig. 13.1 Measure for timing chain wear by stretching the chain and measuring 17 links

Fig. 13.2 Measuring for timing gear wear

Fig. 13.4 Removing the metering needle

Fig. 13.5 The solenoid valve can be removed by turning the body as indicated

Fig. 13.3 Exploded view of 22R engine carburetor

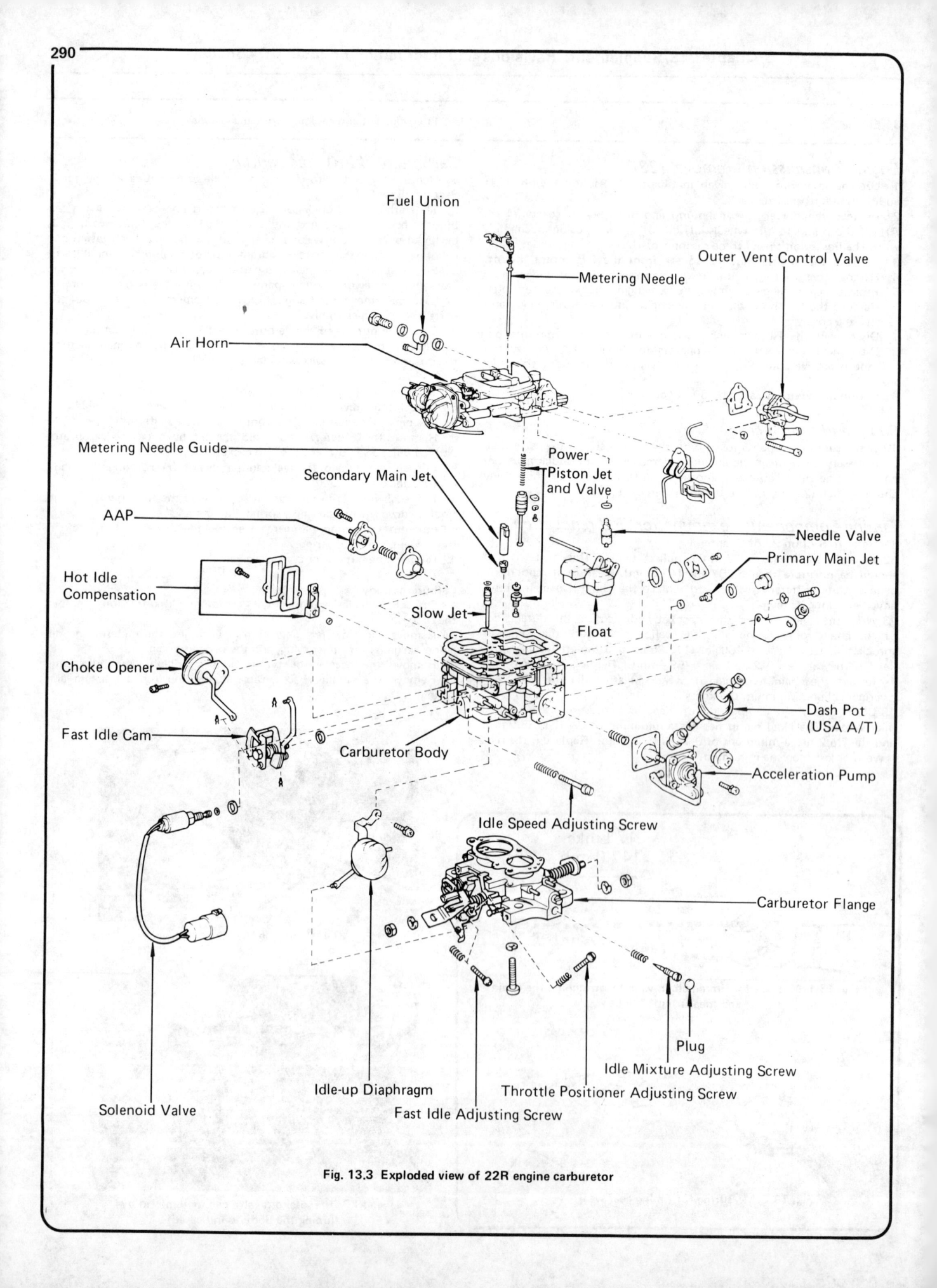

17 Remove the hot idle compensation (HIC) valve cover, the thermostatic valve and rubber valve seat.
18 Disconnect the idle-up diaphragm link and remove the diaphragm.
19 Disconnect the choke opener link and remove the choke opener.
20 Separate the flange from the body.

Carburetor (22R) - Inspection

21 Check all parts for damage and wear. Blow out the jets and air passages to clear them, but do not attempt to use wire to clean jets and orifices, because this can distort them and have a bad effect on fuel consumption.
22 Inspect the vent control valve and valve seats for damage. Check that the rod moves slowly. Measure the resistance between the terminal and solonoid body to be 63 to 73 ohms at 68°F (20°C) with an ohmmeter.
23 Inspect the choke breaker diaphragm by applying vacuum. The choke valve should slightly open and should not lose vacuum pressure immediately.
24 Measure the resistance between the terminal and choke heater housing to be 16 to 20 ohms at 68°F (20°C).
25 Connect a jumper lead from the negative terminal of a battery to the negative side of the fuel cut-out solenoid. When the positive side is connected with another jumper wire to the positive terminal of the battery a "clicking" should be heard.
26 Apply vacuum to the choke opener diaphragm. The vacuum should not drop immediately and the link should move.
27 Apply vacuum to the idle-up diaphragm. Check that the vacuum does not drop immediately and the link moves.

Carburetor (22R) - Reassembly

28 Wash all parts with carburetor cleaner before reassembling. Use new gaskets and as each unit is reassembled, check that any sliding or rotating parts move smoothly. If screws have been peened and filed to remove the peening, they should be discarded and new screws fitted. The new screws should be peened after tightening.

Fig. 13.6 The power piston and spring are retained by a Phillips screw

Fig. 13.7 Removing the power valve with jet (2) the metering needle guide (3), the secondary main jet (4), the plug (5) and the primary main jet (6)

Fig. 13.8 The idle-up diaphragm is secured to the carburetor where shown

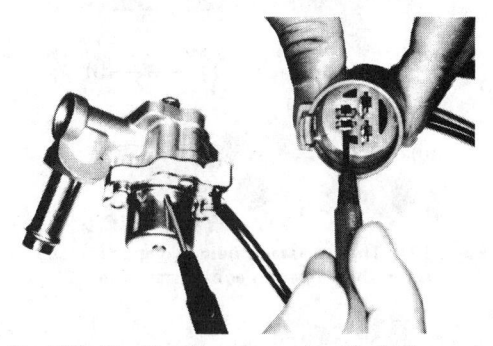

Fig. 13.9 Checking the outer vent control valve

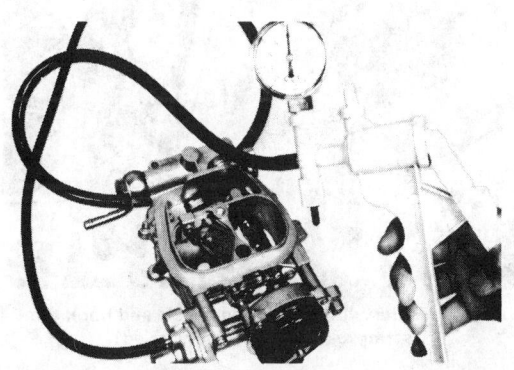

Fig. 13.10 A hand vacuum pump is used to check the choke breaker diaphragm

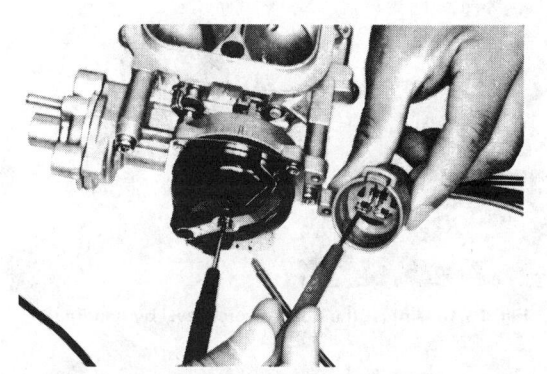

Fig. 13.11 Measuring for resistance between the terminal and choke heater

Chapter 13/Supplement: Revisions and information on later USA models

29 Temporarily install the idle mixture screw if it has been removed.
30 Attach the body to the flange using a new gasket.

Carburetor body
31 Install the primary main jet and plug over a new gasket. Install the metering needle guide, secondary main jet, power valve with jet and the slow jet with a new O-ring.
32 Install the HIC valve, valve seat and cover over a new gasket.
33 Install the auxiliary accelerator pump components in order; diaphragm (with outer gasket), spring then cover and screws
34 Install the accelerator pump components in order; spring, diaphragm (with outer gasket), cover, boot and screws.
35 Install the choke opener, idle-up diaphragm and dash pot (where equipped).

Air horn
36 Install the outer vent control valve on the air horn with a new gasket.
37 Install the power piston spring and piston in the bore. While pushing on the piston, rotate the retainer over the piston and tighten the retainer screw.
38 Install the needle valve and seat. Place the lip of the float under the wire of the needle valve and secure the float with the pivot pin.
39 Let the float hang down. Measure and adjust the clearance between the float top and air horn (without gasket) with a float gauge.
40 Lift the float and measure the distance from the float bottom to the air horn, using a vernier caliper, adjust as necessary.
41 Install the solenoid valve (with new o-ring) into the carburetor body by rotating the body clockwise.

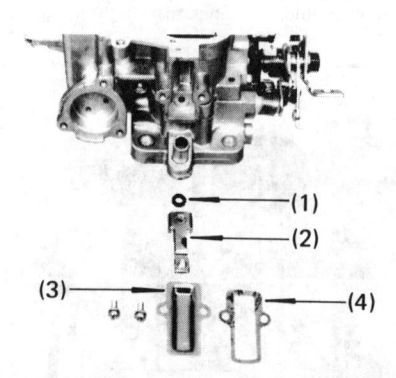

Fig. 13.12 The components of the HIC valve ready for installation

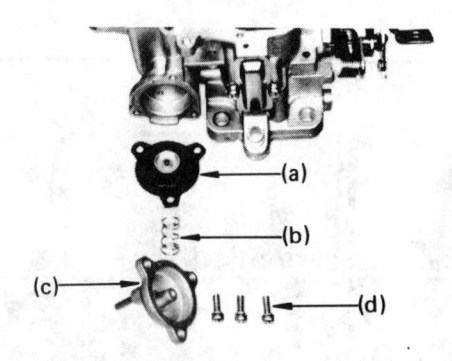

Fig. 13.13 Install the components of the auxiliary accelerator pump (AAP) in the alphabetical order shown

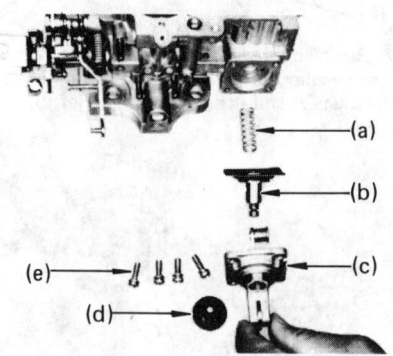

Fig. 13.14 The accelerator pump components shown in the proper order of installation

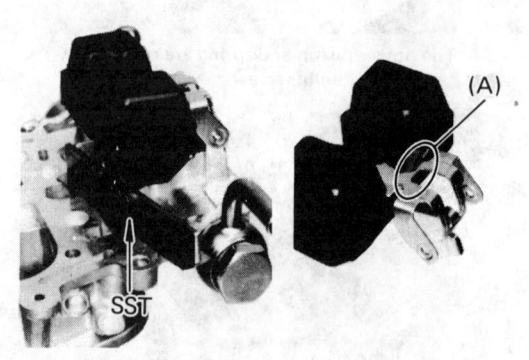

Fig. 13.15 Adjust the float raised level by bending tab (A)

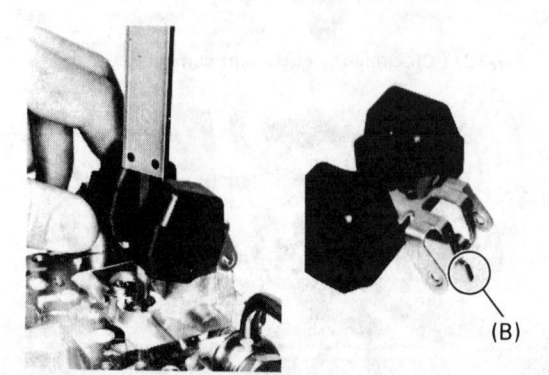

Fig. 13.16 Adjust the float lowered level by bending tab (B)

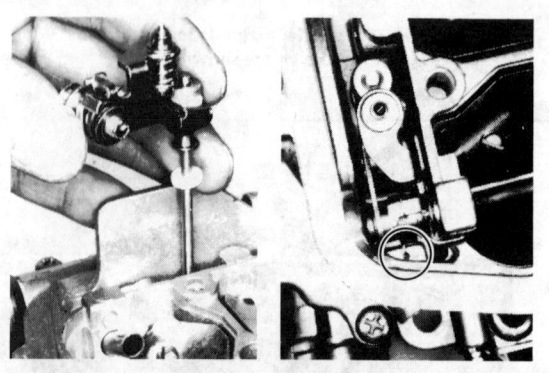

Fig. 13.17 Install the metering needle and hook the spring end into the hole (circled)

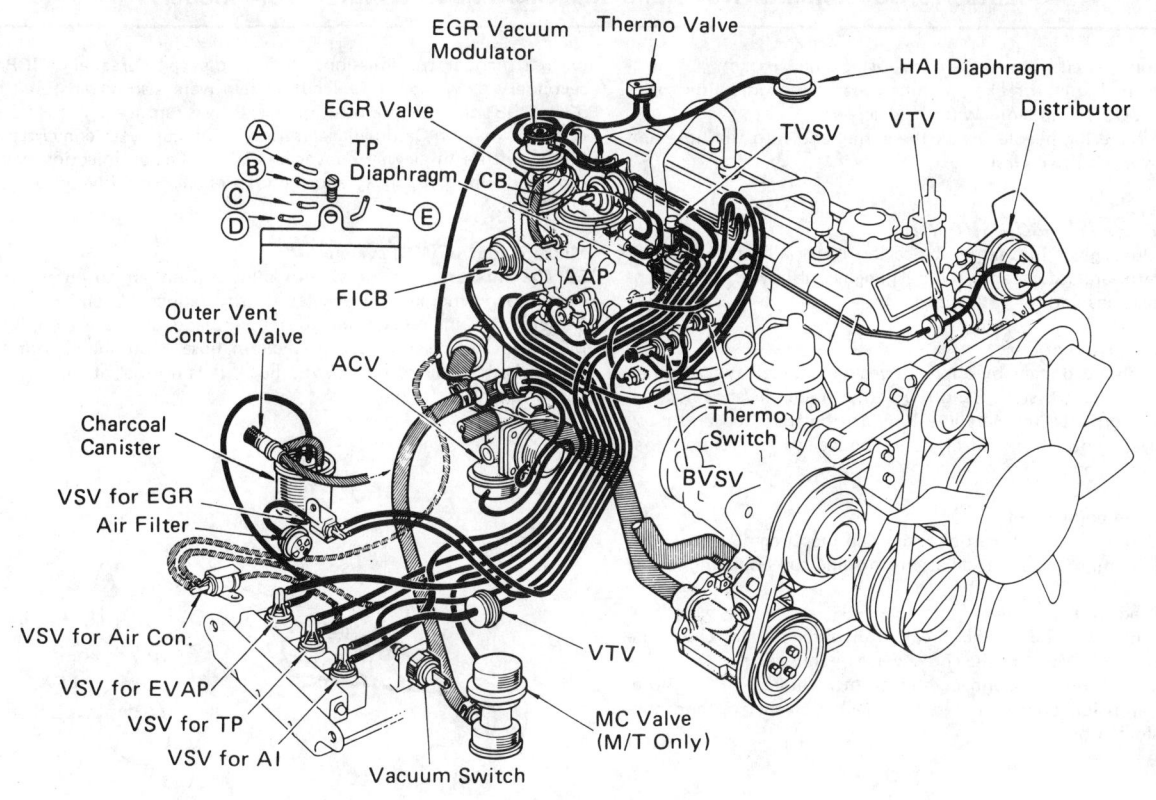

Fig. 13.18 Emission control devices incorporated on 1980 California vehicles
- A Fuel cut port
- B AI port
- C EGR R port
- D EGR port
- E Advancer port

Fig. 13.19 Emission control devices incorporated on 1980 non-California vehicles
- A Fuel cut port
- B AI port
- C EGR R port
- D EGR port
- E Advancer port

42 Carefully join the air horn to the body, using a new gasket.
43 Install the fuel inlet bracket, number plate, fast idle link, wire clamp and the VCU clamp along with the 4 screws.
44 Install the metering needle. Hook the spring end into the hole and tighten the screw with two washers.

Carburetor (22R) - adjustment.

45 The procedures given in Chapter 3, Section 17 for adjusting the 20R engine carburetor are accurate for the 22R engines also, with the exception of specifications and the following.

Choke opener adjustment

46 Adjust the choke opener by applying vacuum. Check that the fast idle cam is on the fourth step. Bend the choke opener lever as needed. Release the vacuum, set the fast idle lever to the first step and close the choke valve. Now check for clearance between the choke opener lever and the fast idle cam.

Idle-up diaphragm adjustment

47 Apply vacuum to the throttle positioner diaphragm. Check the throttle valve opening angle and adjust by turning the adjusting screw.

Choke breaker adjustment

48 Apply vacuum to the choke breaker diaphragm and close the choke valve by hand. Measure the choke valve opening angle.
49 Install the carburetor. Connect the throttle linkage, PCV hose, fuel line and emission control hoses. Put the air cleaner on and connect the remaining hoses.

Emission controls

50 Emission control devices used on 1980 and 1981 cars are very similar to those described in Chapter 3.
51 In 1980 the high altitude compensator (HAC) valve was not used. Also in 1980, a mixture control (MC) system was installed on all cars with a manual transmission. A fast idle cam breaker (FICB) and a secondary slow circuit fuel cut system were also added to all models.
52 In 1981 the throttle positioner (TP) was removed. The HAC became optional on non-California cars only. The catalytic converter was replaced with a three-way converter (TWC). The air injection system was modified with feedback, and a choke opener and idle advance system were added.

Mixture control (MC) system

53 The mixture control system allows fresh air to enter the intake manifold during sudden deceleration to reduce CO and HC emissions.
54 With the engine running, put your hand over the air inlet of the MC valve and disconnect the vacuum hose. You should not feel any vacuum until you reconnect the hose. It is normal at this time for the car to idle rough or die.

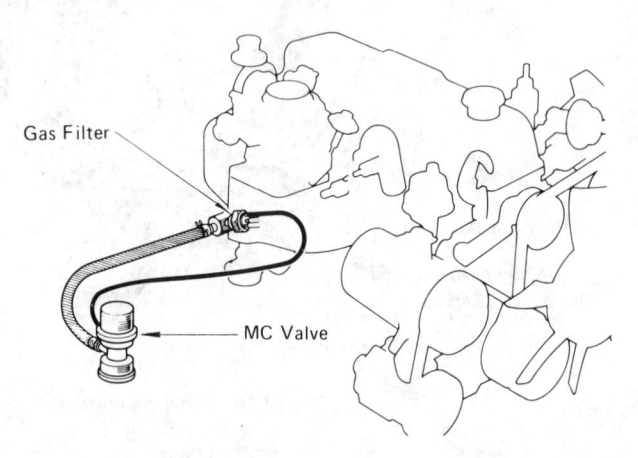

Fig. 13.21 Location of the MC valve

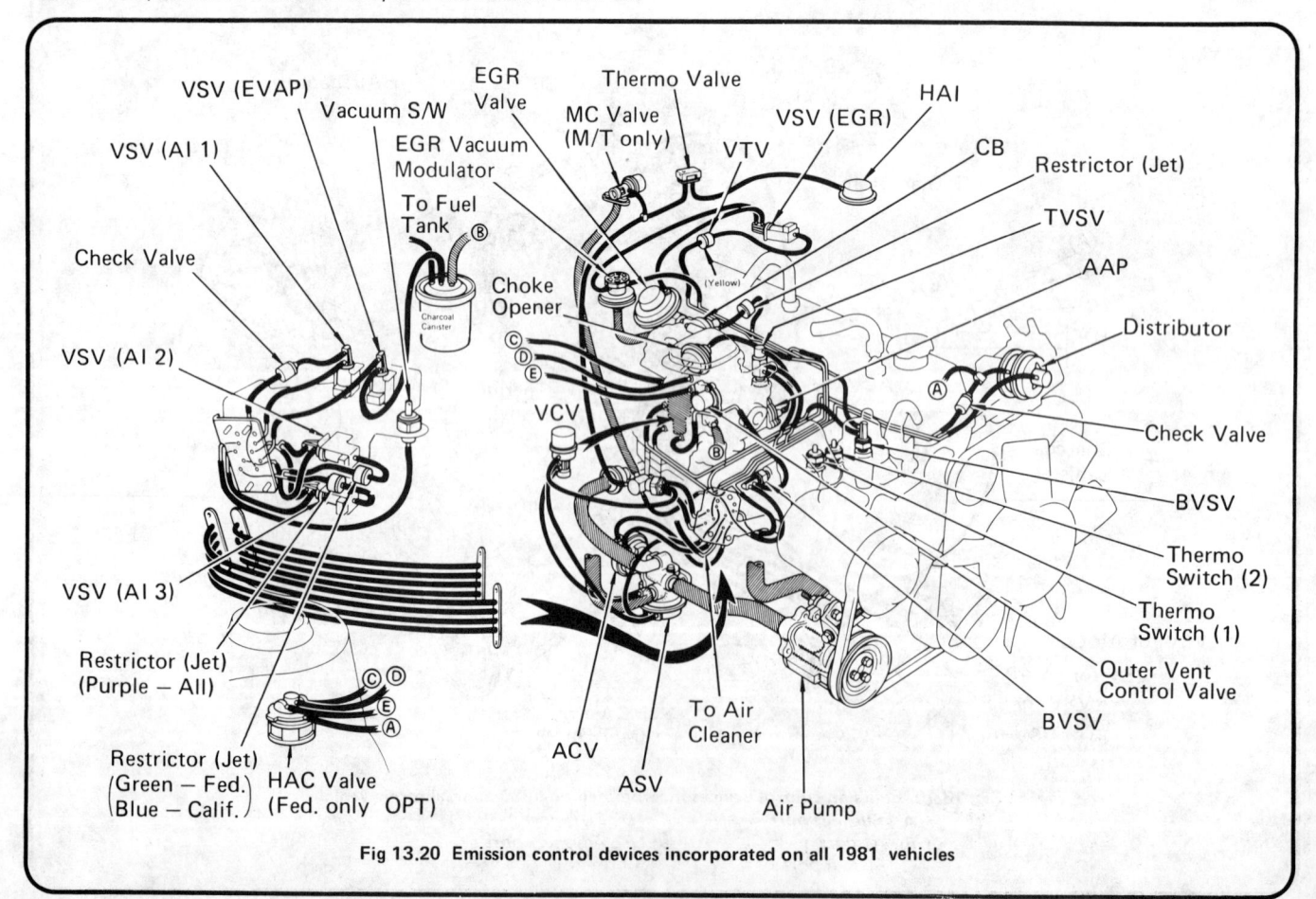

Fig 13.20 Emission control devices incorporated on all 1981 vehicles

Chapter 13/Supplement: Revisions and information on later USA models

Fast idle cam breaker (FICB) system.

55 The FICB lowers the engine speed after warm-up by forcibly releasing the fast idle cam to the third step.

56 Stop the engine after warm-up and disconnect the hose from the fast idle cam breaker.

57 Hold the throttle valve slightly open and set the fast idle cam by pulling up on the FICB linkage and releasing the throttle.

58 Start the engine without touching the accelerator pedal.

59 Reconnect the hose. The fast idle cam should be released to the third step. If not, check the linkage hoses and the TVSV. Also apply vacuum to the FICB diaphragm and check that the linkage moves.

Secondary slow circuit fuel cut system

60 This system prevents dieseling by cutting off part of the fuel in the secondary slow circuit of the carburetor.

61 Measure the stroke of the fuel cut valve by fully opening and closing the throttle valve. This measurement should be between 0.059 and 0.079 in (1.5 to 2.0mm).

62 The stroke can be adjusted by bending the lever (this should be adjusted before the secondary throttle valve opens).

Choke opener system

63 This system prevents an over-rich fuel mixture after warm-up, by releasing the fast idle cam to the fourth step and by forcing the choke valve open.

64 Disconnect the vacuum hose from the choke opener diaphragm while the engine is cold. Push the accelerator pedal down, then release it.

65 Start the engine and reconnect the vacuum hose. The choke linkage should not move.

66 Warm up the engine and disconnect the same vacuum hose.

67 Set the fast idle cam by opening the throttle and pushing the

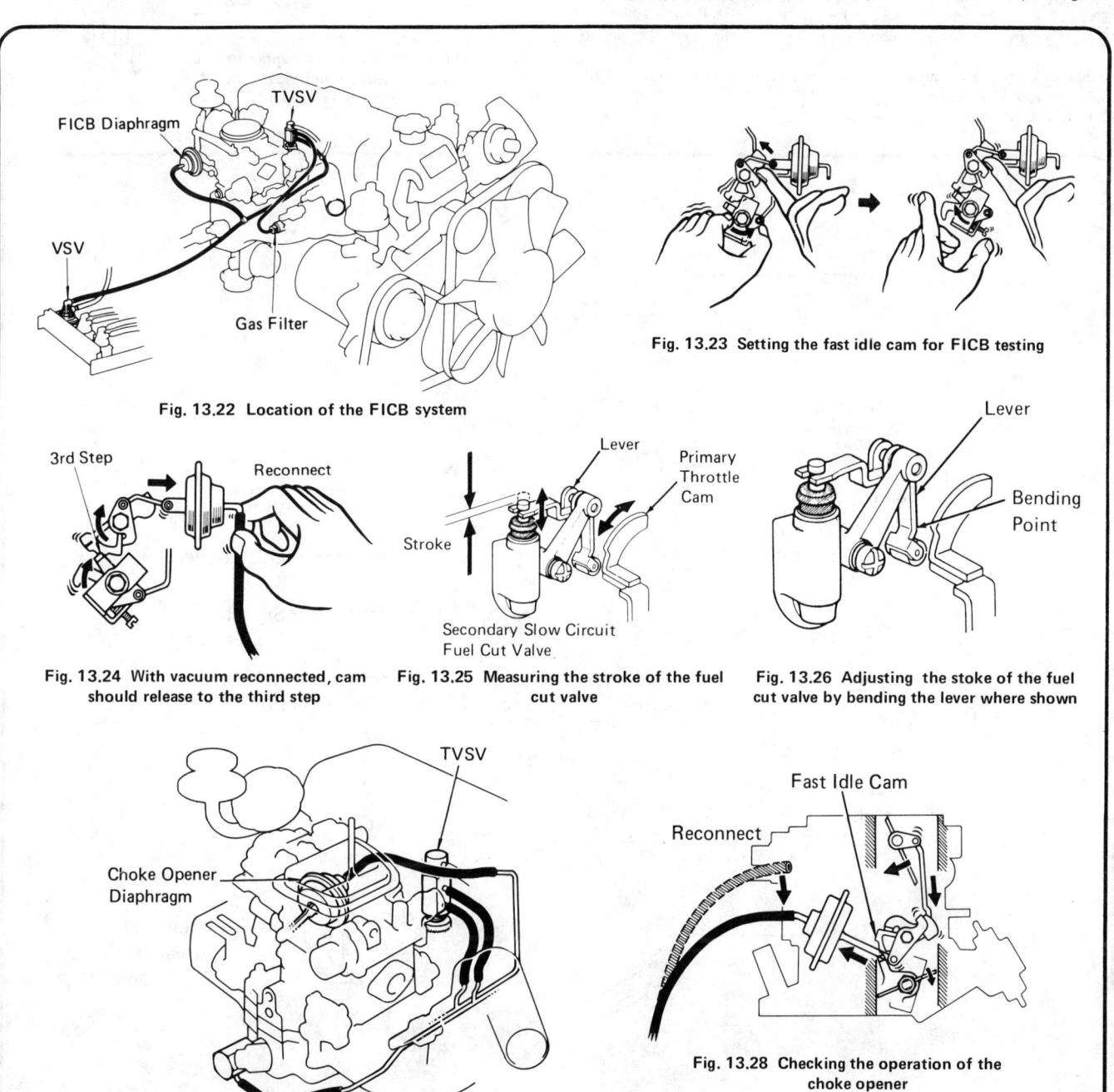

Fig. 13.22 Location of the FICB system

Fig. 13.23 Setting the fast idle cam for FICB testing

Fig. 13.24 With vacuum reconnected, cam should release to the third step

Fig. 13.25 Measuring the stroke of the fuel cut valve

Fig. 13.26 Adjusting the stoke of the fuel cut valve by bending the lever where shown

Fig. 13.27 The choke opener system

Fig. 13.28 Checking the operation of the choke opener

choke valve closed as you release the throttle valve.
68 Start the engine without touching the accelerator.
69 Reconnect the vacuum hose. The fast idle cam should be released to the fourth step and the choke linkage should move.
70 If a problem is encountered, check that the TVSV and the choke opener diaphragm operate the choke linkage when the vacuum is applied.

Idle advance system
71 This system causes the ignition timing to advance, at idle, to improve fuel economy at an idle.
72 Warm the engine to normal operating temperature.
73 Check the ignition timing, as described in Chapter 4, Section 6. At an idle, it should be 15° BTDC.
74 Disconnect the vacuum hose from the distributor sub-diaphragm and plug the hose.
75 Check that the ignition timing has retarded to 8°BTDC at idle.
76 Reconnect the vacuum hose and remove the timing light.
77 If the timing does not change, remove the distributor cap and rotor.
78 Check that the advancer moves when vacuum is applied and repair or replace it if necessary.
79 Install the distributor cap and rotor.

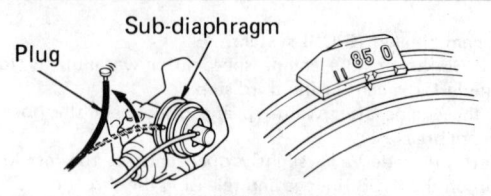

Fig. 13.29 Ignition timing with the idle advance vacuum plugged

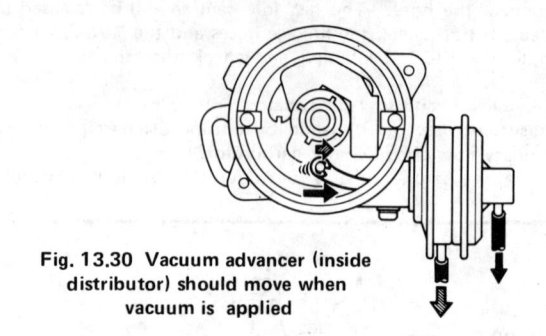

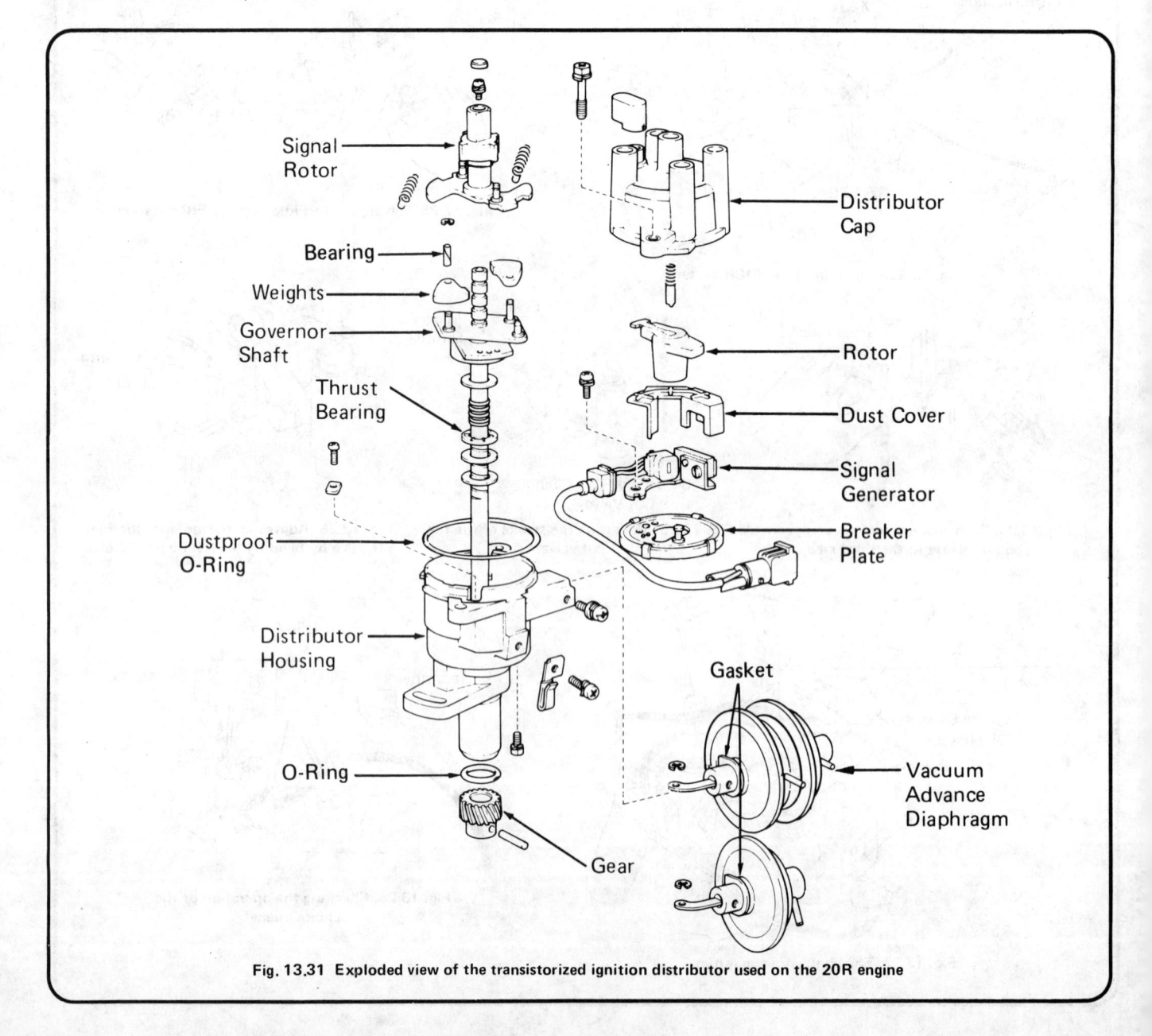

Fig. 13.30 Vacuum advancer (inside distributor) should move when vacuum is applied

Fig. 13.31 Exploded view of the transistorized ignition distributor used on the 20R engine

5 Ignition system

The ignition systems used on 1980 and 1981 models are transistorized. The removal and disassembly procedures are covered in Chapter 4. The only difference is on the 22R engine distributor where the thrust bearing, spring and washers were left off the shaft.

6 Transmission

Manual transmission (W55) - disassembly

1 With the transmission removed from the car as described in Chapter 6, clean off all external dirt.
2 Remove the clutch release mechanism from inside the bellhousing.
3 Unbolt and remove the bellhousing.
4 Unscrew and remove the reversing lamp switch, the speedometer driven gear and the restrictor pins, springs and plugs.
5 Unbolt and remove the gearchange lever retainer, the extension housing and the front bearing retainer.
6 Extract the circlips from the input shaft and layshaft front bearings.
7 Remove the gear casing from the intermediate plate by tapping the intermediate plate away with a soft metal drift.
8 Using a socket wrench, unscrew and remove the plugs from the edge of the intermediate plate and extract the springs and detent balls.
9 Remove the set bolts from the No.1 and No.2 fork shaft and drive out the pin from the No.3 shaft.
10 Remove the snap rings from No.1 and No.2 shafts and the Reverse idle gear shaft stopper.
11 Remove the shift fork shaft No.1 and interlock pins No.1 and No.2.
12 Remove the shift fork shaft No.2 and interlock pin No.3.
13 Pull out the shift fork shaft No.4. Remove the shift fork No.3 and Reverse shift arm with pin.
14 Remove the Reverse idler gear and shaft.
15 Remove the speedometer drive gear and snap ring from the mainshaft.
16 Remove the countershaft rear bearing, spacer, 5th gear and the needle roller bearing using a two-legged puller.
17 Remove the snap ring and clutch hub No.3 with a two-legged puller.
18 Remove the snap ring from the mainshaft, then using a puller, remove the rear bearing and 5th gear.
19 Remove the snap ring and remove Reverse gear from the mainshaft with a puller.
20 Remove the center bearing retainer using a TORX socket wrench.
21 Remove the snap ring and, while tapping the intermediate plate with a mallet, remove the input shaft, mainshaft and counter gear as a unit.

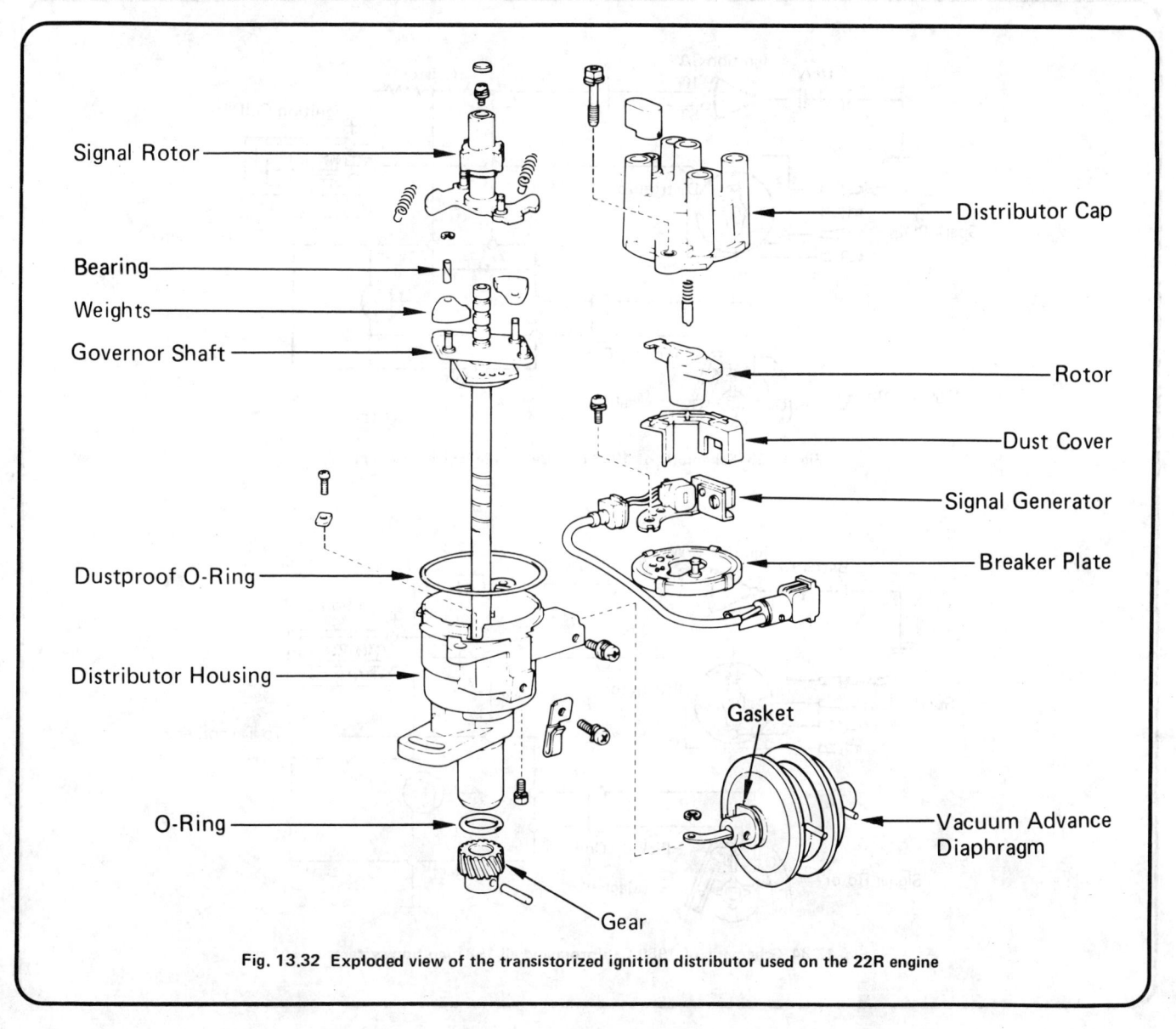

Fig. 13.32 Exploded view of the transistorized ignition distributor used on the 22R engine

Mainshaft overhaul

22 Remove the input shaft from the output shaft.
23 From the rear end of the mainshaft, draw off the bearing. A press will be required for this operation.
24 Remove 1st gear, the needle roller bearing, bearing inner track and synchronizer ring. Take care not to lose the inner track locking ball.
25 Press off 2nd gear complete with synchronizer ring. Reverse gear and 1st/2nd gear synchro unit.
26 Remove the snap ring from the front of the shaft.
27 Remove the clutch hub sleeve No.2, 3rd gear and the synchronizer ring.
28 Clean all components thoroughly and examine for worn or chipped teeth and grooving or scoring of the shaft. The gears should have a running clearance between their internal bores and the shaft of between 0.0004 to 0.0024 in. for 1st, 2nd and counter 5th gear and between 0.0024 to 00.40 in. for 3rd gear.
29 Check the synchronizer units as described in Section 4, Chapter 6 paragraphs 7, 8, 9 and 10. Reassemble the units in accordance with the diagrams.
30 Commence reassembly of the mainshaft by installing the 3rd/4th gear synchronizer ring to 3rd gear and then fitting them to the shaft.
31 Fit the 3rd/4th gear synchronizer unit, positioning it tightly against the mainshaft shoulder. Secure it with a snap ring to give a minimum clearance between snap ring and synchro unit yet will still fit fully into the groove.
32 Snap rings are available in many different thicknesses from a Toyota dealership. Use the snap ring that allows the minimum axial play.

Manual transmission (W55) - reassembly

33 Grip the intermediate plate in a soft-jawed vice and check that the dowel pin projects between 0.24 and 0.32 in (6.0 and 8.0 mm) from the front face of the place.
34 Apply grease to the needle bearing in the recess at the end of the input shaft and fit the input shaft to the front end of the mainshaft.
35 Mesh the teeth of the mainshaft and laygear assemblies and install them simultaneously to the intermediate plate.
36 Fit the retaining snap ring to the mainshaft bearing.
37 Fit the bearing retainer to the intermediate plate.
38 Install Reverse gear and the correct snap ring.
39 Install 5th gear, the output shaft rear bearing and the correct snap ring.
40 Install the clutch hub No.3 and shifting key to the hub sleeve.
41 Install the shifting key springs, positioned so their end gaps will not be in line, under the shifting keys.
42 Install the shifting key retainer using a socket wrench or piece of pipe as a drift.
43 Install the clutch hub No.3 and the correct snap ring.
44 Install the counter 5th gear assembly, aligning the synchronizer

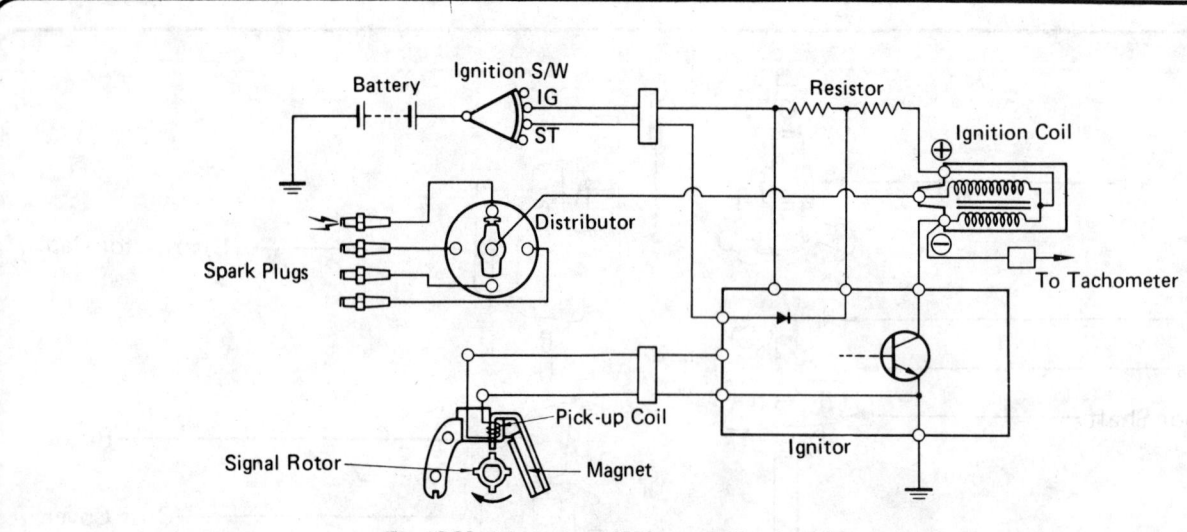

Fig. 13.33 Schematic of 1980 non-California ignition system

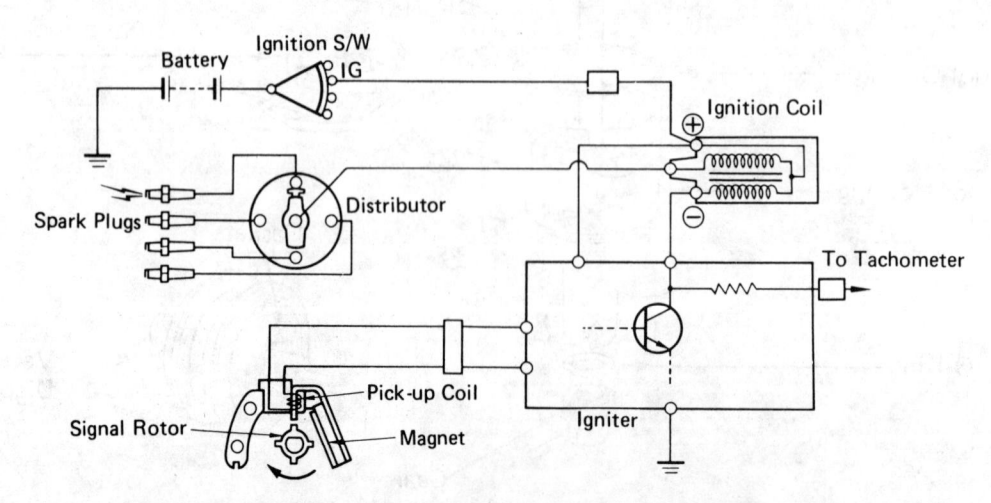

Fig. 13.34 Schematic of 1980 California and all 1981 ignition systems

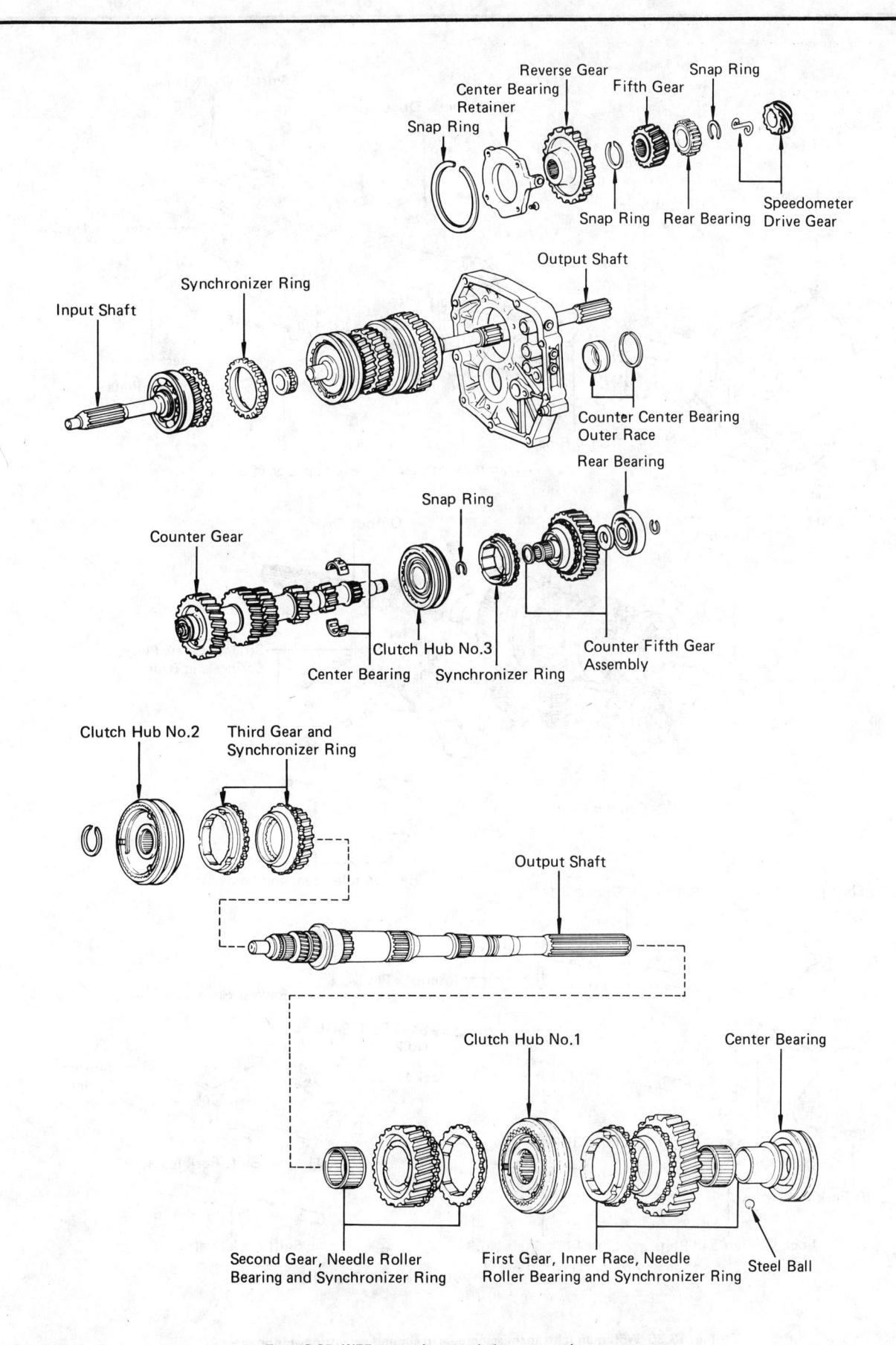

Fig. 13.35 W55 manual transmission gear train components

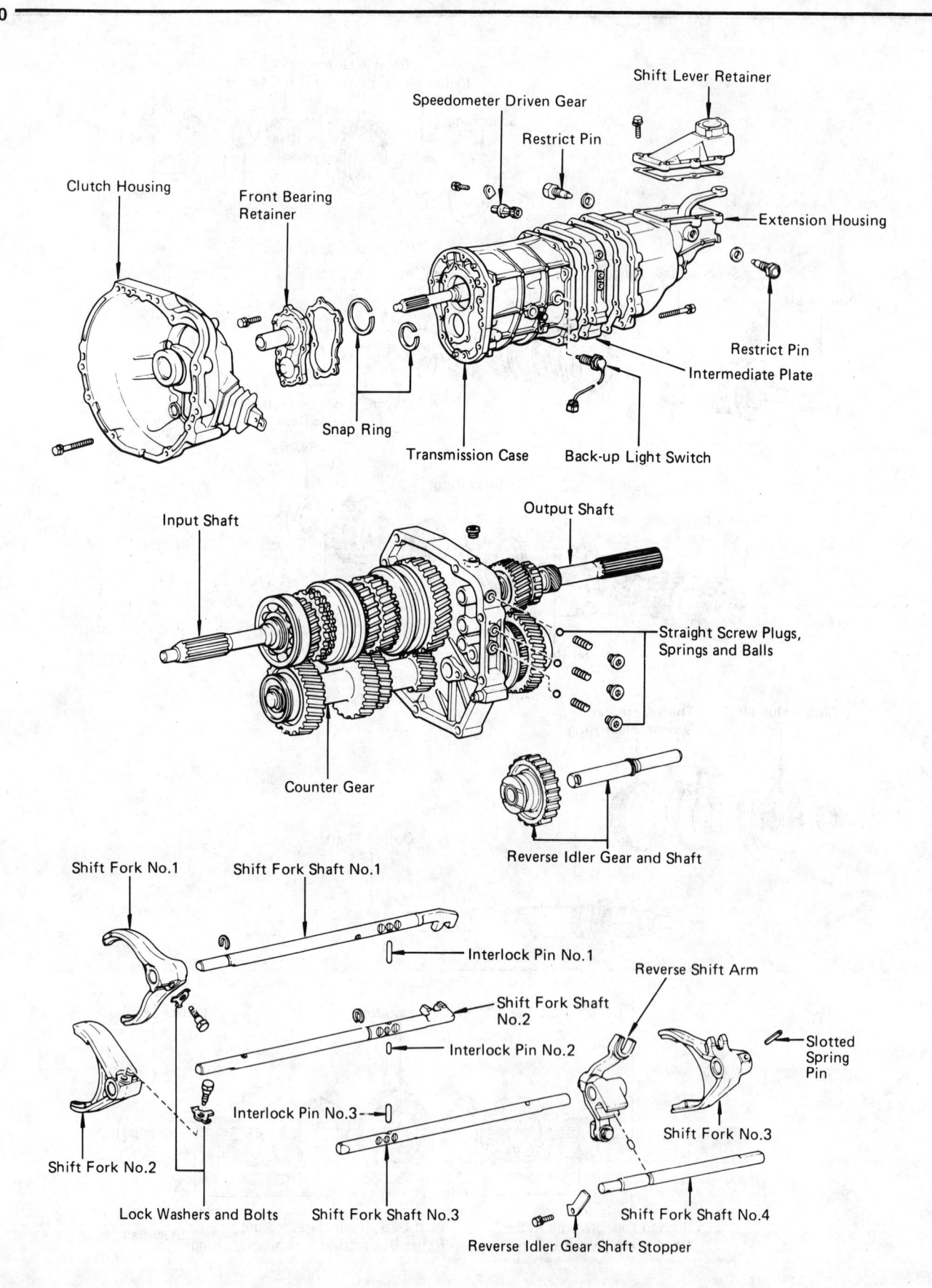

Fig. 13.36 W55 manual transmission gear train and selector components

Chapter 13/Supplement: Revisions and information on later USA models

Fig. 13.37 Measuring for tne correct sized snap ring and then installing it

Fig. 13.38 Installing the output shaft to the intermediate plate

ring slots with the shifting keys.
45 Install the spacer and bearing with a hammer and socket.
46 Install the correct snap ring.
47 Install the speedometer drive gear and clip onto the output shaft.
48 Install the Reverse idler gear and shaft.
49 Insert shift fork shaft No.3 through the Reverse shift arm and shift fork No.3, aligning the shift fork to the hub sleeve No.3 groove.
50 Put the Reverse shift arm into the pivot of the bearing retainer and slip the shift fork shaft into the intermediate plate.
51 Push the pin in the Reverse shift arm hole into the groove of shift fork shaft No.3 and slip shift fork shaft No.4 into the intermediate plate.
52 Drive the slotted spring pin through the pin hole in the fork into the hole in the shaft until it is flush with the fork.
53 Install interlock pin No.3 into the intermediate plate hole.
54 Install shift fork No.2 onto the shift fork shaft No.2 and into the groove of hub sleeve No.2.

Fig.13.39 Installilng the shift fork set bolts and lock washers

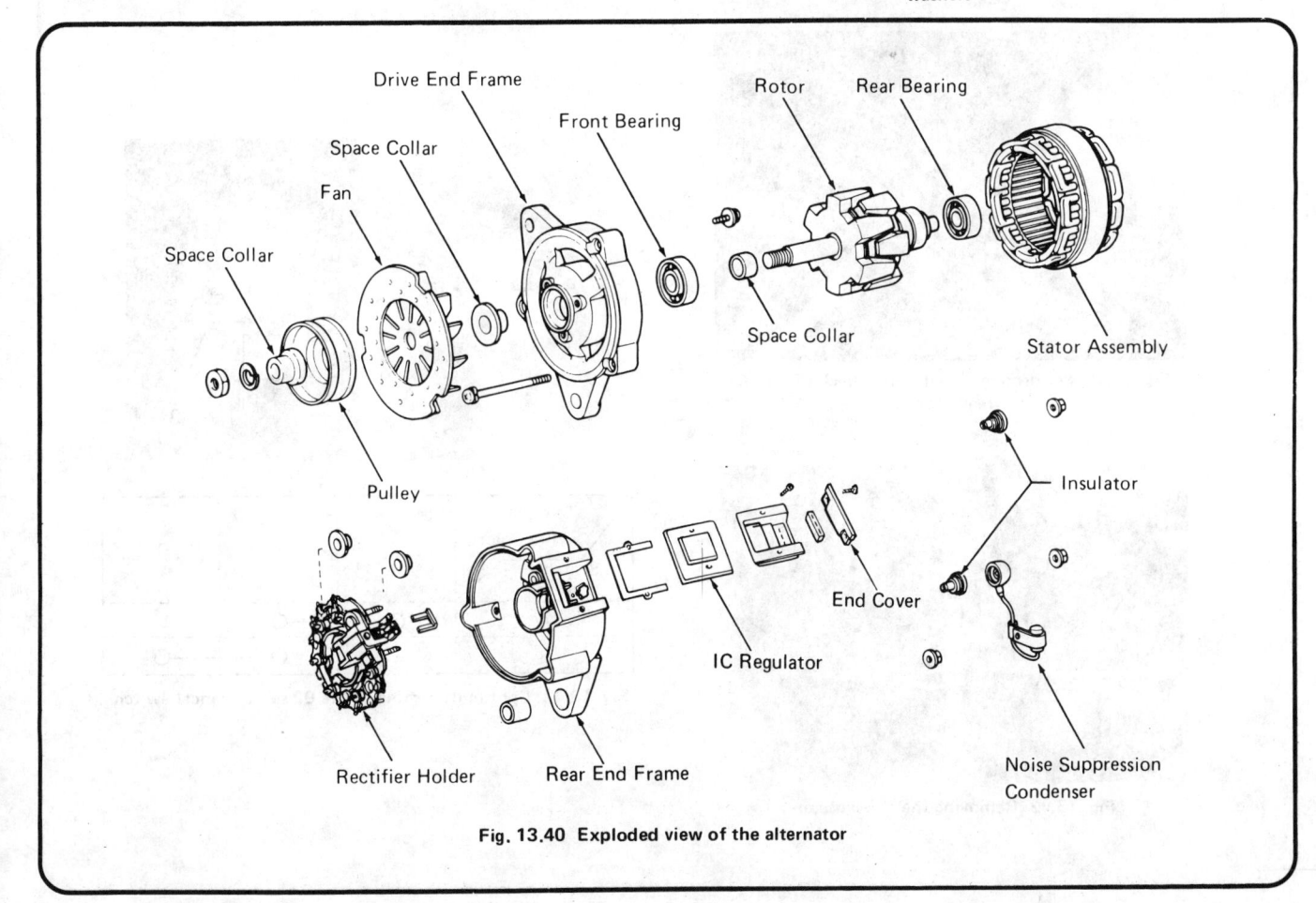

Fig. 13.40 Exploded view of the alternator

55 Install shift fork shaft No.2 to the intermediate plate.
56 Install the fork shaft snap ring No.2 and the interlock pin No.1.
57 Install shift fork No.1 onto the shift fork shaft No.1 and into the groove of hub sleeve No.2.
58 Install shift fork shaft No.1 to the intermediate plate and attach the No.1 shaft snap ring.
59 Install the shift fork set bolts.
60 Insert the detent balls and springs into their holes in the edge of the intermediate plate. Coat the threads of the socket screws with jointing compound and torque them accordingly.
61 Install the Reverse idler gear shaft stopper.
62 Attach the transmission case to the intermediate plate.
63 Install the two bearing snap rings and the front bearing retainer.
64 Fit the extension housing using a new gasket. Turn the remote control rod during the operation so that the rod dog connects with the selector rods.
65 Tighten the extension housing bolts to the specified torque.
66 Install the restrictor pins with a new gasket.
67 Install the shift lever retainer.
68 Install the reverse light switch and wire clamp.
69 Install the speedometer drive gear, bellhousing, release fork and bearing.
70 Fill the transmission with oil after it has been installed in the car.

7 Electrical system

Alternator

1 1980 and 1981 vehicles are equipped with IC regulators built into the alternator.

Disassembly

2 Disassembly of the alternator is the same as described in Chapter 10, Section 8 except that between paragraphs 5 and 6 the IC regulator must be removed.
    a) Remove the end cover
    b) Inside the regulator, remove the three screws on the terminals
    c) Remove the two screws from the top and pull out the regulator
    d) Remove the plastic regulator housing and the rubber seal around the terminals by prying with a small screwdriver

Reassembly

3 During reassembly of the alternator between paragraphs 20 and 21 the IC regulator must be reinstalled.
    a) Install the plastic regulator housing and the rubber seal over the terminals
    b) Install the regulator with the two screws
    c) Install the three screws on the terminals
    d) Install the end cover

IC Regulator - testing

4 Testing is the same as used in Chapter 10, Section 6. If there is a voltage reading greater than 15 volts while testing under no load, the IC regulator needs replacing. If the voltage is less than 13.5 volts, turn the ignition switch to "OFF". Check the continuity, after removing the IC regulator end cover, between the 'L' and 'F' terminals. If there is no continuity, check the alternator. If the continuity is approximately 4 ohms, replace the IC regulator.
5 Turn the ignition switch to "ON" and check the voltage at the 'L' terminal. If it's between 1 and 2 volts or zero check the alternator.

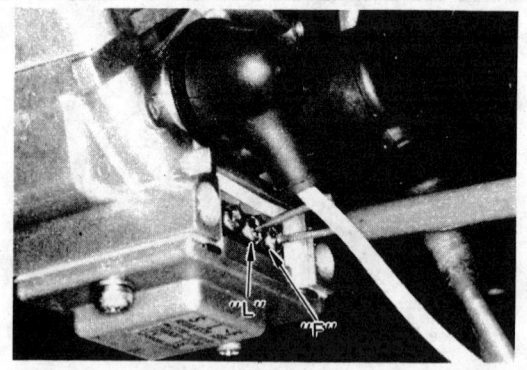

Fig. 13.41 Performing a continuity check of the IC regulator

Fig. 13.42 Removing the IC regulator

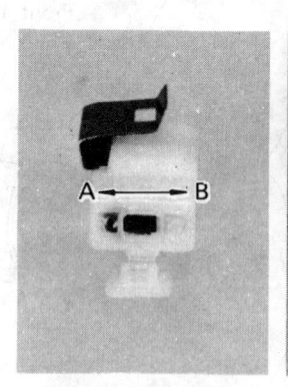

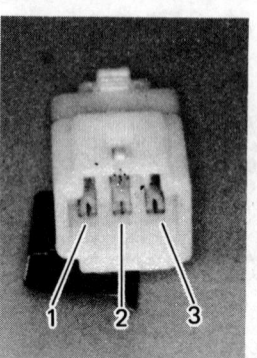

Terminal Wire Color / Switch position	1	2	3
A	O—O		
B		O—O	

Fig. 13.43 Continuity testing of the 02 sensor cancel switch

Fig. 13.44 The two screws which secure the counter assembly lock plate

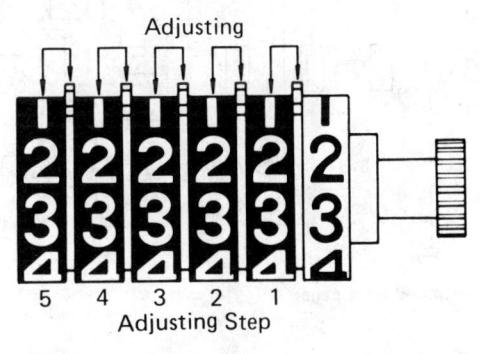

Fig. 13.45 Adjusting the counter assembly

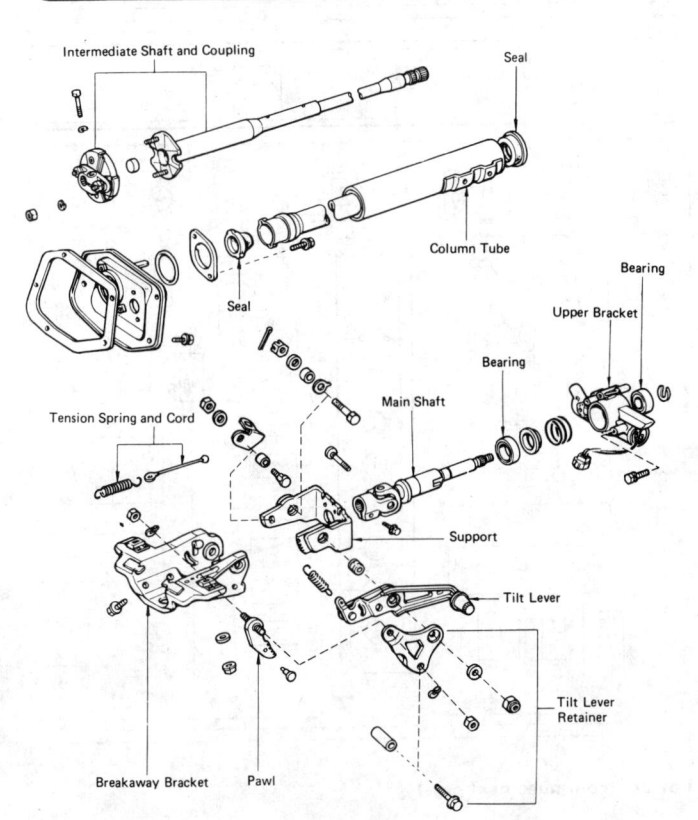

Fig. 13.46 Exploded view of the tilt steering column and components

O_2 sensor - 1981
Maintenance interval counter - reset procedure

6 Remove the speedometer from the combination meter.
7 Pull out the meter needle and remove the counter assembly.
8 Remove the counter lock plate.
9 Align the top side of the meter and adjust from right to left, as seen from the front.
10 Install the lock plate into the 5 clips.
11 Install the meter needle and two screws securing the counter assembly.
12 Install the combination meter into the dashboard.

Cancel switch - inspection
14 Remove the cancel switch from the left-hand kick panel.
15 Using an ohmmeter, check the continuity between the terminals of each switch position.
16 If there is no continuity, replace the switch.

8 Suspension and steering

Tilt steering column and shaft-removel
1 Remove the steering mainshaft as described in Chapter 11, Section 17, paragraphs 1 to 8.
2 Remove the comination switch.
3 With the key in the "Acc." position, push the stop key down with a thin rod and remove the key cylinder.
4 Tilt the mainshaft up and remove the tension springs and cords.
5 Make alignment marks on the intermediate shaft and universal joint.
6 Disconnect the intermediate shaft and mainshaft by removing the joint bolt and four bracket bolts.
7 Disconnect the upper bracket from the tilt steering support.
8 Remove the snap ring and lift the mainshaft from the upper bracket.
9 Remove the tilt lever retainer, the collar and the reclining release pin.
10 Remove the spring, reclining pawl, guide pin bolt and support bolt.
11 Remove the serrated bolt and tilt handle with a mallet.
12 Inspect the components for wear or damage and replace as needed.

Tilt steering column and shaft - installation
13 Reassembly and installation procedures are reversals of removal and dismantling, but make sure that the following operations are carried out.
14 Install the tilt steering support stopper bolt, bracket, washer and nut.
15 Insert the collar, spring and mainshaft into the upper bracket.
16 Press the mainshaft and upper bearing in a soft-jawed vice and install the snap ring.
17 Bolt together the support and upper bracket.
18 Install the column hole cover on the column tube.
19 Connect the breakaway bracket to the column tube.
20 Connect the mainshaft and intermediate shaft, aligning the marks on the joint flange.
21 Connect the springs and cords. Hook the spring to the hanger and the cord end to the support.
22 Hook the cords around the cord guides.
23 Check the mainshaft for play and that it securely locks in all six positions.
24 Install the ignition switch and the combination switch.
25 Connect the combination switch wiring connector and clamp down the wires.
26 Install the steering mainshaft as described in Chapter 11, Section 17.

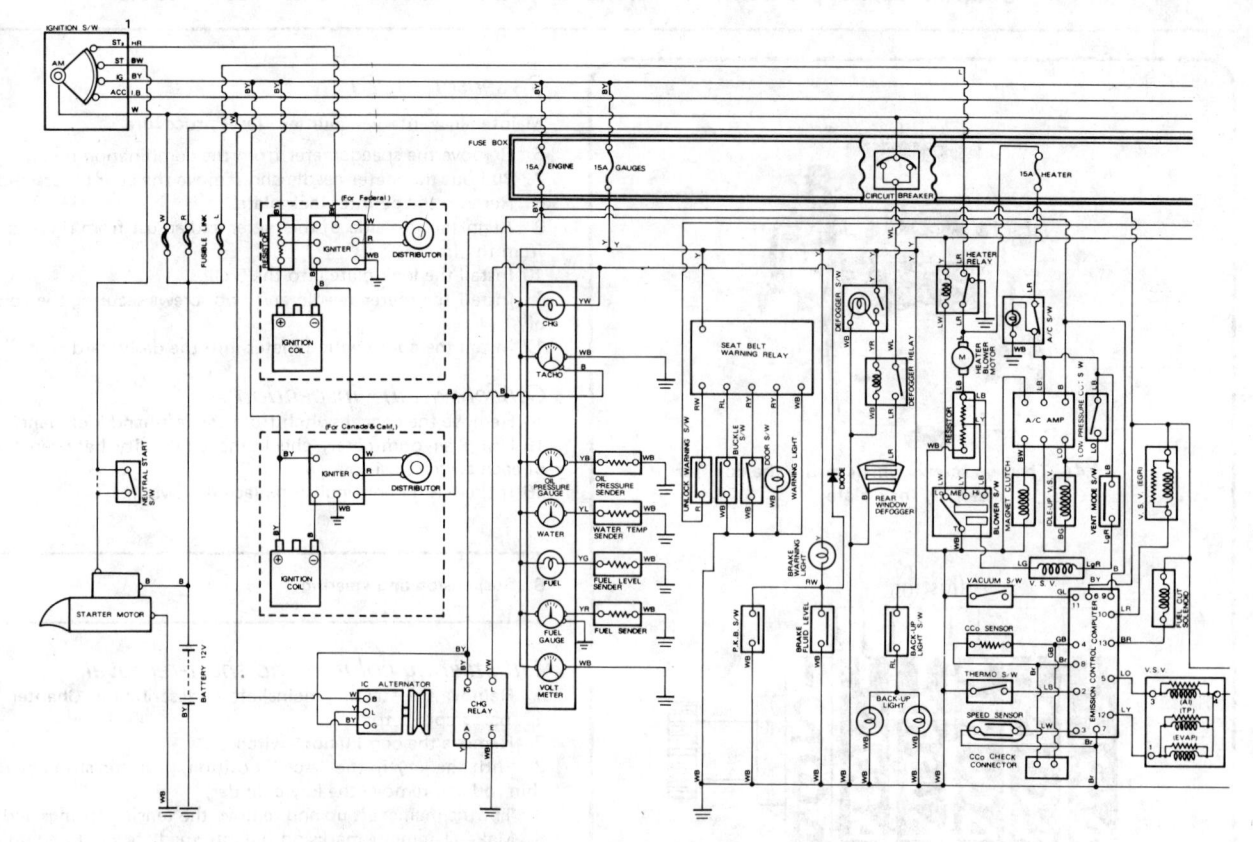

1980 Celica wiring diagram (1 of 2) (continued next page)

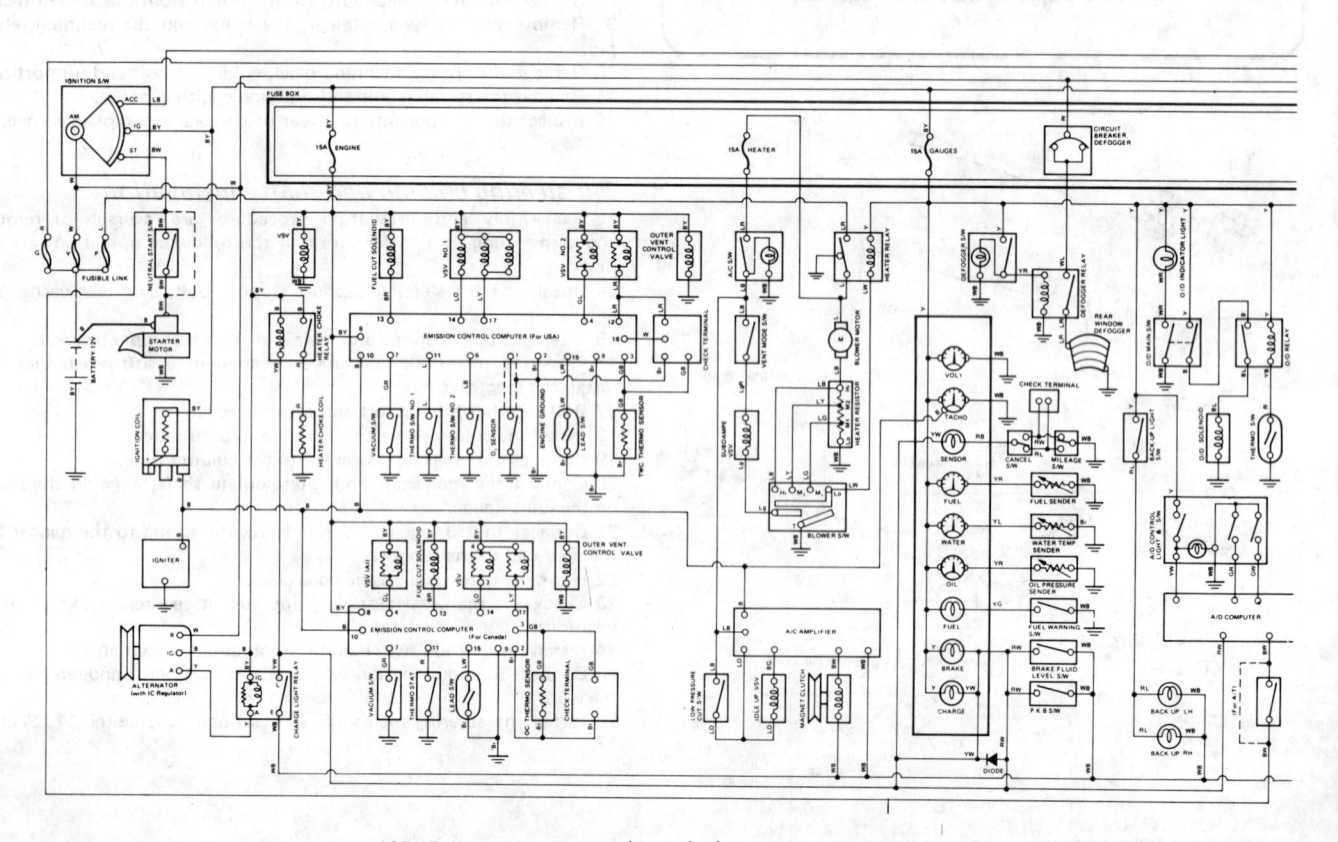

1981 Celica wiring diagram (1 of 2) (continued next page)

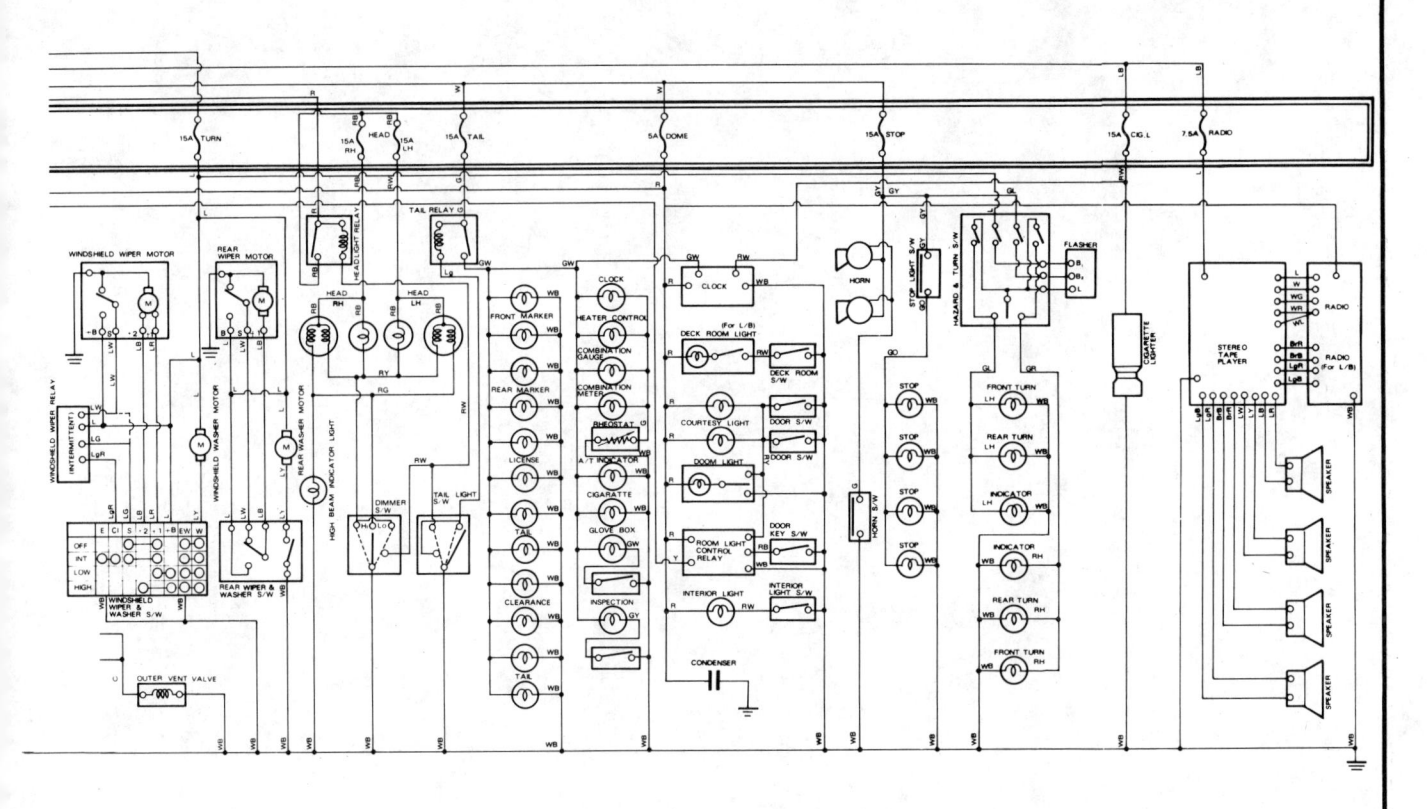

1980 Celica wiring diagram (2 of 2)

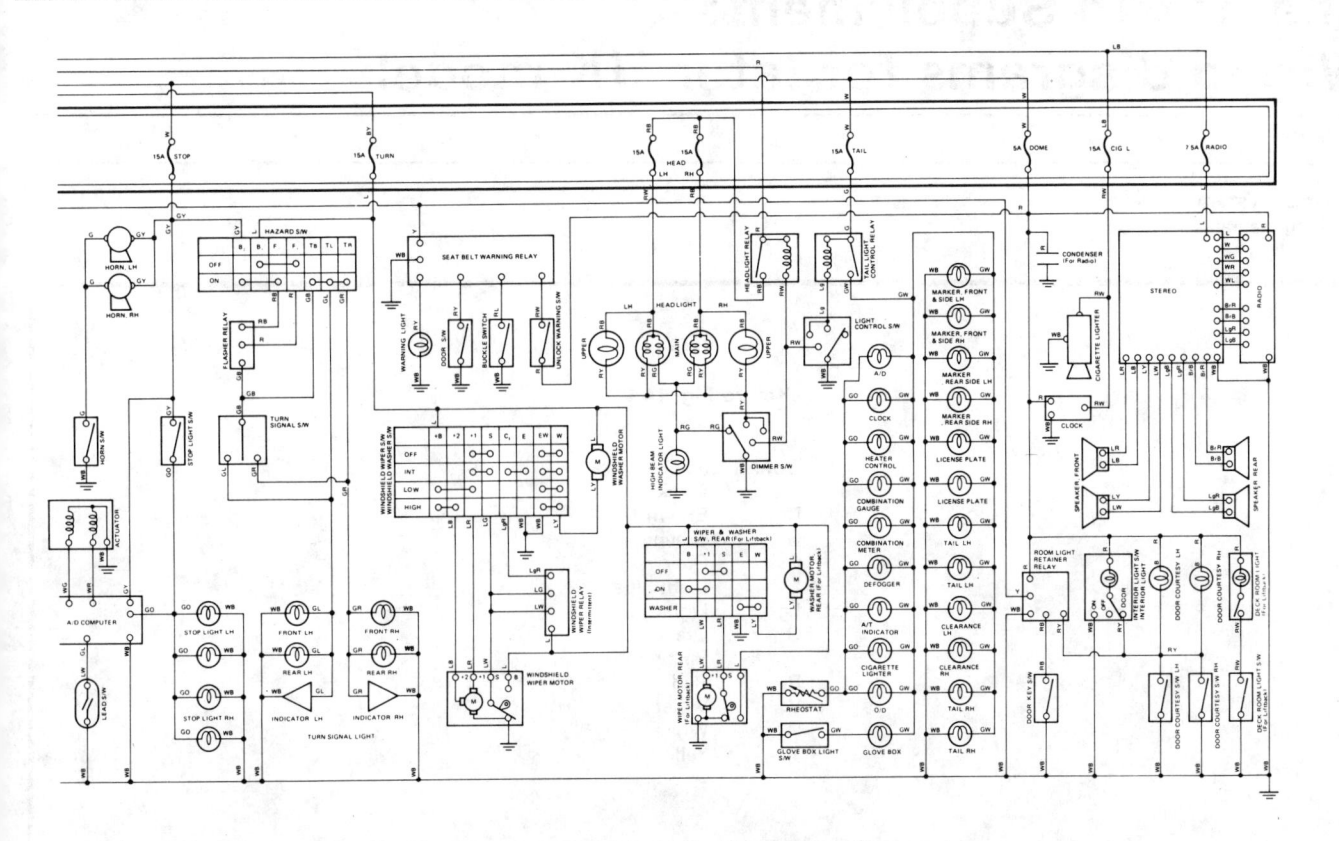

1981 Celica wiring diagram (2 of 2)

Chapter 14 Supplement:
Wiring diagrams for later UK models

Contents

Electrical system
  Wiring diagrams

Key to Fig. 14.1

Colour code
B	Black
Br	Brown
G	Green
Gr	Grey
L	Light blue
Lg	Light green
O	Orange
P	Pink
R	Red
V	Violet
W	White
Y	Yellow

The first letter indicates basic wire colour, the second letter indicates the spiral line colour

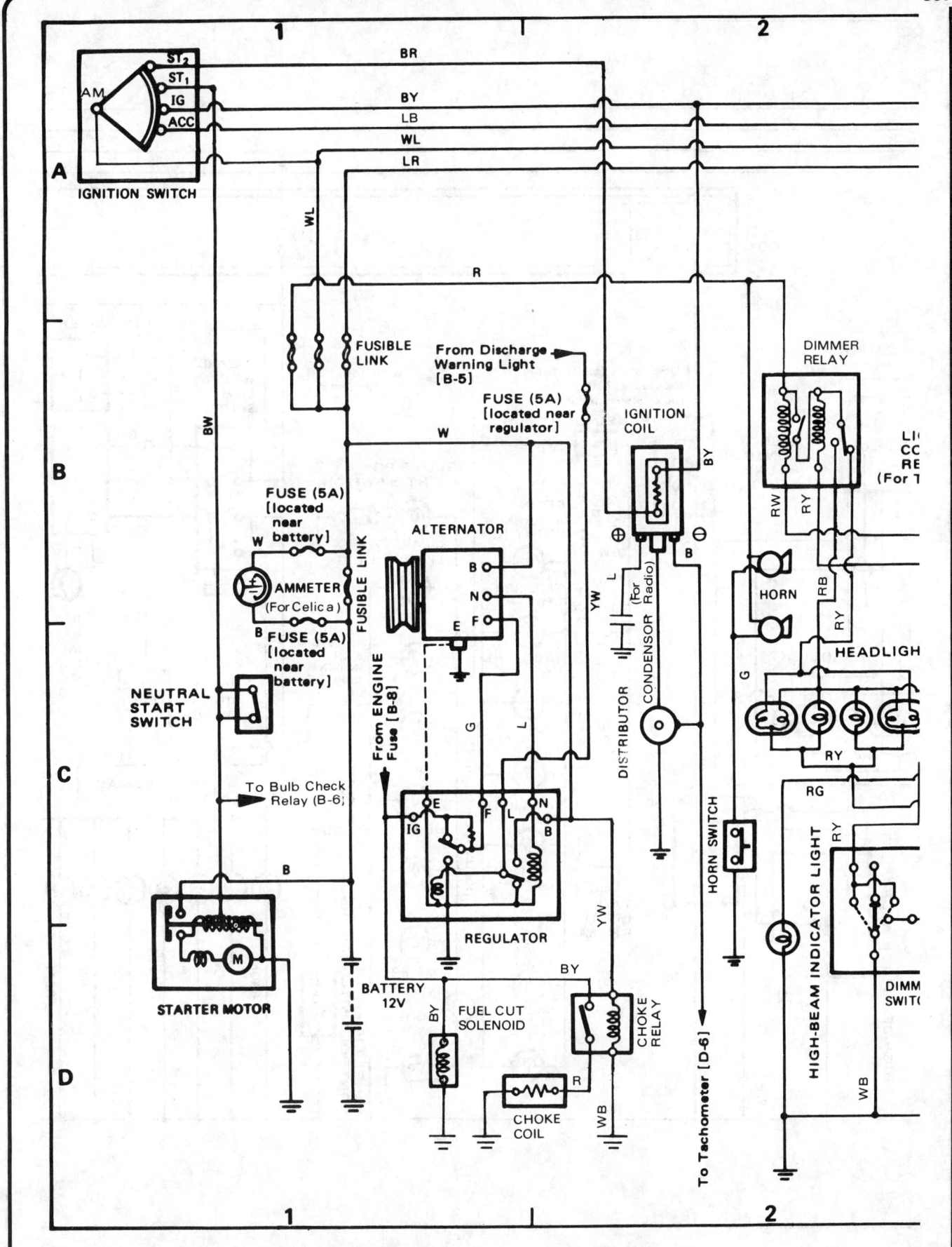

Fig. 14.1 Wiring diagram for 1979/1980 UK models

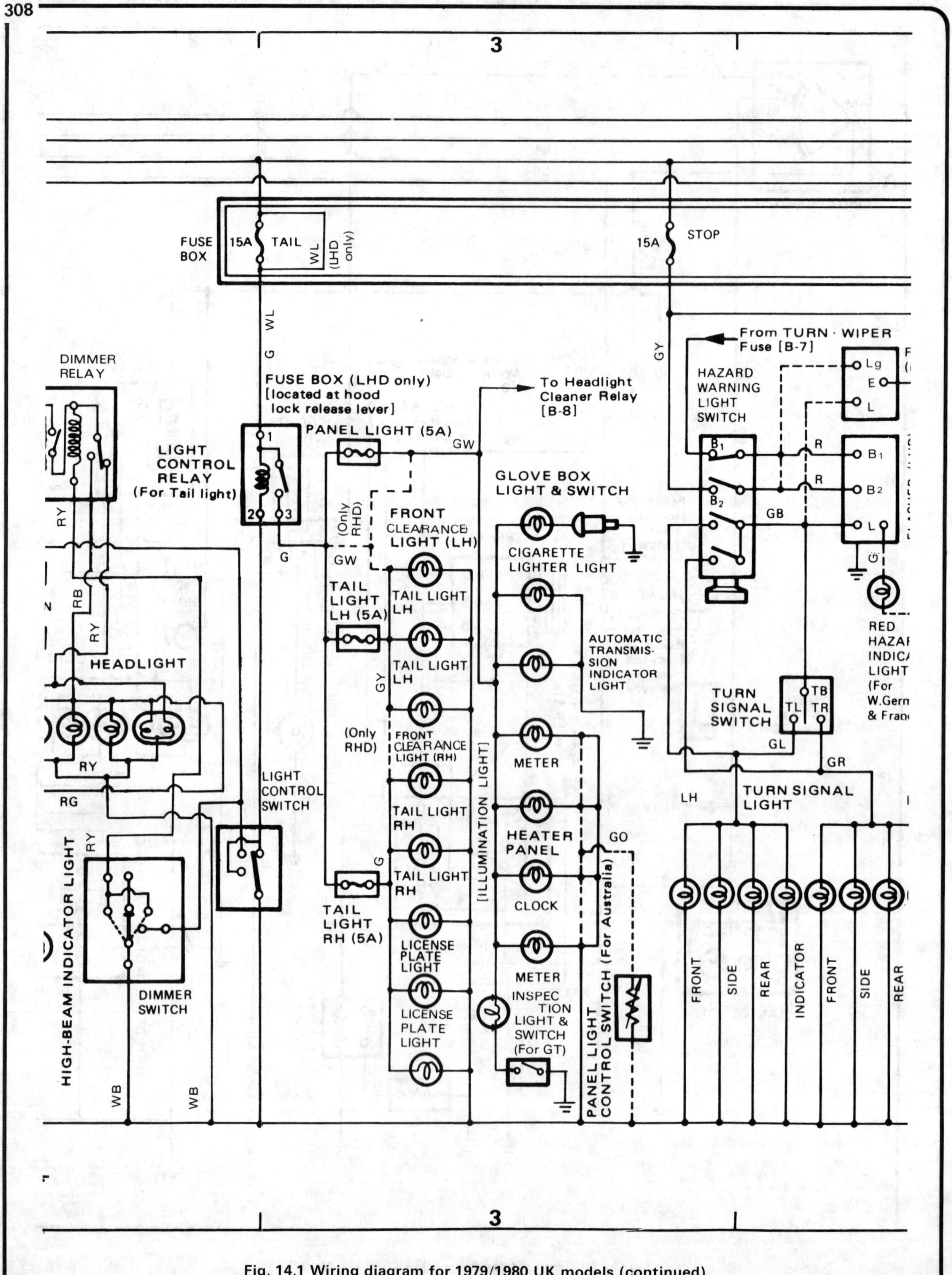

Fig. 14.1 Wiring diagram for 1979/1980 UK models (continued)

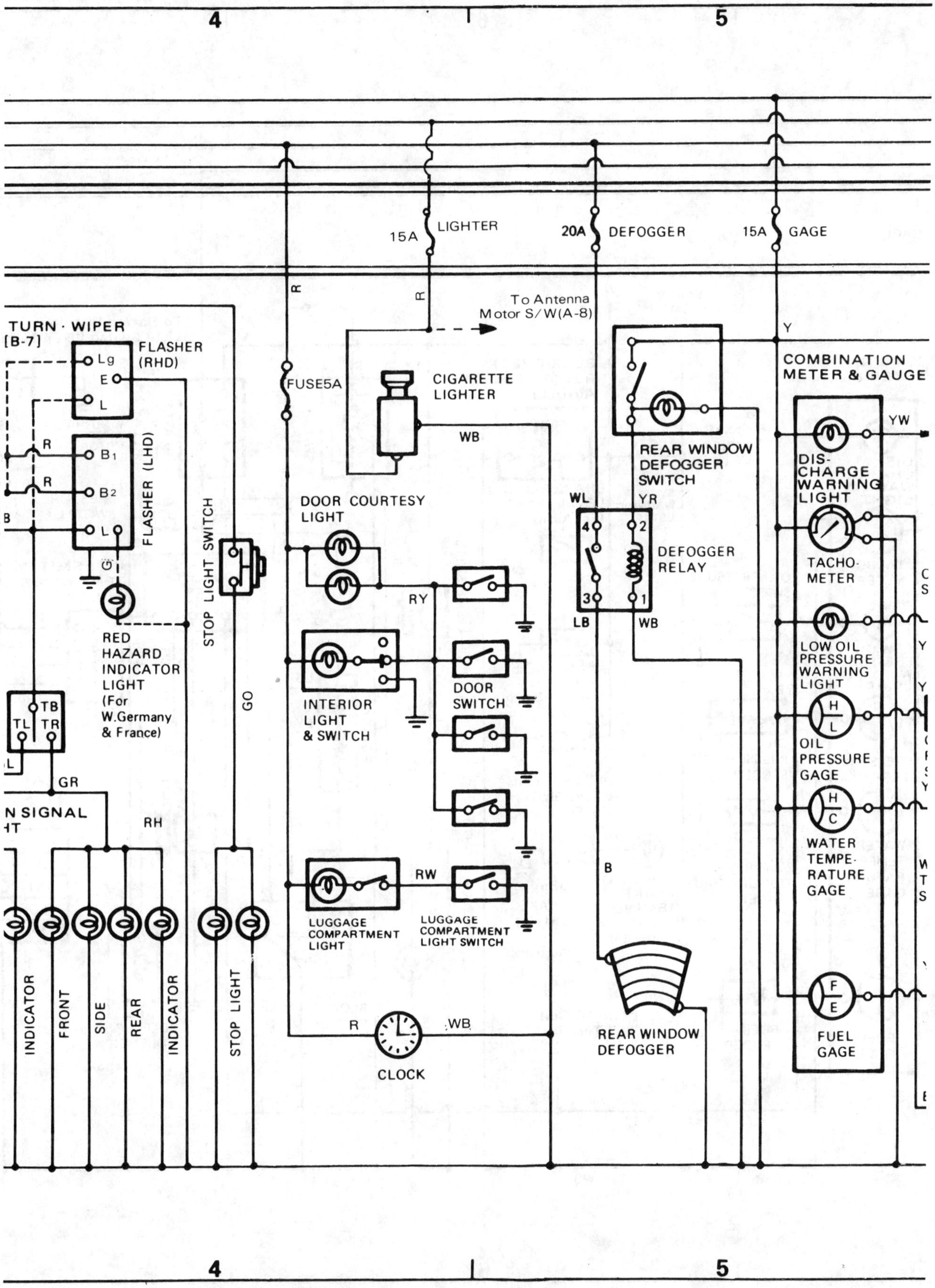

Fig. 14.1 Wiring diagram for 1979/1980 UK models (continued)

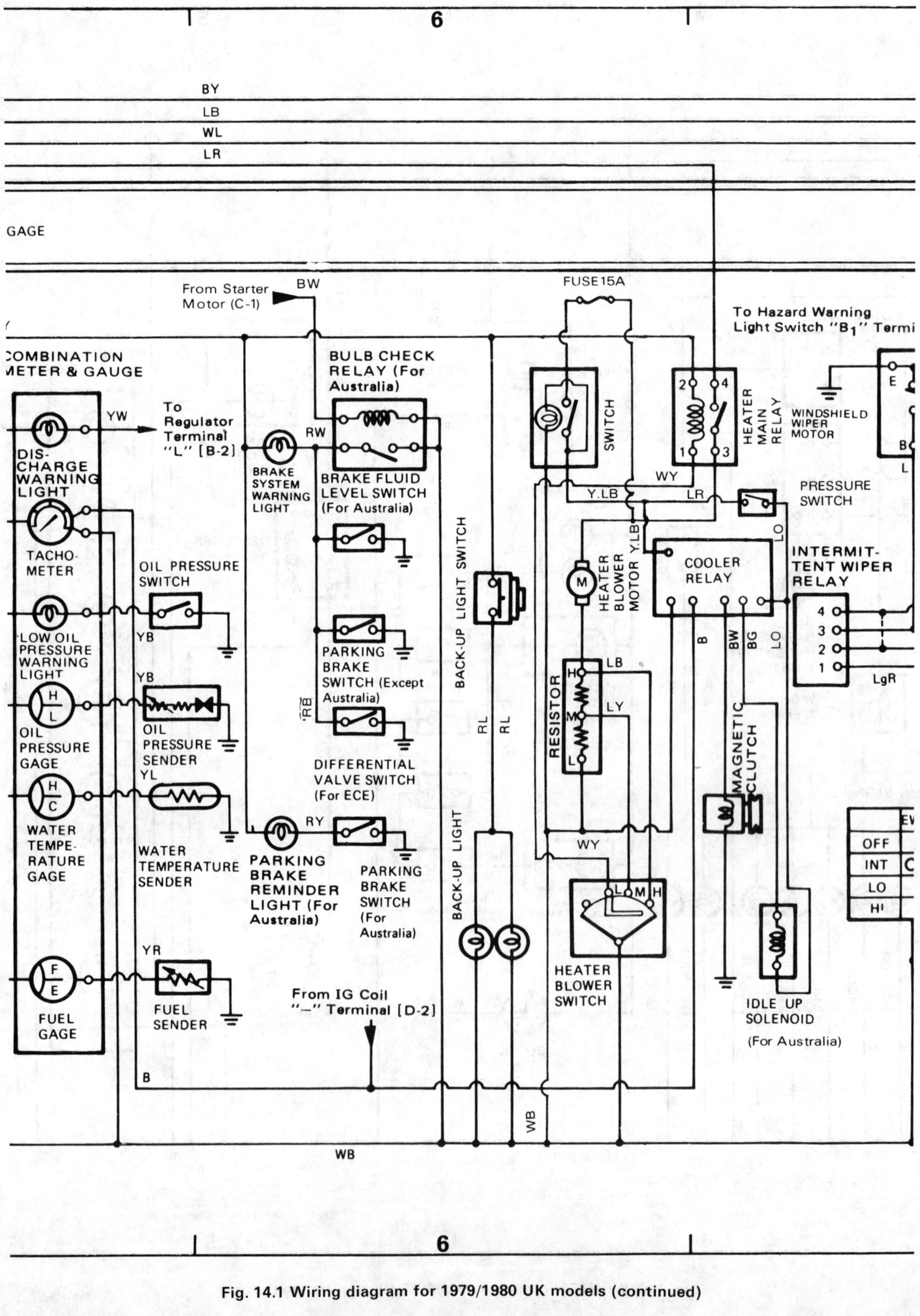

Fig. 14.1 Wiring diagram for 1979/1980 UK models (continued)

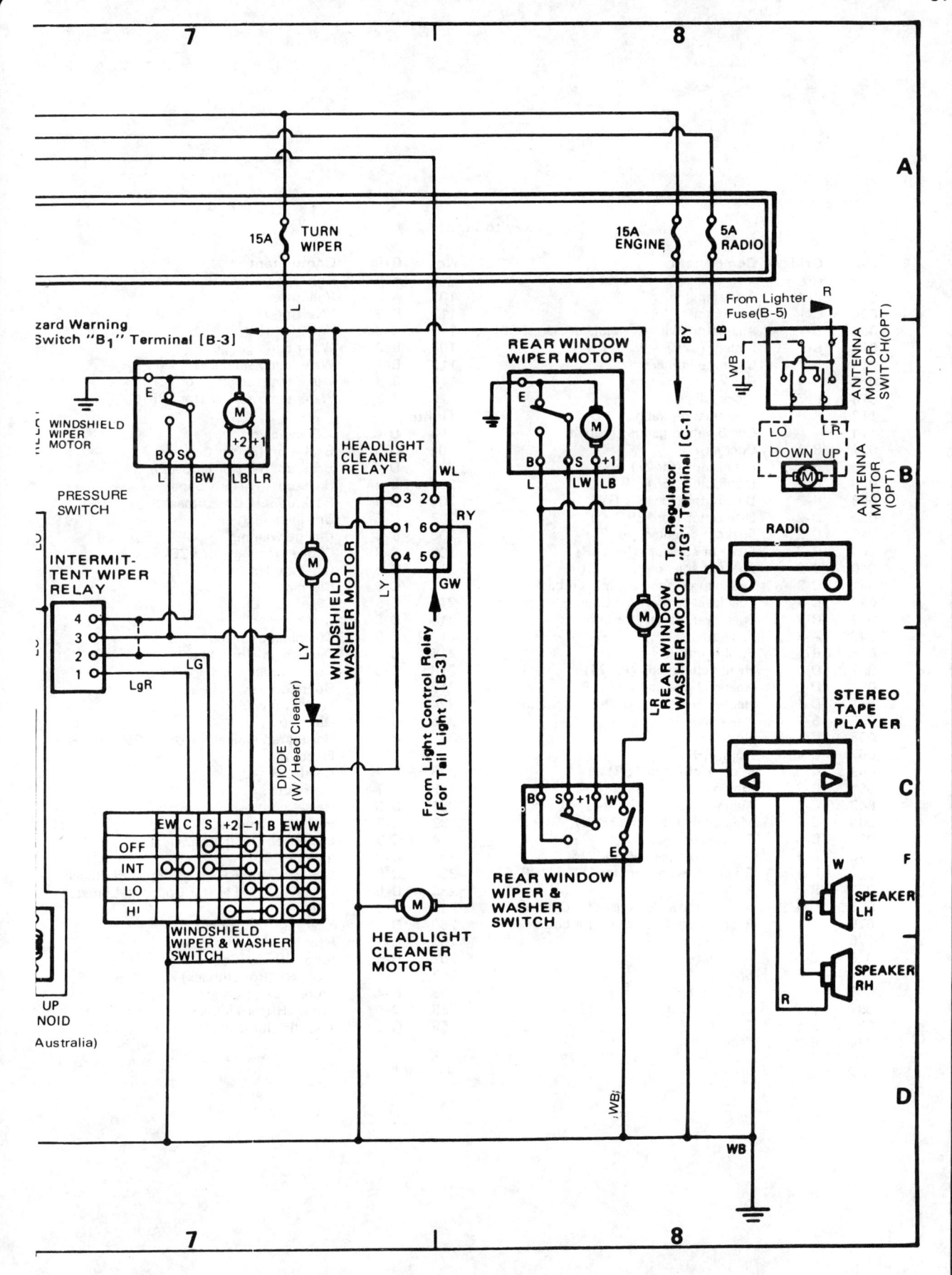

Fig. 14.1 Wiring diagram for 1979/1980 UK models (continued)

Key to Fig. 14.2

No	Grid ref	Component	No	Grid ref	Component
001	B-2	A/C amplifier	103	A-3	Volt meter
004	C-1	Alternator	112	D-3	VSV (for 2T)
005	D-1	Alternator (with IC regulator) (for ECE)	112	D-3	VSV (for 18R)
010	B-1	Battery (12V)	109	B-3	Water temp gauge
013	B-7	Cigarette lighter	110	B-3	Water temp sender (for 2T)
017	D-7	Clock	110	B-3	Water temp sender (for 18R)
018	C-7	Clock (digital)	110	B-3	Water temp sender (for 18R-G)
113	–	Combination gauge (+B)	**Lights:**		
114	–	Combination gauge (ground)	204	D-6	A/T indicator
022	B-2	Condenser	205	D-4	Back-up, RH (for Coupe)
028	B-2	Distributor (for 2T)	205	D-4	Back-up, RH (for Liftback)
028	B-2	Distributor (for 18R)	206	D-4	Back-up, LH (for Coupe)
028	B-2	Distributor (for 18R-G)	206	D-4	Back-up, LH (for Liftback)
039	D-2	Distributor (for 18R & 2T)	207	C-3	Brake warning
040	D-2	Fuel cut solenoid (for 2T)	208	B-3	Charge warning
040	D-2	Fuel cut solenoid (for 18R)	209	C-3	Choke warning (for ECE)
040	D-2	Fuel cut solenoid (for 18R-G)	210	D-6	Cigarette lighter
041	C-3	Fuel gauge (for Coupe & Liftback)	211	C-6	Clearance, RH
042	C-3	Fuel level sender (for Coupe)	212	C-5	Clearance, LH
042	C-3	Fuel level sender (for Liftback)	213	D-6	Clock
047	A-2	Fuse box	214	C-6	Combination meter
048	B-1	Fusible link	215	D-8	Deck room (for Liftback)
051	D-2	Heater choke coil (for 2T)	218	D-8	Door courtesy, RH
051	D-2	Heater choke coil (for 18R)	219	D-8	Door courtesy, LH
051	D-1	Heater choke coil (with IC regulator)	224	C-3	Fuel level warning
052	B-2	Heater resistor	225	C-6	Glovebox
055	C-4	Horn, LH	227	B-5	Hazard red indicator (for W. Germany & France)
057	C-4	Horn, RH			
059	C-2	Idle-up VSV (for 18R)	228	B-7	Head, RH
061	B-2	Ignition coil	229	B-7	Head, LH
063	C-2	Magnet clutch	231	D-6	Heater control
063	C-2	Magnet clutch (for 18R-G)	232	D-7	High beam indicator
069	B-3	Oil pressure gauge	234	C-6	Inspection (for 18R-G & 2T)
070	B-3	Oil pressure sender (for 2T)	235	D-8	Interior
070	B-3	Oil pressure sender (for 18R & 18R-G)	238	C-6	Licence plate, RH (for Coupe & Liftback)
073	B-8	Radio	239	D-6	Licence plate, LH (for Coupe & Liftback)
075	C-4	Rear window defogger (for Coupe)	245	C-3	PKB
075	C-4	Rear window defogger (for Liftback)	250	D-7	Rear fog (for Coupe)
076	C-1	Regulator	250	D-7	Rear fog (for Liftback)
078	C-6	Rheostat	257	D-4	Stop, RH (for Coupe)
087	C-8	Speaker, RH	257	D-4	Stop, RH (for Liftback)
088	C-8	Speaker, LH	258	D-4	Stop, LH (for Coupe)
091	D-2	Speed sensor (for 2T & 18R)	258	D-4	Stop, LH (for Liftback)
093	B-8	Stereo	259	C-6	Tail, RH (for Coupe)
094	B-3	Tachometer			

Key to Fig. 14.2 (continued)

No	Grid ref	Component	No	Grid ref	Component
259	C-6	Tail, RH (for Liftback)	807	C-3	Brake fluid level (for Australia)
260	C-5	Tail, LH (for Coupe)	807	C-3	Brake fluid level (for Sweden)
260	C-5	Tail, LH (for Liftback)	809	C-3	Choke warning (for ECE)
263	C-5	Turn signal, front RH	810	D-8	Deck room light (for Liftback)
264	C-4	Turn signal, front LH	811	B-4	Defogger
265	D-5	Turn signal, indicator RH	812	D-3	Differential (for 2T)
266	D-4	Turn signal, indicator LH	812	D-3	Differential (for 18R & 18R-G)
267	D-5	Turn signal, rear RH (for Coupe)	813	B-7	Dimmer
267	D-5	Turn signal, rear RH (for Liftback)	815	D-8	Door courtesy, front RH
268	D-4	Turn signal, rear LH (for Coupe)	816	D-8	Door courtesy, front LH
268	D-4	Turn signal, rear LH (for Liftback)	819	D-8	Door key
269	C-5	Turn signal, side RH	823	D-7	Rear fog light
270	C-4	Turn signal, side LH	824	C-3	Fuel level warning (for Coupe)
274	D-6	Combination gauge	824	C-3	Fuel level warning (for Liftback)
Motors:			825	C-6	Glove box light
402	B-8	Antenna (for Coupe)	826	A-4	Hazard
402	B-8	Antenna (for Liftback)	827	C-3	Heater blower
406	B-5	Headlight cleaner	828	D-4	Horn
407	B-2	Heater blower	829	A-1	Ignition
415	B-1	Starter (for 2T)	830	C-6	Inspection light (for 18R-G & 2T)
415	B-1	Starter (for 18R & 18R-G)	831	D-8	Interior light
417	C-5	Washer, rear (for Liftback)	834	B-6	Light control
418	C-5	Windshield washer (for 18R)	835	B-2	Low pressure cut
418	C-5	Windshield washer (for 18R-G & 2T)	840	B-1	Neutral start (for A/T)
419	C-5	Windshield wiper	846	C-3	PKB (ex. ECE)
420	D-5	Wiper, rear (for Liftback)	846	C-3	PKB (for Australia)
Relays:			865	D-4	Stop light
604	D-3	Bulb check (for Australia)	874	C-4	Turn signal
610	B-4	Defogger	880	C-5	Washer, rear
611	B-4	Dimmer	882	B-5	Windshield washer
617	C-5	Flasher (for RHD)	883	B-5	Windshield wiper
617	B-4	Flasher (for LHD)	884	C-5	Wiper, rear
622	B-6	Headlight			
623	B-5	Headlight cleaner			
625	B-2	Heater			
626	D-2	Heater choke			
626	D-2	Heater choke (for ECE with IC regulator)			
639	D-8	Interior light control			
644	B-6	Tail light (for RHD)			
644	B-6	Tail light (for LHD)			
650	B-5	Wiper (intermittent)			
Switches:					
801	B-2	A/C			
803	B-7	Antenna motor			
806	D-4	Back-up light			

For colour code see key to Fig. 14.1

Abbreviations
A/C   Air conditioner
A/T   Automatic transmission
ECE   Economic Commission for Europe
LHD   Left-hand drive
M/T   Manual transmission
PKB   Parking brake
RHD   Right-hand drive

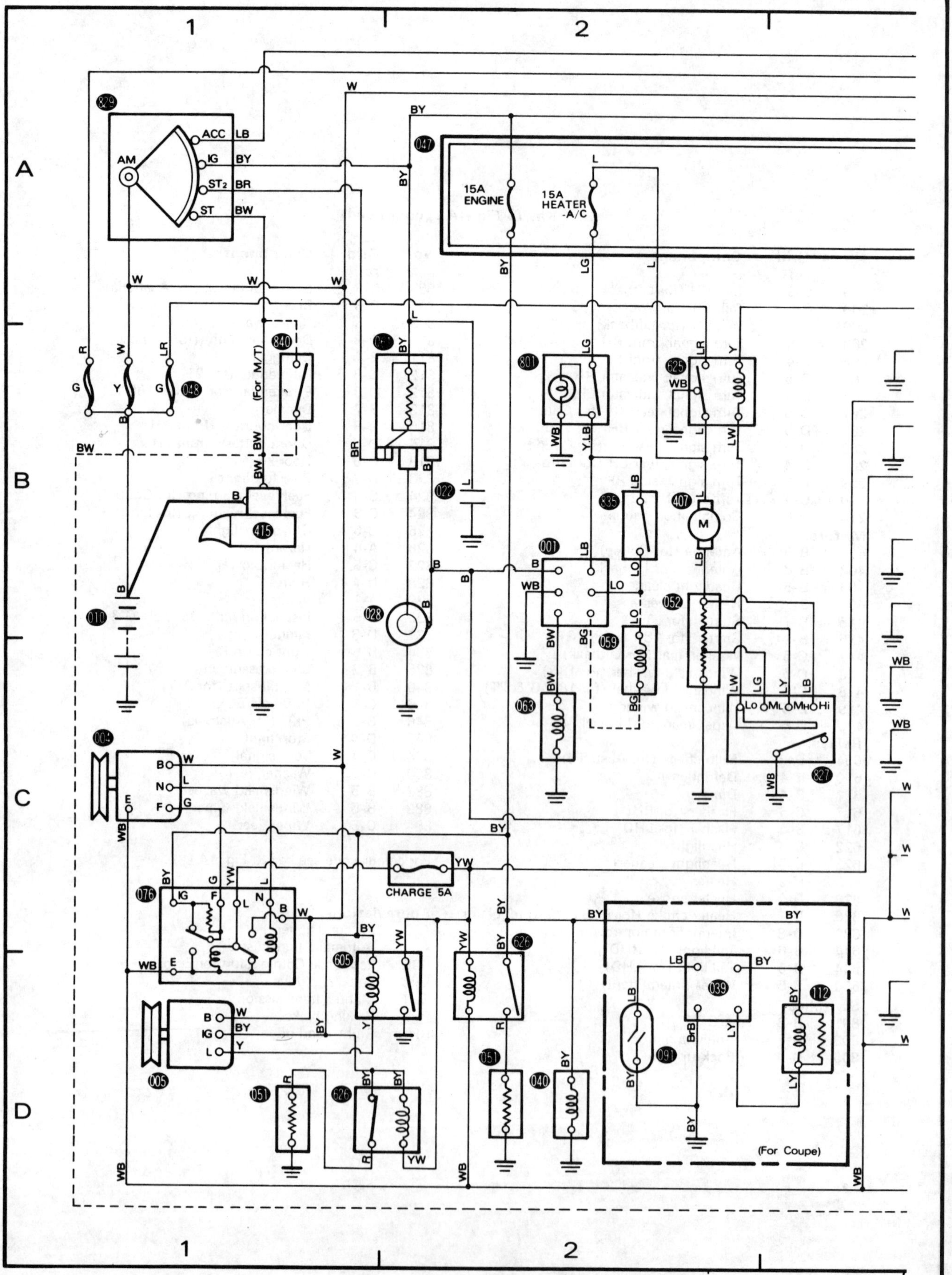

Fig. 14.2 Wiring diagram for 1981/1982 UK models

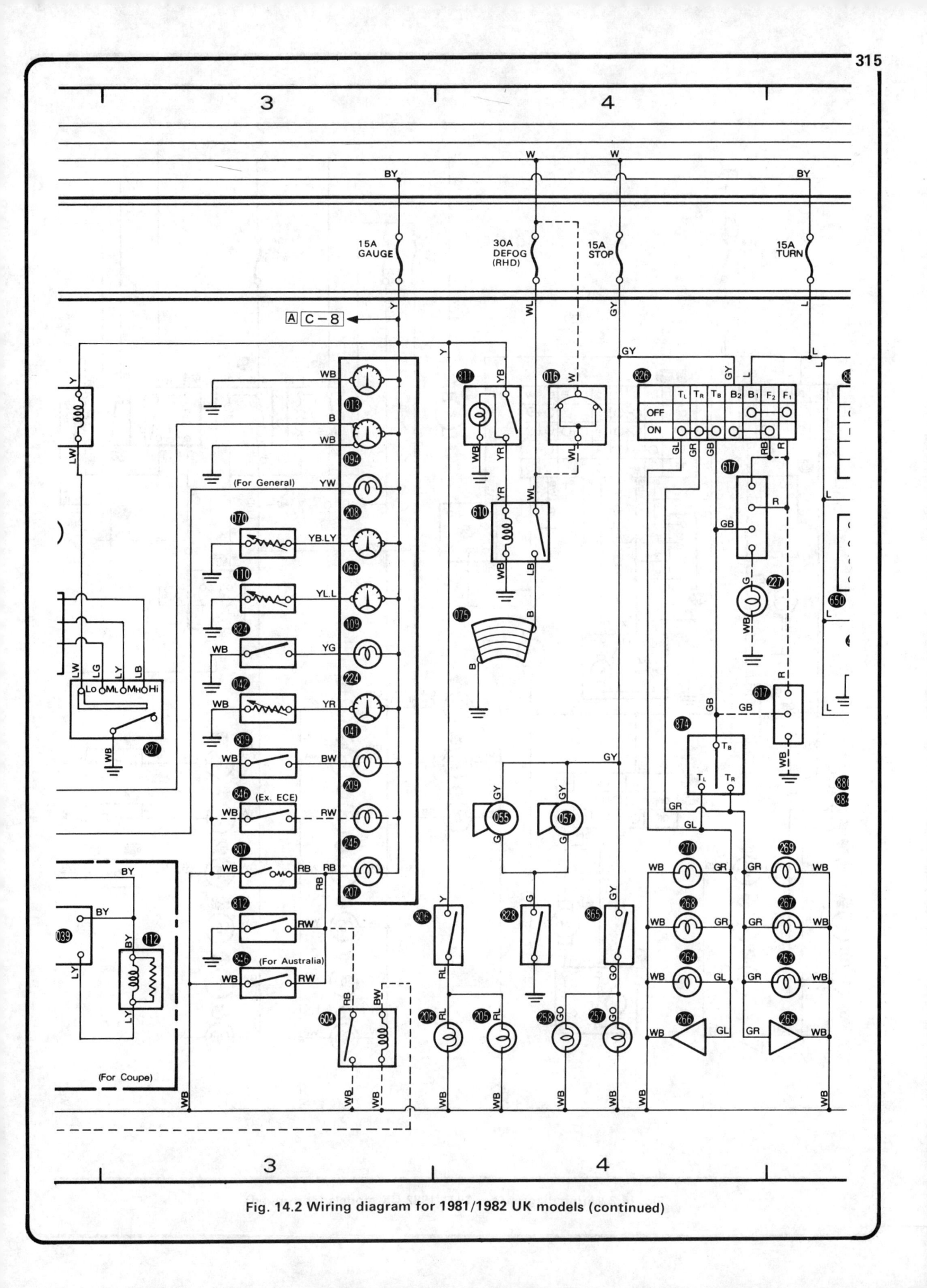

Fig. 14.2 Wiring diagram for 1981/1982 UK models (continued)

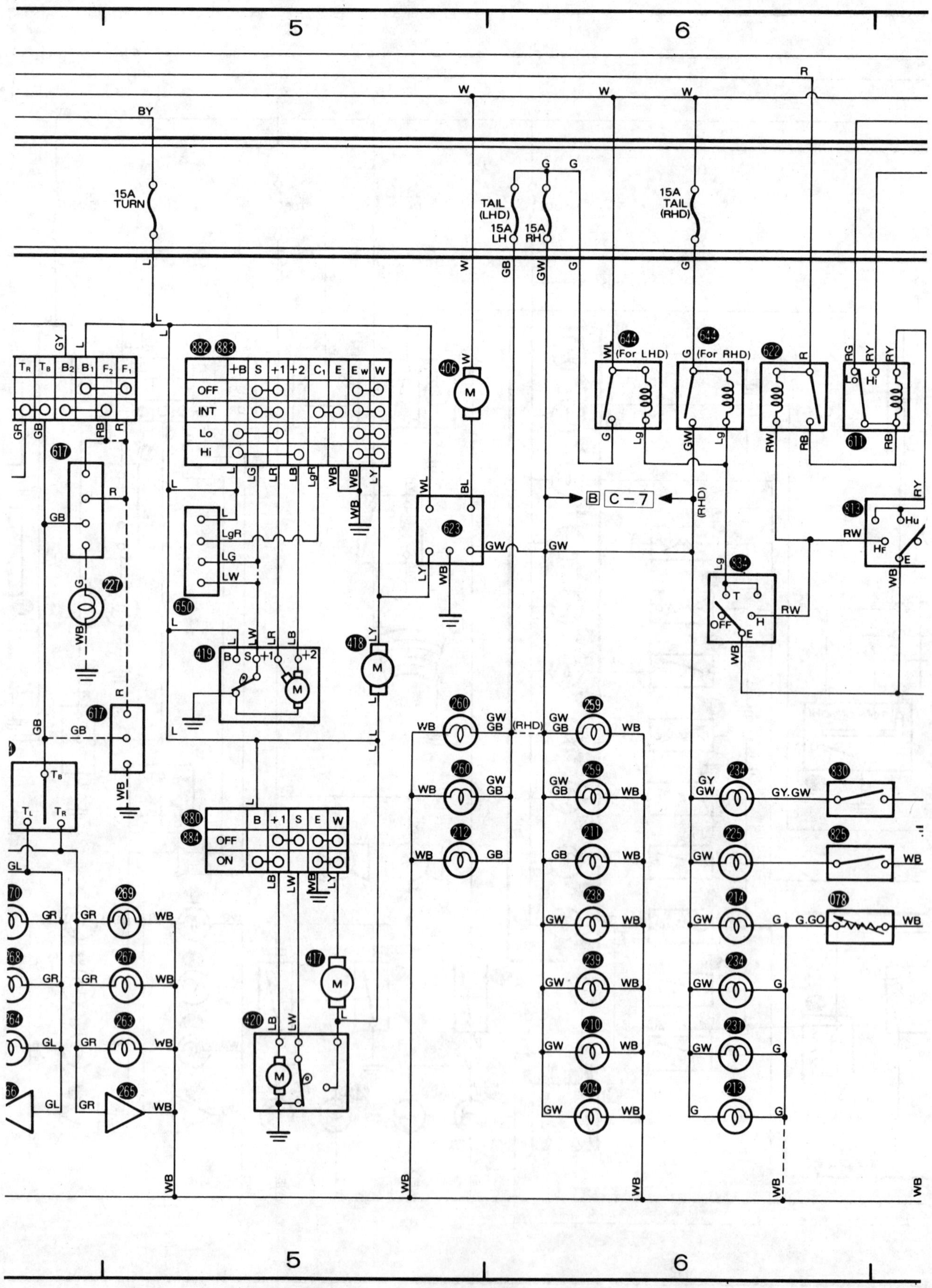

Fig. 14.2 Wiring diagram for 1981/1982 UK models (continued)

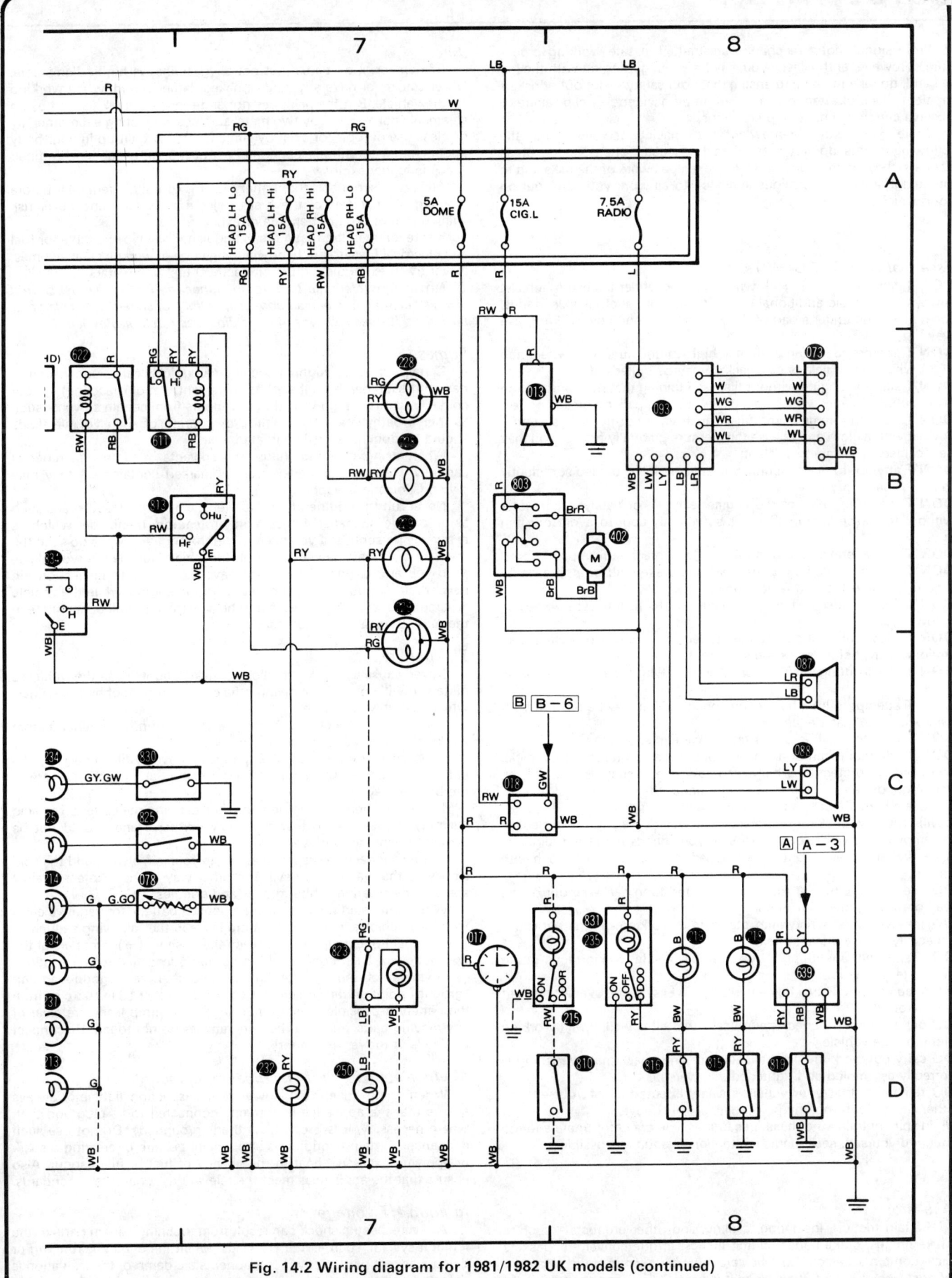

Fig. 14.2 Wiring diagram for 1981/1982 UK models (continued)

Safety first!

Professional motor mechanics are trained in safe working procedures. However enthusiastic you may be about getting on with the job in hand, do take the time to ensure that your safety is not put at risk. A moment's lack of attention can result in an accident, as can failure to observe certain elementary precautions.

There will always be new ways of having accidents, and the following points do not pretend to be a comprehensive list of all dangers; they are intended rather to make you aware of the risks and to encourage a safety-conscious approach to all work you carry out on your vehicle.

Essential DOs and DON'Ts

DON'T rely on a single jack when working underneath the vehicle. Always use reliable additional means of support, such as axle stands, securely placed under a part of the vehicle that you know will not give way.

DON'T attempt to loosen or tighten high-torque nuts (e.g. wheel hub nuts) while the vehicle is on a jack; it may be pulled off.

DON'T start the engine without first ascertaining that the transmission is in neutral (or 'Park' where applicable) and the parking brake applied.

DON'T suddenly remove the filler cap from a hot cooling system – cover it with a cloth and release the pressure gradually first, or you may get scalded by escaping coolant.

DON'T attempt to drain oil until you are sure it has cooled sufficiently to avoid scalding you.

DON'T grasp any part of the engine, exhaust or catalytic converter without first ascertaining that it is sufficiently cool to avoid burning you.

DON'T allow brake fluid or antifreeze to contact vehicle paintwork.

DON'T syphon toxic liquids such as fuel, brake fluid or antifreeze by mouth, or allow them to remain on your skin.

DON'T inhale dust – it may be injurious to health (see *Asbestos* below).

DON'T allow any spilt oil or grease to remain on the floor – wipe it up straight away, before someone slips on it.

DON'T use ill-fitting spanners or other tools which may slip and cause injury.

DON'T attempt to lift a heavy component which may be beyond your capability – get assistance.

DON'T rush to finish a job, or take unverified short cuts.

DON'T allow children or animals in or around an unattended vehicle.

DO wear eye protection when using power tools such as drill, sander, bench grinder etc, and when working under the vehicle.

DO use a barrier cream on your hands prior to undertaking dirty jobs – it will protect your skin from infection as well as making the dirt easier to remove afterwards; but make sure your hands aren't left slippery. Note that long-term contact with used engine oil can be a health hazard.

DO keep loose clothing (cuffs, tie etc) and long hair well out of the way of moving mechanical parts.

DO remove rings, wristwatch etc, before working on the vehicle – especially the electrical system.

DO ensure that any lifting tackle used has a safe working load rating adequate for the job.

DO keep your work area tidy – it is only too easy to fall over articles left lying around.

DO get someone to check periodically that all is well, when working alone on the vehicle.

DO carry out work in a logical sequence and check that everything is correctly assembled and tightened afterwards.

DO remember that your vehicle's safety affects that of yourself and others. If in doubt on any point, get specialist advice.

IF, in spite of following these precautions, you are unfortunate enough to injure yourself, seek medical attention as soon as possible.

Asbestos

Certain friction, insulating, sealing, and other products – such as brake linings, brake bands, clutch linings, torque converters, gaskets, etc – contain asbestos. *Extreme care must be taken to avoid inhalation of dust from such products since it is hazardous to health.* If in doubt, assume that they *do* contain asbestos.

Fire

Remember at all times that petrol (gasoline) is highly flammable. Never smoke, or have any kind of naked flame around, when working on the vehicle. But the risk does not end there – a spark caused by an electrical short-circuit, by two metal surfaces contacting each other, by careless use of tools, or even by static electricity built up in your body under certain conditions, can ignite petrol vapour, which in a confined space is highly explosive.

Always disconnect the battery earth (ground) terminal before working on any part of the fuel or electrical system, and never risk spilling fuel on to a hot engine or exhaust.

It is recommended that a fire extinguisher of a type suitable for fuel and electrical fires is kept handy in the garage or workplace at all times. Never try to extinguish a fuel or electrical fire with water.

Note: *Any reference to a 'torch' appearing in this manual should always be taken to mean a hand-held battery-operated electric lamp or flashlight. It does NOT mean a welding/gas torch or blowlamp.*

Fumes

Certain fumes are highly toxic and can quickly cause unconsciousness and even death if inhaled to any extent. Petrol (gasoline) vapour comes into this category, as do the vapours from certain solvents such as trichloroethylene. Any draining or pouring of such volatile fluids should be done in a well ventilated area.

When using cleaning fluids and solvents, read the instructions carefully. Never use materials from unmarked containers – they may give off poisonous vapours.

Never run the engine of a motor vehicle in an enclosed space such as a garage. Exhaust fumes contain carbon monoxide which is extremely poisonous; if you need to run the engine, always do so in the open air or at least have the rear of the vehicle outside the workplace.

If you are fortunate enough to have the use of an inspection pit, never drain or pour petrol, and never run the engine, while the vehicle is standing over it; the fumes, being heavier than air, will concentrate in the pit with possibly lethal results.

The battery

Never cause a spark, or allow a naked light, near the vehicle's battery. It will normally be giving off a certain amount of hydrogen gas, which is highly explosive.

Always disconnect the battery earth (ground) terminal before working on the fuel or electrical systems.

If possible, loosen the filler plugs or cover when charging the battery from an external source. Do not charge at an excessive rate or the battery may burst.

Take care when topping up and when carrying the battery. The acid electrolyte, even when diluted, is very corrosive and should not be allowed to contact the eyes or skin.

If you ever need to prepare electrolyte yourself, always add the acid slowly to the water, and never the other way round. Protect against splashes by wearing rubber gloves and goggles.

When jump starting a car using a booster battery, for negative earth (ground) vehicles, connect the jump leads in the following sequence: First connect one jump lead between the positive (+) terminals of the two batteries. Then connect the other jump lead first to the negative (−) terminal of the booster battery, and then to a good earthing (ground) point on the vehicle to be started, at least 18 in (45 cm) from the battery if possible. Ensure that hands and jump leads are clear of any moving parts, and that the two vehicles do not touch. Disconnect the leads in the reverse order.

Mains electricity and electrical equipment

When using an electric power tool, inspection light etc, always ensure that the appliance is correctly connected to its plug and that, where necessary, it is properly earthed (grounded). Do not use such appliances in damp conditions and, again, beware of creating a spark or applying excessive heat in the vicinity of fuel or fuel vapour. Also ensure that the appliances meet the relevant national safety standards.

Ignition HT voltage

A severe electric shock can result from touching certain parts of the ignition system, such as the HT leads, when the engine is running or being cranked, particularly if components are damp or the insulation is defective. Where an electronic ignition system is fitted, the HT voltage is much higher and could prove fatal.

Conversion factors

Length (distance)
Inches (in)	X	25.4	= Millimetres (mm)	X 0.0394	= Inches (in)
Feet (ft)	X	0.305	= Metres (m)	X 3.281	= Feet (ft)
Miles	X	1.609	= Kilometres (km)	X 0.621	= Miles

Volume (capacity)
Cubic inches (cu in; in^3)	X	16.387	= Cubic centimetres (cc; cm^3)	X 0.061	= Cubic inches (cu in; in^3)
Imperial pints (Imp pt)	X	0.568	= Litres (l)	X 1.76	= Imperial pints (Imp pt)
Imperial quarts (Imp qt)	X	1.137	= Litres (l)	X 0.88	= Imperial quarts (Imp qt)
Imperial quarts (Imp qt)	X	1.201	= US quarts (US qt)	X 0.833	= Imperial quarts (Imp qt)
US quarts (US qt)	X	0.946	= Litres (l)	X 1.057	= US quarts (US qt)
Imperial gallons (Imp gal)	X	4.546	= Litres (l)	X 0.22	= Imperial gallons (Imp gal)
Imperial gallons (Imp gal)	X	1.201	= US gallons (US gal)	X 0.833	= Imperial gallons (Imp gal)
US gallons (US gal)	X	3.785	= Litres (l)	X 0.264	= US gallons (US gal)

Mass (weight)
Ounces (oz)	X	28.35	= Grams (g)	X 0.035	= Ounces (oz)
Pounds (lb)	X	0.454	= Kilograms (kg)	X 2.205	= Pounds (lb)

Force
Ounces-force (ozf; oz)	X	0.278	= Newtons (N)	X 3.6	= Ounces-force (ozf; oz)
Pounds-force (lbf; lb)	X	4.448	= Newtons (N)	X 0.225	= Pounds-force (lbf; lb)
Newtons (N)	X	0.1	= Kilograms-force (kgf; kg)	X 9.81	= Newtons (N)

Pressure
Pounds-force per square inch (psi; lbf/in^2; lb/in^2)	X	0.070	= Kilograms-force per square centimetre (kgf/cm^2; kg/cm^2)	X 14.223	= Pounds-force per square inch (psi; lbf/in^2; lb/in^2)
Pounds-force per square inch (psi; lbf/in^2; lb/in^2)	X	0.068	= Atmospheres (atm)	X 14.696	= Pounds-force per square inch (psi; lbf/in^2; lb/in^2)
Pounds-force per square inch (psi; lbf/in^2; lb/in^2)	X	0.069	= Bars	X 14.5	= Pounds-force per square inch (psi; lbf/in^2; lb/in^2)
Pounds-force per square inch (psi; lbf/in^2; lb/in^2)	X	6.895	= Kilopascals (kPa)	X 0.145	= Pounds-force per square inch (psi; lbf/in^2; lb/in^2)
Kilopascals (kPa)	X	0.01	= Kilograms-force per square centimetre (kgf/cm^2; kg/cm^2)	X 98.1	= Kilopascals (kPa)
Millibar (mbar)	X	100	= Pascals (Pa)	X 0.01	= Millibar (mbar)
Millibar (mbar)	X	0.0145	= Pounds-force per square inch (psi; lbf/in^2; lb/in^2)	X 68.947	= Millibar (mbar)
Millibar (mbar)	X	0.75	= Millimetres of mercury (mmHg)	X 1.333	= Millibar (mbar)
Millibar (mbar)	X	0.401	= Inches of water (inH$_2$O)	X 2.491	= Millibar (mbar)
Millimetres of mercury (mmHg)	X	0.535	= Inches of water (inH$_2$O)	X 1.868	= Millimetres of mercury (mmHg)
Inches of water (inH$_2$O)	X	0.036	= Pounds-force per square inch (psi; lbf/in^2; lb/in^2)	X 27.68	= Inches of water (inH$_2$O)

Torque (moment of force)
Pounds-force inches (lbf in; lb in)	X	1.152	= Kilograms-force centimetre (kgf cm; kg cm)	X 0.868	= Pounds-force inches (lbf in; lb in)
Pounds-force inches (lbf in; lb in)	X	0.113	= Newton metres (Nm)	X 8.85	= Pounds-force inches (lbf in; lb in)
Pounds-force inches (lbf in; lb in)	X	0.083	= Pounds-force feet (lbf ft; lb ft)	X 12	= Pounds-force inches (lbf in; lb in)
Pounds-force feet (lbf ft; lb ft)	X	0.138	= Kilograms-force metres (kgf m; kg m)	X 7.233	= Pounds-force feet (lbf ft; lb ft)
Pounds-force feet (lbf ft; lb ft)	X	1.356	= Newton metres (Nm)	X 0.738	= Pounds-force feet (lbf ft; lb ft)
Newton metres (Nm)	X	0.102	= Kilograms-force metres (kgf m; kg m)	X 9.804	= Newton metres (Nm)

Power
Horsepower (hp)	X	745.7	= Watts (W)	X 0.0013	= Horsepower (hp)

Velocity (speed)
Miles per hour (miles/hr; mph)	X	1.609	= Kilometres per hour (km/hr; kph)	X 0.621	= Miles per hour (miles/hr; mph)

Fuel consumption*
Miles per gallon, Imperial (mpg)	X	0.354	= Kilometres per litre (km/l)	X 2.825	= Miles per gallon, Imperial (mpg)
Miles per gallon, US (mpg)	X	0.425	= Kilometres per litre (km/l)	X 2.352	= Miles per gallon, US (mpg)

Temperature
Degrees Fahrenheit = (°C x 1.8) + 32   Degrees Celsius (Degrees Centigrade; °C) = (°F - 32) x 0.56

*It is common practice to convert from miles per gallon (mpg) to litres/100 kilometres (l/100km), where mpg (Imperial) x l/100 km = 282 and mpg (US) x l/100 km = 235

Index

A

Accelerator linkage — 115
Air cleaner
    element renewal — 77
Air Injection (AI) system — 105
Alternator
    dismantling, servicing and reassembly — 203, 302
    general description, maintenance and precautions — 202
    removal and refitting — 203
    testing in car — 202, 302
Alternator regulator
    testing and adjustment — 206, 302
Ammeter
    testing — 215
Antifreeze and corrosion inhibiting mixtures — 69
Automatic Hot Air Intake (HAI) system — 105
Automatic transmission
    extension housing oil seal
        renewal — 172
    fault diagnosis — 173
    general description and precautions — 170
    maintenance — 170
    neutral start switch
        adjustment — 172
        removal and refitting — 172
    selector linkage
        adjustment — 172
    specifications — 137
    throttle cable (downshift)
        adjustment — 172

B

Battery
    charging — 202
    maintenance — 202
    removal and installation — 201
Bodywork and fittings
    air conditioning system — 285
    bonnet lid
        adjustment — 268
        removal and installation — 268
    bonnet lock
        adjustment — 268
    boot lid lock
        removal, installation and adjustment — 275
    door glass
        adjustment — 271
    door glass and regulator — 271
    door lock
        removal, refitting and adjustment — 268
    door striker — 268
    doors
        removal, installation and adjustment — 268
        tracing rattles and their rectification — 265
    front bumper — 268
    front wing — 268
    general description — 261
    heater and ventilation system — 279
    heater blower motor — 281
    heater control
        cable adjustment — 285
        removal and installation — 281
    heater unit — 285
    instrument panel
        crash pad — 279
        removal and installation — 279
    liftback rear door damper stays — 279
    maintenance
        bodywork and underframe — 261
        hinges and locks — 265
        upholstery and carpets — 261
    minor body damage repair — 264
    major body damage repair — 265
    sun roof — 273
    windscreen
        removal and refitting — 268
Bodywork and fittings (Coupé)
    boot lid
        removal and installation — 275
    quarter windows — 271
    rear bumper — 279
    rear windows
        removal and installation — 273
Bodywork and fittings (Liftback)
    quarter window — 273
    rear bumper — 279
    rear door — 275
    rear door glass — 279
    rear door lock — 275
Braking system
    booster (vacuum servo unit)
        description — 195
    booster unit
        dismantling and reassembly — 198
        removal and installation — 198
    brake disc
        examination and renovation — 191
    brake drums
        inspection and renovation — 191
    disc caliper — 191
    fault diagnosis — 199
    front disc pads — 188
    general description — 188
    handbrake
        adjustment — 193
        cable renewal — 195
    hydraulic lines — 193
    hydraulic system
        bleeding — 191
    master cylinder — 191
    pedal
        adjustment — 195
        removal and installation — 195
    pressure regulating valve — 195
    rear brake shoes — 190
    rear brake wheel cylinders — 191
    specifications — 186, 286
    warning light — 227
Buying spare parts — 5

Index

C

Camshaft and camshaft bearings — 49
Carburettor (2T–B)
    adjustments after installation — 86
    adjustments during assembly — 84
    dismantling, overhaul and reassembly — 84
    general description — 83
    removal and refitting — 83
Carburettor (18R and 20R)
    adjustments after installation — 95
    adjustments during assembly — 91
    dismantling, overhaul and reassembly — 87
    general description — 86
    removal and refitting — 87
Carburettor (18R–G)
    adjustments after installation — 97
    dismantling, overhaul and assembly — 95
    general description — 95
    removal and refitting — 95
Carburettor (22R)
    adjustments — 294
    dismantling — 289
    inspection — 291
    reassembly — 291
Choke opener system — 294
Clutch
    adjustment — 131
    fault diagnosis — 136
    general description — 131
    hydraulic system
        bleeding — 133
    inspection and renovation — 133
    master cylinder
        dismantling and reassembly — 132
        removal and refitting — 132
    operating cylinder
        dismantling and reassembly — 133
        removal and refitting — 132
    pedal
        removal and refitting — 136
        refitting — 136
    release bearing
        renewal — 136
        removal — 133
    specifications — 131
Connecting rods and bearings
    examination and renovation — 47
Conversion factors — 319
Cooling system
    antifreeze and corrosion inhibiting mixtures — 67
    draining — 67
    drivebelts
        tensioning and renewal — 74
    fault diagnosis — 74
    filling — 69
    flushing — 69
    general description — 67
    radiator — 69
    specifications — 67
    thermostat — 72
    water pump and fluid coupling
        overhaul — 74
    water pump
        removal and refitting — 74
Crankcase ventilation system (2T–B engine) — 33
Crankcase ventilation system (18R engine) — 43
Crankcase ventilation system (20R engine) — 47
Crankshaft and main bearings
    examination and renovation — 47
Cylinder block
    examination and renovation — 51
Cylinder bores
    examination and renovation — 49
Cylinder head and valves
    decarbonising — 49
    examination and renovation — 49

D

Distributor
    dismantling, inspection and reassembly — 125
    removal and refitting — 121
    specifications — 117
    general (22R) — 297

E

Electrical system
    alternator
        dismantling, servicing and assembly — 203, 302
        general description, maintenance and precautions — 202
        removal and refitting — 203
        testing in car — 202, 302
    alternator regulator — 206, 302
    ammeter
        testing — 215
    battery charging — 202
    brake warning light — 227
    bulb replacement
        door courtesy lights — 220
        front flasher, side marker and parking lights — 220
        heater control indicator lamp — 221
        instrument panel lamps — 221
        luggage compartment light — 220
        number plate light — 221
        panel warning lamps — 221
        rear combination light — 220
        rear side marker lamps — 221
        roof light — 220
        transmission indicator lamp — 221
    clock — 227
    ESP system — 229
    fault diagnosis — 229
    fuel level transmitter and gauge
        testing — 212
    fuses and fusible link — 212
    general description — 201
    hazard warning and turn signal lamps — 212
    headlamp
        beam alignment — 220
        removal and refitting — 220
        washers — 227
    heated rear window
        precautions and testing — 224
        switch, relay and choke — 224
    IC regulator testing — 302
    ignition switch — 217
    instrument cluster
        removal and replacement — 217
    neutral safety switch (automatic transmission) — 217
    O_2 sensor
        cancel switch — 303
        resetting — 303
    oil pressure switch and gauge — 215
    radio — 227
    rear window wiper
        dismantling and reassembly — 224
        removal and installation — 224
    relays — 227
    starter motor
        removal and refitting — 206
        testing in car — 206
    starter motor (reduction gear type)
        dismantling — 209
        reassembly — 212
        servicing and testing — 212
    starter motor (standard type)
        dismantling — 206
        reassembly — 209
        servicing and testing — 206
    steering column and facia mounted switches — 217
    specifications — 200, 286
    tachometer — 215

Index

water temperature transmitter and gauge — 212
windscreen washer and rear washer — 224
windscreen wiper — 221
windscreen wiper motor
    dismantling and reassembly — 221
wiring diagrams — 231 to 240, 304-317

Emission control
    fault diagnosis — 116
    general — 104, 294

Engine
    auxiliary driveshaft (18R and 18RG engines) — 51
    camshaft and camshaft bearings — 49
    connecting rods and bearings — 47
    crankshaft and main bearings — 47
    cylinder block — 51
    cylinder bores — 49
    cylinder head and valves
        decarbonising — 49
        examination and renovation — 49
    dismantling the engine (general) — 27
    driveplate (automatic transmission) — 51
    engine/transmission
        refitting — 65
        removing — 26, 289
    engine ancillary components
        refitting — 65
        removing — 27
        renewal — 51
    pistons and piston rings — 49
    removing — 24
    specifications — 14
    timing components — 49

Engine (2T—B engine)
    crankcase ventilation system — 33
    dismantling — 29
    lubrication system, oil pump and filter — 33
    reassembly — 51
    valve clearances
        adjustment — 55

Engine (18R engine)
    crankcase ventilation system — 43
    dismantling — 33
    lubrication system, oil pump and filter — 40
    reassembly — 55
    valve clearances
        adjustment — 62

Engine (18R—G engine)
    dismantling — 37
    lubrication system, oil pump and filter — 40
    reassembly — 62
    valve clearances
        adjustment — 63

Engine (20R engine)
    crankcase ventilation system — 47
    dismantling — 43
    lubrication system, oil pump and filter — 46
    reassembly — 63
    valve clearances
        adjustment — 65

Engine (22R)
    removal — 289
    sump and timing — 289
    timing components — 289

Exhaust Gas Recirculation (EGR) system — 107
Exhaust system
    removal and refitting — 104

F

Fuel and exhaust systems
    accelerator linkage — 115
    air cleaner
        element renewal — 77
    Air Injection (AI) system — 109
    automatic Hot Air Intake (HAI) system — 105
    carburettor (2T—B)
        adjustments after installation — 86
        adjustments during assembly — 84
    carburettor (18R and 20R)
        adjustments after installation — 95
        adjustments during assembly — 91
    carburettor (18R—G engine)
        adjustments after installation — 97
    carburettor (22R)
        dismantling — 289
        inspection — 291
        reassembly — 291
    Catalytic Converter (CCO) system — 112
    Choke Breaker (CB) system — 115
    choke opener system — 294
    deceleration fuel cut system — 115
    emission control — 104, 294
    Exhaust Gas Recirculation (EGR) system — 107
    exhaust system
        removal and refitting — 104
    fault diagnosis — 115, 116
    fuel contents gauge and sender unit — 83
    fuel evaporative emission control system — 115
    fuel filter
        renewal — 77
    fuel pump
        general description — 77
        removal, overhaul and refitting — 83
        testing — 80
    fuel tank
        removal and refitting — 83
    general description — 77
    High Altitude Compensation (HAC) system — 112
    inlet and exhaust manifolds — 100
    Spark Control (SC) system — 105
    specifications — 75, 286
    Throttle Positioner (TP) system — 105

G

Gearbox — see manual transmission
General dimensions — 13, 286

I

Ignition system
    condenser (capacitor)
        removal, testing and refitting — 121
    contact breaker points
        removal and refitting — 121
    contact breaker points and damper spring
        adjustment — 121
    distributor
        dismantling — 125
        inspection and reassembly — 125
        removal and refitting — 121
    fault diagnosis — 130
    general description — 118, 297
    ignition timing
        adjustment — 125
        checking — 125
    spark plugs and leads — 129
    specifications — 117, 286
    transistorized ignition system
        fault finding — 129
        igniter — 129

J

Jacking and towing — 13

L

Lubrication chart — 8

M

Manual transmission
    fault diagnosis – 170
    general description – 138, 297
    removal and installation – 138, 297
    specifications – 137, 286

Manual transmission (Type PSI Gearbox)
    dismantling – 161
    gearcase – 165
    input shaft – 161
    laygear – 161
    reassembly – 165
    reverse idler gear – 165
    synchro-hub – 165

Manual transmission (Type T50 Gearbox)
    dismantling – 141
    gearcase – 145
    input shaft – 145
    laygear – 145
    mainshaft – 141
    reassembly – 147

Manual transmission (Type W50 Gearbox)
    dismantling – 152
    gearcase – 156
    input shaft – 156
    laygear – 156
    mainshaft – 152
    reassembly – 161
    reverse idler gear – 156

Manual transmission (Type W55)
    dismantling – 297
    reassembly – 298
    removal – 297

P

Propeller shaft
    centre bearing
        removal and refitting – 178
    fault diagnosis – 178
    front sliding yoke
        inspection – 178
    general description – 174
    removal and installation – 174
    specifications – 174
    universal joints
        dismantling and reassembly – 178
        inspection – 178

R

Rear axle
    axleshaft, bearings and oil seals
        removal and installation – 179
    fault diagnosis – 185
    final drive unit
        dismantling and reassembly – 184
    general description – 179
    pinion oil seal
        renewal – 184
    removal and installation – 184
    specifications – 179

Routine maintenance – 9

S

Safety first! – 318

Starter motor
    removal and refitting – 206
    specifications – 200
    testing in car – 206

Starter motor (reduction gear type)
    dismantling – 209
    reassembly – 212
    servicing and testing – 212

Starter motor (standard type)
    dismantling – 206
    reassembly – 209
    servicing and testing – 206

Suspension and steering
    fault diagnosis – 260
    front crossmember – 248
    front hubs
        servicing and adjustment – 245
    front anti-roll bar – 248
    front suspension strut – 248
    front suspension radius rod – 248
    front wheel alignment – 259
    general description – 242
    maintenance and inspection – 242
    power steering
        bleeding – 255
        fluid replacement – 255
        maintenance – 252
    power steering pump
        removal and refitting – 255
    rear anti-roll bar – 251
    rear lateral control rod – 251
    rear shock absorbers – 245
    rear suspension
        coil spring – 248
        lower control arm – 248
        upper control arm – 248
    specifications – 241, 286
    steering column lock – 259
    steering column and shaft – 251, 303
    steering linkage
        removal and installation – 251
    steering gear (manual)
        dismantling and reassembly – 255
    steering gear (power-assisted)
        dismantling and reassembly – 255
    steering gear housing – 252
    steering wheel – 251
    suspension lower arm – 248
    tilt steering column
        installation – 303
        removal – 303
    wheels and tyres – 259

T

Tools and working facilities – 6

V

Vehicle identification numbers – 5

W

Wiring diagrams – 231 to 240, 304-317

Printed by
J H Haynes & Co Ltd
Sparkford Nr Yeovil
Somerset BA22 7JJ England